Bayesian Methods for Repeated Measures

Chapman & Hall/CRC Biostatistics Series

Editor-in-Chief

Shein-Chung Chow, Ph.D., Professor, Department of Biostatistics and Bioinformatics,
Duke University School of Medicine, Durham, North Carolina

Series Editors

Byron Jones, Biometrical Fellow, Statistical Methodology, Integrated Information Sciences,
Novartis Pharma AG, Basel, Switzerland

Jen-pei Liu, Professor, Division of Biometry, Department of Agronomy,
National Taiwan University, Taipei, Taiwan

Karl E. Peace, Georgia Cancer Coalition, Distinguished Cancer Scholar, Senior Research Scientist
and Professor of Biostatistics, Jiann-Ping Hsu College of Public Health,
Georgia Southern University, Statesboro, Georgia

Bruce W. Turnbull, Professor, School of Operations Research and Industrial Engineering,
Cornell University, Ithaca, New York

Published Titles

**Adaptive Design Methods in
Clinical Trials, Second Edition**
Shein-Chung Chow and Mark Chang

**Adaptive Designs for Sequential
Treatment Allocation**
Alessandro Baldi Antognini and
Alessandra Giovagnoli

**Adaptive Design Theory and
Implementation Using SAS and R,
Second Edition**
Mark Chang

**Advanced Bayesian Methods for Medical
Test Accuracy**
Lyle D. Broemeling

Advances in Clinical Trial Biostatistics
Nancy L. Geller

Applied Meta-Analysis with R
Ding-Geng (Din) Chen and Karl E. Peace

**Basic Statistics and Pharmaceutical
Statistical Applications, Second Edition**
James E. De Muth

**Bayesian Adaptive Methods for
Clinical Trials**
Scott M. Berry, Bradley P. Carlin,
J. Jack Lee, and Peter Muller

**Bayesian Analysis Made Simple: An Excel
GUI for WinBUGS**
Phil Woodward

**Bayesian Methods for Measures of
Agreement**
Lyle D. Broemeling

Bayesian Methods for Repeated Measures
Lyle D. Broemeling

Bayesian Methods in Epidemiology
Lyle D. Broemeling

Bayesian Methods in Health Economics
Gianluca Baio

**Bayesian Missing Data Problems: EM,
Data Augmentation and Noniterative
Computation**
Ming T. Tan, Guo-Liang Tian,
and Kai Wang Ng

Bayesian Modeling in Bioinformatics
Dipak K. Dey, Samiran Ghosh,
and Bani K. Mallick

**Benefit-Risk Assessment in
Pharmaceutical Research and
Development**
Andreas Sashegyi, James Felli, and
Rebecca Noel

**Biosimilars: Design and Analysis of
Follow-on Biologics**
Shein-Chung Chow

Biostatistics: A Computing Approach
Stewart J. Anderson

**Causal Analysis in Biomedicine and
Epidemiology: Based on Minimal
Sufficient Causation**
Mikel Aickin

**Clinical and Statistical Considerations
in Personalized Medicine**
Claudio Carini, Sandeep Menon, and Mark Chang

Medical Biostatistics, Third Edition
A. Indrayan

Meta-Analysis in Medicine and Health Policy
Dalene Stangl and Donald A. Berry

Mixed Effects Models for the Population Approach: Models, Tasks, Methods and Tools
Marc Lavielle

Modeling to Inform Infectious Disease Control
Niels G. Becker

Modern Adaptive Randomized Clinical Trials: Statistical and Practical Aspects
Oleksandr Sverdlov

Monte Carlo Simulation for the Pharmaceutical Industry: Concepts, Algorithms, and Case Studies
Mark Chang

Multiple Testing Problems in Pharmaceutical Statistics
Alex Dmitrienko, Ajit C. Tamhane, and Frank Bretz

Noninferiority Testing in Clinical Trials: Issues and Challenges
Tie-Hua Ng

Optimal Design for Nonlinear Response Models
Valerii V. Fedorov and Sergei L. Leonov

Patient-Reported Outcomes: Measurement, Implementation and Interpretation
Joseph C. Cappelleri, Kelly H. Zou, Andrew G. Bushmakin, Jose Ma. J. Alvir, Demissie Alemayehu, and Tara Symonds

Quantitative Evaluation of Safety in Drug Development: Design, Analysis and Reporting
Qi Jiang and H. Amy Xia

Randomized Clinical Trials of Nonpharmacological Treatments
Isabelle Boutron, Philippe Ravaud, and David Moher

Randomized Phase II Cancer Clinical Trials
Sin-Ho Jung

Sample Size Calculations for Clustered and Longitudinal Outcomes in Clinical Research
Chul Ahn, Moonseong Heo, and Song Zhang

Sample Size Calculations in Clinical Research, Second Edition
Shein-Chung Chow, Jun Shao and Hansheng Wang

Statistical Analysis of Human Growth and Development
Yin Bun Cheung

Statistical Design and Analysis of Stability Studies
Shein-Chung Chow

Statistical Evaluation of Diagnostic Performance: Topics in ROC Analysis
Kelly H. Zou, Aiyi Liu, Andriy Bandos, Lucila Ohno-Machado, and Howard Rockette

Statistical Methods for Clinical Trials
Mark X. Norleans

Statistical Methods for Drug Safety
Robert D. Gibbons and Anup K. Amatya

Statistical Methods for Immunogenicity Assessment
Harry Yang, Jianchun Zhang, Binbing Yu, and Wei Zhao

Statistical Methods in Drug Combination Studies
Wei Zhao and Harry Yang

Statistics in Drug Research: Methodologies and Recent Developments
Shein-Chung Chow and Jun Shao

Statistics in the Pharmaceutical Industry, Third Edition
Ralph Buncher and Jia-Yeong Tsay

Survival Analysis in Medicine and Genetics
Jialiang Li and Shuangge Ma

Theory of Drug Development
Eric B. Holmgren

Translational Medicine: Strategies and Statistical Methods
Dennis Cosmatos and Shein-Chung Chow

Chapman & Hall/CRC Biostatistics Series

Bayesian Methods for Repeated Measures

Lyle D. Broemeling

Broemeling and Associates, Medical Lake

Washington, USA

CRC Press
Taylor & Francis Group
Boca Raton   London   New York

CRC Press is an imprint of the
Taylor & Francis Group, an **informa** business

A CHAPMAN & HALL BOOK

CRC Press
Taylor & Francis Group
6000 Broken Sound Parkway NW, Suite 300
Boca Raton, FL 33487-2742

First issued in paperback 2018

© 2016 by Taylor & Francis Group, LLC
CRC Press is an imprint of Taylor & Francis Group, an Informa business

No claim to original U.S. Government works

ISBN 13: 978-1-138-89404-4 (pbk)
ISBN 13: 978-1-4822-4819-7 (hbk)

Visit the Taylor & Francis Web site at
http://www.taylorandfrancis.com

and the CRC Press Web site at
http://www.crcpress.com

Contents

Preface

Bayesian methods are being used successfully in many areas of scientific investigation. For example, at the University of Texas MD Anderson Cancer Center, Houston, Texas, Bayesian sequential stopping rules are routinely employed in the design and analysis of clinical trials, and they are also used in diagnostic medicine in estimating the accuracy of diagnostic tests as well as for screening large populations for various chronic diseases. Of course, Bayesian inference is prevalent in other areas of science, including astronomy, engineering, biology, and cryptology.

This book is intended to be a textbook for graduate students in statistics (including biostatistics) and as a reference for consulting statisticians. It will be an invaluable resource especially for those involved in biostatistics, where use of repeated measures is common. The book adopts a practical approach and includes the WinBUGS code necessary to implement posterior analysis. Much of the book is devoted to inferential methods for the analysis of repeated measures for various areas of scientific endeavor. The student should have completed a one-year course in mathematical statistics and a one-year course in regression methods. The analysis of repeated measures is based on linear and nonlinear regression models, which incorporate correlation between observations on the same individual; therefore the student should have completed at least one course in regression analysis.

Consulting statisticians will find that this book is a valuable resource. The book will supplement other books such as *Applied Longitudinal Analysis* by Fitzmaurice, Laird, and Ware and *Statistical Methods for the Analysis of Repeated Measurements* by Davis. These two references are not Bayesian; thus, the present book is unique and presents methods that take advantage of prior information based on previous similar studies. Another feature of the book is that all the computing and analysis is based on the WinBUGS package, which provides the user a platform that efficiently uses prior information. Many of the ideas in this volume go beyond the standard non-Bayesian books. An attractive feature of the book is that the WinBUGS code can be downloaded from http://medtestacc.blogspot.com and executed as one progresses through the book.

Lyle D. Broemeling
Broemeling and Associates

Author

Lyle Broemeling has 30 years of experience as a biostatistician. He has been a professor at the following institutions: The University of Texas Medical Branch at Galveston, Texas; the University of Texas School of Public Health, Dallas, Texas; and the University of Texas MD Anderson Cancer Center in Houston, Texas. His main responsibility was to teach biostatistics and to collaborate with biomedical researchers. His expertise is in the area of Bayesian methods applied to assessing medical test accuracy and inter-rater agreement, and he has written several books on this topic. His latest book is *Bayesian Methods in Epidemiology,* and his present effort is in the analysis of repeated measures.

1

Introduction to the Analysis of Repeated Measures

1.1 Introduction

This book presents the Bayesian approach to the analysis of repeated measures. As such, the book is unique in that it is the only one from a Bayesian viewpoint to present the basic ideas about analyzing repeated measures and associated designs. In repeated measures, measurements of the same experimental unit are taken over time or over different study conditions. In a repeated measure study, the main aim is to determine the average value or mean profile of the individual over the range the measurements are observed. Thus, the focus is on the within-individual change of the average response. Repeated measure studies differ from cross-sectional designs, if the same individual is followed over time; on the other hand, with the cross-sectional design, different individuals appear throughout the observation period. A good example of a cross-sectional study occurs in clinical trials, where one group of subjects receives the treatment under study and another a different treatment (or placebo). Of course, it is possible that the same individual can receive the treatment at various time points, followed by receiving another treatment at later time points.

Often it is not realistic for the same subject to receive different treatments, but there are scenarios when it is practical. An example of a cross-sectional design is provided by Fitzmaurice, Laird, and Ware (p. 3),[1] who describe a study of a group of girls where the percent body fat is followed up to the time of menarche and another group where the percent body fat is measured at menarche and post-menarche. It is thought that the percent body fat increases at menarche but levels off after a period of approximately four years. A more efficient approach is to follow one group beginning some time before menarche followed by annual fat percent measurements until the expected time of leveling out. The latter approach is more efficient because the estimated mean difference between pre- and post-menarche has a smaller variance compared to that in a cross-sectional study. This is true because the repeated measures of percent body fat are not correlated for the cross-sectional design.

Repeated measures are an example of clustered data. In repeated measures, the observations are clustered within individuals. For various types of clustered data, that are not repeated measures of subjects over time, see Aerts et al.[2] For example, developmental toxicology studies with ethylene glycol (antifreeze) represent an illustrative scenario. Consider the example of Price,[3] where timed-pregnant CD-1 mice were dosed by gavage with ethylene glycol in distilled water (see also Aerts et al., p. 14).[2] The various doses were 0, 750, 1500, and 3000 mg/kg/day and were administered during organogenesis and fetal development. For all doses and each of the five National Toxicology Program (NTP) toxic agents (thought to be harmful to the fetus), the number of dams with at least one implant, the number of dams having at least one viable fetus, the mean litter size, and the percentage of malformed fetuses were recorded. Typically, this shows just how similar these types of clustered studies are to the typical repeated measures of subjects over time.

It is important to visualize the symbolic layout of a repeated measure study; thus consider Table 1.1, which is also displayed in Davis (p. 3).[4] The first column is the subject number, the second column is the time the observation is taken, the third column is the symbol for the missing indicator, the fourth column is the response, and the last three columns are the p covariates. In the table, y_{ij} is the response for subject i at the jth occasion, where $i = 1,2,\ldots,N$ and $j = 1,2,\ldots,t_i$.

As for the covariates, x_{ijk} is the kth covariate for subject i at the jth time point, where $k = 1,2,\ldots,p$. When y_{ij} is continuous, the mean will be expressed as a linear or nonlinear function of covariates and unknown parameters, while if the response is categorical, a function of the mean (or link function) will be expressed as a linear or nonlinear function of covariates and unknown parameters. Thus, consider Table 1.1, which is based on information from Davis (p3).

TABLE 1.1

Symbolic Plan for Repeated Measures Study

Subject	Time	Missing Indicator	Response	Covariate	Covariate	Covariate
1	1	ζ_{11}	y_{11}	x_{111}		x_{11p}
1	j	ζ_{1j}	y_{1j}	x_{1j1}	$\ldots$	x_{1jp}
1	t_1	ζ_{1t_1}	y_{1t_1}	x_{1t_11}	$\ldots$	x_{1t_1p}
2	1	ζ_{21}	y_{21}	x_{211}	$\ldots$	x_{21p}
2	j		y_{2j}	x_{2j1}	$\ldots$	x_{2jp}
2	t_2	ζ_{2t_2}	y_{2t_2}	x_{2t_21}	$\ldots$	x_{2t_2p}
N	1	ζ_{N1}	y_{N1}	x_{N11}	$\ldots$	x_{N1p}
N	j	ζ_{Nj}	y_{Nj}	x_{Nj1}	$\ldots$	x_{Njp}
N	t_N	ζ_{Nt_N}	y_{Nt_N}	x_{Nt_N1}	$\ldots$	x_{Nt_Np}

Suppose for subject i and time t_{ij} (the time for the jth occasion), y_{ij} is normally distributed and

$$y_{ij} = \beta_1 + \beta_2 t_{ij} + e_{ij} \tag{1.1}$$

where:
β_1 and β_2 are unknown parameters
the errors $e_i = (e_{i1}, e_{i2}, \ldots, e_{in})$
$i = 1, 2, \ldots, n$ are independent and have a multivariate normal distribution with mean vector zero and arbitrary n by n unknown covariance matrix Σ

The covariance matrix accounts for the positive correlation of the response over the n time points.

Last, the symbol ζ_{ij} is the missing value indicator, where $\zeta_{ij} = 1$ if the response for subject i at the jth time point is observed; otherwise $\zeta_{ij} = 0$. Often in longitudinal studies, there are missing values, and this topic will be presented in depth in Chapter 10. Chapters 2 through 10 illustrate methods that can be used for missing data if the missing value mechanism is understood. If the missing value mechanism is known, one will know what methods are appropriate for the analysis. The topic of missing data is very important and is discussed in many works, for example, Little and Rubin,[5] Schafer,[6] and Daniels and Hogan.[7] The last reference gives a good introduction to the Bayesian approach for the analysis of missing data in longitudinal studies, but the first reference is referred to for the following definitions of missing value mechanisms.

1. Missing completely at random if the chance of a missing value for an observation does not depend on what data is observed and what is missing (or what could have be observed, but isn't).

2. Missing at random if the probability of a missing value for an observation is independent of the missing observations but is not independent of the observed values.

3. Missing not at random or nonignorable if the probability of a missing value for the observation does depend on the missing values.

Suppose in a clinical trial to assess side effects of treatment, the side effects are observed at predetermined time points, and suppose the patient drops out because of severe side effects to the treatment observed at previous visits; the mechanism is missing at random, and regression techniques can be employed to impute missing values for the subsequent analysis.

In what is to follow, the fundamentals of Bayesian inference are presented, WinBUGS examples are explained, and the remaining chapters of the book are summarized.

1.2 Bayesian Inference

Bayesian methods will be employed to design and analyze studies in epidemiology, and this chapter will introduce the theory that is necessary in order to describe Bayesian inference. The Bayes theorem, the foundation of the subject, is first introduced and followed by an explanation of its various components: prior information; information from the sample given by the likelihood function; the posterior distribution, which is the basis of all inferential techniques; and last the Bayesian predictive distribution. A description of the main three elements of inference, namely, estimation, tests of hypotheses, and forecasting future observations, follows.

The remaining sections refer to the important standard distributions for Bayesian inference, namely, the Bernoulli, beta, multinomial, Dirichlet, normal, gamma, normal-gamma, multivariate normal, Wishart, normal-Wishart, and multivariate t distributions.

As will be seen, knowledge of these standard distributions in a discussion of inferential techniques is essential for understanding the analysis of methods used in longitudinal studies.

Of course, inferential procedures can only be applied if there is adequate computing power available. If the posterior distribution is known, often analytical methods are quite sufficient to implement Bayesian inferences and will be demonstrated for the binomial, multinomial, and Poisson populations as well as for several cases of normal populations. For example, when using a beta prior distribution for the parameter of a binomial population, the resulting beta posterior density has well-known characteristics, including its moments. In a similar fashion, when sampling from a normal population with unknown mean and precision and with a vague improper prior, the resulting posterior t distribution for the mean has known moments and percentiles, which can be used for inferences.

Posterior inferences by direct sampling methods are easily done if the relevant random number generators are available. On the other hand, if the posterior distribution is quite complicated and not recognized as a standard distribution, other techniques are needed. To solve this problem, Markov chain Monte Carlo (MCMC) techniques have been developing for the last 25 years and have been a major success in providing Bayesian inferences for quite complicated problems. This has been a great achievement in the field and will be described in Section 1.25.

Minitab, S-Plus, and WinBUGS are packages that provide random number generators for direct sampling from the posterior distribution for many standard distributions, such as binomial, gamma, beta, and t distributions. On occasion, these will be used; however, my preference is for WinBUGS, because it has been well accepted by other Bayesians. This is also true for indirect sampling, where WinBUGS is a good package and is the software of choice for the book; it is introduced in Appendix B. Many institutions provide special purpose software for specific Bayesian routines.

For example, at the MD Anderson Cancer Center, where Bayesian applications are routine, several special purpose programs are available for designing (including sample-size justification) and analyzing clinical trials, and they will be described further. The theoretical foundation of MCMC is introduced in Section 1.19.

Inferences for studies in epidemiology consist of testing hypotheses about unknown population parameters, estimation of those parameters, and forecasting future observations. When a sharp null hypothesis is involved, special care is taken in specifying the prior distribution for the parameters. A formula for the posterior probability of the null hypothesis is derived, via Bayes's theorem, and illustrated for Bernoulli, Poisson, and normal populations. If the main focus is estimation of parameters, the posterior distribution is determined, and the mean, median, standard deviation, and credible intervals found, either analytically or by computation with WinBUGS. For example, when sampling from a normal population with unknown parameters and using a conjugate prior density, the posterior distribution of the mean is a *t* and will be derived algebraically. On the other hand, in observational studies, the experimental results are usually portrayed in a 2 by 2 table that gives the cell frequencies for the four combinations of exposure and disease status, where the consequent posterior distributions are beta for the cell frequencies, and posterior inferences are provided both analytically and with WinBUGS. Of course, all analyses should be preceded by checking to determine if the model is appropriate, and this is where predictive distribution comes into play. By comparing the observed results of the experiment (e.g., a case-control study) with those predicted, the model assumptions are tested. The most frequent use of the Bayesian predictive distribution is for forecasting future observations in time series studies and in cohort studies.

1.3 Bayes's Theorem

Bayes's theorem is based on the conditional probability law:

$$P[A \mid B] = \frac{P[B \mid A]P[A]}{P[B]} \tag{1.2}$$

where:
   $P[A]$ is the probability of A before one knows the outcome of the event B
   $P[B \mid A]$ is the probability of B assuming what one knows about the event A
   $P[A \mid B]$ is the probability of A knowing that event B has occurred
   $P[A]$ is called the prior probability of A, while $P[A \mid B]$ is called the posterior probability of A

Another version of Bayes's theorem is as follows: suppose X is a continuous observable random vector and $\theta \in \Omega \subset R^m$ is an unknown parameter vector, and suppose the conditional density of X given θ is denoted by $f(x|\theta)$. If $x = (x_1, x_2, \ldots, x_n)$ represents a random sample of size n from a population with density $f(x|\theta)$, and $\xi(\theta)$ is the prior density of θ, then Bayes's theorem expresses the posterior density as

$$\xi(\theta \mid x) = c \prod_{i=1}^{i=n} f(x_i \mid \theta)\xi(\theta), x_i \in R \text{ and } \theta \in \Omega \qquad (1.3)$$

where:
  c is the proportionality constant
  the term $\prod_{i=1}^{i=n} f(x_i \mid \theta)$ is called the likelihood function

The density $\xi(\theta)$ is the prior density of θ and represents the knowledge one possesses about the parameter before one observes X. Such prior information is most likely available to the experimenter from other previous related experiments. Note that θ is considered a random variable and that Bayes's theorem transforms one's prior knowledge of θ, represented by its prior density, to the posterior density, and that the transformation is the combining of the prior information about θ with the sample information represented by the likelihood function.

"An Essay towards Solving a Problem in the Doctrine of Chances" by the Reverend Thomas Bayes[8] is the beginning of our subject. He considered a binomial experiment with n trials and assumed that the probability θ of success was uniformly distributed (by constructing a billiard table) and presented a way to calculate $\Pr(a \leq \theta \leq b \mid x=p)$, where x is the number of successes in n independent trials.

This was a first in the sense that Bayes was making inferences via $\xi(\theta \mid x)$, the conditional density of θ given x. Also, by assuming the parameter to be uniformly distributed, he was assuming vague prior information for θ. The type of prior information, where very little is known about the parameter, is called noninformative or vague information.

It can well be argued that Laplace[9] is the greatest Bayesian because he made many significant contributions to inverse probability (he did not know of Bayes), beginning in 1774 with "Memorie sur la probabilite des causes par la evenemens," with his own version of Bayes's theorem, over a period of some 40 years culminating in "Theorie analytique des probabilites." See Stigler[10] and Chapters 9 through 20 of Hald[11] for the history of Laplace's contributions to inverse probability.

It was in modern times that Bayesian statistics began its resurgence, with Lhoste,[12] Jeffreys,[13] Savage,[14] and Lindley.[15] According to Broemeling and Broemeling,[16] Lhoste was the first to justify noninformative priors by invariance principals, a tradition carried on by Jeffreys. Savage's book was a major contribution in that Bayesian inference and decision theory was put on a sound theoretical footing as a consequence of certain axioms of probability and utility, while Lindley's two volumes showed the relevance of Bayesian inference to everyday statistical problems and was quite influential and set the tone and style for later books by, for example, Broemeling,[16] Box and Tiao,[17] and Zellner.[18] Box and Tiao and Broemeling completed essentially works that presented Bayesian methods for the usual statistical problems of the analysis of variance and regression, while Zellner focused Bayesian methods primarily on certain regression problems in econometrics. During this period, inferential problems were solved analytically or by numerical integration. Models with many parameters (such as hierarchical models with many levels) were difficult to use because at that time numerical integration methods had limited capability in higher dimensions. For a good history of inverse probability, see Chapter 3 of Stigler[10] and see Hald,[11] which present a comprehensive history and are invaluable as a reference. Dale[19] gives a complete and very interesting account of Bayes's life.

The last 20 years is characterized by the rediscovery and development of simulation techniques, where samples are generated from the posterior distribution via MCMC methods, such as Gibbs sampling. Large samples generated from the posterior make it possible to make statistical inferences and to employ multilevel hierarchical models to solve complex but practical problems. See Leonard and Hsu,[20] Gelman et al.,[21] Congdon,[22] Carlin and Louis,[23] and Gilks, Richardson, and Spiegelhalter,[24] who demonstrate the utility of MCMC techniques in Bayesian statistics.

1.4 Prior Information

Where do we begin with prior information, a crucial component of Bayes's theorem rule? Bayes assumed the prior distribution of the parameter is uniform, namely

$$\xi(\theta) = 1, 0 \le \theta \le 1$$

where θ is the common probability of success in n independent trials and

$$f(x \mid \theta) = \binom{n}{x} \theta^x (1-\theta)^{n-x} \tag{1.4}$$

where:
   x is the number of successes $= 0,1,2,\ldots,n$

The distribution of X, the number of successes, is binomial and denoted by $X \sim$ binomial (θ,n). The uniform prior was used for many years; however, Lhoste[12] proposed a different prior, namely

$$\xi(\theta) = \theta^{-1}(1-\theta)^{-1}, 0 \le \theta \le 1 \tag{1.5}$$

to represent information that is noninformative and is an improper density function. Lhoste based the prior on certain invariance principles, quite similar to Jeffreys.[13] Lhoste also derived a noninformative prior for the standard deviation σ of a normal population with density

$$f(x \mid \mu, \sigma) = \left(\frac{1}{\sqrt{2\pi}\sigma}\right) \exp-\left(\frac{1}{2\sigma}\right)(x-\mu)^2, \mu \in R \text{ and } \sigma > 0 \tag{1.6}$$

He used invariance as follows: he reasoned that the prior density of σ and the prior density of $1/\sigma$ should be the same, which leads to

$$\xi(\sigma) = \frac{1}{\sigma} \tag{1.7}$$

Jeffreys's approach is similar in that in developing noninformative priors for binomial and normal populations, but he also developed noninformative priors for multiparameter models, including the mean and standard deviation for the normal density as

$$\xi(\mu, \sigma) = \frac{1}{\sigma}, \mu \in R \text{ and } \sigma > 0 \tag{1.8}$$

Noninformative priors were ubiquitous from the 1920s to the 1980s and were included in all the textbooks of that period. For example, see Box and Tiao,[17] Zellner,[18] and Broemeling.[25] Looking back, it is somewhat ironic that noninformative priors were almost always used, even though informative prior information was almost always available. This limited the utility of the Bayesian approach, and people saw very little advantage over the conventional way of doing business. The major strength of the Bayesian way is that it is a convenient, practical, and logical method of utilizing informative prior information. Surely, the investigator knows informative prior information from previous related studies.

How does one express informative information with a prior density? Suppose one has informative prior information for the binomial population. Consider

$$\xi(\theta) = \frac{\Gamma(\alpha+\beta)}{\Gamma(\alpha)\Gamma(\beta)} \theta^{\alpha-1}(1-\theta)^{\beta-1}, 0 \le \theta \le 1 \tag{1.9}$$

as the prior density for θ.

The beta density with parameters α and β has mean $[\alpha/(\alpha+\beta)]$ and variance $[\alpha\beta/(\alpha+\beta)^2(\alpha+\beta+1)]$ and can express informative prior information in many ways. Suppose from a previous cohort study with 20 exposed subjects, there were 6 who developed disease and 14 who did not develop disease, then the probability mass function for the observed number of successes $x = 6$

$$f(6 \mid \theta) = \binom{20}{6} \theta^6 (1-\theta)^{14}, 0 \leq \theta \leq 1 \tag{1.10}$$

As a function of θ (the incidence rate of disease for those exposed), (1.10) is a beta distribution with parameter vector (7,15) and expresses informative prior information, which is combined with (1.10), via Bayes's theorem, in order to make inferences (estimation, tests of hypotheses, and predictions) about the incidence rate θ for the exposed subjects. The beta distribution is an example of a conjugate density, because the prior and posterior distributions for θ belong to the same parametric family. Thus, the likelihood function based on previous sample information can serve as a source of informative prior information. The binomial and beta distributions occur quite frequently in longitudinal studies, and many examples involving a binary response will be presented.

Of course, the normal density also plays an important role as a model in repeated measures. For example, as will be seen in Chapter 9, the normal distribution will model the distribution of observations that occur in repeated measures designs. For example, the measured value of blood glucose can be considered a continuous measurement for diagnosing diabetes. How is informative prior information expressed for the parameters μ and σ (the mean and standard deviation)?

Suppose a previous study has m observations $X = (x_1 x_2, ..., x_m)$, then the density of X given μ and σ is

$$f(x \mid \mu, \sigma) \propto \left(\frac{\sqrt{m}}{\sqrt{2\pi\sigma^2}} \right) \exp - \left(\frac{m}{2\sigma^2} \right)(\bar{x} - \mu)^2$$

$$\left[(2\pi)^{-(m-1)/2} \sigma^{-(m-1)} \right] \exp - \left(\frac{1}{2\sigma^2} \right) \sum_{i-1}^{i=m} (x_i - \bar{x})^2 \tag{1.11}$$

This is a conjugate density for the two-parameter normal family and is called the normal-gamma density. Note it is the product of two function, where the first, as a function of μ and σ, is the conditional density of μ given σ, with mean $\bar{x}$ and variance σ^2/m, while the second is a function of σ only and is an inverse gamma density. Or equivalently, if the normal is parameterized with μ and the precision $\tau = 1/\sigma^2$, the conjugate distribution is as follows: (1) the conditional distribution of μ given τ is normal with mean $\bar{x}$ and precision $m\tau$ and (2) the marginal distribution of τ is gamma with parameters $(m + 1)/2$ and $\sum_{i=1}^{i=m} (x_i - \bar{x})^2/2 = (m-1)S^2/2$, where S^2 is the sample variance. Thus, if one knows the results of a previous experiment (say from related studies for type II diabetes), the likelihood function for μ (the population mean blood glucose) and τ provides informative prior information for the normal population.

1.5 Posterior Information

The preceding section explained how prior information is expressed in an informative or in a noninformative way. Several examples are given and will be revisited as illustrations for the determination of the posterior distribution of the parameters. In the Bayes example, where $X \sim$ binomial (θ, n), a uniform distribution for the incidence rate (of a cohort study) θ is used. What is the posterior distribution? By Bayes's theorem,

$$\xi(\theta \mid x) \propto \binom{n}{x} \theta^x (1-\theta)^{n-x} \tag{1.12}$$

where:

   x is the observed number of subjects with disease among n exposed subjects

Of course, this is recognized as a beta $(x + 1, n - x + 1)$ distribution, and the posterior mean is $(x + 1)/(n + 2)$. On the other hand, if the Lhoste prior density is used, the posterior distribution of θ is beta$(x, n - x)$ with mean x/n, the usual estimator of θ. The conjugate prior (1.9) results in a beta$(x + \alpha, n - x + \beta)$ with mean $(x + \alpha)/(n + \alpha + \beta)$. Suppose the prior is informative in a previous study with 10 subjects with disease among 30 subjects, then $\alpha = 11$ and $\beta = 21$, and the posterior distribution is beta $(x + 11, n - x + 21)$. If the current cohort study has 40 exposed subjects and 15 have disease, the posterior distribution is beta$(26, 46)$ with mean $26/72 = 0.361$, which is the estimated incidence rate compared to a prior estimated incidence rate of 0.343.

   Consider a random sample $X = (x_1, x_2, ..., x_n)$ of size n from a normal $(\mu, 1/\tau)$ population, where $\tau = 1/\sigma^2$ is the inverse of the variance, and suppose the prior information is vague and the Jeffreys–Lhoste prior $\xi(\mu, \tau) \propto 1/\tau$ is appropriate, then the posterior density of the parameters is

$$\xi(\mu, \tau \mid \text{data}) \propto \tau^{n/2-1} \exp{-\left(\frac{\tau}{2}\right)} \left[n(\mu - \bar{x})^2 + \sum_{i=1}^{i=n} (x_i - \bar{x})^2 \right] \tag{1.13}$$

Using the properties of the gamma density, τ is eliminated by integrating the joint density with respect to τ to give

$$\xi(\mu \mid \text{data}) \propto \frac{\left[\Gamma(n/2)n^{1/2} / (n-1)^{1/2} S\pi^{1/2} \Gamma(n - 10/2) \right]}{\left[1 + n(\mu - \bar{x})^2 / (n-1)S^2 \right]^{(n-1+1)/2}} \tag{1.14}$$

which is recognized as a t distribution with $n - 1$ degrees of freedom, location $\bar{x}$, and precision n/S^2. Transforming to $(\mu - \bar{x})\sqrt{n}/S$, the resulting variable has a Student's t distribution with $n - 1$ degrees of freedom. Note the mean of μ is the sample mean, while the variance is $[(n - 1)/n(n - 3)]S^2$, $n > 3$.

Eliminating μ from (1.14) results in the marginal distribution of τ as

$$\xi(\tau \mid S^2) \propto \tau^{[(n-1)/2]-1} \exp-\tau\frac{(n-1)S^2}{2}, \tau > 0 \tag{1.15}$$

which is a gamma density with parameters $(n - 1)/2$ and $(n - 1)S^2/2$. This implies the posterior mean is $1/S^2$ and the posterior variance is $2/(n - 1)S^4$.

The Poisson distribution often occurs as a population for a discrete random variable with mass function

$$f(x \mid \theta) = \frac{e^{-\theta}\theta^x}{x!} \tag{1.16}$$

where the gamma density

$$\xi(\theta) = \left[\frac{\beta^\alpha}{\Gamma(\alpha)}\right]\theta^{\alpha-1}e^{-\theta\beta} \tag{1.17}$$

is a conjugate distribution that expresses informative prior information. For example, in a previous experiment with m observations, the prior density would be (1.17) with the appropriate values of alpha and beta. Based on a random sample of size n, the posterior density is

$$\xi(\theta \mid data) \propto \theta^{\sum_{i=1}^{i=n}x_i+\alpha-1}e^{-\theta(n+\beta)} \tag{1.18}$$

which is identified as a gamma density with parameters $\alpha'=\sum_{j=1}^{i=n}x_i + \alpha$ and $\beta' = n+\beta$. Remember the posterior mean is α'/β', median $(\alpha'-1)/\beta'$, and variance $\alpha'/(\beta')^2$. The Poisson distribution is often assumed to be the distribution of discrete repeated measures and will be illustrated with many examples in Chapter 9, the repeated measures of categorical data.

1.6 Posterior Inference

In a statistical context, by inference one usually means estimation of parameters, tests of hypotheses, and prediction of future observations. With the Bayesian approach, all inferences are based on the posterior distribution of the parameters, which in turn is based on the sample, via the likelihood function and the prior distribution. We have seen the role of the prior density and likelihood function in determining the posterior distribution and presently will focus on the determination of point and interval estimation of the model parameters and later will emphasize how the posterior distribution determines a test of hypothesis. Last, the role of the predictive distribution in testing hypotheses and in goodness of fit will be explained.

When the model has only one parameter, one would estimate that parameter by listing its characteristics, such as the posterior mean, media, and standard deviation and plotting the posterior density. On the other hand, if there are several parameters, one would determine the marginal posterior distribution of the relevant parameters and, as aforementioned, calculate its characteristics (e.g., mean, median, mode, standard deviation) and plot the densities. Interval estimates of the parameters are also usually reported and are called credible intervals.

1.7 Estimation

Suppose we want to estimate θ of the binomial example of the previous section, where the number of people with disease is X, which has a binomial distribution with θ, the incidence rate of those subjects exposed to the risk factor, and the posterior distribution is beta(21, 46) with the following characteristics: mean = 0.361, median = 0.362, standard deviation = 0.055, lower 2½ percent point = 0.254, and upper 2½ percent point = 0.473. The mean and median are the same, while the lower and upper 2½ percent points determine a 95% credible interval of (0.254, 0.473) for θ.

Inferences for the normal (μ, τ) population are somewhat more demanding, because both parameters are unknown. Assuming the vague prior density $\xi(\mu, \tau) \propto 1/\tau$, the marginal posterior distribution of the population mean μ is a t distribution with $n - 1$ degrees of freedom, mean $\bar{x}$, and precision n/S^2; thus, the mean and the median are the same and provide a natural estimator of μ, and because of the symmetry of the t density, a $(1 - \alpha)$ credible interval for μ is $\bar{x} \pm t_{\alpha/2, n-1} S/\sqrt{n}$, where $t_{\alpha/2, n-1}$ is the upper $100\alpha/2$ percent point of the t distribution with $n - 1$ degrees of freedom. To generate values from the $t(n - 1, \bar{x}, n/S^2)$ distribution, generate values from Student's t distribution with $n - 1$ degrees of freedom, multiply each by $S/\sqrt{n}$, and then add $\bar{x}$ to each. Suppose $n = 30$,

$$x = (7.8902, 4.8343, 11.0677, 8.7969, 4.0391, 4.0024, 6.6494, 8.4788, 0.7939,$$
$$5.0689, 6.9175, 6.1092, 8.2463, 10.3179, 1.8429, 3.0789, 2.8470, 5.1471,$$
$$6.3730, 5.2907, 1.5024, 3.8193, 9.9831, 6.2756, 5.3620, 5.3297, 9.3105,$$
$$6.5555, 0.8189, 0.4713)$$

then $\bar{x} = 5.57$ and $S = 2.92$.

Using the same data set, **BUGS CODE 1.1** is used to analyze the problem.

Note that a somewhat different prior was employed here, compared to that used previously, in that μ and τ are independent and assigned proper but noninformative distributions. The corresponding analysis gives Table 1.2.

BUGS CODE 1.1

```
Model;
{for(i in 1:30) {x[i]~dnorm(mu,tau)}
mu~dnorm(0.0,.0001)
tau ~dgamma(.0001,.0001)                                    (4.17)
sigma <- 1/tau}
list(x = c(7.8902,4.8343,11.0677,8.7969,4.0391,4.0024,6.6
494,8.4788,0.7939,5.0689,6.9175,6.1092,8.2463,10.3179,1.8
429,3.0789,2.8470,5.1471,6.3730,5.2907,1.5024,3.8193,9.98
31,6.2756,5.3620,5.3297,9.3105,6.5555,0.8189,0.4713))
list(mu = 0, tau = 1)
```

TABLE 1.2

Posterior Distribution of μ and $\sigma = 1/\sqrt{\tau}$

Parameter	Mean	SD	MC Error	Median	Lower	Upper
μ	5.572	0.5547	0.003566	5.571	4.4790	6.656
σ	9.15	2.570	0.01589	8.733	5.359	15.37

Upper and lower refer to the lower and upper 2½ percent points of the posterior distribution. Note a 95% credible interval for μ is (4.47, 6.65) and the estimation error is 0.003566. See the following section for details on executing the WinBUGS statement.

The program generated 30,000 samples from the joint posterior distribution of μ and σ using a Gibbs sampling algorithm and used 29,000 for the posterior moments and graphs, with a refresh of 100.

1.8 Testing Hypotheses

An important feature of inference is testing hypotheses. Often in longitudinal studies, the scientific hypothesis of a study can be expressed in statistical terms and a formal test is implemented. Suppose $\Omega = \Omega_0 \cup \Omega_1$ is a partition of the parameter space; then the null hypothesis is designated as H: $\theta \in \Omega_0$ and the alternative by A: $\theta \in \Omega_1$, and a test of H versus A consists of rejecting H in favor of A if the observations $x = (x_1, x_2, ..., x_n)$ belong to a critical region C. In the usual approach, the critical region is based on the probabilities of type I errors, namely $\Pr(C|\theta)$, where $\theta \in \Omega_0$, and of type II errors, $1-\Pr(C|\theta)$, where $\theta \in \Omega_1$. This approach to testing a hypothesis was developed formally by Neyman and Pearson and can be found in many of the standard references, such as Lehmann.[26] Lee[27] presents a good elementary introduction to testing and estimation in a Bayesian context. Our approach is to reject the null hypothesis if the

95% credible region for θ does not contain the set of all θ such that $\theta \in \Omega_0$. In the special case that H: $\theta = \theta_0$ versus the alternative A: $\theta \neq \theta_0$, where θ is a scalar, H is rejected when the 95% confidence interval for θ does not include θ_0.

1.9 Predictive Inference

Our primary interest in the predictive distribution is to check for model assumptions. Is the adopted model for an analysis the most appropriate?

What is the predictive distribution of a future set of observations Z? It is the conditional distribution of Z given $X = x$, where x represents the past observations, which when expressed as a density is

$$g(z \mid x) = \int_{\Omega} f(z \mid \theta)\xi(\theta \mid x)d\theta, z \in R^m \tag{1.19}$$

where the integral is with respect to θ and $f(x \mid \theta)$ is the density of $X = (x_1, x_2 ..., x_n)$ given θ. This assumes that given θ, Z and X are independent. Thus the predictive density is posterior average of $f(z \mid \theta)$.

The posterior predictive density will be derived for the binomial and normal populations.

1.10 The Binomial

Suppose the binomial case is again considered, where the posterior density of the binomial parameter θ is

$$\xi(\theta \mid x) = \left[\frac{\Gamma(\alpha + \beta)\Gamma(n + 1)}{\Gamma(\alpha)\Gamma(\beta)\Gamma(x + 1)\Gamma(n - x + 1)} \right] \theta^{\alpha + x - 1}(1 - \theta)^{\beta + n - x - 1} \tag{1.20}$$

a beta with parameters $\alpha + x$ and $n - x + \beta$, and x is the sum of the set of n observations. The population mass function of a future observation Z is $f(z \mid \theta) = \theta^z(1 - \theta)^{1-z}$; thus, the predictive mass function of Z, called the beta-binomial, is

$$g(z \mid x) = \frac{\Gamma(\alpha + \beta)\Gamma(n + 1)\Gamma\left(\alpha + \sum_{i=1}^{i=n} x_i + z\right)\Gamma(1 + n + \beta - x - z)}{\Gamma(\alpha)\Gamma(\beta)\Gamma(n - x + 1)\Gamma(x + 1)\Gamma(n + 1 + \alpha + \beta)} \tag{1.21}$$

where:
  $z = 0, 1$

Note this function does not depend on the unknown parameter, and that the n past observations are known, and that if $\alpha = \beta = 1$, one is assuming a uniform prior density for θ.

1.11 Forecasting from a Normal Population

Moving on to the normal density with both parameters unknown, what is the predictive density of Z, with noninformative prior density?

$$\xi(\mu, \tau) = \frac{1}{\tau}, \mu \in R \text{ and } \sigma > 0$$

The posterior density is

$$\xi(\mu, \tau \mid \text{data}) = \left[\frac{\tau^{n/2-1}}{(2\pi)^{n/2}}\right] \exp\left(\frac{\tau}{2}\right)\left[n(\mu - \bar{x})^2 + (n-1)S_x^2\right]$$

where:
$\bar{x}$ and S_x^2 are the sample mean and variance, based on a random sample of size n, $x = (x_1, x_2, ..., x_n)$

Suppose z is a future observation, then the predictive density of Z is

$$g(z \mid x) = \iint \left[\frac{\tau^{(n)/2-1}}{(2\pi)^{(n)/2}}\right] \exp-(\tau/2)\left[n(\mu - \bar{x})^2 + (n-1)S_x^2\right] \qquad (1.22)$$

where the integration is with respect to $\mu \in R$ and $\sigma > 0$. This simplifies to a t density with $d = n - 1$ degrees of freedom, location $\bar{x}$, and precision

$$p = \frac{n}{(n+1)S_x^2} \qquad (1.23)$$

Recall that a t density with d degrees of freedom, location $\bar{x}$, and precision p has density

$$g(t) = \left\{\frac{\Gamma[(d+1)/2]p^{1/2}}{\Gamma(d/2)(d\pi)^{1/2}}\right\}\left[\frac{1 + (t - \bar{x})^2 p}{d}\right]^{-(d+1)/2}$$

where $t \in R$, the mean is $\bar{x}$, and the variance is $d/(d - 2)p$.

The predictive distribution can be used as an inferential tool to test hypotheses about future observations, to estimate the mean of future observations, and to find confidence bands for future observations. In the context of repeated measures, the predictive distribution for future normal observations will be employed to impute missing blood glucose values.

1.12 Checking Model Assumptions

It is imperative to check model adequacy in order to choose an appropriate model and to conduct a valid study. The approach taken here is based on many sources, including Gelman et al. (Chapter 6),[21] Carlin and Louis (Chapter 5),[23] and Congdon (Chapter 10).[22] Our main focus will be on the likelihood function of the posterior distribution, and not on the prior distribution, and to this end, graphical representations such as histograms, box plots, and various probability plots of the original observations will be compared to those of the observations generated from the predictive distribution. In addition to graphical methods, Bayesian versions of overall goodness of fit type operations are taken to check model validity. Methods presented at this juncture are just a small subset of those presented in more advanced works, including Gelman et al., Carlin and Louis,[23] and Congdon.[22]

Of course, the prior distribution is an important component of the analysis, and if one is not sure of the "true" prior, one should perform a sensitivity analysis to determine the robustness of posterior inferences to various alternative choices of prior information. See Gelman et al. or Carlin and Louis for details of performing a sensitivity study for prior information. Our approach is to use either informative or vague prior distributions, where the former is done when prior relevant experimental evidence determines the prior or the latter is taken if there is none or very little germane experimental studies. In scientific studies, the most likely scenario is that there are relevant experimental studies providing informative prior information.

1.13 Sampling from an Exponential, but Assuming a Normal Population

Consider a random sample of size 30 from an exponential distribution with mean 3. An exponential distribution is often used to model the survival times of a screening test.

$$x = (1.9075, 0.7683, 5.8364, 3.0821, 0.0276, 15.0444, 2.3591, 14.9290,$$
$$6.3841, 7.6572, 5.9606, 1.5316, 3.1619, 1.5236, 2.5458, 1.6693, 4.2076,$$
$$6.7704, 7.0414, 1.0895, 3.7661, 0.0673, 1.3952, 2.8778, 5.8272, 1.5335,$$
$$7.2606, 3.1171, 4.2783, 0.2930)$$

The sample mean and standard deviation are 4.13 and 3.739, respectively.

Assume the sample is from a normal population with unknown mean and variance, with an improper prior density $\xi(\mu, \tau) = 1/\tau$, $\mu \in R$ and $\sigma > 0$, then the posterior predictive density is a univariate t with $n - 1 = 29$ degrees of freedom, mean $\bar{x} = 3.744$, standard deviation $= 3.872$, and precision $p = 0.0645$. This is verified from the original observations x and the formula for the precision. From the predictive distribution, 30 observations are generated:

$$z = (2.76213, 3.46370, 2.88747, 3.13581, 4.50398, 5.09963, 4.39670, 3.24032,$$
$$3.58791, 5.60893, 3.76411, 3.15034, 4.15961, 2.83306, 3.64620, 3.48478,$$
$$2.24699, 2.44810, 3.39590, 3.56703, 4.04226, 4.00720, 4.33006, 3.44320,$$
$$5.03451, 2.07679, 2.30578, 5.99297, 3.88463, 2.52737)$$

which gives a mean of $\bar{z} = 3.634$ and standard deviation $S = 0.975$. The histograms for the original and predicted observations are portrayed in Figures 1.1 and 1.2, respectively.

The histograms obviously are different, where for the original observations, a right skewness is depicted; however, this is lacking for the histogram of the predicted observations, which is for a t distribution. Although the example seems trivial, it would not be for the first time that exponential observations were analyzed as if they were generated from a normal population. The sample statistics do not detect the discrepancy, because they are very similar; the mean and standard deviation for the exponential are 3.744 and 3.872, respectively, but are 3.688 and 3.795 for the predicted observations. Thus, it is important to use graphical techniques to assess model adequacy.

It would be interesting to generate more replicate samples from the predictive distribution in order to see if these conclusions hold firm.

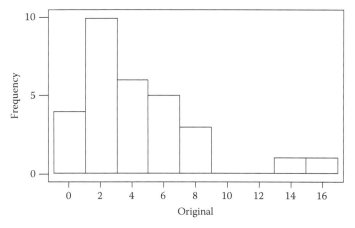

FIGURE 1.1
Histogram of original observations.

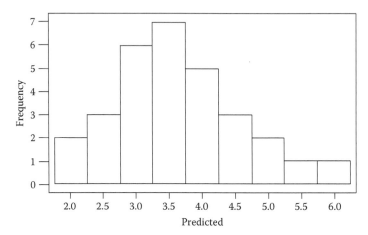

FIGURE 1.2
Histogram of predicted observations.

1.14 Poisson Population

It is assumed the sample is from a Poisson population; however, it is actually generated from a uniform discrete population over the integers from 0 to 10. The sample of size 25 is $x =$ (8, 3, 8, 2, 6, 1, 0, 2, 4, 10, 7, 9, 5, 4, 8, 4, 0, 9, 0, 3, 7, 10, 7, 5, 1), with a sample mean of 4.92 and standard deviation of 3.278. When the population is Poisson, $P(\theta)$, and an uninformative prior

$$\xi(\theta) = \frac{1}{\theta}, \theta > 0$$

is appropriate, the posterior density is gamma with parameters alpha = $\sum_{i=1}^{i=25} x_i = 123$ and beta $= n = 25$. Observations z from the predictive distribution are generated by taking a sample θ from the gamma posterior density, then selecting a z from the Poisson distribution $P(\theta)$. This was repeated 25 times to give $z =$ (2, 5, 6, 2, 4, 3, 5, 3, 2, 3, 3, 6, 7, 5, 5, 3, 1, 5, 7, 3, 5, 3, 6, 4, 5), with a sample mean of 4.48 and standard deviation of 1.896.

The most obvious difference show a symmetric sample from the discrete uniform population, but on the other hand, box plots of the predicted observations reveal a slight skewness to the right. The largest difference is in the interquartile ranges being (2, 8) for the original observations and (3, 5.5) for the predictive sample. Although there are some differences, to therefore declare that the Poisson assumption is not valid might be premature. One should generate more replicate samples from the predictive distribution to reveal additional information.

1.15 Measuring Tumor Size

A study of agreement involving the lesion sizes of five radiologists assumed that the observations were normally distributed. Is this assumption valid? A probability plot of 40 lesion sizes of one replication (there were two) of reader labeled 1 would show the normal distribution use a reasonable assumption.

Is this the implication from the Bayesian predictive density? The original observations are:

$$x = (3.5, 3.8, 2.2, 1.5, 3.8, 3.5, 4.2, 5.4, 7.6, 2.8, 5.0, 2.3, 4.4, 2.5, 5.2, 1.7, 4.5, 4.5,$$
$$6.0, 3.3, 7.0, 4.0, 4.8, 5.0, 2.7, 3.7, 1.8, 6.3, 4.0, 1.5, 2.2, 2.2, 2.6, 3.7, 2.5, 4.8,$$
$$8.0, 4.0, 3.5, 4.8)$$

Descriptive statistics are mean = 3.920 cm, standard deviation = 1.612 cm, and interquartile range of (2.525, 4.800). The basic statistics for the predicted observations are mean = 4.017 and standard deviation = 1.439 cm, with an interquartile range of (4.069, 4.930), and were based on the future observations

$$z = (4.85, 4.24, 3.32, 1.84, 5.56, 3.40, 4.02, 1.38, 4.21, 6.26, 0.55, 4.56, 5.09, 4.51,$$
$$3.28, 3.94, 5.05, 7.23, 4.19, 4.85, 4.24, 2.86, 3.98, 2.00, 2.99, 3.50, 2.53, 1.95,$$
$$6.07, 4.68, 5.39, 1.89, 5.79, 5.86, 2.85, 3.62, 4.95, 4.46, 4.22, 4.33)$$

A comparison of the histograms for the original and predicted observations would show some small differences in the two distributions, but the differences would not be striking; however, the predicted observations appear to be somewhat more symmetric about the mean than those of the original observations. Also, the corresponding sample means, standard deviations, modes, and interquartile ranges are quite alike for the two samples. There is nothing that stands out that implies questioning the validity of the normality assumption. Of course, additional replicates from the posterior predictive density should be computed for additional information about discrepancies between the two.

1.16 Testing the Multinomial Assumption

A good example of a multinomial distribution appears as an outcome of the Shields Heart Study carried out in Spokane Washington from 1993 to 2001 at the Shields Coronary Artery Center where the disease status of coronary artery disease and various risk factors (coronary artery calcium, diabetes, smoking status, blood pressure, and cholesterol, etc.) were measured on each of about 4300 patients. The average age (SD) of 4386 patients was 55.14 (10.757) years,

TABLE 1.3

The Shields Heart Study

Coronary Calcium	+ Disease (Heart Attack)	− Nondisease (No Infarction)
+ (positive)	$\theta_{++}, n_{++} = 119$	$\theta_{+-}, n_{+-} = 2461$
− (negative)	$\theta_{-+}, n_{-+} = 11$	$\theta_{--}, n_{--} = 1798$

with an average age of 55.42 (10.78) years for 2712 males and 56.31 (10.61) years for 1674 females. The main emphasis of this study was to investigate the ability of coronary artery calcium (as measured by computed tomography) to diagnose coronary artery disease. A typical table from this study investigates the association between having a heart attack and the fraction of subjects with a positive reading for coronary calcium. For additional information about the Shield Heart Study, see Mielke, Shields, and Broemeling (Table 1.3).[28]

A multinomial model is valid if the 4386 patients are selected at random from some well-defined population, if the responses of an individual are independent of the other respondents, and if the probability of an event (the disease status and status of coronary calcium) is the same for all individuals for a particular cell of the table.

In some studies, it is difficult to know if the multinomial population is valid, for example, with so-called chart reviews, where medical records are selected, not necessarily at random, but from some population determined by the eligibility criterion of the study. Thus, in the case of epidemiologic studies, such as those discussed above, the best way to determine validity of the multinomial model is to know the details of how the study was designed and conducted. The crucial issue in the multinomial model is independence, that is, given the parameters of the multinomial, the results of one patient are independent of those of another. It is often the case that the details of the study are not available. The other important aspect of a multinomial population is that the probability of a particular outcome is constant over all patients. One statistical way to check is to look for runs in the sequence.

In Table 1.3, one could condition on the row totals, and check if the binomial model is valid for each row. The multinomial distribution will be assumed for categorical repeated measures studies of Chapter 9.

1.17 Computing

This section introduces the computing algorithms and software that will be used for the Bayesian analysis of problems encountered in longitudinal studies. In the previous sections of the chapter, direct methods (noniterative) of computing the characteristics of the posterior distribution were demonstrated with some standard one sample and two sample problems.

An example of this is the posterior analysis of a normal population, where the posterior distribution of the mean and variance is generated from its posterior distribution by the t distribution random number generator in Minitab. In addition to some direct methods, iterative algorithms are briefly explained.

MCMC methods (an iterative procedure) of generating samples from the posterior distribution is introduced, where the Metropolis–Hasting algorithm and Gibb sampling are explained and illustrated with many examples. WinBUGS uses MCMC methods such as the Metropolis–Hasting and Gibbs sampling techniques, and many examples of a Bayesian analysis are given. An analysis consists of graphical displays of various plots of the posterior density of the parameters, by portraying the posterior analysis with tables that list the posterior mean, standard deviation, median, and lower and upper 2½ percentiles, and of other graphics that monitor the convergence of the generated observations.

1.18 Example of a Cross-Sectional Study

The general layout of a cross-sectional study appears in Table 1.4, where a random sample of size $n = n_{++} + n_{+-} + n_{-+} + n_{--}$ subjects is taken from a well-defined population and where the disease status and exposure status of each subject is known. Consider the ++ cell, then θ_{++} is the probability that a subject will have the disease and will be exposed to the risk factor. It is assumed that n is fixed and that the cell frequencies follow a multinomial distribution with mass function

$$f\left(n_{++}, n_{+-}, n_{-+}, n_{--} \mid \theta_{++}, \theta_{+-}, \theta_{-+}, \theta_{--}\right) \propto \theta_{++}^{n_{++}} \theta_{+-}^{n_{+-}} \theta_{-+}^{n_{-+}} \theta_{--}^{n_{--}} \qquad (1.24)$$

where the thetas are between zero and one and their sum is one and the n's are nonnegative integers with a sum equal to the sample size n.

As a function of the thetas, (1.24) is recognized as a Dirichlet density. Usually, the likelihood for the thetas is combined via Bayes's theorem with a prior distribution for the thetas, and the result is a posterior density for thetas. For example, if a uniform prior is used for the thetas, the posterior distribution of the theta is Dirichlet with parameter vector $(n_{++} + 1, n_{+-} + 1, n_{-+} + 1, n_{--} + 1)$, but on the other hand, if the improper prior density

$$f\left(\theta_{++}, \theta_{+-}, \theta_{-+}, \theta_{--}\right) \propto \left(\theta_{++}^{n_{++}} \theta_{+-}^{n_{+-}} \theta_{-+}^{n_{-+}} \theta_{--}^{n_{--}}\right)^{-1} \qquad (1.25)$$

TABLE 1.4

A Cross-Sectional Study

Risk Factor	+ Disease	− Nondisease
+ (positive)	θ_{++}, n_{++}	θ_{+-}, n_{+-}
− (negative)	θ_{-+}, n_{-+}	θ_{--}, n_{--}

is used, the posterior density of the thetas is Dirichlet with parameter vector $(n_{++}, n_{+-}, n_{-+}, n_{--})$. When the latter prior is used, the posterior means of the unknown parameters will be the same as the "usual" estimators. For example, the usual estimator of θ_{++} is n_{++}/n, but the posterior distribution of θ_{++} is beta with parameter vector $(n_{++}, n - n_{++})$; consequently, the posterior mean of θ_{++} is indeed the usual estimator n_{++}/n.

The sampling scheme for a cross-sectional study allows one to estimate the relative risk of disease (the incidence rate of those diseased among those exposed divided by the incidence rate of the diseased among those not exposed to the risk) and the odds of exposure among those diseased versus the odds of exposure among the non-diseased.

A good example of a cross-sectional study is the Shields Heart Study carried out in Spokane, Washington, from 1993 to 2001 at the Shields Coronary Artery Center where the disease status of coronary artery disease and various risk factors (coronary artery calcium, diabetes, smoking status, blood pressure, and cholesterol) were measured on each patient. The average age (SD) of 4386 patients was 55.14 (10.757) years, with an average age of 55.42 (10.78) years for 2712 males and 56.31 (10.61) years for 1674 females. The main emphasis of this study was to investigate the ability of coronary artery calcium (as measured by computed tomography) to diagnose coronary artery disease. A typical table from this study investigates the association between having a heart attack and the fraction of subjects with a positive reading for coronary calcium. For additional information about the Shield Heart Study, see Mielke, Shields, and Broemeling (Table 1.3).[28]

In order to investigate the association between infarction and coronary artery calcium, the relative risk and odds ratio are estimated. From a Bayesian viewpoint, assuming an improper prior distribution for the four parameters, the posterior distribution of the four cell parameters is Dirichlet with hyper parameter vector (119, 2461, 11, 1798), and the relative risk for heart disease is estimated from the parameter

$$\theta_{RR} = \frac{\theta_{++}/(\theta_{++}+\theta_{+-})}{\theta_{-+}/(\theta_{-+}+\theta_{--})} \tag{1.26}$$

Note that the formula for relative risk has a numerator that is the probability of disease among those exposed (have a positive calcium score) to the risk factor, while the denominator is the probability of disease among those not exposed (negative coronary artery calcium). As for the odds ratio, the posterior distribution of

$$\theta_{OR} = \frac{\theta_{pp}\theta_{mm}}{\theta_{mp}\theta_{pm}} \tag{1.27}$$

BUGS CODE 1.2

```
model;
# the cross sectional study
{
# below generates observations from the Dirichlet
distribution
gpp~ dgamma(npp,2)
gpm~dgamma(npm,2)
gmp~dgamma(nmp,2)
gmm~dgamma(nmm,2)
sg<-gpp+gpm+gmp+gmm
thetapp<- gpp/sg
thetapm<-gpm/sg
thetamp<-gmp/sg
thetamm<-gmm/sg
# numerator of RR
nRR<-(thetapp)/(thetapp+thetapm)
# denominator of RR
dRR<-(thetamp)/(thetamp+thetamm)
# the odds ratio
OR<-(thetapp*thetamm)/(thetamp*thetapm)
# the relative rsk
RR<-nRR/dRR
# attributable risk based on the relative risk
ARRR<-(RR-1)/RR
# attributable risk based on the odds ratio
AROR<-(OR-1)/RR
}
# data with improper prior for Shields Heart Study
list(npp = 119,npm = 2461,nmp = 11,nmm = 1798)
# initial values generated from the specification tool
```

will also be determined, and the Bayesian analysis will be executed with **BUGS CODE 1.2** using 55,000 observations generated from the posterior distribution with a burn in of 5,000 and a refresh of 100.

It is interesting that the posterior means of the relative risk and odds ratio are quite similar, as are the posterior means, which implies the disease (coronary infarction) is quite rare, which of course is obvious from Table 2.8, which indicates a disease rate of 2.96%. If the 4389 patients are actually a random sample, then one is confident that 2.9% is an accurate estimate of the "true" disease rate (Table 1.5).

TABLE 1.5

Posterior Distribution for the Shields Heart Study

Parameter	Mean	SD	Error	2½	Median	97½
RR	8.35	2.887	0.0133	4.378	7.805	15.49
OR	8.709	3.308	0.01397	4.529	8.138	16.21
θ_{pp}	0.02711	0.00245	<0.00001	0.0225	0.02704	0.03215
θ_{mm}	0.4096	0.007413	<0.00001	0.395	0.4096	0.4243
θ_{mp}	0.002504	0.007553	<0.000001	0.001246	0.002427	0.004201
θ_{pm}	0.5608	0.0075	<0.00001	0.5461	0.5608	0.5754

1.19 Markov Chain Monte Carlo

MCMC techniques are especially useful when analyzing data with complex statistical models. For example, when considering a hierarchical model with many levels of parameters, it is more efficient to use an MCMC technique such as Metropolis–Hasting or Gibbs sampling iterative procedure in order to sample from the many posterior distributions. It is very difficult, if not impossible, to use non-iterative direct methods for complex models.

A way to draw samples from a target posterior density $\xi(\theta \mid x)$ is to use Markov chain techniques, where each sample depends only on the last sample drawn. Starting with an approximate target density, the approximations are improved with each step of the sequential procedure. Or in other words, the sequence of samples is converging to samples drawn at random from the target distribution. A random walk from a Markov chain is simulated, where the stationary distribution of the chain is the target density, and the simulated values converge to the stationary distribution or the target density. The main concept in a Markov chain simulation is to devise a Markov process whose stationary distribution is the target density. The simulation must be long enough so that the present samples are close enough to the target. It has been shown that this is possible and that convergence can be accomplished. The general scheme for a Markov chain simulation is to create a sequence θ_t, $t = 1,2,...$ by beginning at some value θ_0, and at the t-th stage select the present value from a transition function $Q_t(\theta_t \mid \theta_{t-1})$, where the present value θ_t depends only on the previous one, via the transition function. The value of the starting value θ_0 is usually based on a good approximation to the target density. In order to converge to the target distribution, the transition function much be selected with care. The account given here is a summary of Gelman et al. (Chapter 11),[21] who presents a very complete account of MCMC. Metropolis–Hasting is the general name given to methods of choosing appropriate transition functions, and two special cases of this are the Metropolis algorithm and the Gibbs sampling.

1.20 Metropolis Algorithm

Suppose the target density $\xi(\theta \mid x)$ can be computed, then the Metropolis technique generates a sequence θ_t, $t = 1,2,...$, with a distribution that converges to a stationary distribution of the chain. Briefly, the steps taken to construct the sequence are:

1. Draw the initial value θ_0 from some approximation to the target density.
2. For $t = 1,2,...$, generate a sample θ_* from the jumping distribution

$$G_t(\theta_* \mid \theta_{t-1})$$

3. Calculate the ratio $s = \xi(\theta_* \mid x)\xi(\theta_{t-1} \mid x)$.
4. Let $\theta_t = \theta_*$ with probability min$(s,1)$ or let $\theta_t = \theta_{t-1}$.

To summarize these steps, if the jump given by step 2 increases the posterior density, let $\theta_t = \theta_*$; on the other hand, if the jump decreases the posterior density, let $\theta_t = \theta_*$ with probability s, otherwise let $\theta_t = \theta_{t-1}$. One must show the sequence generated is a Markov chain with a unique stationary density that converges to the target distribution. For more information, see Gelman et al. (p. 325).[16] There is a generalization of the Metropolis algorithm to the Metropolis–Hasting method.

1.21 Gibbs Sampling

Another MCMC algorithm is Gibbs sampling that is quite useful for multidimensional problems and is an alternating conditional sampling way to generate samples from the joint posterior distribution. Gibbs sampling can be thought of as a practical way to implement the fact that the joint distribution of two random variables is determined by the two conditional distributions.

The two-variable case is first considered by starting with a pair (θ_1, θ_2) of random variables. The Gibbs sampler generates a random sample from the joint distribution of θ_1 and θ_2 by sampling from the conditional distributions of θ_1 given θ_2 and from θ_2 given θ_1. The Gibbs sequence of size k

$$\theta_2^0, \theta_1^0 ; \theta_2^1, \theta_1^1 ; \theta_2^2, \theta_1^2 ;...; \theta_2^k, \theta_1^k \qquad (1.28)$$

is generated by first choosing the initial values θ_2^0, θ_1^0 while the remaining are obtained iteratively by alternating values from the two conditional distributions.

Under quite general conditions, for large enough k, the final two values θ_2^k, θ_1^k are samples from their respective marginal distributions. To generate a random sample of size n from the joint posterior distribution, generate the Gibbs sequence n times. Having generated values from the marginal distributions with large k and n, the sample mean and variance will converge to the corresponding mean and variance of the posterior distribution of (θ_1, θ_2).

Gibbs sampling is an example of an MCMC because the generated samples are drawn from the limiting distribution of a 2 by 2 Markov chain. See Casella and George[29] for a proof that the generated values are indeed values from the appropriate marginal distributions. Of course, Gibbs sequences can be generated from the joint distribution of three, four, and more random variables.

1.22 Common Mean of Normal Populations

Gregurich and Broemeling[30] describe the various steps in Gibbs sampling to determine the posterior distribution of the parameters in independent normal populations with a common mean.

The Gibbs sampling approach can best be explained by illustrating the procedure using two normal populations with a common mean θ. Thus, let $y_{ij}, j = 1, 2, \ldots, n_i$ be a random sample of size n_i from a normal population for $i = 1, 2$.

The likelihood function for θ, τ_1, and τ_2 is

$$L(\theta, \tau_1, \tau_2 \mid \text{data}) \propto \tau_1^{n_1/2} \exp{-\frac{\tau_1}{2}\left[(n_1 - 1)s_1^2 + n_1(\theta - \bar{y}_1)^2\right]} \times$$
$$\tau_2^{n_2/2} \exp{-\frac{\tau_2}{2}\left[(n_2 - 1)s_2^2 + n_2(\theta - \bar{y}_2)^2\right]}$$

where:

$$\theta \in \Re, \tau_1 > 0, \tau_2 > 0$$

$$s_1^2 = \sum_{j=1}^{n_1} \frac{\left(y_{1j} - \bar{y}_1\right)^2}{(n_1 - 1)}$$

and

$$s_2^2 = \sum_{j=1}^{n_2} \frac{\left(y_{2j} - \bar{y}_2\right)^2}{(n_2 - 1)}$$

The prior distribution for the parameters $\theta, \tau_1,$ and τ_2 is assumed to be a vague prior defined as

$$g(\theta, \tau_1, \tau_2) \propto \frac{1}{\tau_1} \frac{1}{\tau_2}, \tau_i > 0 \tag{1.29}$$

Then combining the above gives the posterior density of the parameters as

$$P(\theta, \tau_1, \tau_2 \mid \text{data}) \propto \prod_{i=1}^{2} \tau_i^{\frac{n_i-1}{2}} \operatorname{Exp} -\frac{\tau_i}{2}\left[(n_i-1)s_i^2 + n_i(\theta - \bar{y}_i)^2\right]$$

Therefore, the conditional posterior distribution of τ_1 and τ_2 given θ is such that

$$\tau_i \mid \theta \sim \text{Gamma}\left[\frac{n_i}{2}, \frac{(n_i-1)s_i^2 + n_i(\theta - \bar{y}_i)^2}{2}\right] \tag{1.30}$$

for $i = 1,2$ and given θ, τ_1, and τ_2 are independent.

The conditional posterior distribution of θ given τ_1 and τ_2 is normal. It can be shown that

$$\theta \mid \tau_1, \tau_2 \sim N\left[\frac{n_1\tau_1\bar{y}_1 + n_2\tau_2\bar{y}_2}{n_1\tau_1 + n_2\tau_2}, (n_1\tau_1 + n_2\tau_2)^{-1}\right] \tag{1.31}$$

Given the starting values $\tau_1^{(0)}$, $\tau_2^{(0)}$, and $\theta^{(0)}$, where

$$\tau_1^{(0)} = \frac{1}{s_1^2}, \tau_2^{(0)} = \frac{1}{s_2^2}, \text{and } \theta^{(0)} = \frac{n_1\bar{y}_1 + n_2\bar{y}_2}{n_1 + n_2}$$

draw $\theta^{(1)}$ from the normal conditional distribution (44) of θ given $\tau_1 = \tau_1^{(0)}$ and $\tau_2 = \tau_2^{(0)}$. Then draw $\tau_1^{(1)}$ from the conditional gamma distribution (43) given $\theta = \theta^{(1)}$. And at last draw $\tau_2^{(1)}$ from the conditional gamma distribution of τ_2 given $\theta = \theta^{(1)}$. Then generate

$$\theta^{(2)} \sim \theta \mid \tau_1 = \tau_1^{(1)}, \tau_2 = \tau_2^1$$

$$\tau_1^{(2)} \sim \tau_1 \mid \theta = \theta^2$$

$$\tau_2^{(2)} \sim \tau_2 \mid \theta = \theta^2$$

Continue this process until there are t iterations $[\theta^{(t)}, \tau_1^{(t)}, \tau_2^{(t)}]$. For large t, $\theta^{(t)}$ would be one sample from the marginal distribution of θ, $\tau_1^{(t)}$ from the marginal distribution of τ_1, and $\tau_2^{(t)}$ from the marginal distribution of τ_2.

Independently repeating the above Gibbs process m times produces m 3-tuple parameter values $[\theta_j^{(t)}, \tau_{1j}^{(t)}, \tau_{2j}^{(t)}], j = 1,2,\ldots,m,$ which represents a random sample of size m from the joint posterior distribution of (θ, τ_1, τ_2). The statistical inferences are drawn from the m sample values generated by the Gibbs sampler.

The statistical inferences can be drawn from the m sample values generated by the Gibbs sampler. The Gibbs sampler will produce three columns, where each row is a sample drawn from the posterior distribution of (θ, τ_1, τ_2). The first column is the sequence of the sample m, the second column is a random sample of size m from the poly-t distribution of θ, and the third and fourth columns are also random samples of size m but from the marginal posterior distributions of τ_1 and τ_2, respectively.

To find the characteristics of the marginal posterior distribution of a parameter such as the mean and variance, it should be noted that the Gibbs sampler generates a sample of values of a marginal distribution from the conditional distributions without the actual marginal distribution. By simulating a large enough sample, the characteristics of the marginal can be calculated. If m is "large," the sample mean of the column is

$$E(\theta|data) = \sum_{j=1}^{m} \frac{\theta_j^t}{m} = \bar{\theta}$$

and is the mean of the posterior distribution of θ. The sample variance

$$(m-1)^{-1} \sum_{j=1}^{m} \left[\theta_j^t - \bar{\theta}\right]^2$$

is the variance of the posterior distribution of θ.

Additional characteristics such as the median, mode, and the 95% credible region of the posterior distribution of the parameter θ can be calculated from the samples generated by the Gibbs technique. Hypothesis testing can also be performed. Similar characteristics of the parameters $\tau_1, \tau_2, \ldots, \tau_k$ can be calculated from the samples resulting from the Gibbs method.

1.23 An Example

The example is from Box and Tiao (p. 481).[17] It is referred to as "the weighted mean problem." It has two sets of normally distributed independent samples with a common mean and different variances. Samples from the posterior distributions were generated from Gibbs sequences using the statistical software Minitab®. The final value of each sequence was used to approximate the marginal posterior distribution of the parameters $\theta, \tau_1, \ldots, \tau_k$. All Gibbs sequences were generated holding the value of t equal to 50. Each example

TABLE 1.6

Results from Gibbs Sampler for θ Box and Tiao—The Weighted Mean Problem

m	Mean	STD	SEM	95% Credible Region	
				Lower	Upper
250	108.42	1.04	0.07	106.03	110.65
500	108.31	0.94	0.04	106.35	110.21
750	108.31	0.90	0.03	106.64	110.15
1500	108.36	0.94	0.02	106.51	110.26

has the results of the parameters using four different Gibbs sampler sizes, where the sample size m is equal to 250, 500, 750, and 1500.

The "weighted mean problem" has two sets of normally distributed independent observations with a common mean and different variances. The estimated values of θ determined by the Gibbs sampling method are reported in Table 1.6. The mean value of the posterior distribution of θ generated from the 250 Gibbs sequences is 108.42 with 0.07 as the standard error of the mean. The mean value of θ generated from 500 and 750 Gibbs sequences have the same value of 108.31, and the standard errors of the mean equal 0.04 and 0.03, respectively. The mean value of θ generated from 1500 Gibbs sequences is 108.36 and a standard error of the mean of 0.02. Box and Tiao determined the posterior distribution of θ using the t distribution as an approximation to the target density. They estimated the value of θ to be 108.43. This is close to the value generated using the Gibbs sampler method. The exact posterior distribution of θ is the poly-t distribution. The effect of m appears to be minimal indicating that 500 to 750 iterations of the Gibbs sequence are sufficient (Table 1.6).

1.24 Additional Comments about Bayesian Inference

Beginning with Bayes's theorem, introductory material for the understanding of Bayesian inference is presented in this chapter. For example, Bayesian methods for cohort studies and cross-sectional studies are analyzed using the theory and methods unique with the Bayesian approach. Inference for the standard populations is introduced. The most useful population for the case-control and cohort studies is the binomial population, which models the distribution of the number of patients among the exposed who will develop disease. It is shown how to analyze a cohort study with the binomial population, where the posterior distribution of the incidence rate for those exposed is a beta distribution. For the analysis of cross-sectional studies, it is shown that the multinomial distribution models four cell frequencies for the risk and disease status of a subject and that the corresponding posterior distribution of the cell frequencies is a Dirichlet.

There are many books that introduce Bayesian inference and the computational techniques that will execute a Bayesian analysis, and the reader is encouraged to read Ntzoufras.[31] The material found in Ntzoufras is an excellent introduction to Bayesian inference and to WinBUGS, the computing software that is employed in this book. WinBUGS is also introduced in Section 1.25 of this book; thus, together with Ntzoufras, the reader should be able to execute Bayesian analyses for various studies in longitudinal investigations that will be presented in the following chapters.

1.25 WinBUGS

1.25.1 Introduction

WinBUGS is the statistical package that is used for the book, and it is important that the novice be introduced to the fundamentals of working in the language. This is a brief introduction to the package, and for the first-time user, it will be necessary to gain more knowledge and experience by practicing with the numerous examples provided in the download. WinBUGS is specifically designed for Bayesian analysis and is based on MCMC techniques for simulating samples from the posterior distribution of the parameters of the statistical model. It is quite versatile, and once the user has gained some experience, there are many rewards.

Once the package has been downloaded, the essential features of the program are described, first by explaining the layout of the BUGS document. The program itself is made up of two parts, one part for the program statements and the other for the input for the sample data and the initial values for the simulation. Next to be described are the details of executing the program code and what information is needed for the execution. Information needed for the simulation are the sample sizes of the MCMC simulation for the posterior distribution and the number of such observations that will apply to the posterior distribution.

After execution of the program statements, certain characteristics of the posterior distribution of the parameters are computed including the posterior mean, median, credible intervals, and plots of posterior densities of the parameters. In addition, WinBUGS provides information about the posterior distribution of the correlation between any two parameters and information about the simulation. For example, one may view the record of simulated values of each parameter and the estimated error of estimation of the process. These and other activities involving the simulation and interpretation of the output will be explained.

Examples based on repeated measures studies illustrate the use of WinBUGS and include estimation of the true and false positive fractions for the exercise stress test and modeling for the ROC area. Of course, this is only a brief introduction but should be sufficient for the beginner to begin the adventure of analyzing data. Because the book's examples provide the necessary code for all examples, the program can easily be executed by the user. After the book is completed by dedicated students, they will have a good understanding of WinBUGS and the Bayesian approach to measuring test accuracy.

1.25.2 Download

The latest version of WinBUGS can be downloaded from http://www.mrc-bsu.cam.ac.uk/bugs/winbugs/contents.shtml, which you can install in your program files; then you will requested to download a decoder, which will allow one to activate the full capabilities of the package.

1.25.3 Essentials

The Essential feature of the package is a WinBUGS file or document that contains the program statements and space for input information.

1.25.4 Main Body

The main body of the software is a WinBUGS document, which contains the program statements, the major part of the document, and a list statement or statements, which include the data values and some initial values for the MCMC simulation of the posterior distribution. The document is given a title and saved as a WinBUGS file, which can be accessed as needed.

1.25.5 List Statements

List statements allow the program to incorporate certain necessary information that is required for successful implementation. For example, experimental or study information is usually imputed with a list statement. In order to display the list statement, it is contained in **BUGS CODE 1.3**, a program to estimate the mean profile in the following repeated measures study. In order to explain the Bayesian method for estimating the mean profile of two groups, consider the example described by Broemeling[32] which is a study conducted in two regions of Holland, a rural area of Vlagtwedde and more urban industrial area of Vlaardingen. All subjects were followed over time to determine data concerning risk factors for chronic obstructive lung disease. Participants were questioned every 3 years for up to 21 years. The questionnaire inquired about information of respiratory symptoms and smoking status. The main repeated measure was forced expiratory volume, FEV1, which is measured by spirometry, once every 3 years for the first 15 years. FEV1 is measured for 15 years and once at Year 19.

The study will show that a linear association between FEV1 and time for smokers and separately for nonsmokers is appropriate, where the mean profile is modeled as a linear function of time. To be more specific, the sample included in the example consists of 133 subjects aged at least 36 and whose smoking status did not change during the course of the study; thus, smoking status is not a time-dependent covariate.

Therefore for nonsmokers, the proposed mean profile is the linear regression

$$E(Y_{ij}) = \beta_1 + \beta_2 t_{ij} \tag{1.32}$$

and is

$$E(Y_{ij}) = \beta_3 + \beta_4 t_{ij} \tag{1.33}$$

for smokers.

The variance-covariance matrix of the FEV1 values over the various time points is assumed to be unstructured with a noninformative Wishart distribution for the prior distribution.

The execution of the code is explained in terms of the following statements that make up the WinBUGS file or document.

BUGS CODE 1.3

```
model;
{

# prior distribution for the regression coefficients.
               beta1 ~ dnorm(0.0, 0.001)
               beta2 ~ dnorm(0.0, 0.001)
            beta3 ~ dnorm(0.0,.0001)
            beta4 ~ dnorm(0.0,.0001)
# mean profile for nonsmokers
for(i in 1:N1){Y1[i,1:M1]~dmnorm(mu1[],Omega[,])}
for (j in 1:M1){mu1[j]<-beta1+beta2*age[j]}
# noninformative precision matrix
Omega[1:M1,1:M1]~dwish(R[,],7)
Sigma[1:M1,1:M1]<-inverse(Omega[,])
for(i in 1:N2){Y2[i,1:M2]~dmnorm(mu2[],Omega[,])}
# mean profile for smokers
for (j in 1:M2){mu2[j]<-beta3+beta4*age[j]}
# differences compare the mean profile for nonsmokers vs
  smokers
d13<-beta1-beta3
```

```
d23<-beta2-beta4

}

# list statement is for the FEV1 values of nonsmokers


# nonsmokers

list(N1 = 32,N2 = 100,M1 = 7,M2 = 7,Y1 = structure(.Data
    = c(3.4,3.4,3.5,3.2,NA,3.0,2.4,
3.3,3.8,3.5,3.0,3.1,NA,NA,
NA,2.4,2.4,2.4,NA,2.2,2.0,
4.4,4.4,4.1,4.1,4.0,NA,3.6,
3.2,3.4,2.7,3.0,3.1,2.6,1.6,
3.1,NA,NA,2.4,2.5,2.5,2.1,
3.7,3.8,3.9,3.8,3.8,3.5,3.2,
3.8,3.2,3.3,3.2,3.1,3.0,NA,
NA,4.6,4.7,4.7,4.9,NA,4.4,
3.2,NA,2.7,2.8,2.7,2.5,2.5,
3.6,3.7,3.4,3.5,3.3,3.2,3.4,
3.6,4.0,NA,3.6,3.8,NA,3.4,
4.8,4.6,4.7,4.0,3.9,4.2,3.3,
2.6,2.4,2.5,2.6,2.6,2.4,2.3,
NA,NA,3.6,2.7,3.0,3.1,3.2,
NA,NA,2.2,2.1,2.0,1.3,1.5,
3.7,3.3,3.0,3.2,NA,2.9,NA,
3.4,3.3,3.1,3.0,3.0,NA,2.8,
4.1,4.1,3.7,3.8,3.7,3.7,3.6,
NA,2.9,2.8,2.8,2.6,2.6,2.8,
4.1,4.3,3.5,NA,3.3,NA,3.5,
3.2,3.0,2.9,3.0,2.8,2.5,2.6,
NA,3.2,3.2,3.4,3.1,NA,3.2,
3.5,3.6,3.5,3.2,3.0,3.0,3.2,
3.2,3.5,3.4,3.4,3.1,3.0,3.0,
NA,4.9,3.9,3.7,4.0,3.8,3.6,
3.3,3.3,3.2,3.3,2.9,3.2,3.1,
3.1,3.3,3.0,NA,2.6,NA,2.5,
NA,NA,3.7,3.6,3.4,3.4,3.1,
NA,NA,2.7,2.5,2.3,2.2,2.5,
NA,3.2,3.0,2.8,2.8,2.6,2.4,
3.7,3.7,NA,2.7,3.3,3.0,3.1),.Dim = c(32,7)),

# smokers


Y2 = structure(.Data = c(3.1,3.2,3.5,3.0,2.9,NA,NA,
3.6,3.5,3.5,3.1,NA,2.8,2.7
```

```
,NA,2.7,2.9,2.7,2.7,2.5,NA,
3.4,3.3,2.9,2.3,2.5,2.4,NA,
NA,2.5,2.5,2.1,2.4,NA,2.3,
3.9,4.0,4.1,3.8,4.0,NA,NA,
2.7,NA,3.3,2.4,2.2,2.2,2.3,
3.0,2.9,3.0,2.8,2.8,2.7,2.8,
2.8,2.7,2.1,1.9,1.8,NA,1.7,
3.9,3.8,3.5,3.3,3.5,3.2,3.3,
3.2,NA,3.0,2.8,2.9,2.5,2.2,
3.3,3.4,3.4,3.1,3.5,3.0,2.8,
NA,3.1,3.4,3.1,3.2,NA,2.6,
3.2,3.2,3.3,3.1,3.2,2.8,NA,
4.0,3.6,3.7,3.7,3.2,3.0,NA,
3.3,3.6,3.3,3.3,3.0,2.9,NA,
NA,3.0,3.2,3.0,2.9,2.4,NA,
2.8,2.5,2.6,2.5,2.4,2.3,2.2,
3.6,3.9,3.7,NA,3.3,3.1,NA,
3.6,3.4,3.4,NA,3.0,NA,2.7,
NA,2.0,1.8,1.7,1.8,1.6,1.5,
3.6,3.4,3.2,3.3,3.3,NA,2.6,
2.6,3.6,NA,2.9,3.2,3.1,2.2,
3.3,3.0,2.9,3.0,2.9,2.8,2.4,
3.6,3.6,3.4,NA,NA,2.8,2.8,
1.8,1.5,1.7,NA,1.8,1.5,1.2,
3.6,3.8,3.2,3.1,3.2,3.2,2.8,
4.2,3.8,3.5,3.8,3.5,3.4,3.3,
NA,3.9,4.0,NA,3.7,3.3,3.4,
3.7,NA,4.1,2.7,2.9,2.9,2.5,
2.6,2.6,2.7,2.6,2.6,2.4,2.5,
3.5,3.6,3.1,3.0,2.9,3.1,2.7,
2.2,2.3,2.3,2.1,NA,NA,1.5,
NA,3.5,3.3,3.8,3.6,3.9,3.4,
NA,3.1,3.3,3.1,NA,2.8,2.7,
3.7,3.7,3.7,3.3,3.1,NA,NA,
NA,2.4,2.3,2.1,1.9,2.2,2.0,
4.4,4.2,4.1,4.1,NA,4.0,NA,
2.9,2.7,2.8,NA,2.5,NA,2.1,
3.1,2.6,2.7,2.5,2.4,2.1,2.3,
2.5,2.4,2.0,NA,NA,2.1,2.3,
3.3,3.1,3.0,3.2,NA,3.3,2.9,
2.8,2.7,2.7,2.4,NA,2.1,1.8,
2.7,2.4,3.0,1.9,2.6,2.5,2.4,
3.4,3.4,3.4,3.1,2.9,NA,2.5,
3.7,3.4,3.3,3.2,2.7,NA,2.8,
NA,2.5,2.4,2.6,2.0,NA,2.0,
3.6,3.4,3.5,3.5,3.6,3.3,3.2,
2.6,2.7,2.7,2.7,NA,NA,2.4,
```

```
3.4,3.5,3.7,3.4,3.6,NA,3.0,
3.4,3.1,3.0,2.9,NA,2.6,2.4,
3.6,3.3,NA,3.2,3.0,2.7,NA,
4.1,3.6,3.8,3.9,3.6,3.5,NA,
3.7,3.6,3.3,3.1,3.3,3.1,NA,
4.8,3.8,3.8,3.5,3.6,NA,3.3,
3.7,4.2,3.9,3.3,3.2,3.5,3.6,
2.8,2.9,NA,NA,2.7,2.4,2.5,
3.4,3.4,3.3,2.7,3.0,3.1,3.1,
3.0,3.1,2.8,1.7,2.9,2.7,2.2,
3.0,2.5,NA,2.1,2.3,NA,1.9,
3.5,3.4,2.9,3.1,NA,2.8,NA,
2.9,2.5,NA,2.4,2.7,2.3,2.1,
NA,3.5,NA,3.0,3.0,3.2,3.2,
2.3,2.6,2.6,NA,2.4,2.3,2.2,
3.5,3.6,3.4,3.3,NA,2.8,2.8,
3.4,3.3,NA,NA,2.8,2.7,2.6,
3.5,3.1,2.9,2.6,2.6,2.3,NA,
3.8,3.7,3.6,3.4,3.2,3.0,2.8,
3.9,3.1,3.6,3.4,3.3,NA,3.0,
3.5,3.8,4.0,3.8,NA,3.2,3.0,
3.4,3.0,3.1,3.1,NA,2.8,NA,
2.4,2.8,2.7,2.1,NA,2.0,1.9,
1.9,1.9,1.8,1.6,1.6,NA,NA,
3.0,2.8,2.5,NA,NA,2.2,2.2,
2.7,3.0,NA,NA,2.3,1.9,1.6,
2.8,3.4,NA,3.0,2.5,2.5,NA,
NA,3.1,NA,2.8,2.5,2.0,2.7,
NA,4.4,3.5,3.6,NA,3.4,3.2,
4.1,3.7,3.9,2.8,3.4,3.0,3.2,
2.1,2.2,2.4,2.3,2.2,NA,NA,
3.2,2.9,2.8,2.7,2.6,2.4,2.3,
3.5,3.5,3.9,3.3,NA,2.6,2.5,
3.2,3.7,3.4,3.4,3.5,NA,NA,
3.5,3.5,3.2,3.0,3.1,2.8,2.6,
4.3,NA,4.1,4.1,3.8,3.7,3.5,
2.9,3.1,2.7,2.8,NA,2.2,2.2,
2.3,2.2,2.0,2.0,1.9,NA,NA,
3.8,3.4,3.3,3.1,3.1,NA,NA,
4.0,3.3,3.5,NA,3.3,2.8,2.7,
2.8,2.9,2.8,2.4,2.5,2.4,2.3,
3.0,NA,3.2,2.5,2.5,2.7,NA,
NA,2.6,3.0,2.4,2.5,2.1,NA,
3.0,3.1,2.9,2.8,2.5,2.4,2.1,
3.0,2.7,NA,2.4,2.2,2.2,2.3,
2.4,2.3,2.3,2.0,1.9,NA,NA,
NA,2.6,2.5,NA,2.1,2.2,1.7,
```

```
NA,3.1,2.9,NA,2.4,2.7,2.2,
3.7,3.5,3.3,2.9,3.1,NA,2.6,
2.9,3.1,2.9,2.8,2.6,NA,NA,
3.0,3.0,2.7,2.5,2.1,2.0,2.2),.Dim = c(100,7)),

age = c(0,3,6,9,12,15,19),

R = structure(.Data = c(1,0,0,0,0,0,0,0,1,0,0,0,0,0,0,0,0,1
,0,0,0,0,0,0,0,1,0,0,0,0,0,0,0,1,0,0,0,0,0,0,0,
1,0,0,0,0,0,0,0,1),.Dim = c(7,7)))

# initial values
```

The objective of this example is to determine the posterior distribution of the regression coefficients of the mean profile for nonsmokers and smokers and to compare the two groups.

The last list statement contains the initial values or the MCMC simulation.

1.25.6 Executing the Analysis

MCMC can analyze complex statistical models and the following describes the use of drop-down menus from the toolbar for executing the posterior analysis.

1.25.7 Specification Tool

The toolbar of WinBUGS is labeled as follows, from left to right: file, edit, attributes, tools, info, **model, inference**, doodle, maps, text, windows, examples, manuals, and help, and I have highlighted the model and inferences labels. When the user clicks one of the labels, a pop-up menu appears. In order to execute the program, the user clicks **model** and then clicks **specification**, and the specification tool appears (Figure 1.3).

The specification tool is used together with the BUGS document as follows: (1) click the word "model" of the document, (2) click the check model tab

FIGURE 1.3
The specification tool.

of the specification tool, (3) click the compile tab of the specification tool, (4) click the word "list" of the list statement of the document, and (5) click load units tab of the tool. Now close the specification tool.

1.25.8 Sample Monitor Tool

The sample tool is activated by first clicking the inference menu of the toolbar, then click sample, and the sample monitor tool appears as below. Type "beta1" in the node box and click "set," type "beta2" and click "set," type "beta3" and click "set," type "beta4" and click "set," and finally type "*." Type "5000" in the beg box, which means the first 5001 observations generated for the posterior distribution of the beta coefficients. The 5000 observations typed in beg are referred to as the "burn in" (Figure 1.4).

1.25.9 Update Tool

In order to activate the update tool, click on the **model** menu of the toolbar, and then click on updates (see Figure 1.5).

Suppose you want to generate 45,000 observations from the posterior distributions of the four betas, using the statements that are listed in the

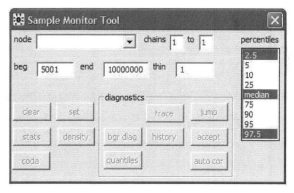

FIGURE 1.4
The sample monitor tool.

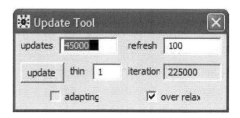

FIGURE 1.5
The update tool.

document above; key in "45000" in the updates box and "100" in refresh. In order to execute the simulation using the program statements in the document, click on update of the update tool.

1.25.10 Output

On clicking the history box, the values of 45,000 observations from the joint density of the betas are displayed, and the output for the posterior analysis will look like this Table 1.7.

The Bayesian analysis reveals the mean profile for nonsmokers, where the posterior means of β_1 and β_2 are 3.523 and −0.033770, respectively, while the mean profile for smokers is defined by posterior means for β_3 and β_4 given as 3.263 and −0.3828, respectively. Figure 1.6 displays the posterior density of $d13=\beta_1-\beta_3$ and implies that the two intercept are not the same.

1.25.11 Another Example

In order to illustrate the use of WinBUGS and to gain additional insight, consider the following example of a linear relation between the main response and time was assumed for the mean profile of two groups of subjects. From experience, one would expect the mean profile to not be too complex and

TABLE 1.7

Posterior Analysis for the FEV1 Mean Profiles

Parameter	Mean	SD	Error	2½	Median	97½
β_1	3.523	0.1007	0.000576	3.326	3.524	3.721
β_2	−0.033770	0.00327	0.0000201	−0.04021	−0.03376	−0.027365
β_3	3.263	0.05863	0.000315	3.148	3.263	3.379
β_4	−0.03828	0.001924	0.0000127	−0.04206	−0.03829	−0.034485
d_{13}	0.2606	0.1157	0.000601	0.0344	0.2607	0.4879
d_{24}	0.00451	0.00378	0.0000212	−0.002943	0.00454	0.011895

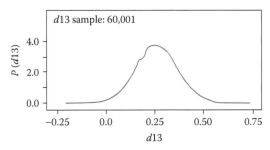

FIGURE 1.6
Posterior density of $\beta_1-\beta_3$.

that simple polynomials or linear shift point models should be able to indeed model the mean profile. For our next case, a quadratic will be proposed to model the mean profile of two groups. A good example of this is the study of where the hematocrit of hip-replacement patients is measured on four occasions. Also measured is the age of each subject at the beginning of the study, and there are two groups, male and female patients. Hematocrit is a measure of how much volume the red blood cells take up in the blood and is measured as a percentage; thus, an observed value of 33 means red blood cells comprise 33% by volume of the blood. For our study, one follows the hematocrit of hip-replacement patients in order to see if they are becoming anemic. The data for Crowder and Hand is found in the list statement of **BUGS CODE 1.4**.

This is an example of a quadratic mean profile for two groups (males and females), and for males, the expectation is

$$E(Y_{ij}) = \beta_1 + \beta_2 t_{ij} + \beta_3 t_{ij}^2 \qquad (1.34)$$

and for females, the expectation is

$$E(Y_{ij}) = \beta_1 + \beta_4 + (\beta_2 + \beta_5)t_{ij} + (\beta_3 + \beta_6)t_{ij}^2 \qquad (1.35)$$

The covariance matrix of the hematocrit values over the four time points is assumed to be unstructured and to have a noninformative Wishart distribution.

Note that there are 13 male and 17 female hip-replacement patients in the Crowder and Hand (p. 79) study. Also note that not all the patients have a complete set of four measurements. Indeed, some have four, some three, and some two, but all have at least two repeated values.

BUGS CODE 1.4

```
        model;
        {
                beta1 ~ dnorm(0.0, 0.001)
                beta2 ~ dnorm(0.0, 0.001)
        beta3 ~ dnorm(0.0,.0001)
        beta4 ~ dnorm(0.0,.0001)
        beta5 ~ dnorm(0.0,.0001)
        beta6 ~ dnorm(0.0,.0001)
        # Y1 is hematocrit for males
        for(i in 1:N1){Y1[i,1:M1]~dmnorm(mu1[],Omega[,])}
        for (j in 1:M1){mu1[j]<-beta1+beta2*age[j]+beta3*age[j]*
        age[j]}

        Omega[1:M1,1:M1]~dwish(R[,],4)
        Sigma[1:M1,1:M1]<-inverse(Omega[,])
```

```
# Y2 is hematocrit for females
for(i in 1:N2){Y2[i,1:M2]~dmnorm(mu2[],Omega[,])}
for (j in 1:M2){mu2[j]<-beta1+beta4+(beta2+beta5)*age[j]+
(beta3+beta6)*age[j]*age[j]}


}

# Males

list(N1 = 13,N2 = 17,M1 = 4,M2 = 4,

Y1 = structure(.Data =

c(47.1,31.1,NA,32.8,
44.1,31.5,NA,37.0,
39.7,33.7,NA,24.5,
43.3,18.4,NA,36.6,
37.4,32.3,NA,29.1,
45.7,35.5,NA,39.8,
44.9,34.1,NA,32.1,
42.9,32.1,NA,NA,
46.1,28.8,NA,37.8,
42.1,34.4,34.0,36.1,
38.3,29.4,32.9,30.5,
43.0,33.7,34.1,36.7,
37.8,26.6,26.7,30.6),.Dim = c(13,4)),

# females

Y2 = structure(.Data =

c(37.3,26.5,NA,38.5,
NA,28.0,NA,33.9,
27.0,32.5,NA,32.0,
38.4,32.3,NA,37.9,
38.8,32.6,NA,26.9,
44.7,32.2,NA,34.2,
38.0,27.1,NA,37.9,
34.0,23.2,NA,26.0,
44.8,37.2,NA,29.7,
46.0,29.1,NA,26.7,
41.9,32.0,37.1,37.6,
38.0,31.7,38.4,35.7,
42.2,34.0,32.9,33.3,
39.7,33.5,26.6,32.7,
37.5,28.2,28.8,30.3,
34.6,31.0,30.1,28.7,
35.5,24.7,28.1,29.8),.Dim = c(17,4)),
```

```
age = c(1,2,3,4),
R = structure(.Data = c(1,0,0,0,
                        0,1,0,0,
                        0,0,1,0,
                        0,0,0,1),.Dim = c(4,4)))

# initial values

list(beta1 = 0,beta2 = 0,beta3 = 0,beta4 = 0,beta5 = 0,
beta6 = 0)
```

To begin the analysis, click the model menu of the toolbar and pull down the specification tool:

1. Click the word "model" of the BUGS document.
2. Click the check model tab of the specification tool (see Figure 1.3).
3. Activate the word "list" of the first list statement.
4. Click the compile tab of the specification tool.
5. Click the word "list" of the third list statement of the document. If a mistake is made, the user will be notified, but you are now ready to execute the analysis.

In order to continue the process, pull down from the inference menu the sample monitor tool (see Figure 1.4) and type "beta1," followed by clicking the set box, and then repeat the operation for nodes beta2, beta3, beta4, beta5, and beta6. For the final operation, put an * in the node box, and type "5000" in the beg box for the burn in.

Pull down the update tool (see Figure 1.5) from the model menu and type "65000" in the updates box, put "100" in the refresh box, and click the update box. The simulation now begins with 65,000 observations generated form the posterior distribution of the six beta coefficients of the sample monitor tool, and the following output will appear (Table 1.8).

TABLE 1.8

Posterior Analysis for Hip-Replacement Study

Parameter	Mean	SD	Error	2½	Median	97½
β_1	61.17	3.1	0.178	54.65	61.36	66.7
β_2	−22.62	2.901	0.1737	−27.78	−22.81	−16.54
β_3	3.892	0.5597	0.03345	2.735	3.927	4.888
β_4	−8.559	4.084	0.2361	−15.8	−8.791	0.5215
β_5	6.402	3.769	0.2267	−2.198	6.676	12.8
β_6	−1.156	0.7173	0.04317	−2.368	−1.205	0.4907

This example will be revisited in Chapter 5, where the mean profile of the two groups will be compared.

1.25.12 Summary

This section introduces the reader to WinBUGS, and the novice should be able to begin with this chapter and learn the main topic, namely, how a Bayesian analyzes repeated measures studies. To gain additional experience, refer to the manual and to the numerous examples that come with the downloaded version of the package. Practice is the key to understanding the importance of analyzing actual data with a Bayesian approach. There are many references about WinBUGS, including Broemeling,[25] Woodworth,[33] and the WinBUGS link,[34] which in turn refer to many books and other resources about the package.

1.26 Preview

To follow is a brief preview of the various chapters in the book. Only the important topics will be highlighted.

Chapter 2 is a review of Bayesian regression techniques. As regression is the way to analyze repeated measures studies, it is important that students have a good foundation in the subject. The first topic to be reviewed is the Bayesian approach using logistic regression for binary variables. Using one covariate and then expanding to several, the methods of logistic regression are illustrated with an example from heart disease. Logistic regression gets its name from the logistic transformation, the log of the odds of the probability of success of the binary variable. The logistic function is expressed as a linear function of the independent variables, where a prior distribution is assigned to the unknown regression coefficients. Also discussed are goodness-of-fit techniques that provide evidence of how well the logistic model fits the data. Logistic regression is followed by a review of simple and multiple linear regression for a continuous dependent variable following a normal distribution with constant variance. The mean of the dependent variable is expressed as a linear function of the independent variables with unknown coefficients. Using an example from public health with multiple independent variables, the Bayesian analysis is performed with WinBUGS, assuming noninformative priors for the regression coefficients. Relaxing the assumption of a constant variance for the dependent variable, weighted regression techniques are introduced. Finally, examples of nonlinear regression are analyzed with the Bayesian approach. As the analysis of repeated measures is based on a regression methodology, this chapter is pivotal for students.

Chapter 3 introduces several preliminary concepts and techniques that should always be part of the analysis for repeated measures. First to be introduced for continuous variables is the notation used to express a repeated measures experiment with a regression model. In order to define a suitable model, descriptive statistics and graphical techniques should always be performed. Descriptive statistics consist of computing the mean, standard deviation, and median for the response at each time point. Scatter plots should accompany the descriptive statistics to gain additional information about trends over time and about the heterogeneity between units (subjects). Also briefly reviewed are descriptions of sources of variation, measurement error, and experimental error. The second part of the chapter focused on the preliminary concepts required to understand repeated measures with categorical variable. An example involving a categorical variable following a multinomial distribution reveals the important analytical techniques of computing the descriptive statistics and the variances, covariances, and correlations between the three possible values of a response with a multinomial distribution. Using uninformative prior distributions for the parameters of a multinomial distribution, a drug study is examined via Bayesian techniques.

Chapter 4 introduces the most important concept for repeated measures studies, namely the idea of linear models. Notation for the linear model is introduced to make it easier to model the mean profile and to model the covariance structure of the response between the time points of the study. Two historical approaches are examined, namely, (1) univariate analysis of variance and (2) multivariate analysis of variance. Using the notation for a linear model, Bayesian inference for the parameters (those defining the mean profile and those determining the covariance structure) are discussed. The important components (the prior, likelihood function, and posterior distribution) of a Bayesian analysis are explained and illustrated within example, expressed in terms of the linear model. Additional examples stress the modeling of the mean profile and the covariance structure and include computing descriptive statistic and graphical methods. The chapter concludes with explaining various patterns (unstructured, autoregressive, and compound symmetry) for the covariance matrix and how to estimate the parameters involved with a Bayesian approach.

Chapter 5 presents an in-depth explanation of estimating the mean profile for repeated measures studies. Of course there are many ways to express the mean profile (the expected value of the response over time) of a repeated measures response, and one of the most popular is a polynomial (linear, quadratic, cubic, and polynomial spline) representation. In particular, the linear, quadratic, and a linear–linear spline (with one join point) are used to estimate the mean profile. The second part of the chapter describe the same topics but for categorical variables, and two examples (one involving plasma inorganic phosphate and the other involving a hospital study if a dietary regime) reveal the power of the Bayesian approach.

Of course there are two important sets of parameters to estimate in a repeated measures study, the mean profile and the correlation study.

Although the main emphasis is usually on the mean profile, an accurate determination of the correlation structure is essential for successful estimation of the mean profile.

Chapter 6 is exclusively focused on choosing and estimating the covariance structure of the response. For a repeated study with N subjects and n time points (with the same time points for each subject), the covariance matrix consists of the pair-wise correlations of the response between the n time points. The various patterns defined are (1) unstructured, (2) autoregressive, (3) compound symmetry, and (4) Toeplitz matrix. It is a challenge to determine the appropriate form of the covariance matrix, and Chapter 6 examines the various strategies for choosing the "correct" pattern. Unstructured patterns have a large number of parameters compared to the autoregressive and compound symmetry, and it is a challenge to choose the most appropriate structure for the covariance matrix.

Chapter 7 expands the representation of a repeated measure to that called general mixed linear models. The word *mixed* means that the model includes both fixed and random effects. Random effects are intended to account for the subject–subject variability inherent in a repeated measures study. The inclusion of random factors in the model induces a particular pattern for the covariance pattern, and it also makes it more difficult to interpret the mean profile of the subjects. There are several ways to use random effects in the model. Suppose the fixed part of the mean profile is linear in time, then one could use a random factor to measure the subject–subject heterogeneity in the intercept, one could use a random factor to measure the subject–subject heterogeneity in the slope, or one could use both. For the three alternatives, a covariance pattern is induced. Several examples illustrate the Bayesian analysis with a model including random factors. The latter part of the chapter discusses diagnostic procedures that assess the fit of the model to the data. Discussed are transformed residuals of the general mixed linear model that are employed as a diagnostic procedure, and the methods are explained in terms of an air-quality study from the Netherlands.

The response in a repeated measures study is often categorical, and Chapter 8 introduces the reader to the Bayesian analysis under various scenarios. First is when the categorical variable exhibits a pattern that can be represented as a multinomial distribution. The Bayesian analysis is based on the Dirichlet distribution with uninformative priors for the parameters of the Dirichlet. This approach is applicable if there are no covariates. The second approach with categorical data with covariates and a Bayesian version of generalized estimating equations is employed, and the method is illustrated with a Poisson distribution for the categorical response. The third alternative presented is the generalized mixed linear model for categorical data, and presented are techniques for binary data with a Bernoulli distribution, count data with a Poisson distribution, and count data with a multinomial distribution. Of course, the generalized mixed linear model has two components: (1) a distribution for the response and (2) a link function that expressed the mean of the response as a linear function of various covariates and unknown parameters and random factors. For example, for a binary outcome, the response is assumed to follow a Bernoulli distribution with parameter θ, whose logic is expressed as a linear function of time, other covariates, and unknown parameters, including random factors.

Chapter 9 is a description of the Bayesian analysis for repeated measures when the mean profile is nonlinear. First to be considered is when the response is continuous and the covariance matrix is unstructured, and this is followed by the case when the correlation structure is autoregressive, and finally the case when random effects are part of the model. The Bayesian analysis is demonstrated with blood glucose values using the Michaelis–Menten model. For the case when the response is categorical, the Bayesian analysis is illuminated with several examples: (1) when the response is binary and describes the pill-dissolution times, (2) the timber load study, with categorical data, and (3) the heart rate investigation. Recall for categorical data that the model is specified by a distribution for the response and a link function that expresses the mean of the response as a nonlinear function of time, covariates, random factors, and unknown parameters. For the heart rate data, the response is categorized as an ordinal variable assumed to have a multinomial distribution, and the link function is the cumulative proportional odds model.

The book ends with Chapter 10, a Bayesian approach to missing values in the response variable. Examples are based on previous examples with no missing values, and then missing values are designated in such a way that the mechanism is missing at random. With WinBUGS, missing values are considered unknown parameters, and the software automatically estimates the missing values the usual way, with a posterior mean, standard deviation, MCMC error, and a 95% credible interval. Thus, if the simulation size is 75,000, it is used to compute the posterior characteristics of the missing values. When there are missing covariate values, one must assume a distribution for the covariate, and then WinBUGS treats the missing value as an unknown parameter. Continuous responses with linear and nonlinear mean profiles are analyzed with the Bayesian approach, and the techniques are repeated for linear and nonlinear models with categorical responses.

Exercises

1. For the beta density with parameters α and β, show that the mean is $[\alpha / (\alpha + \beta)]$ and the variance is $[\alpha\beta / (\alpha + \beta)^2(\alpha + \beta + 1)]$.

2. Consider a normal distribution. If the normal distribution is parameterized with μ and the precision $\tau = 1/\sigma^2$, the conjugate distribution is as follows: (1) the conditional distribution of μ given τ is normal with mean $\bar{x}$ and precision τ and (2) the marginal distribution of τ

is gamma with parameters $(n-1)/2$ and $\sum_{i=1}^{i=n}(x_i-\bar{x})^2/2 = (n-1)S^2/2$, where S^2 is the sample variance.

3. Verify Table 1.2, which reports the Bayesian analysis for the parameters of a normal population.

4. Verify the following statement: To generate values from the $t(n-1,\bar{x},n/S^2)$ distribution, generate values from Student's t distribution with $n-1$ degrees of freedom and multiply each by $S/\sqrt{n}$ and then add $\bar{x}$ to each.

5. Verify (1.22), the predictive density of a future observation Z from a normal population with both parameters unknown.

6. Suppose $x_1, x_2, ..., x_n$ are independent and that $x_i \sim$ gamma(α_i, β), and show that $y_i = x_i/(x_1 + x_2 + ... + x_n)$ jointly have a Dirichlet distribution with parameter $(\alpha_1, \alpha_2, ..., \alpha_n)$. Describe how this can be used to generate samples from the Dirichlet distribution.

7. Derive the conditional posterior distribution of the two precisions given the common mean, that is, verify (1.30).

8. Derive the conditional posterior distribution of the common mean given the two precision, that is, verify (1.31).

9. Suppose $(X_1, X_2, ..., X_k)$ is multinomial with parameters n and $(\theta_1, \theta_2, ..., \theta_k)$, where $\sum_{i=1}^{i=k} X_i = n$, $0 < \theta_i < 1$, and $\sum_{i=1}^{i=k} \theta_i = 1$. Show that $E(X_i) = n\theta_i$, Var$(X_i) = n\theta_i(1-\theta_i)$, and Cov$(X_i, X_j) = -n\theta_i\theta_j$. What is the marginal distribution of θ_i?

10. Suppose $(\theta_1, \theta_2, ..., \theta_k)$ is Dirichlet with parameters $(\alpha_1, \alpha_2, ..., \alpha_k)$, where $\alpha_i > 0$, $\theta_i > 0$, and $\sum_{i=1}^{i=k} \theta_i = 1$. Find the mean and variance of θ_i and covariance between θ_i and θ_j, $i \neq j$.

11. Show that the Dirichlet family is conjugate to the multinomial family.

12. Suppose $(\theta_1, \theta_2, ..., \theta_k)$ is Dirichlet with parameters $(\alpha_1, \alpha_2, ..., \alpha_k)$. Show that the marginal distribution of θ_i is beta and give the parameters of the beta. What is the conditional distribution of θ_i given θ_j?

13. For the exponential density $f(x|\theta) = \theta\exp-\theta x$, $x > 0$, where x is positive and θ is a positive unknown parameter, suppose the prior density of θ is $g(\theta) \propto \theta^{\alpha-1}\exp-\beta\theta$, $\theta > 0$, what is the posterior density of θ? In epidemiology, the exponential distribution is often used as the distribution of survival times in screening tests.

2

Review of Bayesian Regression Methods

2.1 Introduction

Regression is a powerful method to study the association between a dependent variable and various independent variables and is the main approach to evaluate repeated measures. When several covariates (independent variables) are available, regression models are ideal and a standard tool for the analysis of data. Regression analysis is an important topic in statistics, and many textbooks are accessible. For a good introduction, see Chatterjee and Price[1]; for a Bayesian approach, see Ntzoufras.[2]

Several types of regression models are reviewed: (1) logistic regression models, (2) simple and multiple linear regression models, (3) nonlinear models, and (4) a simple repeated measures model. The types of models differ in regard to the type of dependent variable. A regression model has a dependent variable and at least one independent variable. When the dependent variable is binary (two values), the association between the dependent variable and other covariates can be modeled by a logistic regression. If the dependent variable is continuous or quantitative, normal-theory simple linear and multiple regression models are applicable, but if the dependent variable is categorical (several values), multinomial regression models are appropriate.

As an example of a logistic model, a heart disease study will be analyzed, where the dependent variable is the disease status (myocardial infarction) and the independent variables are age and systolic blood pressure.

In this case, the independent variables are binary (take on two values), but they could be regarded as continuous if one uses the actual age and blood pressure values. Using the logistic regression model, the odds ratios are used for measuring association between the disease status and the two dependent variables age and systolic blood pressure. For a good introduction to logistic regression, and for many examples, see Hosmer and Lemeshow.[3]

Simple linear and multiple regression models are appropriate when the dependent variable (an indicator of disease status) is quantitative (often called continuous) and considered to have a normal distribution. A simple linear regression model has one independent variable, while a multiple linear regression model has more than one covariate. For example, for a population of diabetics, one might want to study the association between blood glucose values and age. One could use the simple linear regression model where blood glucose values constitute the dependent variable and age the independent variable, but if another variable such as gender is included as an independent variable, one would have a multiple linear regression model with two independent variables, where age and gender are the covariates. The term *linear regression* refers to the fact that the average value of the dependent variable is a linear function of the regression coefficients. Obviously, a model does not have to be linear but can be nonlinear, a topic that will be explained in more detail at the end of Section 2.5. For additional information about normal theory regression models, see Chatterjee and Price,[1] a work that will be frequently referred throughout this chapter.

For categorical regression models, the dependent variable can have a small number of values. For example, disease status for cancer could be labeled as local and confined to the primary tumor, metastasized to the lymph nodes, or metastasized beyond the lymph nodes, in which case the dependent variable assumes three values. Special models are used in this situation, which are also referred to as multinomial regression models. The independent variables can be either continuous or categorical. As before, the dependent variable is an indicator of disease status, while the independent variables are risk factors or confounders. For example, for cancer, possible independent variables are age, sex, type of treatment, and so on, and one wants to study their effect on the stage of disease.

Additional references are Agresti,[4] where categorical regression models are described, and Congdon,[5] where a Bayesian approach to the general subject of regression analysis is presented.

2.2 Logistic Regression

2.2.1 Introduction

For n subjects, suppose for the ith response y_i is binary, that is, $y_i = 0$ or 1, and y_i is distributed as a Bernoulli with parameter θ_i, that is to say, $\theta_i = P(y_i = 1)$ and $1 - \theta_i = P(y_i = 0)$. In addition, suppose there is one independent variable x that assumes the value x_i for subject i; then the logistic model is defined as

$$\text{logit}(\theta_i) = \alpha + \beta x_i \tag{2.1}$$

where

$$\text{logit}(\theta_i) = \ln \frac{\theta_i}{1 - \theta_i} \tag{2.2}$$

The term *logistic* comes from the fact that (3.2) is equivalent to the logistic transformation

$$\theta_i = \frac{\exp(\alpha + \beta x_i)}{1 + \exp(\alpha + \beta x_i)} \tag{2.3}$$

Of course, additional independent variables can be added as covariates, in which case

$$\text{logit}(\theta_i) = \alpha + \beta_1 x_{1i} + \beta_2 x_{2i} + \cdots + \beta_p x_{ip} \tag{2.4}$$

Unknown regression coefficients in the model are $\alpha, \beta_1, \beta_2, ..., \beta_p$, which are estimated from the data (x_i, y_i), and for the Bayesian, are estimated from their joint posterior distribution. The model is linear in the regression coefficients on the logit scale (2.4), but not for the probability parameter (3.3). Note also that the odds that $y_i = 1$ is the antilog of (2.1) or

$$\frac{\theta_i}{(1 - \theta_i)} = \exp(\alpha + \beta x_i) \tag{2.5}$$

If x is a categorical variable with $x_i = 0$ or 1, then the odds that $y_i = 1$ when $x_i = 1$ divided by the odds that $y_i = 1$ when $x_i = 0$ is the odds ratio $\exp(\beta)$.

Formally, we have that the odds ratio expressed as

$$\frac{\theta_i/(1 - \theta_i)|x_i = 1}{\theta_i/(1 - \theta_i)|x_i = 0} = \exp(\beta) \tag{2.6}$$

or that the log of the odds ratio is β. On the other hand, if x is continuous, then $\exp(\beta)$ is the odds ratio for a unit increase in x, that is, when x increases from a value of x to a value of $x + 1$.

The same interpretation applies for the multiple linear logistic model (2.4). Note that for a unit increase in x_{1i}, the odds that $y_i = 1$ given $x_{1i} = 1$ divided by the odds that $y_i = 1$ given $x_{1i} = 0$ is $\exp(\beta_1)$, assuming the other $p - 1$ variables are constant, for all possible values of the remaining independent variables. It is important to remember the restriction that $\exp(\beta_1)$ is the odds ratio, assuming the other variables are constant for all possible values of those other variables. Bayesian inferences about the odds will be based on the posterior distribution of β_1, which induces the posterior distribution of the odds $\exp(\beta_1)$.

2.2.2 Example of Heart Disease

Consider the example of a heart study, where the association between heart attack and two independent variables age and systolic blood pressure is to

TABLE 2.1

Myocardial Infarction versus Systolic Blood
Pressure by Age: Hypothetical Study

	MI	No MI	Total
Age ≥60			
SBP ≥140	3	28	31
SBP <140	0	19	19
Total	3	47	50
Age <60			
SBP ≥140	4	149	153
SBP <140	5	142	147
Total	9	291	300

be determined. Age is categorized as age <60 and age ≥60 as well as systolic blood pressure (SBP), where SBP ≥140 or SBP <140. The occurrence
of heart attack is modeled as $y = 0$ for no attack and $y = 1$ for heart attack.
In a similar way, let $x_1 = 1$ if age is age ≥ 60 or 0 otherwise, and let $x_2 = 1$
if SBP ≥ 140 mm Hg or 0 otherwise. For the logistic regression, we have a
model with two independent variables, where both are binary; thus, β_1 is
the odds of a heart attack for patients over age 60 divided by the odds
of a heart attack for patients under the age of 60, for all values of blood
pressure.

Consider Table 2.1, which portrays the results of a hypothetical study.
There are a total of 350 subjects, of which 12 had a heart attack and 338
did not. Logistic regression will be used to assess the association between
systolic blood pressure and heart attack, adjusting for age. There are two
levels of age, blood pressure, and disease status. The Bayesian analysis
is based on **BUGS CODE 2.1** and is executed with 55,000 observations
for the simulation, with a burn in of 5,000 and a refresh of 100. The code
closely follows the various formulas for logistic regression given by (2.1)
through (2.6). Note that the data is given by the first list statement in the
code, where y is the column of values for the occurrence of a heart attack,
x_1 is the column of age indicators with a 1 indicating age greater than 60,
while x_2 is the column indicating low and high ($x_2 = 1$) systolic blood
pressure.

A glance of the study results show that for age less than 60, the number
of subjects who had a heart attack and a blood pressure at least 140 is 4.
On the other hand, the number of subjects who had a heart attack who are
greater than 60 years of age and have a blood pressure greater than 140 is 3.
A Bayesian posterior analysis is executed on the basis of **BUGS CODE 2.1**,
and 65,000 observations are generated for the simulation with a burn in of
5,000 and a refresh of 100.

BUGS CODE 2.1

```
model;
{
# Bernoulli distribution for the observations
# theta is the probability of a heart attack
for (i in 1:400){y[i]~dbern(theta[i])}


for (i in 1:400){z[i]~dbern(theta[i])}

# logistic regression of theta on age and systolic blood
pressure
for (i in 1:400){logit(theta[i])<-alpha+beta[1]*x1[i]+bet
a[2]*x2[i]}
# prior distributions for the regression coefficients
# uninformative priors
alpha~ dnorm(0.0000,.00001)
beta[1]~dnorm(0.0000,.00001)
beta[2]~dnorm(0.0000,.00001)
# the odds ratio for age
ORage<-exp(beta[1])
ORsbp<-exp(beta[2])
}
# y is the occurrence of a heart attack
# x1 is age
# x2 is systolic blood pressure
list(y = c(1,1,
      1,
      0,0,0,0,0,0,0,0,0,0,0,0,0,0,0,0,0,0,0,0,0,0,0,0,0,0,
      0,0,0,0,0,0,0,0,0,0,0,0,0,0,0,0,0,0,0,0,0,
      1,1,1,1,
      1,1,1,1,1,
      0,0,0,0,0,0,0,0,0,0,0,0,0,0,0,0,0,0,0,0,0,0,0,0,0,0,
      0,0,0,0,0,0,0,0,0,0,0,0,0,0,0,0,0,0,0,0,0,0,0,0,0,0,
      0,0,0,0,0,0,0,0,0,0,0,0,0,0,0,0,0,0,0,0,0,0,0,0,0,0,
      0,0,0,0,0,0,0,0,0,0,0,0,0,0,0,0,0,0,0,0,0,0,0,0,0,0,
      0,0,0,0,0,0,0,0,0,0,0,0,0,0,0,0,0,0,0,0,0,0,0,0,0,0,
      0,0,0,0,0,0,0,0,0,0,0,0,0,0,0,0,0,0,0,0,0,0,0,0,0,0,

      0,0,0,0,0,0,0,0,0,0,0,0,0,0,0,0,0,0,0,0,0,0,0,0,0,0,
      0,0,0,0,0,0,0,0,0,0,0,0,0,0,0,0,0,0,0,0,0,0,0,0,0,0,
      0,0,0,0,0,0,0,0,0,0,0,0,0,0,0,0,0,0,0,0,0,0,0,0,0,0,
      0,0,0,0,0,0,0,0,0,0,0,0,0,0,0,0,0,0,0,0,0,0,0,0,0,0,
      0,0,0,0,0,0,0,0,0,0,0,0,0,0,0,0,0,0,0,0,0,0,0,0,0,0,
      0,0,0,0,0,0,0,0,0,0,0,0,0,0,0,0,0,0,0,0,0,0,0,0,0,0,
      0,0,0,0,0,0,0,0,0,0,0,0,0,0,0,0,0,0,0,0,0,0,0,0,0,0,
      0,0,0,0,0,0,0,0,0,0,0,0,0,0,0,0,0,0,0,0,0,0,0,0,0,0,
```

```
       0,0,0,0,0,0,0,0,0,0,0,0,0,0,0,0,0,0,0,0,0,0,0,0,0,0,0,0,
       0,0,0,0,0,0,0,0,0,0,0,0,0,0,0,0,0,0,0,0),


x1 = c(1,1,
       1,
       1,1,1,1,1,1,1,1,1,1,1,1,1,1,1,1,1,1,1,1,1,1,1,1,1,1,1,1,
       1,1,1,1,1,1,1,1,1,1,1,1,1,1,1,1,1,1,1,1,1,1,1,1,
       0,0,0,0,
       0,0,0,0,0,
       0,0,0,0,0,0,0,0,0,0,0,0,0,0,0,0,0,0,0,0,0,0,0,0,0,0,0,0,
       0,0,0,0,0,0,0,0,0,0,0,0,0,0,0,0,0,0,0,0,0,0,0,0,0,0,0,0,
       0,0,0,0,0,0,0,0,0,0,0,0,0,0,0,0,0,0,0,0,0,0,0,0,0,0,0,0,
       0,0,0,0,0,0,0,0,0,0,0,0,0,0,0,0,0,0,0,0,0,0,0,0,0,0,0,0,
       0,0,0,0,0,0,0,0,0,0,0,0,0,0,0,0,0,0,0,0,0,0,0,0,0,0,0,0,
       0,0,0,0,0,0,0,0,0,0,0,0,0,0,0,0,0,0,0,0,0,0,0,0,0,0,0,0,

       0,0,0,0,0,0,0,0,0,0,0,0,0,0,0,0,0,0,0,0,0,0,0,0,0,0,0,0,
       0,0,0,0,0,0,0,0,0,0,0,0,0,0,0,0,0,0,0,0,0,0,0,0,0,0,0,0,
       0,0,0,0,0,0,0,0,0,0,0,0,0,0,0,0,0,0,0,0,0,0,0,0,0,0,0,0,
       0,0,0,0,0,0,0,0,0,0,0,0,0,0,0,0,0,0,0,0,0,0,0,0,0,0,0,0,
       0,0,0,0,0,0,0,0,0,0,0,0,0,0,0,0,0,0,0,0,0,0,0,0,0,0,0,0,
       0,0,0,0,0,0,0,0,0,0,0,0,0,0,0,0,0,0,0,0,0,0,0,0,0,0,0,0,
       0,0,0,0,0,0,0,0,0,0,0,0,0,0,0,0,0,0,0,0,0,0,0,0,0,0,0,0,
       0,0,0,0,0,0,0,0,0,0,0,0,0,0,0,0,0,0,0,0,0,0,0,0,0,0,0,0,
       0,0,0,0,0,0,0,0,0,0,0,0,0,0,0,0,0,0,0,0,0,0,0,0,0,0,0,0,
       0,0,0,0,0,0,0,0,0,0,0,0,0,0,0,0,0,0,0,0,0,0,0,0,0,0,0,0,
       0,0,0,0,0,0,0,0,0,0,0,0,0,0,0,0,0,0,0),

x2 = c(1,1,
       0,
       1,1,1,1,1,1,1,1,1,1,1,1,1,1,1,1,1,1,1,1,1,1,1,1,1,1,1,1,
       1,1,1,
       0,0,0,0,0,0,0,0,0,0,0,0,0,0,0,0,0,0,0,0,0,
       1,1,1,1,
       0,0,0,0,0,
       1,1,1,1,1,1,1,1,1,1,1,1,1,1,1,1,1,1,1,1,1,1,1,1,1,1,1,1,
       1,1,1,1,1,1,1,1,1,1,1,1,1,1,1,1,1,1,1,1,1,1,1,1,1,1,1,1,
       1,1,1,1,1,1,1,1,1,1,1,1,1,1,1,1,1,1,1,1,1,1,1,1,1,1,1,1,
       1,1,1,1,1,1,1,1,1,1,1,1,1,1,1,1,1,1,1,1,1,1,1,1,1,1,1,1,
       1,1,1,1,1,1,1,1,1,1,1,1,1,1,1,1,1,1,1,1,1,1,1,1,1,1,1,1,
       1,1,1,1,1,1,1,1,1,1,1,1,1,1,1,1,1,1,1,1,1,1,1,1,1,1,1,1,

       0,0,0,0,0,0,0,0,0,0,0,0,0,0,0,0,0,0,0,0,0,0,0,0,0,0,0,0,
       0,0,0,0,0,0,0,0,0,0,0,0,0,0,0,0,0,0,0,0,0,0,0,0,0,0,0,0,
       0,0,0,0,0,0,0,0,0,0,0,0,0,0,0,0,0,0,0,0,0,0,0,0,0,0,0,0,
       0,0,0,0,0,0,0,0,0,0,0,0,0,0,0,0,0,0,0,0,0,0,0,0,0,0,0,0,
       0,0,0,0,0,0,0,0,0,0,0,0,0,0,0,0,0,0,0,0,0,0,0,0,0,0,0,0,
       0,0,0,0,0,0,0,0,0,0,0,0,0,0,0,0,0,0,0,0,0,0,0,0,0,0,0,0,
```

```
      0,0,0,0,0,0,0,0,0,0,0,0,0,0,0,0,0,0,0,0,0,0,0,0,0,0,0,
      0,0,0,0,0,0,0,0,0,0,0,0,0,0,0,0,0,0,0,0,0,0,0,0,0,0,0,
      0,0,0,0,0,0,0,0,0,0,0,0,0,0,0,0,0,0,0,0,0,0,0,0,0,0,0,
      0,0,0,0,0,0,0,0,0,0,0,0,0,0,0,0,0,0))

# below is the data with the actual values of age and
systolic blood pressure
list( y = c(1,1,
      1,
      0,0,0,0,0,0,0,0,0,0,0,0,0,0,0,0,0,0,0,0,0,0,0,0,0,0,0,
      0,0,0,
      0,0,0,0,0,0,0,0,0,0,0,0,0,0,0,0,0,0,0,0,0,
      1,1,1,1,
      1,1,1,1,1,1,
      0,0,0,0,0,0,0,0,0,0,0,0,0,0,0,0,0,0,0,0,0,0,0,0,0,0,0,
      0,0,0,0,0,0,0,0,0,0,0,0,0,0,0,0,0,0,0,0,0,0,0,0,0,0,0,
      0,0,0,0,0,0,0,0,0,0,0,0,0,0,0,0,0,0,0,0,0,0,0,0,0,0,0,
      0,0,0,0,0,0,0,0,0,0,0,0,0,0,0,0,0,0,0,0,0,0,0,0,0,0,0,
      0,0,0,0,0,0,0,0,0,0,0,0,0,0,0,0,0,0,0,0,0,0,0,0,0,0,0,
      0,0,0,0,0,0,0,0,0,0,0,0,0,0,0,0,0,0,0,0,0,0,0,0,0,0,0,

      0,0,0,0,0,0,0,0,0,0,0,0,0,0,0,0,0,0,0,0,0,0,0,0,0,0,0,
      0,0,0,0,0,0,0,0,0,0,0,0,0,0,0,0,0,0,0,0,0,0,0,0,0,0,0,
      0,0,0,0,0,0,0,0,0,0,0,0,0,0,0,0,0,0,0,0,0,0,0,0,0,0,0,
      0,0,0,0,0,0,0,0,0,0,0,0,0,0,0,0,0,0,0,0,0,0,0,0,0,0,0,
      0,0,0,0,0,0,0,0,0,0,0,0,0,0,0,0,0,0,0,0,0,0,0,0,0,0,0,
      0,0,0,0,0,0,0,0,0,0,0,0,0,0,0,0,0,0,0,0,0,0,0,0,0,0,0,
      0,0,0,0,0,0,0,0,0,0,0,0,0,0,0,0,0,0,0,0,0,0,0,0,0,0,0,
      0,0,0,0,0,0,0,0,0,0,0,0,0,0,0,0,0,0,0,0,0,0,0,0,0,0,0,
      0,0,0,0,0,0,0,0,0,0,0,0,0,0,0,0,0,0,0,0,0,0,0,0,0,0,0,
      0,0,0,0,0,0,0,0,0,0,0,0,0,0,0,0,0,0),

x1 = c(77,85,87,74,56,76,69,71,74,83,67,69,52,77,72,76,
77,76,70,67,70,74,72,74,57,88,79,64,67,67,69,77,68,86,58,
77,79,86,70,80,60,72,63,92,56,79,76,65,70,76,55,56,48,54,
50,49,45,36,54,47,52,47,47,49,46,39,50,51,39,59,46,44,52,
53,49,53,52,46,49,46,41,58,61,50,45,58,51,58,44,50,57,60,
48,51,56,47,44,53,41,39,50,36,55,47,59,43,51,43,40,43,39,
38,62,49,46,45,42,50,53,43,47,31,40,43,45,51,51,55,56,54,
49,46,53,49,34,43,60,44,56,50,46,44,49,46,53,45,46,56,43,
53,45,49,49,41,46,44,43,48,39,59,55,46,42,38,40,42,29,41,
38,52,51,54,50,48,47,47,47,55,44,50,55,50,43,54,42,55,42,
38,52,49,50,56,50,39,51,49,51,54,47,42,59,54,52,47,52,44,
44,45,50,49,48,35,43,45,49,46,47,47,59,48,52,44,48,50,58,
47,49,43,56,50,42,48,41,52,41,43,47,43,47,54,52,50,62,47,
56,42,46,53,43,43,50,56,54,46,51,49,47,42,42,54,53,45,35,
48,52,45,47,41,48,40,42,50,41,56,56,49,53,34,46,50,54,42,
```

```
48,51,51,56,49,56,44,47,51,51,36,46,58,58,51,40,53,53,44,
44,51,47,48,61,40,50,49,39,47,36,45,47,57,44,44,51,40,44,
51,60,48,51,46,51,47,48,57,48,39,49,40,58,50,45,41,47,49,
41,64,36,47,49,58,59,45,45,53,37,43,43,50,44,49,43,53,48,
54,53,44,47,55,52,40,43,47,47,54,46,51,65,54,53,56,53,49,
47,38,59,55,44,52,41,55,43,47,46,55,38,45,42,54,52,42,56,
48,44,40,47,37,44,51,37,50,47,46,45,46,44,44,52,45,38,48,
54,46,41,41,43,44,44,49,48,41,57,45,46,46,35,50,43,41,52,
43,54,46,40,50,61,49,44,50,40,50,38,52,46,49,53),

x2 = c(158,164,150,157,161,165,166,147,169,159,150,166,
161,161,167,164,167,163,166,159,162,166,145,163,161,162,
156,160,178,164,154,115,114,128,117,123,122,114,116,114,119,
112,123,119,122,130,119,128,120,115,163,166,162,154,167,163,
155,160,157,154,149,159,167,162,159,160,162,157,158,168,
168,164,157,171,150,158,150,163,163,159,171,165,148,156,
164,168,158,167,153,167,149,158,159,144,167,151,156,150,
154,164,158,147,151,167,151,151,160,161,160,160,156,168,
162,171,155,150,159,156,147,162,163,170,158,163,168,153,151,
150,156,162,164,164,159,152,164,159,158,163,162,152,162,157,
158,162,149,166,142,151,160,154,168,165,161,152,151,159,
155,169,169,168,166,163,158,150,151,150,152,167,155,150,157,
152,158,168,158,152,162,165,154,168,173,158,164,157,155,
161,160,160,159,155,162,164,160,158,164,157,168,159,169,159,
155,161,150,119,114,120,114,128,126,116,119,109,121,114,
121,120,122,112,113,116,126,117,109,120,111,123,113,119,
110,112,120,127,116,124,111,134,137,113,124,127,125,123,
120,119,121,115,119,110,123,123,124,126,128,119,122,118,
115,121,114,125,117,125,124,133,115,121,120,127,120,118,
136,125,119,120,111,119,121,111,125,130,117,131,127,109,
124,114,120,112,130,127,118,119,121,119,117,117,117,123,
122,121,119,125,113,115,131,126,116,126,120,119,125,113,
124,127,116,119,128,119,129,125,122,121,113,118,113,130,
119,117,125,130,120,115,118,131,121,117,130,103,136,127,
118,122,125,121,123,117,110,126,118,125,123,128,129,116,
127,109,118,127,134,123,124,115,113,126,120,116,122,112,
109,121,110,114,133,131,123,122,122,116,120,104,120,125,
117,121,125,121,121,116,115,121,126,121,113,126,128,112,
112,126,120,122,113,126,119,131,118,112,126,131,124,116,
122,120,115,104,115,127,122,123,134,120,124,104,113,110,
118,117,118,128,121,120,116,121,116,121,116,128,115,123,
118,114,123,121,116,111,118,119,127,129,114,122))

# Initial values
list(alpha = 0,beta = c(0,0))
```

TABLE 2.2

Bayesian Analysis of Logistic Regression for a Heart Study with Binary Values for All Variables

Parameter	Mean	SD	Error	2½	Median	97½
Odds ratio—age	2.776	2.11	0.02174	0.4575	2.207	8.318
Odds ratio—systolic blood pressure	1.292	0.8591	0.01157	0.316	1.069	3.58
α	−3.765	0.4535	0.005423	−4.729	−3.731	−2.963
β_1	0.7673	0.7367	0.008129	−0.7819	0.7917	2.118
β_2	0.06499	0.6179	0.006504	−1.152	0.06713	1.275

The analysis of the association between myocardial infarction and systolic blood pressure, adjusted for age, is reported in Table 2.2.

The interpretation of the parameters is as follows. For the odds ratio of age with a posterior median of 2.207, the implication is that the odds of a heart attack for a person older than 60 years is 2.207 times the odds of a heart attack for a person younger than 60. For the odds ratio of systolic blood pressure, the median is 1.069, implying that there is very little difference in the odds of a heart attack for a person with a blood pressure greater than 140 mm Hg and the odds of a heart attack for a person with a blood pressure less than 140. The 95% credible intervals for the two odds ratios include the value 1, implying that age and systolic blood pressure have very little effect on the occurrence of a heart attack. It is of interest to note that the Markov chain Monte Carlo (MCMC) errors are not as small as in earlier examples. For example, the error of 0.02174 indicates that the posterior mean of 2.776 is within 0.02174 units of the "true" posterior mean. The posterior distribution of the two odds ratios are highly skewed; therefore, I use the posterior median (see Figure 2.1).

It is interesting to analyze the heart study data from Table 2.1 using the actual values for age and systolic blood pressure; see the second list statement of **BUGS CODE 2.1**, which contains the actual values. I performed a Bayesian analysis

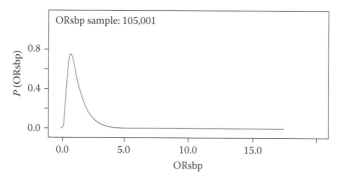

FIGURE 2.1
Skewed posterior density of the odds ratio.

TABLE 2.3

Bayesian Analysis for a Heart Study with Actual Age and Systolic Blood Pressure Values

Parameter	Mean	SD	Error	2½	Median	97½
Odds ratio—age	1.041	0.02495	0.0004006	0.9908	1.041	1.089
Odds ratio—blood pressure	1.087	0.03363	0.0009797	1.034	1.082	1.167
α	−18.29	5.089	0.1478	−30.42	−17.65	−10.2
β_1	0.03692	0.02402	0.0003863	−0.009285	0.04021	0.08518
β_2	0.08256	0.03054	0.000466	0.03362	0.0786	0.1546

with the actual data, using 55,000 observations for the simulation with a burn in of 5,000 and a refresh of 100, and the analysis is reported in Table 2.3.

Table 2.2 should be compared to Table 2.3, because the differences are quite interesting. Recall from Table 2.2 that when age and blood pressure are coded with binary values, there is a small hint that age and systolic blood pressure can impact the probability of a heart attack, but referring to Table 2.3 when the actual values of age and blood pressure are used, one sees that there is strong evidence that age and blood pressure have absolutely no effect on the probability of a heart attack. Of course, the analysis with the actual values of age and blood pressure should be preferred, because coding age and systolic blood pressure with binary values involves a loss of information. For example, all age values greater than 60 are treated the same.

2.2.3 Example with Several Independent Variables

An example exhibiting four independent variables is provided by Hosmer and Lemeshow (p. 48),[3] where the disease is coronary heart disease, and the four variables correspond to four races. The data appears in the first list statement of **BUGS CODE 2.2**, where the y column (5,20,15,10) is the number of white, black, Hispanic, and other subjects, respectively, with the disease.

In addition, the vector n is (25,30,25,20) and is the total number of white, black, Hispanic, and other subjects, respectively, in the study.

Thus, among 25 whites, 5 have coronary heart disease, and among 20 of another race (not black or Hispanic), 10 have heart disease; therefore, the odds of having heart disease among whites relative to that among other races is estimated as $(5/20)/(10/10) = 0.25$. This information is analyzed with the logistic model

$$\text{logit}(\theta_i) = \alpha + \beta_1 x_{1i} + \beta_2 x_{2i} + \beta_3 x_{i3} \qquad (2.7)$$

where:

θ_i is the probability of disease for the ith race, $i = 1,2,3$

The four racial groups are coded as indicator variables $x_1 = (1,0,0,0)$, $x_2 = (0,1,0,0)$, and $x_3 = (0,0,1,0)$; thus, the group corresponding to other races is coded as $x_{14} = x_{24} = x_{34} = 0$. This coding makes group 4 (the one with all other races) as the reference group. In order to perform a Bayesian analysis for this study, consider the following. Note that this information appears in the first list statement of **BUGS CODE 2.2**.

BUGS CODE 2.2

```
model;
{
# binomial distribution for the number of respondents

for(i in 1:4){y[i]~dbin(theta[i],n[i])}
# logistic regression of theta

for(i in 1:4){logit(theta[i])<-alpha+beta[1]*x1[i]+beta[2
]*x2[i]+beta[3]*x3[i]}
# prior distributions
alpha~dnorm(.0000,.0001)
for(i in 1:3){beta[i]~dnorm(.0000,.00001)}
# odds ratio of first group versus fourth
OR1<-exp(beta[1])
OR2<-exp(beta[2])
OR3<-exp(beta[3])
for( in 1:4){z[i]~dbin(theta[i],n[i])}

}
# data from Hosmer and Lemeshow page 48
# below is the number with CHD
list(y = c(5,20,15,10),
# n is the total number in each group
n = c(25,30,25,20),
# indictor for white
x1 = c(1,0,0,0),
# indicator for black
x2 = c(0,1,0,0),
# indicator for Hispanic
x3 = c(0,0,1,0))

# data below Table 2.1
list(y = c(2,1,4,5),
# n is the number in each group
n = c(31,19,153,247),
x1 = c(1,0,0,0),
x2 = c(0,1,0,0),
x3 = c(0,0,1,0))

# initial values
list(alpha = 0, beta = c(0,0,0))
```

TABLE 2.4

Posterior Analysis for Association between Race and Coronary Artery Disease

Parameter	Mean	SD	Error	2½	Median	97½
OR1	0.2932	0.2239	0.001455	0.05672	0.2346	0.8732
OR2	2.477	1.694	0.01225	0.6358	2.084	6.851
OR3	1.85	1.287	0.009219	0.4571	1.523	5.158
alpha	$<10^{-20}$	0.458	0.00399	−0.9015	$<10^{-20}$	0.9008
beta[1]	−1.464	0.6965	0.004732	−2.87	−1.45	−0.1352
beta[2]	0.7211	0.6057	0.004637	−0.4529	0.7171	1.924
beta[3]	0.4225	0.6184	0.00469	−0.783	0.421	1.641
theta[1]	0.2001	0.07883	<0.0001	0.07162	0.1916	0.3752
theta[2]	0.6671	0.08462	<0.0001	0.4924	0.6707	0.8214
theta[3]	0.5999	0.09587	<0.0001	0.4065	0.6025	0.7775
theta[4]	0.5	0.1089	<0.0001	0.2887	0.4998	0.7111

Based on **BUGS CODE 2.2**, a posterior analysis is conducted with 55,000 observations generated for the simulation, with a burn in of 5,000 and a refresh of 100, and the calculations are reported in Table 2.4.

This is an interesting analysis because the posterior distributions of the odds ratio are skewed. For example, the posterior mean of the odds ratio for group 1 (whites) is 0.2932, but the posterior median is 0.2346. Consider the "usual" estimate of $(5/20)/(10/10) = 0.25$ for the odds ratio for coronary artery disease of whites relative to the group consisting of other races; then neither the posterior mean nor the median is equal to the usual estimate. I would use 0.2346 as the estimate, because the posterior distribution is skewed.

2.2.4 Goodness of Fit

Goodness of fit of regression models is a vast topic and can be read by referring to Ntzoufras[2] and Hosmer and Lemeshow[3]; thus, what is presented here on the topic is based on comparing the actual number with disease to the number predicted by the model. Recall for the logistic model that the number y_i is the number of diseased for the ith group (or individual) and has a binomial distribution with parameters θ_i and n_i, and the Bayesian analysis involves determining the posterior distribution of the θ_i.

The Bayesian predictive distribution is the conditional distribution of a future observation given the data y_i and will be used to generate future observations z_i corresponding to y_i. WinBUGS easily generates the future observations.

Consider the example of the effect of race on the probability of coronary heart disease, where **BUGS CODE 2.2** is used for the Bayesian analysis of estimating the odds of the various racial groups relative to some reference group.

TABLE 2.5

Predicted Number with Coronary Artery Disease

Parameter	Mean	SD	Error	2½	Median	97½
$z[1]$	5.007	2.768	0.008715	1	5	11
$z[2]$	20.0	3.596	0.010081	13	20	26
$z[3]$	15.01	3.393	0.01056	8	15	21
$z[4]$	9.99	3.093	0.01906	4	10	16

Table 2.5 reports the Bayesian analysis with the posterior distribution of the odds ratios and other relevant parameters. The last statement

```
for(i in 1:4){z[i]~dbin(theta[i],n[i])}
```

before the first list statement of the code produces the future number of diseased subjects for each group. Returning to the Hosmer and Lemeshow example of four racial groups, the posterior distribution of the four predicted distributions is listed in Table 2.5.

The posterior mean of the future observations is almost equal to the exact number of people with disease; thus, the logistic model appears to give an excellent fit. Some uncertainty exists for the four predicted observations. For example, for the first observation, the posterior standard deviation is 2.768, and this is demonstrated with Figure 2.2. A Bayesian analysis is performed, via **BUGS CODE 2.2**, with 106,000 observations for the simulation, with a burn in of 5,000 and a refresh of 100.

In summary, in order to test how well the logistic model fits the data, generate a future observation for each group and compare to the actual number with disease.

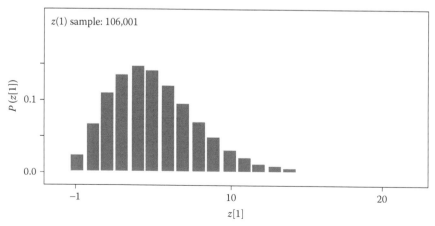

FIGURE 2.2
Predictive density of the number of Caucasians with coronary heart disease.

2.3 Linear Regression Models

2.3.1 Introduction

Simple and multiple linear regression models are used in medicine and epidemiology to determine associations between a dependent and several independent variables, and the subject is vast. If one refers to the latest issue of the *American Journal of Epidemiology* or the *Annals of Internal Medicine*, one will mostly likely find a regression analysis that is employed to find some type of association between disease and various covariates representing exposure to various risk factors. The models considered in this chapter differ from the logistic model in that the dependent variable is quantitative. Quantitative variables assume values similar to those one would encounter in measuring blood glucose values, where in principle, between any two values one can find another blood glucose value; thus, measurements such as age, weight, and systolic blood pressure are examples of a quantitative variable.

The review of regression analysis is initiated with the definition of a simple linear regression model, which has one dependent variable and one independent variable, with the goal being to establish an association between the two. For example, the dependent variable might be systolic blood pressure, and the independent variable indicating two groups, where the subjects are with and without coronary artery disease. Simple linear regression has very strict assumptions, such as the dependent variable must be normally distributed, and the variance of the dependent must be constant over all values of the independent variable. These assumptions will be relaxed to some extent by allowing for unequal variances, where a weighted regression is appropriate. Of course, the approach is Bayesian, where the posterior distribution of the regression coefficients (intercept and slope) and the variance about the regression line is determined. Several examples, relevant to longitudinal studies, are presented. One problem to be explained is that of interpreting the estimated value of the regression coefficient.

Simple linear regression models are generalized to multiple linear regression models, where the goal is to establish an association between one quantitative dependent variable and several (more than one) independent variables. For example, the dependent variables might be blood glucose values, and the dependent variables might be age, weight, gender, and subjects with and without diabetes. The goal is to estimate the effect of age, weight, gender, and diabetes, on the average blood glucose value. One challenge with such a regression analysis is the interpretation of the regression coefficients of the model, and this will be carefully explained with many examples relevant to epidemiology.

One uses the term *linear regression* with the emphasis on linear, which means that the average value of the dependent variable is linear in the unknown regression coefficients, which are to be estimated from their posterior distribution. In practice, the dependent and independent variables are often transformed to achieve the linear assumption and the assumption of constant variance.

Often, one must use nonlinear regression models to assess the association between disease and various risk factors, and this will be presented and explained in Chapter 9.

2.3.2 Simple Linear Regression

The definition of simple linear regression is as follows:

$$y_i = \alpha + \beta x_i + e_i \qquad (2.8)$$

where values of the dependent variable y and independent variable x are paired as (x_i, y_i) for the ith individual, for $i = 1,2,...,n$. If one plots the n pairs of observations, one would expect a linear association to develop; however, the relationship would not appear exactly linear because of the error term e with n values e_i, which are assumed to be independent and normally distributed with mean zero and unknown variance σ^2. This implies that the average value of the dependent variable y is

$$\text{Avg}(y_i) = \alpha + \beta x_i \qquad (2.9)$$

thus, the average value of y is linear in x, namely $\alpha + \beta x_i$ for $i = 1,2,...,n$.

With simple linear regression, there are three unknown parameters, the intercept term α, the slope β, and the variance of the error term σ^2, which is also the variance of the dependent variable y.

Consider Table 2.6, which gives 15 pairs of observations for the independent variable y and dependent variable x. This is the first example of simple linear regression, and the values are hypothetical.

These pairs are plotted in Figure 2.3, and the red line is the fitted simple linear regression line. This fitted line is the estimated average value of the dependent variable plotted against the three values 1, 2, 3 of the independent variable. The line is fitted by a procedure called least squares, and a comparable Bayesian technique will be employed and explained in what is to follow. This is an example of replication, where the independent variable occurs with the same value five times, at $x = 1$, 2, and 3. There are five corresponding y values for $x = 1$, 2, and 3; one can see the variability of the y values for each x values, and the variance σ^2 is assumed the same for $x = 1$, 2, and 3. Thus, σ^2 is called the variance about the regression line, and its posterior distribution will be determined.

In order to estimate the three unknown parameters, **BUGS CODE 2.3** is executed with 55,000 observations for the simulation, with a burn in of 5,000 and a refresh of 100.

The code is self-explanatory, where alpha is the intercept, beta is the slope, sigsq is the variance about the regression line, and tau is the precision about the regression line, that is, $\sigma^2 = 1/\tau$. Note that the first list statement is the

TABLE 2.6

(x,y) Pairs for Simple Linear Regression

x	y
1	3.2
1	2.9
1	3.0
1	3.3
1	2.8
2	5.3
2	4.7
2	4.9
2	5.2
2	5.1
3	7.4
3	6.9
3	7.2
3	6.8
3	7.0

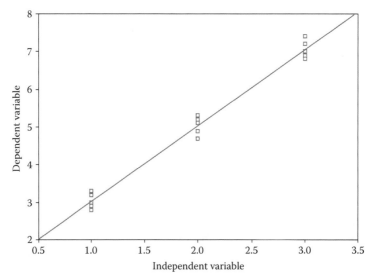

FIGURE 2.3
Simple linear regression.

data from Table 2.7 and the first and second statements of the code make up the linear regression of y on x. See Table 2.7 for the Bayesian analysis of the data. I used uninformative prior distributions for the slope and intercept with a normal distribution with mean 0.0 and variance $1/0.0001 = 10{,}000$. As for the precision parameter τ, I used a gamma with mean 1 and variance 100,000.

BUGS CODE 2.3

```
model;
{
for(i in 1:n){y[i]~dnorm(mu[i], tau)}
for(i in 1:n){z[i]~dnorm(mu[i], tau)}
for(i in 1:n){mu[i]<- alpha+beta*x[i]}
# alpha is intercept
# uninformative priors
alpha~dnorm(0,.0001)
# beta is slope


beta~dnorm(0,.0001)
tau~dgamma(.00001,.00001)
# sigsq is variance about regression line
sigsq<-1/tau
}
# hypothetical example Table 4.6
list(n = 15,x = c(1,1,1,1,1,2,2,2,2,2,3,3,3,3,3),
y = c(3.2,2.9,3.0,3.3,2.8,5.3,4.7,4.9,5.2,5.1,7.4,6.9,
    7.2,6.8,7.0))
# data from table 3.8
list(n = 30,
x = c(14,12,16,13,11,21,17,15,17,20,21,17,23,12,16,15,19,24,
    24,12,23,23,20,13,18,18,19,24,25,22),
y = c(97,100,100,106,119,111,106,117,111,126,115,118,124,
    126,116,131,125,125,128,130,138,131,129,127,131,126,
    142,165,157,163))


# initial values
list(alpha = 0,beta = 0,tau = 1)
```

TABLE 2.7

Posterior Analysis for Linear Regression

Parameter	Mean	SD	Error	2½	Median	97½
α	1.029	0.1661	0.002606	0.7002	1.029	1.36
β	2.009	0.07002	0.001206	1.855	2.008	2.16
σ^2	0.05801	0.02769	<0.0001	0.02573	0.05143	0.1282
τ	20.42	8.02	0.04845	7.8	19.44	38.87

Thus, the intercept is estimated as 1.029 with the posterior mean and median and the slope as 2.009 with the posterior mean and median, and last, the variance about the regression line is estimated as 0.05801 with the posterior mean. The interpretation of the slope is that for each unit increase in the independent variable, the average value of the dependent variable increases by 2.008 units.

Refer to Figure 2.3, and the estimated intercept and slope appear to be very reasonable estimates. Of course, the hypothetical values are designed to give a slope of 2 and an intercept of 1, and one can conclude the Bayesian analysis provides "correct" estimates of the unknown parameters. Simulation errors are "small" where the MCMC error for the intercept implies that the estimate of 1.029 is within 0.002606 of the "true" posterior mean. Also note that the estimated value of 0.05143 for the variance about the regression line is the estimate of the "scatter" of the y values when $x = 1, 2$, or 3.

2.3.3 Another Example of Simple Linear Regression

Consider another example of simple linear regression that examines the effect of age x on systolic blood pressure y, and the data appears in Table 2.8.

TABLE 2.8

Age and Systolic Blood Pressure for 30 Patients

Age	Systolic Blood Pressure	Weight	Subject
14	97	94	1
12	100	95	2
16	100	104	3
13	106	107	4
11	119	108	5
21	111	116	6
17	106	117	7
15	117	117	8
17	111	122	9
20	126	124	10
21	115	125	11
17	118	125	12
23	124	125	13
12	126	128	14
16	116	128	15
15	131	129	16
19	125	134	17
24	125	134	18
24	128	136	19
12	130	137	20

(Continued)

TABLE 2.8 (*Continued*)

Age and Systolic Blood Pressure for 30 Patients

Age	Systolic Blood Pressure	Weight	Subject
23	138	137	21
23	131	138	22
20	129	139	23
13	127	141	24
18	131	141	25
18	126	143	26
19	142	150	27
24	165	171	28
25	157	172	29
22	163	172	30

Source: Woolson, R.F., *Statistical Methods for the Analysis of Biomedical Data*, John Wiley & Sons, New York, 1987.

This example is taken from Woolson (p. 298)[6] and is a subset of a larger study investigating the effect of weight on systolic blood pressure adjusted for age.

The dependent variable is the systolic blood pressure y, and the independent variable is age x; a plot of blood pressure versus age appears in Figure 2.4, which includes a plot of the regression line of systolic blood pressure on age. It appears that there is a linear relationship between age and systolic blood pressure, but it also seems as if there is a lot of variation in the y variable for each value of x. **BUGS CODE 2.4** is employed to perform the Bayesian regression analysis, where the slope, intercept, and variance

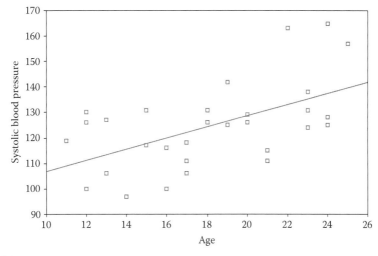

FIGURE 2.4
Systolic blood pressure versus age.

BUGS CODE 2.4

```
model;
{
for(i in 1:n){y[i]~dnorm(mu[i], tau)}
for(i in 1:n){z[i]~dnorm(mu[i], tau)}
for(i in 1:n){mu[i]<- alpha+beta[1]*x1[i]+beta[2]*x2[i]}
# alpha is intercept
# uninformative priors
alpha~dnorm(0,.0001)
# regression coefficients for x1 and x2
beta[1]~dnorm(0,.0001)
beta[2]~dnorm(0,.0001)


# prior for precision
tau~dgamma(.00001,.00001)
# sigsq is variance about regression line
sigsq<-1/tau
}
# data from table 2.8.
# y is systolic blood pressure
# x1 is age
#x2 is weight
list(n = 30,
x1 = c(14,12,16,13,11,21,17,15,17,20,21,17,23,12,16,15,19,
24,24,12,23,23,20,13,18,18,19,24,25,22),
y = c(97,100,100,106,119,111,106,117,111,126,115,118,124,
126, 116,131,125,125,128,130,138,131,129,127,131,126,142,
165,157,163),
x2 = c(94,95,104,107,108,116,117,117,122,124,125,125,125,
128,128,129,134,134,136,137,137,138,139,141,141,143,150,1
71,172,172))
# cigarette data below
# x1 is income
# x2 is price per pack
# y is consumption


list(n = 51, x1 = c(2948,4644,3665,2878,4493,3855,4917,45
24,5079,3738,3354,4623,
3290,4507,3772,3751,3853,3112,3090,3302,4309,4340,4180,
3859,
2626,3781,3500,3789,4563,3737,4701,3077,4712,3252,3086,
4020,
3387,3719,3971,3959,2990,3123,3119,3606,3227,
3468,3712,4053,
3061,3812,3815),
```

```
x2 = c(42.70,41.80,38.50,38.80,39.70,31.10,45.50,41.30,
32.60,43.80,
35.80,36.70,33.60,41.40,32.20,38.50,38.90,30.10,39.30,
38.80,
34.20,41.00,39.20,40.10,37.50,36.80,43.70,34.70,44.00,34.10,
41.70,41.70,41.70,29.40,38.90,38.10,39.80,29.00,44.70,
40.20,
34.30,38.50,41.60,42.00,36.60,39.50,30.20,40.30,41.60,40.20,
34.40),

y = c(90,121,115,100,123,125,120,155,200,124,110,82,102,1
25,135,109,
114,156,116,129,124,124,129,104,93,121,111,108,190,266,
121,90,
119,172,94,122,108,157,107,124,104,93,100,106,66,123,124,97,
115,106,132))
# initial values
list(alpha = 0,beta = c(0,0,0))
```

about the regression line are estimated. The second list statement contains the information from Table 2.8, and I would expect the goodness of fit to not be as good as with the previous example. Using 55,000 observations for the simulation with a burn in of 5,000 and a refresh of 100, the Bayesian analysis is reported in Table 2.9.

Alpha, the intercept, is estimated as 84.94 with the posterior mean and a 95% credible interval (60.98, 108.7), and the slope is estimated as 2.19 with (0.9094, 3.478) as the 95% credible interval for that parameter. Thus, for every year increase in age, the average systolic blood pressure is estimated to increase by 2.19 mm Hg, but the average increase in systolic blood pressure can vary from 0.9094 to 3.478 mm Hg, with 95% confidence. Of course, the surprise is the estimate of 207.7 with the posterior median for the variance about the regression line, which implies a less than good

TABLE 2.9

Bayesian Analysis for Association between Blood Pressure and Age

Parameter	Mean	SD	Error	2½	Median	97½
α	84.94	12.09	0.2181	60.98	84.02	108.7
β	2.19	0.6497	0.01172	0.9094	2.191	3.478
σ^2	218.3	63.01	0.2658	127.5	207.7	371.1
τ	0.004936	0.001324	<0.000001	0.002695	0.004814	0.007844

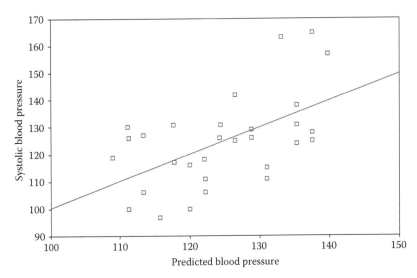

FIGURE 2.5
Systolic blood pressure versus predicted values.

fit of the model to the data. This is somewhat confirmed by Figure 2.5, which is a graph of the predicted values of systolic blood pressure versus the actual blood pressure values y of Table 2.8. Figure 2.5 is an informal way to assess how well the model fits the data, but there are more formal approaches including calculating the correlation between the observed and predicted values of systolic blood pressure. One may show the correlation between the observed and predicted values is $R = 0.551$, corresponding to $R^2 = 0.3036$, where R^2 is the traditional way to assess the fit of a linear regression model.

There is a linear association between the two, but there is a lot of variation in the predicted values for each value of the actual blood pressure values. One assumption that needs to be checked is that the variance of the dependent variables is the same for all values of the independent variable x. For this example, the residuals (systolic blood pressure minus the predicted blood pressure) are plotted against the independent variable age.

The absolute value of the residual for a given age is an approximate estimate of the variance of the dependent variable systolic blood pressure, and Figure 2.6 does not reveal any surprises. Note that the length of the residual is the absolute value of the residual. The lowess curve portrays no discernible trend in the residuals; thus, there is no sufficient information to question the assumption of constant variance for all possible ages.

2.3.4 More on Multiple Linear Regression

The study result that relates systolic blood pressure to age and weight is portrayed in Table 2.8, and a simple linear regression was employed to assess the

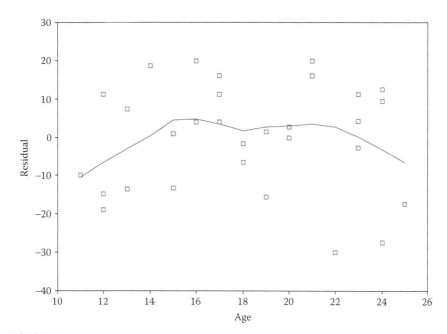

FIGURE 2.6
Residual versus age.

association between age and systolic blood pressure, based on information from 30 subjects.

Age can be viewed as a confounder; thus, a multiple linear regression will be performed in order to assess the association between weight and blood pressure adjusted for age—therefore, consider the model:

$$y_i = \alpha + \beta_1 x_{1i} + \beta_2 x_{2i} + e_i \qquad (2.10)$$

where:

the independent variable x_1 is the age of the ith individual
x_{2i} is the corresponding weight

Model (2.10) implies the average value of the dependent variable y is a linear function of the regression coefficients, and that for a unit increase of one pound in weight, the average value of systolic blood pressure increases by β_2 holding the value of age constant (and for all possible ages). As with simple linear regression, the assumptions are as follows: (1) the n observations are independent random variables; (2) for each value of x_1 and x_2, the variance of y is the same; and (3) the errors are normally distributed with a mean of zero and a variance of σ^2.

A Bayesian analysis produces the joint posterior distribution of the four unknown parameters: α, β_1, β_2, and σ^2, and the analysis is based on the information from Table 2.8.

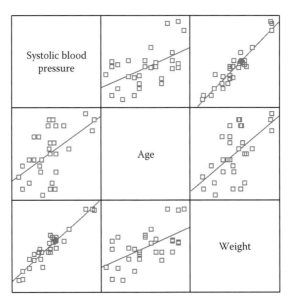

FIGURE 2.7
Systolic blood pressure versus age and weight.

BUGS statements in **BUGS CODE 2.4** are self-explanatory and closely follow the notation of the defined model (2.10). Before analyzing the data, one should plot the data, as for example in Figure 2.7.

In order to show trends, a linear regression line is shown for each plot. All possible relations are shown: (1) systolic blood pressure versus age, (2) blood pressure versus weight, and (3) age versus weight. It is seen that blood pressure versus weight is more linear than blood pressure versus age.

BUGS CODE 2.4 is executed with 55,000 observations for the simulation, with a burn in of 5,000 and a refresh of 100, and produced the following posterior analysis for the multiple linear regression (Table 2.10).

One is interested in assessing the association between systolic blood pressure and weight adjusted for age. The effect weight is the increase in the

TABLE 2.10

Bayesian Analysis for Multiple Linear Regression: Systolic Blood Pressure versus Weight and Age

Parameter	Mean	SD	Error	2½	Median	97½
α	19.09	7.127	0.2839	5.169	19.01	33.4
β_1	−0.05125	0.3119	0.009947	−0.6557	−0.05085	0.578
β_2	0.8173	0.06849	0.002954	0.68	0.8181	0.9498
σ^2	31.96	9.481	0.08804	18.55	30.34	54.78
τ	0.0338	0.009188	<0.00001	0.01826	0.03296	0.05391

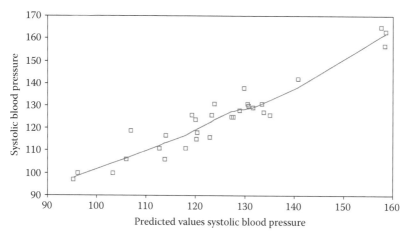

FIGURE 2.8
Systolic blood pressure versus predicted values, adjusted for age and weight.

average blood pressure by 0.8173 mm Hg for an increase of one pound, for all ages. The 95% credible interval for β_2 is (0.68, 0.9489), suggesting a real effect on blood pressure, but that for age is (−0.6557, 0.578), suggesting that age can be eliminated as a predictor of blood pressure.

How well does the model (2.10) fit the data? In order to answer this question, one should plot the predicted blood pressure values versus the corresponding actual blood pressure values y_i, $i = 1,2,...,30$.

Figure 2.8 demonstrates that the multiple linear regression model is a good fit to the data and in fact appears to be a linear relationship. In addition, the R^2 value is 0.9025 giving further evidence of an excellent fit to the data.

One should check to see if the variance of the dependent variable is constant for all pairs (x_{1i}, x_{2i}), $i = 1,2,...,n$. A plot of the residuals versus age and weight is presented in Figure 2.9. Can you detect a trend for the residuals when plotted against age and plotted against weight?

The lowess curve is plotted, and no discernible trend is perceived for residuals versus age and residuals versus weight.

2.3.5 Example for Public Health

One of the major health problems in the United States is cigarette smoking, and the dangerous effects of smoking are well known, especially for the development of lung cancer. Epidemiologists took an active role in promoting no smoking and providing evidence from scientific studies of the deleterious effects of smoking.

The following data set is taken from Chatterjee and Price (p. 265),[1] which reports the 1970 consumption of cigarettes and certain social demographic factors for the 50 states and the District of Columbia, including the median

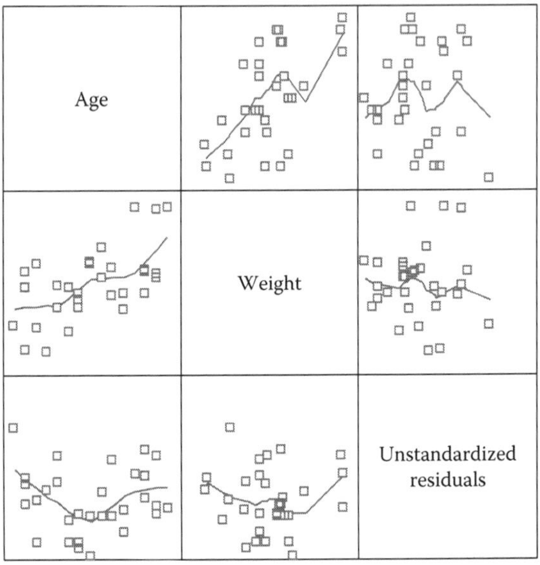

FIGURE 2.9
Residual versus age versus weight.

age of people living in each state, the percentage of people over 25 years of age who completed high school, the per capita income, the percentage of blacks, the percentage of females, the average price of a pack of cigarettes, and the number of pack sold on a per capita basis. Such information is valuable to public health officials, to state attorneys general, and to insurance companies. This is a good example of aggregate data, where the various state variables are averaged over the relevant population of the state.

For example, what is the effect of the level of education on cigarette consumption? What is the average price per capita? If one knew the effect of price per pack on consumption, one could adjust the price (via taxes) by increasing it and decreasing the consumption. It is well known that price per pack and income are the most important variables in predicting cigarette consumption (number of packs sold in a state per capita) and that the other factors such as education, % blacks, and % females do not play an important role in cigarette consumption (Table 2.11).

A multiple linear regression model

$$y_i = \alpha + \beta_1 x_{1i} + \beta_2 x_{2i} + e_i \tag{2.11}$$

will be utilized to model the association between cigarette consumption and income and price.

One should plot consumption versus income and price and verify the linear association. The Bayesian analysis is executed using the information in Table 2.11, which is also listed as the third list statement of **BUGS CODE 2.4**.

TABLE 2.11

Cigarette Consumption by State in 1970

State	Age	Education	Income	% of Blacks	% of Females	Price	Consumption
AL	27.0	41.3	2948	26.2	51.7	42.7	89.8
AK	22.9	66.7	4644	3.0	45.7	41.8	121.3
AZ	26.3	58.1	3665	3.0	50.8	38.5	115.2
AR	29.1	39.9	2878	18.3	51.5	38.8	100.3
CA	28.1	62.6	4493	7.0	50.8	39.7	123.0
CO	26.2	63.9	3855	3.0	50.7	31.1	124.8
CT	29.1	56	4917	6.0	51.5	45.5	120.0
DE	26.8	54.6	4524	14.3	51.3	41.3	155.0
DC	28.4	55.2	5079	71.1	53.5	32.6	200.4
FL	32.3	52.6	3738	15.3	51.8	43.8	123.6
GA	25.9	40.6	3354	25.9	51.4	35.8	109.9
HI	25.0	61.9	4623	1.0	48.0	36.7	82.1
ID	26.4	59.5	3290	0.3	50.1	33.6	102.4
IL	28.6	52.6	4507	12.8	51.5	41.4	124.8
IN	27.2	52.9	3772	6.9	51.3	32.2	134.6
IO	28.8	59.0	3751	1.2	51.4	38.5	108.5
KA	28.7	59.9	3853	4.8	51.0	38.9	114.0
KY	27.5	38.8	3112	7.2	50.9	30.1	155.8
LA	24.8	42.2	3090	29.8	51.4	39.3	115.9
ME	28.6	54.7	3302	0.3	51.3	38.8	128.5
MD	27.1	52.3	4309	17.8	51.1	34.2	123.5
MA	29.0	58.5	4340	3.1	52.2	41.0	124.3
MI	26.3	52.8	4180	11.2	51.0	39.2	128.6
MN	26.8	57.6	3859	0.9	51.0	40.1	104.3
MS	25.1	41.0	2626	36.8	51.6	37.5	93.4
MO	29.4	48.8	3781	10.3	51.8	36.8	121.3
MT	27.1	59.2	3500	0.3	50.0	43.7	111.2
NB	28.6	59.3	3789	2.7	51.2	34.7	108.1
NV	27.8	65.2	4563	5.7	49.3	44.0	189.5
NH	28.0	57.6	3737	0.3	51.1	34.1	265.7
NJ	30.1	52.5	4701	10.8	51.7	41.7	120.7
NM	23.9	55.2	3077	1.9	50.7	41.7	90.0
NY	30.3	52.7	4712	11.9	52.2	41.7	119.0
NC	26.5	38.5	3252	22.2	51.0	29.4	172.4
ND	26.4	50.3	3086	0.4	49.5	38.9	93.8
OH	27.7	53.2	4020	9.1	51.5	38.1	121.6
OK	29.4	51.6	3387	6.7	51.3	39.8	108.4
OR	29.0	60.0	3719	1.3	51.0	29.0	157.0
PA	30.7	50.2	3971	8.0	52.0	44.7	107.3

(Continued)

TABLE 2.11 (*Continued*)

Cigarette Consumption by State in 1970

State	Age	Education	Income	% of Blacks	% of Females	Price	Consumption
RI	29.2	46.4	3959	2.7	50.9	40.2	123.9
SC	24.8	37.8	2990	30.5	50.9	34.3	103.6
SD	27.4	53.3	3123	0.3	50.3	38.5	92.7
TN	28.1	41.8	3119	15.8	51.6	41.6	99.8
TX	26.4	47.4	3606	12.5	51.0	42.0	106.4
UT	23.1	67.3	3227	0.6	50.6	36.6	65.5
VT	26.8	57.1	3468	0.2	51.1	39.5	122.6
VA	26.8	47.8	3712	18.5	50.6	30.2	124.3
WA	27.5	63.5	4053	2.1	50.3	40.3	96.7
WV	30.0	41.6	3061	3.9	51.6	41.6	114.5
WI	27.2	54.5	3812	2.9	50.9	40.2	106.4
WY	27.2	62.9	3815	0.8	50.0	34.4	132.2

Source: Chatterjee, S., and Price, B., *Regression Analysis by Example*, John Wiley & Sons, New York, 1991.

I used 55,000 observations for the simulation, with a burn in of 5,000 and a refresh of 5,000, and the posterior distribution for the parameters are reported in Table 2.12. I deleted the value 265.7 for the consumption of NH, the state of New Hampshire, which is considered an outlier and not used in the Bayesian analysis.

The effect of price (dollars/pack) is estimated by the posterior mean as −2.17, implying that for fixed income (regardless of the income value), the average annual decrease in consumption (number of packs sold per capita) is −2.17 with a 95% credible interval of (−3.528, −0.7488). The relatively long interval is a reflection of the standard deviation.

On the other hand, the effect of price per pack is estimated as 0.02168, implying that consumption increases by 0.02335 packs (packs sold per capita) for a one cent increase in price, for a given value of income (and for all income levels). MCMC errors are small. And one had confidence that 55,000 observations are enough for the simulation.

TABLE 2.12

Bayesian Analysis for Cigarette Consumption Data

Parameter	Mean	SD	Error	2½	Median	97½
α	120.2	28.81	1.325	64.68	120.7	175.9
β_1	0.02168	0.004872	0.0001583	0.01238	0.0216	0.0315
β_2	−2.171	0.7161	0.03237	−3.528	−2.188	−0.7488
σ^2	449.5	97.37	0.8215	298.2	436.6	677.1
τ	0.002324	0.000482	<0.000001	0.0001477	0.00229	0.003354

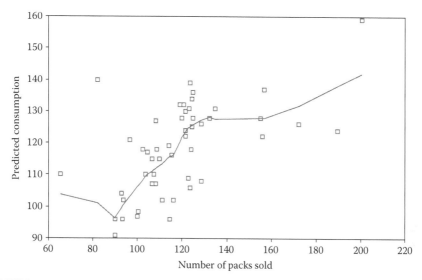

FIGURE 2.10
Predicted consumption versus consumption.

I am beginning to suspect that the model is only a fair fit to the data, but one should plot the predicted values of consumption versus the observed consumption values. This plot is revealed in Figure 2.10.

The lowess curve shows an increasing trend in that as the consumption increases so does the predicted cigarette consumption, and a reasonable conclusion is that the multiple linear regression model with price and income as independent variables has a reasonably good fit to the data. Incidentally, the Pearson correlation between observed and predicted consumption is only 0.576.

2.4 Weighted Regression

When the assumption of equal variance of the dependent variable (for all values of the independent variables) is not reasonable, weighted regression models will be implemented in order to estimate the regression coefficients of the model. Often, the larger the measurement, the larger the variance. Consider an example from Chatterjee and Price,[1] which is analyzed with a weighted regression. A weighted regression is appropriate if the variance of the dependent variable is not constant for all observations. The information from Table 2.13 is state expenditure data for education for all 50 states and is of interest to the administrators at the state medical and public health schools, including departments

TABLE 2.13

State Expenditure for Education

Case	State	URB70	SE75	PI73	Y74
1	ME	508	235	3944	325
2	NH	564	231	4578	323
3	VT	322	270	4011	328
4	MA	846	261	5233	305
5	RI	871	300	4780	303
6	CT	774	317	5889	307
7	NY	856	387	5663	301
8	NJ	889	285	5759	310
9	PA	715	300	4894	300
10	OH	753	221	5012	324
11	IN	649	264	4908	329
12	IL	830	308	5753	320
13	MI	738	379	5439	337
14	WI	659	342	4634	328
15	MN	664	378	4921	330
16	IA	572	232	4869	318
17	MO	701	231	4672	309
18	ND	443	246	4782	333
19	SD	446	230	4296	330
20	NB	615	268	4827	318
21	KS	661	337	5057	304
22	DE	722	344	5540	328
23	MD	766	330	5331	323
24	VA	631	261	4715	317
25	WV	390	214	3828	310
26	NC	450	245	4120	321
27	SC	476	233	3817	342
28	GA	603	250	4243	339
29	FL	805	243	4647	287
30	KY	523	216	3967	325
31	TN	588	212	3946	315
32	AL	584	208	3724	332
33	MS	445	215	3448	358
34	AR	500	221	3680	320
35	LA	661	244	3825	355
36	OK	680	234	4189	306
37	TX	797	269	4336	335
38	MT	534	302	4418	335
39	ID	541	268	4323	344

(Continued)

TABLE 2.13 (*Continued*)

State Expenditure for Education

Case	State	URB70	SE75	PI73	Y74
40	WY	605	323	4813	331
41	CO	785	304	5046	324
42	NM	698	317	3764	366
43	AZ	796	332	4504	340
44	UT	804	315	4005	378
45	NV	809	291	5560	330
46	WA	726	312	4989	313
47	OR	671	316	4697	305
48	CA	609	332	5438	307
49	AK	484	546	5613	386
50	HI	831	311	5309	333

Source: Chatterjee, S., and Price, B., *Regression Analysis by Example*, John Wiley & Sons, New York, p. 131, 1991.

of epidemiology. The variables (columns) are defined as follows: URB70 is the number of residents per thousand living in urban areas in 1970, SE75 per capita expenditures for education in 1975, PI73 is 1973 per capita income, and Y74 is the number of year 1974 residents per thousand under 18 years of age.

Is an unweighted regression reasonable?

Using **BUGS CODE 2.5**, the information from Table 2.13, and the multiple linear regression model

$$y_i = \alpha + \beta_1 x_{1i} + \beta_2 x_{2i} + \beta_3 x_{3i} + e_i \qquad (2.12)$$

for state $i = 1,2,\ldots,50$, the residuals are computed. In this model, the dependent variable y_i is the projected 1975 expenditure for education for state i, x_{1i} is the number of residents per thousand living in an urban area for state i, x_{2i} is the 1973 per capita income, and x_{3i} is the number of residents per thousand under 18 years of age.

BUGS CODE 2.5

```
model;
{
for(i in 1:n){y[i]~dnorm(mu[i], tau)}
for(i in 1:n){z[i]~dnorm(mu[i], tau)}
for(i in 1:n){mu[i]<-alpha+beta[1]*x1[i]+beta[2]*x2[i]+be
ta[3]*x3[i]}
```

```
# alpha is intercept
# un informative priors
# y is the projected per capita expenditure for education
# x1 is the number of residents per thousand living in
urban areas

# x2 is the 1973 per capita income
# x3 is the number of residents per thousand under 18
# regression coefficients
beta[1]~dnorm(0,.0001)
beta[2]~dnorm(0,.0001)
beta[3]~dnorm(0,.0001)
alpha~dnorm(0.0,.0001)
# prior for precision
tau~dgamma(.000001,.000001)
# sigsq is variance about regression line
sigsq<-1/tau
}
# data for state education expenditure data
list(n = 50,
# URB70
x1 = c(508,564,322,846,871,774,856,889,715,753,649,
830,738,659,664,
572,701,443,446,615,661,722,766,631,390,450,
476,603,805,523,
588,584,445,500,661,680,797,534,541,605,785,
698,796,804,809,
726,671,609,484,831),
# state expenditure education
y = c(235,231,270,261,300,317,387,285,300,221,
264,308,379,342,378,
232,231,246,230,268,337,344,330,261,214,245,
233,250,243,216,
212,208,215,221,244,234,269,302,268,323,304,
317,332,315,291,
312,316,332,NA,311),
# PI 73
x2 = c(
3944,4578,4011,5233,4780,5889,5663,5759,4894,
5012,4908,5753,
5439,4634,4921,4869,4672,4782,4296,4827,5057,
5540,5331,4715,
3828,4120,3817,4243,4647,3967,3946,3724,3448,
3680,3825,4189,
4336,4418,4323,4813,5046,3764,4504,4005,5560,
4989,4697,5438,
```

```
5613,5309),
# Y74
x3 = c(325,323,328,305,303,307,301,310,300,324,
329,320,337,328,330,
318,309,333,330,318,304,328,323,317,310,321,
342,339,287,325,
315,332,358,320,355,306,335,335,344,331,
324,366,340,378,330,
313,305,307,386,333))

# weighted observations
# log transformation
list(n = 50,
# URB70
x1 = c(508,564,322,846,871,774,856,889,715,753,649,
830,738,659,664,
572,701,443,446,615,661,722,766,631,390,450,
476,603,805,523,
588,584,445,500,661,680,797,534,541,605,785,
698,796,804,809,
726,671,609,484,831),
# state expenditure education
y = c(5.46,5.44,5.60,5.56,5.70,5.76,5.96,5.65,5.70,5.40,
5.58,5.73,5.94,5.83,5.93,5.45,5.44,5.51,5.44,5.59,5.82,
5.84,5.80,5.56,5.37,5.50,5.45,5.52,5.49,5.38,5.36,5.34,
5.37,5.40,5.50,5.46,5.59,5.71,5.59,5.78,5.72,5.76,5.81,
5.75,5.67,5.74,5.76,5.81,NA,5.74),
# PI 73
x2 = c(
3944,4578,4011,5233,4780,5889,5663,5759,4894,
5012,4908,5753,
5439,4634,4921,4869,4672,4782,4296,4827,5057,
5540,5331,4715,
3828,4120,3817,4243,4647,3967,3946,3724,3448,
3680,3825,4189,
4336,4418,4323,4813,5046,3764,4504,4005,5560,
4989,4697,5438,
5613,5309),
# Y74
x3 = c(325,323,328,305,303,307,301,310,300,324,329,
320,337,328,330,
318,309,333,330,318,304,328,323,317,310,321,
342,339,287,325,
315,332,358,320,355,306,335,335,344,331,324,
366,340,378,330,
313,305,307,386,333))
```

```
# square root transformation
# state expenditure education
# weighted observations
list(n = 50,
# URB70
x1 = c(508,564,322,846,871,774,856,889,715,753,649,
830,738,659,664,
572,701,443,446,615,661,722,766,631,390,450,
476,603,805,523,
588,584,445,500,661,680,797,534,541,605,785,
698,796,804,809,
726,671,609,484,831),

y = c(15.33,15.20,16.43,16.16,17.32,17.80,19.67,
16.88,17.32,14.87,
16.25,17.55,19.47,18.49,19.44,15.23,15.20,15.68,15.17,16.37,
18.36,18.55,18.17,16.16,14.63,15.65,15.26,15.81,15.59,14.70,
14.56,14.42,14.66,14.87,15.62,15.30,16.40,17.38, 16.37,17.97,
17.44,17.80,18.22,17.75,17.06,17.66,17.78, 18.22,NA,17.64),
# PI 73
x2 = c(
3944,4578,4011,5233,4780,5889,5663,5759,4894,5012,4908,5753,
5439,4634,4921,4869,4672,4782,4296,4827,5057,5540,5331,4715,
3828,4120,3817,4243,4647,3967,3946,3724,3448,3680,3825,4189,
4336,4418,4323,4813,5046,3764,4504,4005,5560,4989,4697,5438,
5613,5309),
# Y74
x3 = c(325,323,328,305,303,307,301,310,300,324,329,320,337,
328,330,
318,309,333,330,318,304,328,323,317,310,321,342,339,287,325,
315,332,358,320,355,306,335,335,344,331,324,366,340,378,330,
313,305,307,386,333))

# initial values
list(alpha = 0,beta = c(0,0,0), tau = 1)
```

The Bayesian analysis is executed with **BUGS CODE 2.5** and 65,000 observations for the simulation, with a burn in of 5,000 and a refresh of 100. The dependent variable is the projected per capita state expenditure for education, and the independent variables are x_1 (the number of residents per thousand living in urban areas), x_2 (1973 per capita income), and x_3 (the number of residents per thousand under 18 years of age).

The first list statement is the data from Table 2.13, the second list statement is data for a weighted regression when the dependent variables have been transformed with a log, while the third list statement contains the data for a weighted regression when the dependent variable is the square root of y.

TABLE 2.14

Bayesian Analysis for Multiple Linear Regression: State Expenditures for Education

Parameter	Mean	SD	Error	2½	Median	97½
α	−93.12	86.05	4.672	−257.2	−95.89	76.72
β_1	0.06355	0.04903	0.001588	−0.03248	0.06329	0.1615
β_2	0.041	0.01012	0.000414	0.02089	0.04135	0.06055
β_3	0.4318	0.2328	0.01238	−0.04971	0.4391	0.8851
σ^2	1399	311.7	3.802	917.3	1357	2127
τ	0.000748	0.0001593	0.00000191	0.0004702	0.000737	0.00109

The Bayesian analysis for the unweighted regression is as follows (Table 2.14). Thus, the effect of per capita income is 0.041, that is, as per capita income increases by one dollar, the average per capita expenditure increases by 0.040 dollars. Note that the 95% credible intervals for β_1 and β_3 include zero, implying that perhaps variables x_1 (the number of residents per thousand living in urban areas) and x_3 (the number of residents per thousand under 18 years of age) do not impact the average per capita state expenditure for education.

Figure 2.11 is a plot of the absolute residuals versus the observed state education expenditures, and the lowess curve reveals an increasing trend, that is, as state expenditures for education increase, so do the absolute values of the residuals. The residual is the difference between the observed and predicted expenditures.

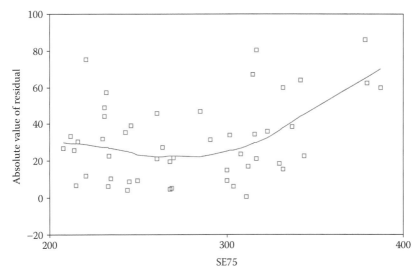

FIGURE 2.11
Absolute residuals versus state expenditures.

The absolute value of the residual is an estimator of the variance of the observed expenditures; thus, it appears that the variance is not constant and a weighted linear regression is called for. Referring to **BUGS CODE 2.5**, note that the first statement is the regression of y on three independent variables, the second generates the predicted values of y, and the third is the regression function, which is linear in three independent variables.

Recall that the value of the predicted expenditure is the mean of the posterior distribution of the projected expenditure listed in Table 2.13. The absolute residual is the absolute value of the predicted expenditure minus the observed expenditure. What are plotted are the absolute values of the residuals versus the state expenditures for education. The absolute residual is a surrogate of the variance of the observations.

Notice that the code is for a multiple linear regression model (2.11) with three independent variables, where the unknown parameters are given uninformative prior distributions, namely, normal distributions for the regression coefficients and the precision about the regression line. The analysis to follow is for an unweighted regression, which is executed in order to compute the predicted values z and the residuals (observed minus predicted state expenditures). The code will be modified for the weighted regression analysis. How should the observations be weighted?

Weighted regression is accomplished by transforming the dependent variable and performing a multiple regression on the independent variables. The type of transformation depends on the association between the variance and the mean of the dependent variable. Table 2.15 will guide the user in how to transform the dependent variable in order so that approximately the variance of the dependent variable is constant for all combination of values of the independent variables.

Figure 2.10 suggests using a square root transformation; thus, **BUGS CODE 2.5** is executed again, but using the second list statement with a square root

TABLE 2.15

Variance Stabilizing Transformations for Y

Relationship	Transformation
Variance is constant	No transformation
Variance proportional to mean	Square root of y
Variance proportional of the square of the mean	Log (log) of y
Variance is proportional to the cube of the mean	1 divided by the square root of y
Variance is proportional to the fourth power of the mean	Reciprocal of y

Source: Montgomery, D.C. et al., *Introduction to Linear Regression Analysis,* John Wiley & Sons, New York, p. 174, 2001.

TABLE 2.16

Bayesian Weighted Regression for State Education Expenditures, Square Root Transformation

Parameter	Mean	SD	Error	2½	Median	97½
α	−0.2643	4.104	0.2452	−7.693	0.3235	8.551
β_1	0.001698	0.001427	0.0000466	−0.001099	0.001697	0.004576
β_2	0.001505	0.000336	0.00004156	0.0008312	0.001513	0.002122
β_3	0.02548	0.0104	0.0006148	0.004202	0.02548	0.04636
σ^2	1.175	0.2608	0.00305	0.7721	1.14	1.784
τ	0.8911	0.1886	0.002062	0.5607	0.8769	1.295

transformation for the dependent variable y, the projected state expenditures for education and the results appear in Table 2.16.

Comparing Table 2.14 with Table 2.16 reveals that for weighted regression, the effect of three independent variables has less effect than the corresponding effects for the unweighted regression. This is true because the dependent variable (square root of the original) is smaller than the dependent variable for unweighted regression. Did this transformation have an effect on the variance of the dependent variable? Consider a plot of the absolute residuals versus the predicted values of the square root of the dependent variable.

Figure 2.12 shows that the variance of the dependent variable (the square root of y) has somewhat stabilized, compared to the trend shown in Figure 2.10. The variance (as measured by the absolute value of the residual) still depends to some extent on the predicted values but not to the extreme as the variance of the original observations; thus, the square root transformation has done its job.

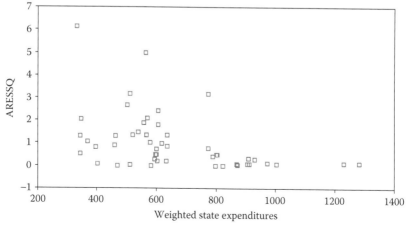

FIGURE 2.12

Absolute residuals versus state expenditures, square root transformation.

2.5 Nonlinear Regression

Nonlinearity is a feature of many studies involving repeated measures. A nonlinear regression model is defined as

$$Y_n = f(x_n, \theta) + e_n \tag{2.13}$$

where:

the nth observation of the dependent variable Y is Y_n

x_n is the corresponding observation of the q by one independent variable x

e_n is the corresponding error term corresponding to the nth observation

It is assumed that the N error terms e_n are independent random variables with a normal distribution, mean zero, and unknown variance σ^2. The N values of Y_n and the vector x_n are known, but the vector

$$\theta = (\theta_1, \theta_2, ..., \theta_p) \tag{2.14}$$

is assumed to be unknown.

Using a Bayesian approach, the objective is to examine the effect of exposure to q risk factors on the dependent variable Y. A prior distribution is placed on the unknown parameters θ and σ^2, and then via Bayes's theorem, the posterior distributions of θ and σ^2 are determined. As before, the posterior analysis will be executed using WinBUGS.

In order to illustrate nonlinear regression, several examples will be presented as a five-step procedure:

1. Plots of the independent variables versus the dependent variable will be portrayed.
2. A model will be defined based on the plots of 1.
3. A prior distribution will be assumed for the unknown parameters θ and σ^2.
4. Based on WinBUGS, the posterior distribution of θ and σ^2 will be determined.
5. The goodness of fit of the model will be assessed by plotting the predicted values of Y (those predicted by the model) versus the corresponding actual Y_n values.

In order to investigate the polychlorinated biphenyl (PCB) concentration in fish from Cayuga Lake, New York, Bache et al.[8] conducted a study to see the effect of the age of the fish on PCB concentration. As the fish are annually stocked as yearlings and distinctly marked as to year class, the ages of the fish are accurately known. The fish is mechanically chopped, ground,

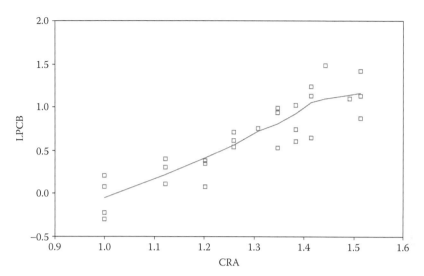

FIGURE 2.13
Log of PCB versus cube root of age.

and mixed, and 5-gram samples are taken. Age is recorded in years and PCB concentration expressed as ppm (parts per million). PCB is a toxin, and it is important for public health to know its concentration in the environment. The data is listed below in the first list statement of **BUGS CODE 2.6**.

In order to investigate the effect of age on PCB, the plot of the log of the PCB values versus the cube root of age reveals a linear association depicted in Figure 2.13.

Based on the plot, the regression model is assumed to be

$$\ln(Y_n) = \theta_1 + \theta_2\, x_n + e_n \tag{2.15}$$

where:

$n = 1, 2, \ldots, 28$, and 28 is the total number of observations

The dependent variable is the natural log of PCB $= Y_n$, the independent variable is cube root of age $= x_n$, and the association is linear with unknown regression coefficients $\theta = (\theta_1, \theta_2)$; thus, it can be analyzed like a simple normal linear regression model.

However, it should be noted that the association between PCB and age can be expressed as nonlinear regression, namely,

$$Y_n = \exp\left(\beta_1 + \beta_2 \sqrt[3]{x_n}\right) \tag{2.16}$$

A Bayesian analysis is based on (3.15) and **BUGS CODE 2.6**.

Note that the first list statement contains two columns, one for the PCB data and one for the age of the fish.

BUGS CODE 2.6

```
model;
{
# nonlinear regression of PCB on cube root of age
for(i in 1:N){y[i]~dnorm(vu[i], tauy)}
for(i in 1:N){vu[i]<-exp(beta[1]+beta[2]*x[i])}
# predicted values of PCB
for(i in 1:N){u[i]~dnorm(vu[i], tauy)}
# linear regression of log PCB on cube root of age
for(i in 1:N){z[i]~dnorm(mu[i], tau)}
# z is natural log of y
for(i in 1:N){z[i]<-log(y[i])}
for(i in 1:N){mu[i]<-theta[1]+theta[2]*x[i]}
# x is the cube root of age
for(i in 1:N){x[i]<-pow(age[i],.333)}
# prior distribution of the beta
for(i in 1:2){beta[i]~dnorm(0.000,0.0001)}
for(i in 1:2){theta[i]~dnorm(0.000,0.0001)}
# prior distribution for tau
tau~dgamma(.0001,.0001)
tauy~dgamma(.0001,.0001)
# predicted values of log PCB
for(i in 1:N){w[i]~dnorm(mu[i], tau)}
# sigma is the inverse of the precision tau
sigma<-1/tau
}

# PCB data
list(N = 28,
y = c(.6,1.6,.5,1.2,2,1.3,2.5,2.2,2.4,1.2,3.5,4.1,5.1,5.7,
3.4,9.7,8.6,
4.0,5.5,10.5,17.5,13.4,4.5,30.4,12.4,13.4,26.2,7.4),
age = c(1,1,1,1,2,2,2,3,3,3,4,4,4,5,6,6,6,7,7,7,8,8,8,9,1
1,12,12,12))

# initial values
list(beta = c(0,0), theta = c(0,0), tauy = 1,tau = 1)
```

A Bayesian analysis is executed with 55,000 observations for the simulation, with a burn in of 5,000 and a refresh of 100, and results are reported in Table 2.17.

The simple linear regression of log PCB on the cube root of age has an estimated (posterior mean) intercept of −2.398 and 95% credible interval of (−3.21, −1.581), while the slope is estimated as 2.307 (posterior mean) and 95% credible interval of (1.831, 2.778); thus, as the cube root of age increases one unit, the average log PCB increases 2.307 units. The *W* values are the

TABLE 2.17

Posterior Analysis for PCB Study: Concentration of PCB in Lake Cayuga Trout

Parameter	Mean	SD	Error	2½	Median	97½
theta[1]	−2.398	0.4148	0.01064	−3.21	−2.4	−1.581
theta[2]	2.307	0.2406	0.0061	1.831	2.307	2.778
sigma	0.267	0.08064	0.000503	0.1528	0.2523	0.4635
W[1]	−0.0889	0.552	0.005062	−1.18	−0.08778	0.9973
W[2]	−0.09105	0.552	0.005108	−1.181	−0.08926	0.991
W[3]	−0.09073	0.5507	0.005058	−1.177	−0.09177	0.99
W[4]	−0.0898	0.5523	0.005026	1.173	0.08976	1.003
W[5]	0.5047	0.5375	0.003725	−0.5582	0.5063	1.56
W[6]	0.5075	0.537	0.003669	−0.5554	0.5118	1.57
W[7]	0.5078	0.5346	0.00373	−0.5473	0.5094	1.555
W[8]	0.924	0.5259	0.00280	−0.1065	0.9226	1.965
W[9]	0.9275	0.5299	0.00294	−0.1233	0.9334	1.969
W[10]	0.9289	0.5303	0.00294	−0.1231	0.9334	1.969
W[11]	1.263	0.5264	0.00252	0.2201	1.265	2.309
W[12]	1.264	0.5302	0.00240	0.2336	1.264	2.315
W[13]	1.265	0.527	0.00217	0.232	1.265	2.307
W[14]	1.541	0.5297	0.00225	0.486	1.542	2.579
W[15]	1.79	0.5288	0.00239	0.7396	1.791	2.834
W[16]	1.791	0.5259	0.00239	0.7554	1.789	2.849
W[17]	1.792	0.5276	0.00252	0.75	1.794	2.831
W[18]	2.014	0.5283	0.00257	0.9642	2.013	3.064
W[19]	2.007	0.5279	0.00262	0.9633	2.008	3.045
W[20]	2.012	0.5316	0.00261	0.9637	2.011	3.062
W[21]	2.213	0.5312	0.00286	1.162	2.214	3.268
W[22]	2.214	0.5313	0.00263	1.16	2.213	3.267
W[23]	2.213	0.5309	0.00294	1.168	2.211	3.265
W[24]	2.396	0.5352	0.00330	1.339	2.392	3.454
W[25]	2.727	0.5391	0.00386	1.657	2.729	3.793
W[26]	2.876	0.5465	0.00418	1.797	2.876	3.596
W[27]	2.878	0.545	0.00402	1.792	2.877	3.95
W[28]	2.878	0.5443	0.003869	1.801	2.877	3.954

predicted values of PCB concentration corresponding to the observed concentration values of PCB.

Consider an alternative way of assessing the association between PCB and age directly using the nonlinear regression model

$$Y_n = \exp\left(\beta_1 + \beta_2 \sqrt[3]{x_n}\right) + e_n \tag{2.17}$$

where the thetas have been replaced by betas and an error term is added.

BUGS CODE 2.6 is executed again with 55,000 observations for the simulation, with a burn in of 5,000 and a refresh of 100, to give the following results for the nonlinear regression of PCB of the cube root of age (Table 2.18).

From Table 2.18, the posterior mean of the intercept is −1.67 with a 95% credible interval of (−4.042, 0.05629), and the posterior mean of the slope is 1.981 with a 95% credible interval of (1.151, 3.063). From Figure 2.14, which is a plot of the posterior density of the slope, it appears that the distribution is symmetric about the posterior mean 1.981.

TABLE 2.18

Posterior Analysis for Nonlinear Regression of PCB on the Cube Root of Age: PCB Concentration in Lake Cayuga Trout

Parameter	Mean	SD	Error	2½	Median	97½
beta[1]	−1.67	1.047	0.01268	−4.042	−1.565	0.05629
beta[2]	1.981	0.4899	0.00583	1.151	1.939	3.063
tauy	0.0362	0.0101	0.0000598	0.01911	0.0352	0.0585
$U[1]$	1.55	5.543	0.02417	−9.351	1.532	12.6
$U[2]$	1.618	5.553	0.02446	−9.279	1.566	12.62
$U[3]$	1.559	5.521	0.02608	−9.281	1.513	12.54
$U[4]$	1.559	5.514	0.02703	−9.338	1.566	12.44
$U[5]$	2.475	5.552	0.02601	−8.529	2.454	13.46
$U[6]$	2.472	5.573	0.0254	−8.551	2.462	13.56
$U[7]$	2.476	5.581	0.02662	−8.512	2.453	13.5
$U[8]$	3.463	5.563	0.02647	−7.588	3.483	14.45
$U[9]$	3.488	5.588	0.02711	−7.545	3.49	14.51
$U[10]$	3.502	5.594	0.02824	−7.512	3.504	14.52
$U[11]$	4.544	5.609	0.02973	−6.587	4.566	5.6
$U[12]$	4.538	5.613	0.02673	−6.591	4.52	15.6
$U[13]$	4.547	5.593	0.02744	−6.52	4.533	15.64
$U[14]$	5.568	5.612	0.03007	−5.498	5.697	16.7
$U[15]$	7.031	5.615	0.02692	−4.418	7.018	18.06
$U[16]$	7.003	5.63	0.02964	−4.094	7.011	18.13
$U[17]$	6.992	5.624	0.03191	−4.165	6.998	18.07
$U[18]$	8.835	5.631	0.03129	−2.862	8.433	19.46
$U[19]$	8.411	5.656	0.03055	−2.741	8.393	19.56
$U[20]$	8.435	5.607	0.028	−2.751	8.437	19.48
$U[21]$	9.933	5.646	0.02855	−1.201	9.91	21.07
$U[22]$	9.919	5.641	0.03111	−1.246	9.953	20.96
$U[23]$	9.292	5.625	0.02733	−1.31	9.953	21.02
$U[24]$	11.64	5.623	0.02987	0.5627	11.65	2.71
$U[25]$	15.51	5.833	0.03418	4.03	15.51	27.04
$U[26]$	17.68	6.049	0.03663	5.75	17.7	29.62
$U[27]$	17.7	6.048	0.03822	5.756	17.72	29.56
$U[28]$	17.64	6.047	0.03573	5.671	17.67	29.57

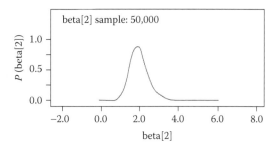

FIGURE 2.14
Posterior density of slope, nonlinear regression.

Note the $U[2]$ is the predicted value the PCB for the second observation of 1.60. The intercept and slope of the simple linear regression of log PCB on the cube root of age (2.15) are estimated (by the posterior mean) as −2.398 and 2.307, respectively. These are different than those estimated from the nonlinear regression (2.15), but not too different, and note that the two models are not the same.

How well does the nonlinear regression model fit the data? To investigate this, plot the 28 U values of Table 2.14 versus the actual PCB values. For a good fit, one would expect the plot to be linear with a slope of one going through the origin; thus, one might conclude the nonlinear regression model is a "bad" fit. How would you improve the fit to the PCB study?

Regressing the log PCB on the cube root yields the following graph of the residuals versus the cube root of age (Figure 2.15). Can you detect a pattern?

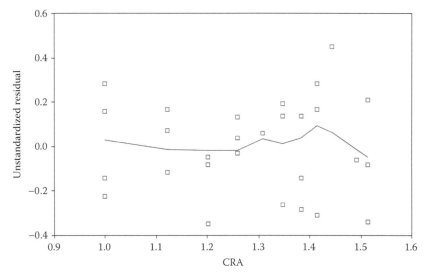

FIGURE 2.15
Residuals versus cube root of age.

2.6 Repeated Measures Model

This example of repeated measures introduces the subject of this book. It should be noted that the approach taken here is very specialized and is just one of many ways the analysis could be accomplished.

Our first encounter with repeated measures is an example involving Alzheimer's disease in a study by Hand and Taylor,[9] where two groups of patients were compared. One group was placebo and the other group received lecithin. Each of the 26 patients in the placebo group and 22 in the treatment group were measured at five times, where the measurement was the number of words the subject could recall from a list of words. Note that the same measurement is repeated on the same subject for a fixed number of occasions, and one would expect the measurements to be correlated. The unique aspect of a repeated measures study is the presence of correlation between measurements on the same subject. From a statistical point of view, this correlation is taken into account when estimating the other parameters of the model. The data of this study appear in the first list statement of **BUGS CODE 2.7**.

BUGS CODE 2.7

```
model;
{

for(i in 1:n){for(j in 1:p){y[i,j]~dnorm(mu[i,j],tau)}}
for(i in 1:n){for(j in 1:p)
{mu[i,j]<- theta+alpha[i]+beta[j]}}

for(i in 1:n){alpha[i]~dnorm(0,tau.alpha)}
for(j in 1:p){beta[j]~dnorm(0,tau.beta)}
for(i in 1:n){for(j in 1:p){z[i,j]~dnorm(mu[i,j],tau)}}
# prior distributions
        theta~dnorm(0,.0001)
tau.alpha~dgamma(.0001,.0001)
tau.beta~dgamma(.0001,.0001)
tau~dgamma(.0001,.0001)

sigma.alpha<-1/tau.alpha
sigma.beta<-1/tau.beta
sigma<-1/tau

}

# example with placebo group only
list(n = 26, p = 5, y = structure(.Data =
```

```
c(20,19,20,20,18,
  14,15,16,9,6,
  7,5,8,8,5,
  6,10,9,10,10,
  9,7,9,4,6,
  9,10,9,11,11,
  7,3,7,6,3,
  18,20,20,23,21,
  6,10,10,13,14,
  10,15,15,15,14,
  5,9,7,3,12,
  11,11,8,10,9,
  10,2,9,3,2,
  17,12,14,15,13,
  16,15,13,7,9,
  7,10,4,10,5,
  5,0,5,0,0,
  16,7,7,6,10,
  5,6,9,5,6,
  2,1,1,2,2,
  7,11,7,5,11,
  9,16,17,10,6,
  2,5,6,7,6,
  7,3,5,5,5,
  19,13,19,17,17,
  7,5,8,8,6),.Dim = c(26,5)))

# initial values
list(alpha = c(0,0,0,0,0,0,0,0,0,0,0,0,0,0,0,0,0,0,0,0,0,
0,0,0,0,0),
beta = c(0,0,0,0,0), tau.alpha = 1, tau.beta = 1, tau = 1,
theta = 0)
```

In order to analyze the Alzheimer's information, the following model is adopted.

Let the observation for the *i*th subject on occasion *j* be

$$y_{ij} = \theta + \alpha_i + \beta_j + e_{ij} \tag{2.18}$$

where:

$i = 1,2,...,n$, where n is the number of subjects
$j = 1,2,...,p$, where p is the number of time points

It is assumed that θ is a constant,

$$\alpha_i \sim nid(0, \tau_\alpha), i = 1, 2, \ldots, n \tag{2.19}$$

$$\beta_j \sim nid(0, \tau_\beta), j = 1, 2, \ldots, p \tag{2.20}$$

and

$$e_{ij} \sim nid(0, \tau) \tag{2.21}$$

The variance of the α_i is $\sigma_\alpha^2 = 1 / \tau_\alpha$, of β is $\sigma_\beta^2 = 1 / \tau_\beta$, and of e_{ij} is $\sigma^2 = 1/\tau$, where the three tau's are positive. The variance component $\sigma_\alpha^2 = 1/\tau_\alpha$ measures the variability of the observations between the various subjects, while the component $\sigma_\beta^2 = 1 / \tau_\beta$ measures the variability between the several times (occasions) and $\sigma^2 = 1 / \tau$ measures the overall variability of the $y(i, j)$ observations. Note that the θ parameter measures the overall mean of the observations.

Note that

$$\mathrm{cov}\left(y_{ij}, y_{ij'}\right) = \sigma_\alpha^2 \tag{2.22}$$

and

$$\mathrm{cov}\left(y_{ij}, y_{ij}\right) = \sigma_\alpha^2 + \sigma_\beta^2 + \sigma^2 \tag{2.23}$$

that is, that observations of the same subject are correlated with covariance σ_α^2 and that the common variance is $\sigma_\alpha^2 + \sigma_\beta^2 + \sigma^2$, which implies that the correlation between measurements of the same subject is

$$\rho = \frac{\sigma_\alpha^2}{\sigma_\alpha^2 + \sigma_\beta^2 + \sigma^2} \tag{2.24}$$

This pattern of correlation could be criticized in that the correlations are constant across time, that is, regardless of the time between observations, the correlation is the same. In Chapters 4 through 10, determining the "correct" way to model the correlation will be discussed in depth. There are other patterns of correlation between observations of the same subject that will be introduced in Chapter 6.

For the first example, consider a Bayesian analysis for the placebo group of the Hand and Taylor[9] Alzheimer's study. See **BUGS CODE 2.7** that closely follows formulas (2.18) through (2.24). Figure 2.16 portrays the trend of the placebo and treatment groups. The vertical axis is the number of correctly recalled words, and the horizontal axis denotes the time periods at times 0, 1, 2, 4, and 6. The two lowess curves corresponding to the two groups can be used to compare the two groups. The green

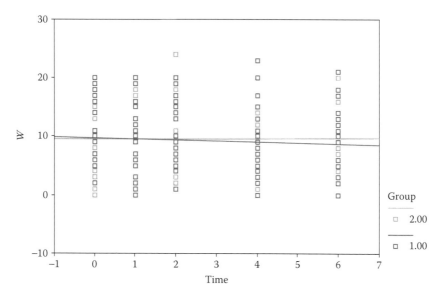

FIGURE 2.16
Words correctly recalled versus time by group.

curve corresponds to the treatment group, while the red denotes the placebo cohort.

BUGS CODE 2.7 is applicable only for the placebo group.

The analysis is executed with 55,000 observations for the simulation, with a burn in of 5,000 and a refresh of 100. Note that vague noninformative prior distributions are placed on the model parameters. A normal distribution for the theta parameter and a gamma for the three precision parameters is specified (Table 2.19).

TABLE 2.19

Posterior Analysis of Alzheimer's Study, Placebo Group

Parameter	Mean	SD	Error	2½	Median	97½
ρ	0.7536	0.06426	0.000403	0.6195	0.7578	0.8629
σ^2	6.978	0.9967	0.00585	5.296	6.889	9.202
σ_α^2	23.2	7.537	0.04688	12.56	21.88	41.51
σ_β^2	0.1256	0.6324	0.01072	0.0000977	0.005952	0.9147
θ	9.315	0.9729	0.0238	7.41	9.317	11.22
Z[1,1]	18.82	2.884	0.01297	13.16	18.84	24.5
Z[1,2]	18.78	2.874	0.01169	13.13	18.8	24.4
Z[1,3]	18.9	2.883	0.01315	13.22	18.9	24.54
Z[1,4]	18.72	2.907	0.0136	13.02	18.72	24.39
Z[1,5]	18.69	2.882	0.01401	12.99	18.68	24.36

(Continued)

TABLE 2.19 (*Continued*)

Posterior Analysis of Alzheimer's Study, Placebo Group

Parameter	Mean	SD	Error	2½	Median	97½
Z[2,1]	11.91	2.89	0.01263	6.272	11.9	17.54
Z[2,2]	11.8	2.887	0.01157	6.103	11.82	17.43
Z[2,3]	11.94	2.873	0.01424	6.287	11.96	17.57
Z[2,4]	11.77	2.87	0.01288	6.177	11.76	17.44
Z[2,5]	11.74	2.878	0.01239	6.077	11.75	17.4
⋮						
Z[26,1]	7.015	2.891	0.01264	1.3	7.025	12.66
Z[26,2]	6.938	2.88	0.0139	1.315	6.926	12.64
Z[26,3]	7.075	2.892	0.01323	1.41	7.063	12.75
Z[26,4]	6.88	2.901	0.01274	1.18	6.87	12.55
Z[26,5]	6.902	2.888	0.01324	1.247	6.902	12.6

Bayesian inferences for the Alzheimer's placebo cohort:

1. The correlation between observations of the same patient is estimated as 0.753 via the posterior mean with a 95% credible interval of (0.6195, 0.8629).
2. The variance of the main observations (the number of correctly recalled words) is estimated as 6.978.
3. The variance of the observations between individuals is estimated as 23.2 with a 95% credible interval of (12.56, 41.51).
4. σ_β^2 measures the variability of the observations between time periods, and the posterior mean gives 0.1256.
5. Last, the average number of correctly recalled words is estimated at 9.315 with a 95% credible interval of (7.41, 11.22).

The Z values of Table 2.19 are the predicted number of correctly recalled words for the placebo group for patients 1, 2, and 26, with five values for each patients corresponding to the five time points 0, 1, 2, 4, and 6.

How well does the model (2.18) fit the data for the placebo group? Figure 2.17 depicts the association between the observed and predicted number of recalled words. The lowess curve shows a close association implying a very good fit of the model to the data.

In order to compare the placebo to the treatment group, refer to Figure 2.15, which reveals very little difference in the two trend curves of the two groups. Thus, it appears that the average value of the number of correctly recalled words is about the same for the placebo and treatment groups. The predicted values z are computed by executing **BUGS CODE 2.7** and reported in Table 2.19.

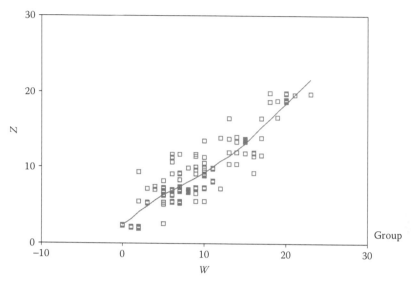

FIGURE 2.17
Predicted number of words versus actual number.

2.7 Remarks about Review of Regression

Chapter 2 has presented a brief review of Bayesian methods to analyze associations between variables. Such techniques will be useful when performing an analysis for repeated measures. Regression analysis based on the linear model will be the foundation for the approach taken in this book.

The first model to be introduced is the logistic regression model with several covariates, and the Bayesian analysis is revealed with a hypothetical example where the dependent variable is the occurrence (yes or no) of a myocardial infarction and the two independent variables are age and systolic blood pressure. The analysis is presented in two ways: (1) where age and systolic blood pressure are dichotomized and (2) where the actual values of age and systolic blood pressure are used. A Bayesian analysis employing the logistic model with three covariates (there are four races), and with coronary artery disease being the dependent variable, also illustrates the use of WinBUGS for the computations necessary for the posterior analysis. See Broemeling[10] for additional information.

The next class of models to be explained are those with a continuous dependent variable and several independent variables. The average value of the dependent variable is a linear function of the coefficients of the model, and two examples are provided. The first is based on hypothetical data with an exact linear relation to one independent variable.

A second example involves two independent variables, age and weight, where systolic blood pressure is the dependent variable. A Bayesian analysis estimates three regression coefficients and the residual error with their posterior distribution; thus, one can estimate the effect of each independent variable on the average systolic blood pressure taking into account the "other" independent variable. In order to assess how well the model fits the data, predicted values of systolic blood pressure versus actual values are plotted. Other diagnostic procedures included plotting the residuals versus age and weight.

Often with linear regression, the variance of the observations increases with values of the independent variables, which violate the standard assumption of constant variance. If one knows the relation between the mean and variance of the dependent variable, one can transform the observations and execute a weighted regression where the variance of the transformed dependent variable is approximately constant over all values of the independent variables. Bayesian methodology for weighted regression is demonstrated with a public health example involving cigarette consumption as a function of price per pack.

Often in repeated measures, the average value of a continuous variable is not linear in the unknown parameters of the regression model; thus, this chapter introduces a brief introduction to nonlinear regression where the Bayesian analysis is illustrated with an example taken from wildlife biology. The dependent variable is the amount of PCB measure on a fish and the independent variable is the age of the fish, and the relation between the average amount of PCB and age is nonlinear in unknown parameters. A Bayesian posterior analysis shows that age does indeed have an effect on the average amount of PCB. Of course, this is of interest to public health officials because PCB is a toxin that could pose a hazard to the public. For additional information about nonlinear regression, see Bates and Watts[11] and Denison et al.[12] The latter reference is Bayesian, while the former is a general introduction to nonlinear regression from a non-Bayesian viewpoint.

It should be noted that the models assume an independent dependent variable, which is not the case for a repeated measures situation. Finally, the last model to be considered involves an Alzheimer study, where the dependent variable is the number of correctly recalled words measures for each patient on five occasions. A random model is assumed for the relation between the dependent variable and time (measured at the five occasions at 0, 1, 2, 4, and 6). One would expect the dependent values of the five occasions to be correlated. The random model induces a correlation pattern with a constant correlation between the observations. That is, the correlation between times 0 and 1 is the same as that between times 0 and 4. As will be seen, this is somewhat unrealistic and will be relaxed. For more information about nonlinear repeated measures, see Davidian and Giltinan.[13]

Exercises

1. Using **BUGS CODE 2.1** and the information from Table 2.1, perform a Bayesian analysis with 55,000 observations for the simulation with a burn in of 5,000 and a refresh of 100.

    a. Verify the posterior analysis of Table 2.2.

    b. Verify the plot of the posterior density of the odds ratio for blood pressure.

    c. Is there an association between heart attack and systolic blood pressure, adjusted for age?

    d. Is age a confounder? Does age affect the association between heart attack and systolic blood pressure?

2. Refer to problem 1, the Bayesian analysis for the study of Table 2.1, where the odds ratios for the age and systolic blood pressure factors are estimated with the logistic model. An alternative model is logistic with an interaction term, namely,

$$\text{logit}(\theta_i) = \alpha + \beta_1 x_{1i} + \beta_2 x_{2i} + \beta_3 x_{1i} x_{2i} \qquad (2.25)$$

where:
θ_i is the probability of a heart attack for the ith subject
α is the intercept
(x_{1i}, x_{2i}) are the values of the independent variables for the ith subject

The first variable is the age, where $x_{1i} = 0$ if age <60, otherwise it is 1, and $x_{2i} = 0$ if the systolic blood pressure <140, otherwise its value is 1. The third regression coefficient β_3 is the effect of the interaction on the logit scale, where the interaction is the product of age and blood pressure. I used this model to reanalyze the study results of Table 2.1 using **BUGS CODE 2.1 Alternative**. Does the inclusion of an interaction term change the posterior analysis of Table 2.2?

    a. Execute a Bayesian analysis similar to that of Table 2.2 and base the analysis on **BUGS CODE 2.1 Alternative**. Use the study results of Table 2.1, and for the simulation, generate 55,000 observations with a burn in of 5,000 and a refresh of 100.

    b. The main focus of this analysis is on the interaction coefficient β_3. What is the 95% credible interval for β_3?

    c. Does the inclusion of the interaction term in the model (2.12) alter the analysis reported in Table 2.12?

    d. What is the most appropriate model? The one with or without interaction?

Table 2.1.1 reports the posterior analysis for the heart study.

BUGS CODE 2.1 ALTERNATIVE

```
model;
{
# Bernoulli distribution for the observations

# theta is the probability of a heart attack

for (i in 1:400){y[i]~dbern(theta[i])}
# logistic regression of theta on age and systolic blood
pressure
# with interaction
for (i in 1:400){logit(theta[i])<-alpha+beta[1]*x1[i]+bet
a[2]*x2[i]+beta[3]*x1[i]*x2[i]}

# prior distributions for the regression coefficients
# uninformative priors
alpha~ dnorm(0.0000,.00001)
for(i in 1:3){beta[i]~dnorm(.0000,.00001)}
ORage<-exp(beta[1])
ORsbp<-exp(beta[2])
}
# y is the occurrence of a heart attack
# x1 is age
# x2 is systolic blood pressure
# age and blood pressure are coded with binary values
list(y=c(1,1,
       1,
       0,0,0,0,0,0,0,0,0,0,0,0,0,0,0,0,0,0,0,0,0,0,0,0,0,0,0,
       0,0,0,
       0,0,0,0,0,0,0,0,0,0,0,0,0,0,0,0,0,0,0,0,
       1,1,1,1,
       1,1,1,1,1,
       0,0,0,0,0,0,0,0,0,0,0,0,0,0,0,0,0,0,0,0,0,0,0,0,0,0,0,
       0,0,0,0,0,0,0,0,0,0,0,0,0,0,0,0,0,0,0,0,0,0,0,0,0,0,0,
       0,0,0,0,0,0,0,0,0,0,0,0,0,0,0,0,0,0,0,0,0,0,0,0,0,0,0,
       0,0,0,0,0,0,0,0,0,0,0,0,0,0,0,0,0,0,0,0,0,0,0,0,0,0,0,
       0,0,0,0,0,0,0,0,0,0,0,0,0,0,0,0,0,0,0,0,0,0,0,0,0,0,0,
       0,0,0,0,0,0,0,0,0,0,0,0,0,0,0,0,0,0,0,0,0,0,0,0,0,0,0,

       0,0,0,0,0,0,0,0,0,0,0,0,0,0,0,0,0,0,0,0,0,0,0,0,0,0,0,
       0,0,0,0,0,0,0,0,0,0,0,0,0,0,0,0,0,0,0,0,0,0,0,0,0,0,0,
       0,0,0,0,0,0,0,0,0,0,0,0,0,0,0,0,0,0,0,0,0,0,0,0,0,0,0,
       0,0,0,0,0,0,0,0,0,0,0,0,0,0,0,0,0,0,0,0,0,0,0,0,0,0,0,
       0,0,0,0,0,0,0,0,0,0,0,0,0,0,0,0,0,0,0,0,0,0,0,0,0,0,0,
       0,0,0,0,0,0,0,0,0,0,0,0,0,0,0,0,0,0,0,0,0,0,0,0,0,0,0,
       0,0,0,0,0,0,0,0,0,0,0,0,0,0,0,0,0,0,0,0,0,0,0,0,0,0,0,
       0,0,0,0,0,0,0,0,0,0,0,0,0,0,0,0,0,0,0,0,0,0,0,0,0,0,0,
```

```
           0,0,0,0,0,0,0,0,0,0,0,0,0,0,0,0,0,0,0,0,0,0,0,0,0,0,0,
           0,0,0,0,0,0,0,0,0,0,0,0,0,0,0,0,0,0,0,0),

x1 = c(1,1,
       1,
       1,1,1,1,1,1,1,1,1,1,1,1,1,1,1,1,1,1,1,1,1,1,1,1,1,1,1,
       1,1,1,
       1,1,1,1,1,1,1,1,1,1,1,1,1,1,1,1,1,1,1,1,
       0,0,0,0,
       0,0,0,0,0,
       0,0,0,0,0,0,0,0,0,0,0,0,0,0,0,0,0,0,0,0,0,0,0,0,0,0,0,
       0,0,0,0,0,0,0,0,0,0,0,0,0,0,0,0,0,0,0,0,0,0,0,0,0,0,0,
       0,0,0,0,0,0,0,0,0,0,0,0,0,0,0,0,0,0,0,0,0,0,0,0,0,0,0,
       0,0,0,0,0,0,0,0,0,0,0,0,0,0,0,0,0,0,0,0,0,0,0,0,0,0,0,
       0,0,0,0,0,0,0,0,0,0,0,0,0,0,0,0,0,0,0,0,0,0,0,0,0,0,0,
       0,0,0,0,0,0,0,0,0,0,0,0,0,0,0,0,0,0,0,0,0,0,0,0,0,0,0,

       0,0,0,0,0,0,0,0,0,0,0,0,0,0,0,0,0,0,0,0,0,0,0,0,0,0,0,
       0,0,0,0,0,0,0,0,0,0,0,0,0,0,0,0,0,0,0,0,0,0,0,0,0,0,0,
       0,0,0,0,0,0,0,0,0,0,0,0,0,0,0,0,0,0,0,0,0,0,0,0,0,0,0,
       0,0,0,0,0,0,0,0,0,0,0,0,0,0,0,0,0,0,0,0,0,0,0,0,0,0,0,
       0,0,0,0,0,0,0,0,0,0,0,0,0,0,0,0,0,0,0,0,0,0,0,0,0,0,0,
       0,0,0,0,0,0,0,0,0,0,0,0,0,0,0,0,0,0,0,0,0,0,0,0,0,0,0,
       0,0,0,0,0,0,0,0,0,0,0,0,0,0,0,0,0,0,0,0,0,0,0,0,0,0,0,
       0,0,0,0,0,0,0,0,0,0,0,0,0,0,0,0,0,0,0,0,0,0,0,0,0,0,0,
       0,0,0,0,0,0,0,0,0,0,0,0,0,0,0,0,0,0,0,0,0,0,0,0,0,0,0,
       0,0,0,0,0,0,0,0,0,0,0,0,0,0,0,0,0,0),

x2 = c(1,1,
       0,
       1,1,1,1,1,1,1,1,1,1,1,1,1,1,1,1,1,1,1,1,1,1,1,1,1,1,1,
       1,1,1,
       0,0,0,0,0,0,0,0,0,0,0,0,0,0,0,0,0,0,0,0,
       1,1,1,1,
       0,0,0,0,0,
       1,1,1,1,1,1,1,1,1,1,1,1,1,1,1,1,1,1,1,1,1,1,1,1,1,1,1,
       1,1,1,1,1,1,1,1,1,1,1,1,1,1,1,1,1,1,1,1,1,1,1,1,1,1,1,
       1,1,1,1,1,1,1,1,1,1,1,1,1,1,1,1,1,1,1,1,1,1,1,1,1,1,1,
       1,1,1,1,1,1,1,1,1,1,1,1,1,1,1,1,1,1,1,1,1,1,1,1,1,1,1,
       1,1,1,1,1,1,1,1,1,1,1,1,1,1,1,1,1,1,1,1,1,1,1,1,1,1,1,
       1,1,1,1,1,1,1,1,1,1,1,1,1,1,1,1,1,1,1,1,1,1,1,1,1,1,

       0,0,0,0,0,0,0,0,0,0,0,0,0,0,0,0,0,0,0,0,0,0,0,0,0,0,0,
       0,0,0,0,0,0,0,0,0,0,0,0,0,0,0,0,0,0,0,0,0,0,0,0,0,0,0,
       0,0,0,0,0,0,0,0,0,0,0,0,0,0,0,0,0,0,0,0,0,0,0,0,0,0,0,
       0,0,0,0,0,0,0,0,0,0,0,0,0,0,0,0,0,0,0,0,0,0,0,0,0,0,0,
       0,0,0,0,0,0,0,0,0,0,0,0,0,0,0,0,0,0,0,0,0,0,0,0,0,0,0,
```

```
        0,0,0,0,0,0,0,0,0,0,0,0,0,0,0,0,0,0,0,0,0,0,0,0,0,0,0,0,0,
        0,0,0,0,0,0,0,0,0,0,0,0,0,0,0,0,0,0,0,0,0,0,0,0,0,0,0,0,0,
        0,0,0,0,0,0,0,0,0,0,0,0,0,0,0,0,0,0,0,0,0,0,0,0,0,0,0,0,0,
        0,0,0,0,0,0,0,0,0,0,0,0,0,0,0,0,0,0,0,0,0,0,0,0,0,0,0,0,0,
        0,0,0,0,0,0,0,0,0,0,0,0,0,0,0,0,0,0,0))


# actual values of age and blood pressure
list( y = c(1,1,
        1,
        0,0,0,0,0,0,0,0,0,0,0,0,0,0,0,0,0,0,0,0,0,0,0,0,0,0,0,0,
        0,0,0,
        0,0,0,0,0,0,0,0,0,0,0,0,0,0,0,0,0,0,0,0,0,
        1,1,1,1,
        1,1,1,1,1,
        0,0,0,0,0,0,0,0,0,0,0,0,0,0,0,0,0,0,0,0,0,0,0,0,0,0,0,0,0,
        0,0,0,0,0,0,0,0,0,0,0,0,0,0,0,0,0,0,0,0,0,0,0,0,0,0,0,0,0,
        0,0,0,0,0,0,0,0,0,0,0,0,0,0,0,0,0,0,0,0,0,0,0,0,0,0,0,0,0,
        0,0,0,0,0,0,0,0,0,0,0,0,0,0,0,0,0,0,0,0,0,0,0,0,0,0,0,0,0,
        0,0,0,0,0,0,0,0,0,0,0,0,0,0,0,0,0,0,0,0,0,0,0,0,0,0,0,0,0,
        0,0,0,0,0,0,0,0,0,0,0,0,0,0,0,0,0,0,0,0,0,0,0,0,0,0,0,0,0,

        0,0,0,0,0,0,0,0,0,0,0,0,0,0,0,0,0,0,0,0,0,0,0,0,0,0,0,0,0,
        0,0,0,0,0,0,0,0,0,0,0,0,0,0,0,0,0,0,0,0,0,0,0,0,0,0,0,0,0,
        0,0,0,0,0,0,0,0,0,0,0,0,0,0,0,0,0,0,0,0,0,0,0,0,0,0,0,0,0,
        0,0,0,0,0,0,0,0,0,0,0,0,0,0,0,0,0,0,0,0,0,0,0,0,0,0,0,0,0,
        0,0,0,0,0,0,0,0,0,0,0,0,0,0,0,0,0,0,0,0,0,0,0,0,0,0,0,0,0,
        0,0,0,0,0,0,0,0,0,0,0,0,0,0,0,0,0,0,0,0,0,0,0,0,0,0,0,0,0,
        0,0,0,0,0,0,0,0,0,0,0,0,0,0,0,0,0,0,0,0,0,0,0,0,0,0,0,0,0,
        0,0,0,0,0,0,0,0,0,0,0,0,0,0,0,0,0,0,0,0,0,0,0,0,0,0,0,0,0,
        0,0,0,0,0,0,0,0,0,0,0,0,0,0,0,0,0,0,0,0,0,0,0,0,0,0,0,0,0,
        0,0,0,0,0,0,0,0,0,0,0,0,0,0,0,0,0,0,0),

x1 = c(77,85,87,74,56,76,69,71,74,83,67,69,52,77,72,76,77,
76,70,67,70,74,72,74,57,88,79,64,67,67,69,77,68,86,58,77,
79,86,70,80,60,72,63,92,56,79,76,65,70,76,55,56,48,54,50,
49,45,36,54,47,52,47,47,49,46,39,50,51,39,59,46,44,52,53,49,
53,52,46,49,46,41,58,61,50,45,58,51,58,44,50,57,60,48,51,
56,47,44,53,41,39,50,36,55,47,59,43,51,43,40,43,39,38,62,
49,46,45,42,50,53,43,47,31,40,43,45,51,51,55,56,54,49,46,
53,49,34,43,60,44,56,50,46,44,49,46,53,45,46,56,43,53,45,
49,49,41,46,44,43,48,39,59,55,46,42,38,40,42,29,41,38,52,
51,54,50,48,47,47,47,55,44,50,55,50,43,54,42,55,42,38,52,49,
50,56,50,39,51,49,51,54,47,42,59,54,52,47,52,44,44,45,50,
49,48,35,43,45,49,46,47,47,59,48,52,44,48,50,58,47,49,43,
56,50,42,48,41,52,41,43,47,43,47,54,52,50,62,47,56,42,46,
```

```
53,43,43,50,56,54,46,51,49,47,42,42,54,53,45,35,48,52,45,47,
41,48,40,42, 50,41,56,56,49,53,34,46,50,54,42,48,51,51,56,49,
56,44,47,51,51,36,46,58,58,51,40,53,53,44,44,51,47,48,61,40,
50,49,39,47,36,45,47,57,44,44,51,40,44,51,60,48,51,46,51,47,
48,57,48,39,49,40,58,50,45,41,47,49,41,64,36,47,49,58,59,45,
45,53,37,43,43,50,44,49,43,53,48,54,53,44,47,55,52,40,43,47,
47,54,46,51,65,54,53,56,53,49,47,38,59,55,44,52,41,55,43,47,
46,55,38,45,42,54,52,42,56,48,44,40,47,37,44,51,37,50,47,46,
45,46,44,44,52,45,38,48,54,46,41,41,43,44,44,49,48,41,57,45,
46,46,35,50,43,41,52,43,54,46,40,50,61,49,44,50,40,50,38,
52,46,49,53),

x2 = c(158,164,150,157,161,165,166,147,169,159,150,166,161,
161,167,164,167,163,166,159,162,166,145,163,161,162,156,
160,178,164,154,115,114,128,117,123,122,114,116,114,119,112,
123,119,122,130,119,128,120,115,163,166,162,154,167,163,
155,160,157,154,149,159,167,162,159,160,162,157,158,168,168,
164,157,171,150,158,150,163,163,159,171,165,148,156,164,
168,158,167,153,167,149,158,159,144,167,151,156,150,154,
164,158,147,151,167,151,151,160,161,160,160,156,168,162,
171,155,150,159,156,147,162,163,170,158,163,168,153,151,
150,156,162,164,164,159,152,164,159,158,163,162,152,162,
157,158,162,149,166,142,151,160,154,168,165,161,152,151,
159,155,169,169,168,166,163,158,150,151,150,152,167,155,
150,157,152,158,168,158,152,162,165,154,168,173,158,164,
157,155,161,160,160,159,155,162,164,160,158,164,157,168,
159,169,159,155,161,150,119,114,120,114,128,126,116,119,
109,121,114,121,120,122,112,113,116,126,117,109,120,111,
123,113,119,110,112,120,127,116,124,111,134,137,113,124,
127,125,123,120,119,121,115,119,110,123,123,124,126,128,
119,122,118,115,121,114,125,117,125,124,133,115,121,120,
127,120,118,136,125,119,120,111,119,121,111,125,130,117,
131,127,109,124,114,120,112,130,127,118,119,121,119,117,
117,117,123,122,121,119,125,113,115,131,126,116,126,120,
119,125,113,124,127,116,119,128,119,129,125,122,121,113,
118,113,130,119,117,125,130,120,115,118,131,121,117,130,
103,136,127,118,122,125,121,123,117,110,126,118,125,123,
128,129,116,127,109,118,127,134,123,124,115,113,126,120,
116,122,112,109,121,110,114,133,131,123,122,122,116,120,
104,120,125,117,121,125,121,121,116,115,121,126,121,113,
126,128,112,112,126,120,122,113,126,119,131,118,112,126,131,
124,116,122,120,115,104,115,127,122,123,134,120,124,104,113,110,
118,117,118,128,121,120,116,121,116,121,116,128,115,123,118,
114,123,121,116,111,118,119,127,129,114,122))

list(alpha = 0,beta = c(0,0,0))
```

TABLE 2.1.1

Posterior Distribution of Heart Study: Logistic Regression with Interaction

Parameter	Mean	SD	Error	2½	Median	97½
Odds ratio—age	2.808	3.671	0.06011	0.0541	1.605	12.85
Odds ratio—blood pressure	1.301	1.063	0.01729	0.2384	1.013	4.124
α	−3.752	0.477	0.00704	−4.776	−3.718	−2.912
β_1	0.2902	1.392	0.03692	−2.917	0.4734	2.554
β_2	0.005863	0.7201	0.01135	−1.434	0.01283	1.417
β_3	0.5249	1.704	0.04159	−2.613	0.4451	4.14

3. Using the actual values of age and systolic blood pressure, perform a Bayesian analysis with the logistic model containing an interaction term. The second list statement of **BUGS CODE 2.1 Alternative** contains the actual age and systolic blood pressure values. Generate 55,000 observations for the simulation, with a burn in of 5,000 and a refresh of 100.

   a. Confirm the Bayesian analysis reported in Table 2.3.2.

   b. Compare Table 2.3.2 with Tables 2.10 and 2.3.

   c. After comparing Table 2.3.2 with Table 2.3, would you conclude an interaction term is needed?

   d. After comparing Table 2.3.2 with Table 2.10, is it better to use the actual values of age and systolic blood pressure? Explain.

4. Refer to Table 2.3 and **BUGS CODE 2.2** for the analysis of the association between race and coronary heart disease. The study results are provided by Hosmer and Lemeshow (p. 48).[3] Verify the results of Table 2.3 with a simulation using 55,000 observations, with a burn in of 5,000 and a refresh of 100.

   a. The posterior mean and median of the odds ratio of coronary heart disease of blacks relative to that of the group consisting of other races are 2.477 and 2.084, respectively. Is the posterior

TABLE 2.3.2

Bayesian Analysis for Heart Study Information: Logistic Regression with Interaction with Actual Values of Age and Systolic Blood Pressure

Parameter	Mean	SD	Error	2½	Median	97½
Odds ratio—age	1.266	0.3549	0.008029	0.6845	1.227	2.039
Odds ratio—blood pressure	1.167	0.1324	0.003044	0.9139	1.158	1.45
α	−28.7	18.02	0.423	−64.97	−28.39	6.536
β_1	0.1971	0.2823	0.006742	−0.3791	0.2046	0.7142
β_2	0.1484	0.1126	0.002613	−0.07116	0.1466	0.3713
β_3	−0.001001	0.001771	0.0000418	−0.004821	−0.001052	0.002591

distribution skewed? In order to estimate the odds ratio, what is your estimate? The usual estimate is $(20/10)/(10/10) = 2$.

b. Suppose one specifies the first group (whites), the reference group, for our analysis. Then one would have to code the independent variables as:

$x_1 = (0,1,0,0)$, $x_2 = (0,0,1,0)$, and $x_3 = (0,0,0,1)$

Using this representation, perform a Bayesian analysis based on **BUGS CODE 2.2** with 55,000 observations for the simulation, with a burn in of 5,000 and a refresh of 100. What is the odds ratio of coronary artery disease of Hispanics relative to that of whites? Find the posterior mean and median of the odds ratio.

5. Repeat the analysis for the Heart Study of Table 2.1 with **BUGS CODE 2.2**. Note that there are four groups of patients, thus relative to the notation used in **BUGS CODE 2.2**, there will be four thetas; therefore, code the groups with three vectors $x_1, ..., x_3$ and make the last group, group 4, the reference group. The reference group consists of subjects with age less than 60 and systolic blood pressure <140. Note that each vector will have four components. Compare the results of your Bayesian analysis with Table 2.2, which is based on the results of Table 2.1, but where the Bayesian analysis is based on 400 observations.

6. With **BUGS CODE 2.2**, verify Table 2.5, the Bayesian analysis for predicting the number with coronary artery disease for four racial groups.

7. Table 2.7.3 gives the posterior analysis for the 15 predicted y values of Table 2.6. Verify this table by using **BUGS CODE 2.3** with 55,000

TABLE 2.7.3

Predicted Values

Parameter	Mean
$z[1]$	3.036
$z[2]$	3.037
$z[3]$	3.030
$z[4]$	3.04
$z[5]$	3.037
$z[6]$	5.048
$z[7]$	5.046
$z[8]$	5.047
$z[9]$	5.047
$z[10]$	5.046
$z[11]$	7.055
$z[12]$	7.056
$z[13]$	7.055
$z[14]$	7.054
$z[15]$	7.055

for the observations, with a burn in of 5,000 and a refresh of 100. The first list statement of the code gives the data for Table 2.6, and the second statement generates the predicted y values. This is only part of the output because I did not include the standard deviation, the MCMC error, the median, and the upper and lower 21/2 percentiles of the posterior distribution.

a. Plot the predicted values y versus the actual values x.

b. Is the simple linear regression model a good fit to the data?

c. Plot the posterior density of $z[1]$.

d. What is the 95% credible interval for $z[1]$?

8. Verify Table 2.7, the Bayesian analysis for the association between blood pressure and age, using **BUGS CODE 2.2**.

a. What is your estimate of the intercept?

b. What is your estimate of the slope?

c. Is the model a good fit to the data?

9. Using **BUGS CODE 2.4** and the information about cigarette consumption, verify Table 2.12, the Bayesian analysis for the multiple regression of consumption on income and price per pack. Use 55,000 observations with a burn in of 5,000 and a refresh of 100. Note that the second statement of the code generates the 51 predicted values of cigarette consumption. You must use z as the node and the predicted values will be generated. Use the posterior mean of the predicted values as the predicted values. I deleted the value 265.7 for the consumption of NH, the state of New Hampshire, which is considered an outlier and is not used in the Bayesian analysis.

a. Does the model provide a good fit to the observed cigarette consumption?

b. What is the 95% credible interval for the variance about the regression line?

c. What is the interpretation for the estimated intercept term of the model?

d. Verify Figure 2.10, a plot of predicted consumption versus observed consumption.

10. Based on **BUGS CODE 2.1**, perform a Bayesian regression analysis with 55,000 observations for the simulation, with a burn in of 5,000 and a refresh of 100. Recall that the first three statements of the code are relevant for the regression analysis, with y as the dependent variable (binary for the occurrence of disease) and x_1 and x_2 as independent variables. The z vector consists of the predicted systolic blood pressure values.

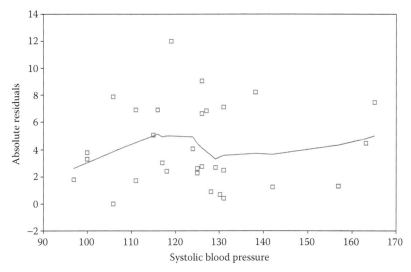

FIGURE 2.10.1
Absolute value of residuals versus blood pressure.

a. Compute the residuals, that is, the differences: systolic blood pressure minus the corresponding predicted values z.

b. Graph the absolute value of the residuals versus the predicted value (Figure 2.10.1). The graphs should look similar to Figure 2.12.

c. Is there a trend in absolute residuals as the predicted values increase?

d. If there is a trend, perform a weighted regression by transforming the dependent variable (systolic blood pressure) by the appropriate transform. In order to choose the appropriate transformation, refer to Table 2.15 for selecting the transformation of the dependent variable.

e. Does the transformation stabilize the variance of the dependent variable?

f. Is the weighted regression model a good fit to the data?

11. a. Verify Table 2.14, the Bayesian analysis for the state education expenditure data for 50 states. Use **BUGS CODE 2.5** to execute the regression of state expenditure on three independent variables. Note that the first list statement is the information for the unweighted regression. Use 65,000 observations for the simulation, with a burn in of 5,000 and a refresh of 100.

b. Verify Table 2.15, the Bayesian weighted regression with the square root of the dependent variable (state education expenditure data)

TABLE 2.12.4

Myocardial Infarction versus Systolic Blood
Pressure by Age: A Previous Study

	MI	No MI	Total
Age ≥60			
SBP ≥140	1	15	16
SBP <140	0	8	8
Total	2	23	24
Age <60			
SBP ≥140	2	80	82
SBP <140	3	118	121
Total	5	198	203

on three independent variables. The second list statement gives the data for the weighted regression. Use 65,000 observations for the simulation, with a burn in of 5,000 and a refresh of 100.

c. Did the square root transformation stabilize the variance? Why?

d. Is the weighted regression model a good fit to the data? Plot the predicted values of the dependent variable versus the dependent variable.

12. Table 2.12.4 is a previous study that was carried out before the study reported in Table 2.1.

Using **BUGS CODE 2.1**, execute a Bayesian analysis with 55,000 observations using Table 2.12.4 as prior information and the information in Table 2.1. Use a burn in of 5,000 and a refresh of 100.

a. What is the posterior distribution of the odds ratio for age?

b. What is the posterior distribution of the odds ratio for systolic blood pressure?

c. Compare the results with those reported in Table 2.2.

d. How much does the standard deviation of the posterior distribution for the odds ratio decrease?

3

Foundation and Preliminary Concepts

3.1 Introduction

Since regression analysis will be used to analyze repeated measures, Chapter 2 reviewed Bayesian regression techniques for standard models, including logistic for binary responses and simple and multiple linear regression for continuous responses. Also reviewed was the analysis of nonlinear models and last to be presented was an example of repeated measures. As was discussed, the standard regression model assumes a constant variance for the dependent variable and that the observations are independent. For a repeated measures model, correlation is present, and the variance of observations is not the same. In order to execute Bayesian analysis, correlation must be taken into account. Our first example of repeated measures is based on a clinical trial of epilepsy conducted by Lippik[1] and analyzed by Thall and Vail.[2]

The chapter consists of two parts: the first nine sections that are appropriate for quantitative responses and the remaining two sections to responses that have only a few possible values.

3.2 An Example

Patients having epileptic seizures were involved in a randomized study of progabide versus placebo, where 28 were assigned to placebo and 31 to progabide; they were all receiving standard chemotherapy for epilepsy. The information from the clinical trial is given in Table 3.1, where $y1$, $y2$, $y3$, and $y4$ are the successive two-week seizure counts. There are two treatments (Trt = 0 is placebo and Trt = 1 indicates progabide); base is the eight-week seizure count before assignment of treatment and the last column is the subject's age.

The seizure count was taken over an eight-week period with four two-week records of the number of seizures for the previous two weeks. It should be noted that this is a two-period crossover study, where the patients were switched over to the other treatment after eight weeks postrandomization;

TABLE 3.1

Successive Two-Week Seizure Counts for 59 Epileptic Subjects

y1	y2	y3	y4	Trt	Base	Age
5	3	3	3	0	11	31
3	5	3	3	0	11	30
2	4	0	5	0	6	25
4	4	1	4	0	8	36
7	18	9	21	0	66	22
5	2	8	7	0	27	29
6	4	0	2	0	12	31
40	20	23	12	0	52	42
5	6	6	5	0	23	37
14	13	6	0	0	10	28
26	12	6	22	0	52	36
12	6	8	4	0	33	24
4	4	6	2	0	18	23
7	9	12	14	0	42	36
16	24	10	9	0	87	26
11	0	0	5	0	50	26
0	0	3	3	0	18	28
37	29	28	29	0	111	31
3	5	2	5	0	18	32
3	0	6	7	0	20	21
3	4	3	4	0	12	29
3	4	3	4	0	9	21
2	3	3	5	0	17	32
8	12	2	8	0	28	25
18	24	76	25	0	55	30
2	1	2	1	0	9	40
3	1	4	2	0	10	19
13	15	13	12	0	47	22
11	14	9	8	1	76	18
8	7	9	4	1	38	32
0	4	3	0	1	19	20
3	6	1	3	1	10	30
2	6	7	4	1	19	18
4	3	1	3	1	24	24
22	17	19	16	1	31	30
5	4	7	4	1	14	35
2	4	0	4	1	11	27
3	7	7	7	1	67	20
4	18	2	5	1	41	22
2	1	1	0	1	7	28
0	2	4	0	1	22	23

(Continued)

TABLE 3.1 (*Continued*)

Successive Two-Week Seizure Counts for 59 Epileptic Subjects

y1	y2	y3	y4	Trt	Base	Age
5	4	0	3	1	13	40
11	14	25	15	1	46	33
10	5	3	8	1	36	21
19	7	6	7	1	38	35
1	1	2	3	1	7	25
6	10	8	8	1	36	26
2	1	0	0	1	11	25
102	65	72	63	1	151	22
4	3	2	4	1	22	32
8	6	5	7	1	41	25
1	3	1	5	1	32	35
18	11	28	13	1	56	21
6	3	4	0	1	24	41
3	5	4	3	1	16	32
1	23	19	8	1	22	26
2	3	0	1	1	15	21
0	0	0	0	1	13	36
1	4	3	2	1	23	37

Source: Thall, P.F., and Vail, S.C., (Table 2) *Biometrics*, 46, 657–671, 1990.

thus what is reported in Table 3.1 is the first part of the crossover study. This study will be used to illustrate the remaining topics of this chapter. For example, the various sources of variation in the epilepsy study will be identified, and the basic descriptive statistics (means, standard deviations, and correlations) computed. This example will also provide a demonstration of various graphical procedures that explore just how the seizure count varies over the four time periods.

Next to be introduced is the notation that is used for the analysis of repeated measures studies, which is followed by some Bayesian inference for the epilepsy study.

3.3 Notation

The following notation is standard for repeated measures. It is assumed that there are N subjects where the response is measured on the same subject the same number of times. Of course, there are many studies where there are missing values—the times at which the responses measured are not equally spaced. In fact missing values and unequally spaced observations are the usual situation with repeated measures studies, and this case will be presented in the later chapters.

Let Y_{ij} be the response for the *i*th subject on occasion *j*, where $i = 1,2,...,N$ and $j = 1,2,...,n$. Suppose for subject *i* that

$$Y_i = \left(Y_{i1}, Y_{i2},..., Y_{in}\right)' \tag{3.1}$$

is the *n* by 1 vector of *n* responses.

Corresponding to Y_{ij}, there is a px1 vector of covariates

$$X_{ij} = \left(X_{ij1}, X_{ij2},..., X_{ijp}\right)' \tag{3.2}$$

and it is important to examine the effect of the various covariates on the main response.

The Y_{ij} are random variables and the corresponding observed values are denoted by y_{ij}, and Table 3.2 presents the layout of observations for a repeated measures study and is appropriate for complete observations on each subject.

Descriptive statistics are important features in describing a repeated measures study and are usually expressed as means, standard deviations, and correlations.

For the mean of Y_{ij}, let

$$\mu_{ij} = E\left(Y_{ij}\right) \tag{3.3}$$

where the expectation is taken with respect to the hypothetical repeated replications of the repeated measures study. Often the main response will depend on the covariates and the expectation takes into account the variation of the covariates. Such expectations are often referred to as conditional expectations, see Fitzmaurice, Laird, and Ware (p. 28).[3] By investigating the association between the mean (3.3) and time *j*, the dependence between the main response and time over an individual can be revealed.

TABLE 3.2

Observed Response Values for *N* Subjects and *n* Occasions

Subject	1	2	...	...	N
1	y_{11}	y_{12}	...	....	y_{1n}
2	y_{21}	y_{22}	...	....	y_{2n}
⋮					
N	y_{N1}	y_{N2}	...	....	y_{Nn}

If the expectation is the same for all individuals, one writes the expectation as

$$\mu_j = E(Y_{ij}) \tag{3.4}$$

for $j = 1,2,\ldots,n$.

In repeated studies, the observations between responses at different times of the same individual are expected to be correlated, thus one defines conditional covariance between Y_{ij} and Y_{ik} as

$$\sigma_{jk} = E\left[(Y_{ij} - \mu_{ij})(Y_{ik} - \mu_{ik})\right] \tag{3.5}$$

for all i and j ranging between 1 and n. The subscript i does not appear in the covariance expression (3.5), implying that the covariance between observations at different times is the same for all individuals.

Note that when $j = k$, (3.5) reduces to the variance of observation $j = k$. Having defined the covariance, the conditional correlation between observations j and k is defined as

$$\rho_{jk} = \frac{\sigma_{jk}}{\sqrt{\sigma_{jj}\sigma_{kk}}} \tag{3.6}$$

where j and k range over the integers from 1 to n.

For individual i, the $n \times n$ correlation matrix is

$$\text{Corr}(Y_i) = (\rho_{jk})$$

where the jkth entry is ρ_{jk} and $\rho_{jk} = 1$ when $j = k$.

The main interest in repeated measures studies is the association between the response and time and the effect of the covariates on that association. The correlation is not usually of principle interest but needs to be taken into account when estimating the coefficients of the regression model and when using diagnostic tests to examine the goodness of fit of the model.

3.4 Descriptive Statistics

Descriptive statistics reveal the underlying structure of a repeated measures study by computing the characteristics of the distribution of responses at each time point averaged over all individuals measured at that occasion. For example, in the epileptic study, there are two groups (placebo and those receiving progabide) and five time points (baseline, 2, 4, 6, and 8 weeks); for each combination of treatment and time, the mean, median, and standard deviation appear in Tables 3.3 and 3.4.

This is a randomized study, and the mean and standard deviation at baseline for the two groups are similar. Notice the skewness of the distribution of the number of seizures at each time point, where the sample median is always less than the corresponding mean. Knowing the nature of the distribution of responses will play an important role when choosing a regression model to analyze the data. A cursory examination of Tables 3.3 and 3.4 imply a similar response profile for the average number of seizures for the two groups. Graphical techniques will give us additional information when comparing the two groups with regard to the number of seizures.

Also important is the structure of the correlation between the five responses for each group. There are five time points, and for the placebo group, the 5×5 matrix of sample correlations appears in Table 3.5.

TABLE 3.3

Descriptive Statistics of Placebo Group: Number of Seizures for Epileptic Study

Time	Mean	Standard Deviation	Median	Minimum	Maximum
0	7.70	6.526	4.75	2	28
2	9.36	10.137	5	0	40
4	8.29	8.164	4.5	0	29
6	8.79	14.673	5	0	76
8	7.96	7.628	5	0	29

TABLE 3.4

Descriptive Statistics of Progabide Group: Number of Seizures for Epileptic Study

Time	Mean	Standard Deviation	Median	Minimum	Maximum
0	7.91	6.984	5.75	2	38
2	8.58	18.241	4	0	102
4	8.42	11.86	5	0	65
6	8.13	13.894	4	0	72
8	6.71	11.264	4	0	63

TABLE 3.5

Sample Correlations between Five Points of Placebo Group: Number of Seizures for Epileptic Study

Time (weeks)	0	2	4	6	8
0	1	0.744	0.831	0.493	0.818
2		1	0.782	0.507	0.675
4			1	0.661	0.780
6				1	0.676
8					1

It is important to identify the correlation pattern in a repeated measures study, because correlation plays an important role in estimating the response profile of subjects over time. Generally speaking, one would expect the correlations to be positive and to become smaller as the time lag increases. One usually does not expect to see the estimated correlations decrease to zero and is unlikely to see correlations close to one.

See Thall and Vail[2] for additional information on the structure of the correlation matrix for the epileptic study.

3.5 Graphics

In addition to the descriptive statistics, graphical techniques reveal the underlying structure of the repeated measures study. Of primary interest in comparing the two groups of the epileptic clinical trial, a scatter plot with lowess curves is appropriate.

Consider Figure 3.1, which depicts two lowess curves that differentiate the two groups of epileptic patients. It shows very little difference in the two, which corroborates the information in Tables 3.3 and 3.4.

Graphical procedures that accompany the estimated correlations are called matrix plots. For example, Figure 3.2 shows bivariate scatter plots

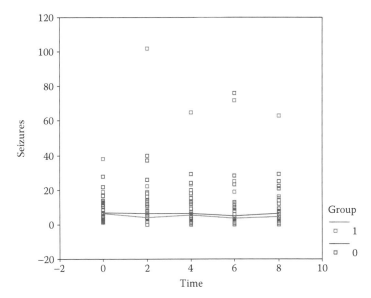

FIGURE 3.1
Seizures versus time with lowess curves for two groups of epileptic patients.

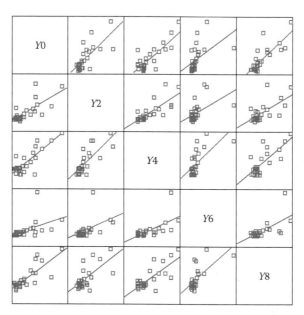

FIGURE 3.2
Scatter plot of number of seizures for placebo group.

with linear regressions between the number of seizures taken at times 0, 2, 4, 6, and 8 weeks. For the placebo group, z0, z2, z4, z6, and z8 are the vectors of the number of seizures observed at times 0, 2, 4, 6, and 8 weeks, while the corresponding vectors for the progabide group are denoted by y0, y2, y4, y6, and y8, respectively. Note that all bivariate plots are over the same set of individuals of that particular group, and the trends are represented by linear regressions. Recall that the estimated correlation between baseline and time 2 weeks is 0.744, which is represented graphically by the first scatter plot of Figure 3.2.

One should examine the scatter plots of Figure 3.2 carefully. What does a scatter plot between the values of two occasions tell one about the correlation between the responses of the two time points? The more the points cluster about a straight line, the higher the correlation. One should compare each plot with the estimated correlation given in Tables 3.6 and 3.7.

Similar matrix scatter plots are portrayed for the progabide group in Figure 3.3.

The distribution of the number of seizures at a given time should be examined. This is important because the identification of the distribution of the dependent variable helps determine the type of regression model employed to examine the profile of individuals over time. For example, for the placebo group, the histogram of the number of seizures at baseline is shown in Figure 3.4 and demonstrates definite skewness, which is also apparent with Table 3.1. The sample mean is 7.70, but the median

TABLE 3.6

Sample Correlations between Five Points of Progabide Group:
Number of Seizures for Epileptic Study

Time (weeks)	0	2	4	6	8
0	1	0.855	0.848	0.839	0.877
2		1	0.907	0.912	0.971
4			1	0.925	0.947
6				1	0.952
8					1

TABLE 3.7

Mean and Standard Deviation of the Estimated Slopes by Group

Group	Mean	Standard Deviation	N
0 (placebo)	−0.00179	0.9306	28
1 (progabide)	−0.14403	0.5541	31

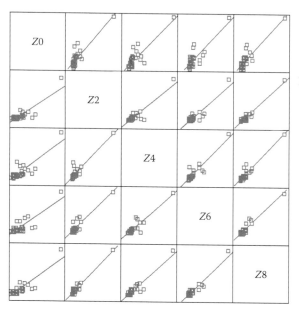

FIGURE 3.3
Scatter plots for progabide group.

is 4.75, which implies right skewness and is corroborated by Figure 3.4. It is left as an exercise for students to elucidate the skewness at other time points of the two groups. As was stated earlier, the choice of the regression model depends on the distribution of the response variable, thus the model should be appropriate for a categorical response, which

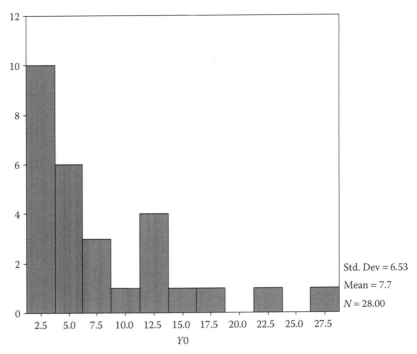

FIGURE 3.4
Histogram of number of seizures for placebo at baseline.

is highly skewed to the right. In this case, a categorical regression model appears to be appropriate.

3.6 Sources of Variation

Suppose the repeated measures study is replicated an infinite number of times, then one would not expect the same values appearing in Table 3.1 to appear again in any future replication. The investigator must rely on only one replication to make inferences about the individual profiles, but the estimated parameters (such as the mean and variance) at each time point refer to those hypothetical replications. What accounts for the variation between the replications of the repeated measures study?

Several authors, such as Fitzmaurice, Lair, and Ware,[3] discuss the various sources of variation identified as (1) variation between individuals, (2) within biological variation, and (3) measurement error.

Between individual variation refers to the differences in the responses over time of the various subjects in the study. Some individuals tend to score higher values than others and some tend to score lower than others. And of course, other individuals tend to score somewhere near the "average" compared to others.

For these two subjects, subject 8 has higher scores than subject 1. Also note that higher scores tend to be followed by higher for subject 8 and lower scores tend to be followed by lower scores for subject 1. How can one estimate the interindividual variation? One approach is to compute the absolute value of the difference in the response at each time point and sum the absolute differences over the five time points (baseline, 2, 4, 6, and 8 weeks). If this sum is zero, then two individuals have identical repeated measures (Figure 3.5).

Of course one would expect to see a large variety of scatter plots taken over all pairs of subjects, varying from pairs like subjects 8 and 1, where the difference is obvious, to pairs where it is not so obvious just how the two individuals do differ.

Within individual variation refers to the variation of the responses over time of a given individual. If the individual responses are repeated over time

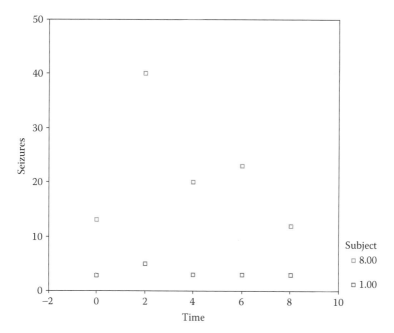

FIGURE 3.5
Number of seizures over time for subjects 8 and 1.

without intervention, one has something similar to replication. For example, in the epileptic study, a subject is randomized to the placebo group or to the treatment group administered with progabide; thus, the repeated measures for a subject assigned to the one group are similar to replications. What accounts for the variation between the responses within individuals? Why does the number of seizures vary from time period to time period in the epileptic study?

Biological variation contributes to these differences. Another example is what individuals measuring their blood glucose values say on a daily basis. An example of biological variation is that the glucose value varies due to the metabolism of individuals, which is affected by their energy and any medicines intake.

Another source of variation is measurement error, which is always present in repeated measures observations. Consider the epileptic study of the number of seizures observed at five time points. How does one measure the number of seizures in a two-week period? One can see that the observed number depends on someone making the count and recording it at each seizure. What is the definition of a seizure and who is responsible for reporting the seizure count? Does the subject keep a record or does another person count the number of seizures? It is possible that different individuals could count the number of seizures differently.

There are other types of response where measurement error is not as important. For example, height and weight measurements usually do not have the amount of variation as others. Usually, it is impossible to measure the measurement error in repeated measures studies. When is it possible to estimate the measurement error? Suppose one took a blood sample at a particular time and divided the sample in two parts, and measured the blood glucose value the same way; one could then estimate the variation between the two values, giving one an estimate of the measurement error.

In experimental design courses, measurement error is an important statistical topic, and the interested reader will find the Fuller[4] account very informative and useful. Refer to Figure 3.5 for the seizure accounts of two subjects and focus on the observed counts on the graph. It is easy to see the variation between the two individuals, but how does one visualize the measurement error? One would have to imagine another point corresponding to each observation and imagine this point as the "true" point without measurement error. For the epileptic study, see Figure 3.6 where the first curve are the number of seizures with measurement error and the other curve without. The larger values correspond to subject 8 and the smaller to subject 1. From Figure 3.6, one can discern the three sources of variation: between individual, within biological variation, and measurement error.

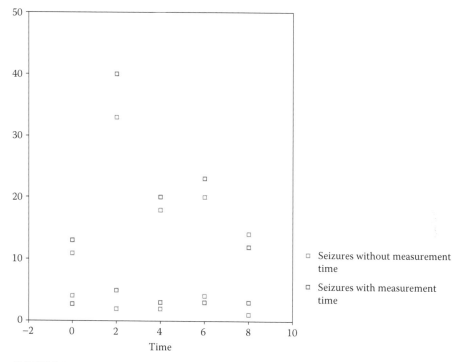

FIGURE 3.6
Number of seizures for subjects 8 and 1, with and without measurement error.

3.7 Bayesian Inference

Recall Bayesian inference is the approach taken in this book, and the reader is referred to Chapter 1 for an introduction to the subject. The principle upon which the approach is based is common sense in that Bayes's theorem combines information from the data with information from similar previous scientific investigations; thus, the Bayesian analyst must become very involved with the principal investigator in order to optimally use Bayesian inference. Using information from previous related experiments allows the analyst to specify the prior information about the unknown parameters in the likelihood function (the statistical model).

For example, in order to present this type of scenario, we consider the epileptic study of Lippik[1] with the experimental results reported in Table 3.1. Thall and Vail[2] analyzed the data for this study, and the paper reported several other studies concerning the use of progabide for the treatment of epilepsy.

For example, Crawford and Chadwick[5] describe various drug treatments for epilepsy including progabide and valproate, while Chadwick[6] gives an in-depth review of many drug treatments (including progabide). Later in Chapter 5, this information will be used to determine the prior information in the regression model for the profile of individuals over time of the Lippik study.

3.8 Summary Statistics

One approach to comparing two groups of repeated measures is to use summary statistics, that is, a statistic that summarizes a subject's profile over time. This should only be done once the nature of the profile is ascertained by graphical techniques. For example, with the epilepsy study, Figure 3.1 reveals a flat response for both groups (the placebo and progabide) with the response of the placebo group appearing slightly larger than the treatment group. For each individual, a summary statistic is computed, and then the groups are compared.

In the epilepsy study, the average number of seizures is computed for each individual, and it can be shown that the distribution of the average number of seizures per individual does not appear to be normally distributed. The histogram of the individual average number of seizures is definitely skewed to the right. On the other hand, when the slope of the regression of the number of seizures versus time is estimated by least squares for each individual, the sample histogram for each group appear to be normally distributed. See Figure 3.7, which depicts the sample histogram of the least squares estimated slopes for the progabide group. As the estimated slopes appear to be normally distributed, the two groups can be compared using a two-sample t test. Note that the descriptive statistics for the two groups are as shown in Figure 3.7.

The t test for the difference in the population mean is 0.474, indicating no significant difference in the slopes of the two treatment groups. It is assumed that the variances of the two populations are the same.

What is the Bayesian approach to this problem? I will assume the estimated slopes are a random sample from a normal population assuming vague prior distributions for the two parameters, namely the population mean and precision. Recall that it is assumed that the responses from the individuals are independent; thus, the estimated slopes are assumed to be independent observations.

For a normal distribution with mean μ and precision τ, the density is

$$f(x \mid \mu, \tau) \propto \tau^{1/2} \exp{-\left(\frac{\tau}{2}\right)(x - \mu)^2} \tag{3.7}$$

where:
x is any real number
the precision $\tau > 0$

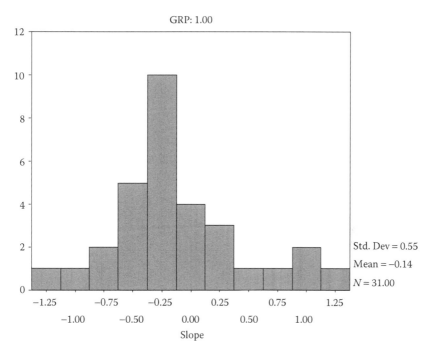

GRP: 1.00

Std. Dev = 0.55
Mean = −0.14
N = 31.00

Slope

FIGURE 3.7
Sample histogram of slope for the progabide group, group 1.00.

The precision is the inverse of the variance, and the Bayesian analysis will be based on vague prior distribution for the unknown parameters, that is, it will be assumed that μ has a normal distribution with mean 0 and precision 0.0001 and that the precision τ has a gamma distribution with parameters 0.0001 and 0.0001. **BUGS CODE 3.1** will implement the Bayesian analysis. In the epileptic example, x is the number of seizures for a two-week period.

A Bayesian analysis is executed with 55,000 observations for the simulation, with 1,000 initial observations and a refresh of 100, and the results are reported in Table 3.8.

Bayesian inferences are based on the posterior distribution of diff = mu0 − mu2, where the posterior mean is 0.1426 and a 95% credible interval

TABLE 3.8

Posterior Analysis of Epilepsy Study

Parameter	Mean	SD	Median	2½	97½	MCMC
diff	0.1426	0.2087	0.1419	−0.2681	0.5527	0.000917
mu0	−0.002133	0.182	−0.001625	−0.36311	0.3545	0.0007783
mu1	−0.1447	0.1036	−0.1447	−0.348	0.06171	0.000487
tau0	1.155	0.315	1.128	0.6227	1.847	0.001393
tau1	3.252	0.8147	3.177	1.826	5.104	0.00378

BUGS CODE 3.1

```
model;

{
# normal distribution for the slopes
for (i in 1:n0){y0[i]~dnorm(mu0,tau0)}
for (j in 1:n1){y1[j]~dnorm(mu1,tau1)}

# prior distributions
# vague priors for the means and precisions
Mu0~dnorm(0,.0001)
Mu1~dnorm(0,.0001)
Tau0~dgamma(.0001,.0001)
Tau1~dgamma(.0001,.0001)
# diff is the main parameter
diff<-mu0-mu1

}

#data
list(n0 = 28,n1 = 31,
# slopes for placebo epileptic study
y0 = c(-.075,.025,.250,.050,.550,.175,-.400,-.950,
      -.025,-.650,-.100,-.625,-.150,.600,-1.575,-1.300,
      .000,-.325,.000,.350,.100,.175,.125,-.200,4.025,-
      .125,.000,.025),

# slopes for progabide epileptic study
y1 = c(-1.200,-.500,-.325,-.050,.175,-.450,.675,
      .150,.025,-.725,-.625,-.225,-.350,-.275,1.050,
      -.450,-.990,.175,.000,-.375,1.025,-.250,-.475,-
      .300,.400,-.700,-.050,1.150,-.375,-.325,-.275))
# information for Alzheimer's study
list(n0 = 26,n1 = 22,
y0 = c(-.224,-1.595,-.121,.474,-.603,.345,-.371,
      .552,1.216,.414,.578,-.276,-1.017,-.284,-1.422,-
      .241,-.690,-.629,  -.126,.095,.190,-1.069,.578,
      -.086,0,.026),

y1 = c(.586,1.681,-.060,-.422,-.371,.241,-.543,.328,-
      .121,1.121,1.198,.483,.052,-.259,.241,-.500,-.759,.259,
      -.043,-.483,-.069,-.776))

# initial values
list(mu0 = 0,mu1 = 0,tau0 = 1,tau1 = 1)
```

of (−0.2681, 0.5527), implying very little difference in the average slope of the two groups. Referring to Figure 3.1, one can see a flat response to the two groups over the five time periods, noting that the posterior mean of the slope for the placebo group is −0.0021 and is −0.1447 for the progabide group. The latter estimated slope is more negative than that for placebo, implying that progabide is having a beneficial effect in reducing seizures; however, as seen by the 95% credible interval for the difference in mu0 − mu2, to declare progabide as efficacious is somewhat problematic! The slope measures the rate of change in the average number of seizures. and the time unit is two weeks. Comparing Tables 3.7 and 3.8, one can see the similarity between the estimated slope for placebo and posterior mean of mu0. One would expect to see such similarity, because the Bayesian analysis employed vague-type priors for the mean and precision for the normal population of the placebo group. For the Bayesian analysis reported in Table 3.8, the Markov chain Monte Carlo (MCMC) errors are very small, which indicates that the computed posterior means are very close to the "true" posterior means.

3.9 Another Example

The next example will illustrate the various ideas presented earlier in this chapter, namely: (1) descriptive statistics, (2) graphical techniques (scatter plots and histograms), (3) discussion of prior information, and (4) a Bayesian analysis based on summary statistics.

This example involves patients with Alzheimer's disease. The study analyzed by Hand and Taylor[7] consists of 48 patients, of which 26 received placebo and the remainder lecithin, which is a therapy intended to improve memory. The information also appears as Table A.10 in Hand and Crowder.[8] In Table 3.9, the main response is the number of correctly remembered words measured at times 0, 1, 2, 4, and 6 units. Is there any difference in the groups, that is, is there a group time interaction?

In order to arrive at an initial impression of the time profile of the two groups, a scatter plot with linear regression reveals that the two groups are somewhat similar. The vertical axis is the number of words and the horizontal the time periods 0, 1, 2, 4, and 6. Linear regression shows that there is a slight increase in the average number of correctly recalled words for the treatment group compared to the placebo group. There appears to be a slight decrease in the average number of correctly recalled words for placebo; however, the difference in the two slopes reveals very little difference; thus, I doubt that lecithin is efficacious in increasing the number of correctly

TABLE 3.9

Number of Correctly Remembered Words by Alzheimer's
Patients by Group

Group	Time	Word	Subject
1.00	0.00	20.00	1.00
1.00	1.00	19.00	1.00
1.00	2.00	20.00	1.00
1.00	4.00	20.00	1.00
1.00	6.00	18.00	1.00
1.00	0.00	14.00	2.00
1.00	1.00	15.00	2.00
1.00	2.00	16.00	2.00
1.00	4.00	9.00	2.00
1.00	6.00	6.00	2.00
1.00	0.00	7.00	3.00
1.00	1.00	5.00	3.00
1.00	2.00	8.00	3.00
1.00	4.00	8.00	3.00
1.00	6.00	5.00	3.00
1.00	0.00	6.00	4.00
1.00	1.00	10.00	4.00
1.00	2.00	9.00	4.00
1.00	4.00	10.00	4.00
1.00	6.00	10.00	4.00
1.00	0.00	9.00	5.00
1.00	1.00	7.00	5.00
1.00	2.00	9.00	5.00
1.00	4.00	4.00	5.00
1.00	6.00	6.00	5.00
1.00	0.00	9.00	6.00
1.00	1.00	10.00	6.00
1.00	2.00	9.00	6.00
1.00	4.00	11.00	6.00
1.00	6.00	11.00	6.00
1.00	0.00	7.00	7.00
1.00	1.00	3.00	7.00
1.00	2.00	7.00	7.00
1.00	4.00	6.00	7.00
1.00	6.00	3.00	7.00
1.00	0.00	18.00	8.00
1.00	1.00	20.00	8.00
1.00	2.00	20.00	8.00
1.00	4.00	23.00	8.00
1.00	6.00	21.00	8.00

(Continued)

TABLE 3.9 (*Continued*)

Number of Correctly Remembered Words by Alzheimer's
Patients by Group

Group	Time	Word	Subject
1.00	0.00	6.00	9.00
1.00	1.00	10.00	9.00
1.00	2.00	10.00	9.00
1.00	4.00	13.00	9.00
1.00	6.00	14.00	9.00
1.00	0.00	10.00	10.00
1.00	1.00	15.00	10.00
1.00	2.00	15.00	10.00
1.00	4.00	15.00	10.00
1.00	6.00	14.00	10.00
1.00	0.00	5.00	11.00
1.00	1.00	9.00	11.00
1.00	2.00	7.00	11.00
1.00	4.00	3.00	11.00
1.00	6.00	12.00	11.00
1.00	0.00	11.00	12.00
1.00	1.00	11.00	12.00
1.00	2.00	8.00	12.00
1.00	4.00	10.00	12.00
1.00	6.00	9.00	12.00
1.00	0.00	10.00	13.00
1.00	1.00	2.00	13.00
1.00	2.00	9.00	13.00
1.00	4.00	3.00	13.00
1.00	6.00	2.00	13.00
1.00	0.00	17.00	14.00
1.00	1.00	12.00	14.00
1.00	2.00	14.00	14.00
1.00	4.00	15.00	14.00
1.00	6.00	13.00	14.00
1.00	0.00	16.00	15.00
1.00	1.00	15.00	15.00
1.00	2.00	13.00	15.00
1.00	4.00	7.00	15.00
1.00	6.00	9.00	15.00
1.00	0.00	7.00	16.00
1.00	1.00	10.00	16.00
1.00	2.00	4.00	16.00
1.00	4.00	10.00	16.00
1.00	6.00	5.00	16.00

(*Continued*)

TABLE 3.9 (*Continued*)

Number of Correctly Remembered Words by Alzheimer's Patients by Group

Group	Time	Word	Subject
1.00	0.00	5.00	17.00
1.00	1.00	0.00	17.00
1.00	2.00	5.00	17.00
1.00	4.00	0.00	17.00
1.00	6.00	0.00	17.00
1.00	0.00	16.00	18.00
1.00	1.00	7.00	18.00
1.00	2.00	7.00	18.00
1.00	4.00	6.00	18.00
1.00	6.00	10.00	18.00
1.00	0.00	5.00	19.00
1.00	1.00	6.00	19.00
1.00	2.00	9.00	19.00
1.00	4.00	5.00	19.00
1.00	6.00	6.00	19.00
1.00	0.00	2.00	20.00
1.00	1.00	1.00	20.00
1.00	2.00	1.00	20.00
1.00	4.00	2.00	20.00
1.00	6.00	2.00	20.00
1.00	0.00	7.00	21.00
1.00	1.00	11.00	21.00
1.00	2.00	7.00	21.00
1.00	4.00	5.00	21.00
1.00	6.00	11.00	21.00
1.00	0.00	9.00	22.00
1.00	1.00	16.00	22.00
1.00	2.00	17.00	22.00
1.00	4.00	10.00	22.00
1.00	6.00	6.00	22.00
1.00	0.00	2.00	23.00
1.00	1.00	5.00	23.00
1.00	2.00	6.00	23.00
1.00	4.00	7.00	23.00
1.00	6.00	6.00	23.00
1.00	0.00	7.00	24.00
1.00	1.00	3.00	24.00
1.00	2.00	5.00	24.00
1.00	4.00	5.00	24.00
1.00	6.00	5.00	24.00

(Continued)

TABLE 3.9 (*Continued*)

Number of Correctly Remembered Words by Alzheimer's
Patients by Group

Group	Time	Word	Subject
1.00	0.00	19.00	25.00
1.00	1.00	13.00	25.00
1.00	2.00	19.00	25.00
1.00	4.00	17.00	25.00
1.00	6.00	17.00	25.00
1.00	0.00	7.00	26.00
1.00	1.00	5.00	26.00
1.00	2.00	8.00	26.00
1.00	4.00	8.00	26.00
1.00	6.00	6.00	26.00
2.00	0.00	9.00	27.00
2.00	1.00	11.00	27.00
2.00	2.00	14.00	27.00
2.00	4.00	11.00	27.00
2.00	6.00	14.00	27.00
2.00	0.00	6.00	28.00
2.00	1.00	7.00	28.00
2.00	2.00	9.00	28.00
2.00	4.00	12.00	28.00
2.00	6.00	16.00	28.00
2.00	0.00	13.00	29.00
2.00	1.00	18.00	29.00
2.00	2.00	24.00	29.00
2.00	4.00	20.00	29.00
2.00	6.00	14.00	29.00
2.00	0.00	13.00	29.00
2.00	1.00	18.00	29.00
2.00	2.00	24.00	29.00
2.00	4.00	20.00	29.00
2.00	6.00	14.00	29.00
2.00	0.00	9.00	30.00
2.00	1.00	10.00	30.00
2.00	2.00	9.00	30.00
2.00	4.00	8.00	30.00
2.00	6.00	7.00	30.00
2.00	0.00	6.00	31.00
2.00	1.00	7.00	31.00
2.00	2.00	4.00	31.00
2.00	4.00	5.00	31.00
2.00	6.00	4.00	31.00

(Continued)

TABLE 3.9 (*Continued*)

Number of Correctly Remembered Words by Alzheimer's
Patients by Group

Group	Time	Word	Subject
2.00	0.00	11.00	32.00
2.00	1.00	11.00	32.00
2.00	2.00	5.00	32.00
2.00	4.00	10.00	32.00
2.00	6.00	12.00	32.00
2.00	0.00	7.00	33.00
2.00	1.00	10.00	33.00
2.00	2.00	11.00	33.00
2.00	4.00	8.00	33.00
2.00	6.00	5.00	33.00
2.00	0.00	8.00	34.00
2.00	1.00	18.00	34.00
2.00	2.00	19.00	34.00
2.00	4.00	15.00	34.00
2.00	6.00	14.00	34.00
2.00	0.00	3.00	35.00
2.00	1.00	3.00	35.00
2.00	2.00	3.00	35.00
2.00	4.00	1.00	35.00
2.00	6.00	3.00	35.00
2.00	0.00	4.00	36.00
2.00	1.00	10.00	36.00
2.00	2.00	9.00	36.00
2.00	4.00	17.00	36.00
2.00	6.00	10.00	36.00
2.00	0.00	11.00	37.00
2.00	1.00	10.00	37.00
2.00	2.00	5.00	37.00
2.00	4.00	15.00	37.00
2.00	6.00	16.00	37.00
2.00	0.00	1.00	38.00
2.00	1.00	3.00	38.00
2.00	2.00	2.00	38.00
2.00	4.00	2.00	38.00
2.00	6.00	5.00	38.00
2.00	0.00	6.00	39.00
2.00	1.00	7.00	39.00
2.00	2.00	7.00	39.00
2.00	4.00	6.00	39.00
2.00	6.00	7.00	39.00

(*Continued*)

TABLE 3.9 (*Continued*)

Number of Correctly Remembered Words by Alzheimer's
Patients by Group

Group	Time	Word	Subject
2.00	0.00	0.00	40.00
2.00	1.00	3.00	40.00
2.00	2.00	2.00	40.00
2.00	4.00	0.00	40.00
2.00	6.00	0.00	40.00
2.00	0.00	18.00	41.00
2.00	1.00	19.00	41.00
2.00	2.00	15.00	41.00
2.00	4.00	17.00	41.00
2.00	6.00	20.00	41.00
2.00	0.00	15.00	42.00
2.00	1.00	15.00	42.00
2.00	2.00	15.00	42.00
2.00	4.00	14.00	42.00
2.00	6.00	12.00	42.00
2.00	0.00	14.00	43.00
2.00	1.00	11.00	43.00
2.00	2.00	8.00	43.00
2.00	4.00	10.00	43.00
2.00	6.00	8.00	43.00
2.00	0.00	6.00	44.00
2.00	1.00	6.00	44.00
2.00	2.00	5.00	44.00
2.00	4.00	5.00	44.00
2.00	6.00	8.00	44.00
2.00	0.00	10.00	45.00
2.00	1.00	10.00	45.00
2.00	2.00	6.00	45.00
2.00	4.00	10.00	45.00
2.00	6.00	9.00	45.00
2.00	0.00	4.00	46.00
2.00	1.00	6.00	46.00
2.00	2.00	6.00	46.00
2.00	4.00	4.00	46.00
2.00	6.00	2.00	46.00
2.00	0.00	4.00	47.00
2.00	1.00	13.00	47.00
2.00	2.00	9.00	47.00
2.00	4.00	8.00	47.00
2.00	6.00	7.00	47.00

(*Continued*)

TABLE 3.9 (*Continued*)

Number of Correctly Remembered Words by Alzheimer's
Patients by Group

Group	Time	Word	Subject
2.00	0.00	14.00	48.00
2.00	1.00	7.00	48.00
2.00	2.00	8.00	48.00
2.00	4.00	10.00	48.00
2.00	6.00	6.00	48.00

Source: Hand, D.J., and Taylor, C.C., *Multivariate Analysis of Variance and Repeated Measures*, Chapman & Hall, London, 1987; Hand, D., and Crowder, M., *Practical Longitudinal Data Analysis*, Chapman & Hall, Boca Raton, FL, 1999.

recalled words in Alzheimer's disease. Later in this chapter, this will be examined in more detail with a Bayesian approach.

Descriptive statistics provide additional information about comparing the two groups and complement Figure 3.8.

The baseline measurements are not the same, but for each group the means do not vary much. For example, for placebo, the mean varies from 8.73 at time 6 to 10.07 at time 2, and for the treatment group, the mean varies from 8.347 at baseline to 10.130 at time 2 (Table 3.10).

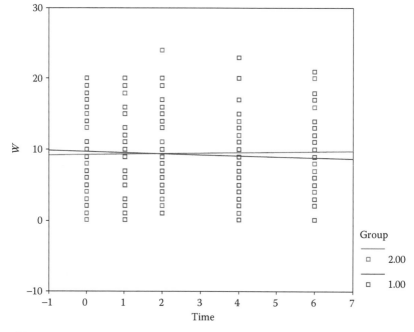

FIGURE 3.8
Two groups of Alzheimer's patients.

TABLE 3.10

Descriptive Statistics for Alzheimer's Study

Time	Mean	SD	Median
Placebo Group			
0	9.6538	5.184	8
1	9.230	5.452	10
2	10.0769	5.106	9
4	8.923	5.585	8
6	8.730	5.250	7.5
Lecithin Group			
0	8.347	4.744	8
1	10.130	4.883	10
2	9.521	6.309	8
4	9.913	5.751	10
6	9.2609	5.145	8

TABLE 3.11

Estimated Pearson Correlations

Time	0	1	2	4	6
Placebo Group					
0	1	0.693	0.790	0.679	0.646
1		1	0.843	0.804	0.790
2			1	0.798	0.705
4				1	0.845
6					1
Treatment Group					
0	1	0.718	0.574	0.706	0.663
1		1	0.877	0.861	0.738
2			1	0.813	0.620
4				1	0.841
6					1

Information about the pair-wise correlations between the word counts at times 0, 1, 2, 4, and 6 are needed in order to use regression techniques to compare the two groups. For the placebo group, the correlations are portrayed in Table 3.11.

For the treatment group correlations, the smallest is 0.584 for the correlation between time 0 and time 1, while the largest is 0.877, the estimated Pearson correlation coefficient of the word count between times 1 and 2 units. The estimated correlations appear to not die out and are somewhat similar for the two groups. One can "see" these correlations depicted by scatter matrix plots, one for the placebo group and one for lecithin patients (see Figure 3.2).

In summary, the descriptive statistics for the Alzheimer's study are given in two parts: (1) the mean, median, and standard deviation by Table 3.10 and (2) the estimated pair-wise Pearson correlations by Table 3.11.

How should the Alzheimer's study be analyzed? As in the previous example, for each individual, I computed the slope by simple least squares, and then assuming the slopes were a random sample from each group, a Bayesian approach compared the two groups. Estimated least squares slopes for the placebo group and the treatment group are portrayed in Table 3.12.

It is assumed that the estimated least squares slopes are random samples from normal distributions, where the placebo group has a mean slope of

TABLE 3.12

Least Squares Slopes

Subject	Slope	Subject	Slope
26 Placebo Subjects			
1	−0.224	14	−0.284
2	−1.595	15	−1.422
3	−0.121	16	−0.241
4	0.474	17	−0.690
5	−0.603	18	−0.629
6	0.345	19	−0.026
7	−0.371	20	0.095
8	0.552	21	0.190
9	1.216	22	−1.066
10	0.414	23	0.578
11	0.578	24	−0.086
12	−0.276	25	0
13	−1.017	26	0.026
22 Lecithin Subjects			
27	0.586	38	0.483
28	1.681	39	0.052
29	−0.060	40	−0.259
30	−0.422	41	0.241
31	−0.371	42	−0.500
32	0.241	43	−0.759
33	−0.542	44	0.259
34	0.328	45	−0.043
35	−0.121	46	−0.483
36	1.121	47	−0.069
37	1.198	48	−0.776

TABLE 3.13

Posterior Analysis for Alzheimer Study

Parameter	Mean	SD	Error	2½	Median	97½
Diff	−0.02453	0.1957	0.0009299	−0.6308	−0.2448	0.1398
mu1	−0.1648	0.1339	0.0005819	−0.4298	−0.1645	0.09957
mu2	0.08408	0.1444	0.0006715	−0.204	0.08059	0.3681
tau1	2.328	0.6595	0.003239	1.217	2.268	3.787
tau2	2.436	0.7539	0.003445	1.191	2.355	4.142

mu1, and for lecithin, the mean slope is mu2. From the histogram for the estimated slopes of the placebo group, I verified that one could assume that they are normally distributed. The corresponding precisions are tau1 and tau2 for placebo and treatment groups, respectively, and the analysis is executed with 55,000 observations for the MCMC simulation, with 5000 initial values and a refresh of 100.

Of primary interest is the parameter difference, which is the difference between mu1 and mu2, which has posterior mean of −0.02453, posterior standard deviation of 0.1957, and 95% credible interval (−0.6308, 0.1398). As the 95% credible interval contains zero, one is reluctant to declare a difference in the two mean slopes, implying that the treatment lecithin is not efficacious. Note the symmetry in the posterior distributions of mu1 and mu2, which is not the case for the posterior distributions of the two precisions tau1 and tau2.

With regard to the Alzheimer's study, one should base the prior distribution of the Bayesian analysis on relevant previous experiments. For example, Little et al.[9] review the general area of treatments for Alzheimer's disease and conclude that lecithin has not shown a beneficial effect, which agrees with that for the present study (Table 3.13).

3.10 Basic Ideas for Categorical Variables

3.10.1 Introduction

As with the previous sections devoted to quantitative variables, the following sections will focus on repeated measures whose possible values are few in number. The basic concepts including descriptive statistics, graphical techniques, and the fundamentals of a Bayesian analysis will be presented for categorical variables.

The simplest case is when the response takes on only two possible values, and the approach is to consider the patterns of response over time as samples of a multinomial distribution. This allows one to determine the distribution of the response at each time point.

3.10.2 An Example

Davis[10] provides the first example, which involves 46 subjects given three drugs A, B, and C, where each subject gives a favorable (F) or unfavorable (U) response to each drug. See Table 3.14 for the results of the study.

For the drug study, the first task is to determine the descriptive statistics, which consists of two parts, namely the percentage of patients with a favorable response to drugs A, B, and C.

TABLE 3.14

Response F or U to Three Drugs A, B, and C

Subject	A	B	C
1	F	F	U
2	U	U	U
3	U	U	F
4	F	F	U
5	U	U	U
6	F	F	U
7	F	F	F
8	F	F	U
9	F	U	U
10	U	U	F
11	F	F	U
12	U	F	U
13	F	F	F
14	F	F	U
15	U	F	F
16	F	U	F
17	U	U	U
18	F	F	U
19	F	U	U
20	U	U	F
21	F	F	F
22	F	F	U
23	F	F	U
24	U	F	U
25	F	F	U
26	U	U	U
27	F	U	U
28	U	U	F
29	U	U	U
30	F	F	U
31	F	F	F
32	F	U	F

(Continued)

TABLE 3.14 (Continued)

Response F or U to Three Drugs A, B, and C

Subject	A	B	C
33	F	F	U
34	U	F	F
35	U	F	U
36	F	F	U
37	F	F	U
38	F	F	F
39	F	U	U
40	U	U	F
41	F	F	U
42	U	U	U
43	U	U	F
44	F	F	U
45	F	F	F
46	U	F	U

Source: Davis, C.S., *Statistical Methods for the vAnalysis of Repeated Measurements*, Springer-Verlag, New York, p. 185, 2002.

TABLE 3.15

Profiles of 46 Subjects to Eight Responses

Profile	1	2	3	4	5	6	7	8
	F	F	F	F	U	U	U	U
	F	F	U	U	F	F	U	U
	F	U	F	U	F	U	F	U
Number of subjects	6	16	2	4	2	4	6	6

There are eight distinct patterns of two events U and F over three time periods at which the drugs are administered to the subjects, summarized in Table 3.15.

Therefore, of the 46 subjects, 6 had a response depicted by profile 1, while 16 had a F response to drug A, a F response to drug B, and a U response to drug C. It is assumed that among the 46 subjects, the number of subjects among the eight categories follow a multinomial distribution.

Now a subject responses favorably to drug A if a subject belongs to profile 1, 2, 3, or 4. In addition, a subject responds with an F to drug B if the subject follows profile 1, 2, 5, or 6, and finally a subject responds favorably to drug C if the subject is in profile 1, 3, 5, or 7. Note that this study is a clustered study somewhat different than the repeated format of studies reported earlier. The "repeated" factor, drug, is not ordered.

3.10.3 Multinomial Distribution of Response to Drugs

Let X_i denote the number of subjects in the ith category or profile, then $(X_1, X_2, X_3, X_4, X_5, X_6, X_7, X_8)$ has a multinomial distribution with parameters n and $(\theta_1, \theta_2, \theta_3, \theta_4, \theta_5, \theta_6, \theta_7, \theta_8)$, where θ_i is the probability that a subject will belong to category $i = 1, 2, \ldots, 8$, the sum of the thetas is 1, and n is the total number of observations.

The properties of the multinomial mass function

$$f(x) \propto \prod_{i=1}^{i=8} \theta_i^{x_i} \tag{3.8}$$

where $0 \leq x_i \leq n$, $0 < \theta_i < 1$ are easily found.

Also $\sum_{i=1}^{i=8} x_i = n$ and $\sum_{i=1}^{i=8} \theta_i = 1$. It should be noted that the normalizing constant for the mass function (3.8) is the multinomial coefficient $(n! / x_1! x_2! \ldots x_8!)$.

The properties of the multinomial distribution are well known; for example, the mean of the number of observations in category i is

$$E(X_i) = n\theta_i \tag{3.9}$$

the variance by

$$\text{Var}(X_i) = n\theta_i(1 - \theta_i) \tag{3.10}$$

and the covariance is

$$\text{Cov}(\theta_i, \theta_j) = -n(\theta_i\theta_j) \tag{3.11}$$

Our ultimate goal is to determine the correlation between observations in one category and those of another. In particular, one is interested in the number of subjects who score favorably F to drug A. From Table 3.15, this occurs for subjects who have profiles 1, 2, 3, or 4, thus, let

$$Y_1 = X_1 + X_2 + X_3 + X_4 \tag{3.12}$$

In a similar fashion, let

$$Y_2 = X_1 + X_2 + X_5 + X_6 \tag{3.13}$$

then Y_2 is the number of subjects responding favorably to drug B.

Last, the number of subjects who react favorably to drug C is

$$Y_3 = X_1 + X_3 + X_5 + X_7 \tag{3.14}$$

Using the equations for $Y_1, Y_2,$ and Y_3 allows one to determine the percentage of a favorable response to the three drugs: (1) For drug A, there are 28 favorable response among 46 patients, thus comprising 60.86% of the patients;

(2) for drug B, there are again 28/46 = 0.6086; while (3) for drug C, there are 16/46 = 0.3487, or 34.87% who respond favorably. Also, note that for drug A, 28/46 = Y_1/n; for drug B, 28/46 = Y_2/n; and for drug C, 16/46 = Y_3/n.

The variances and covariances among $Y_1, Y_2,$ and Y_3 allow one to find the correlations between the number of responses favorable to the three drugs. Using well-known formulas for the variances and covariances involving sums of random variables (e.g., $Y_1, Y_2,$ and Y_3), the relevant expressions appear in (3.15) through (3.21).

In order to determine the pair-wise correlations between $Y_1, Y_2,$ and Y_3, the variances and covariances for the eight profiles will be

3.10.4 Descriptive Statistics

The basic information about the favorable response to the three drugs from the drug study is presented in Table 3.16.

It is seen that the favorable response to drugs A and B is the same at 60.87% and is 34.78% for drug C.

3.10.5 Variances, Covariances, and Correlations between Y_1, Y_2, and Y_3

$$\text{Cov}(Y_1, Y_2) = \text{Var}(X_1) + \text{Cov}(X_1, X_2) + \text{Cov}(X_1, X_5) + \text{Cov}(X_1, X_6)$$
$$+ \text{Cov}(X_2, X_1) + \text{Var}(X_2) + \text{Cov}(X_2, X_5) + \text{Cov}(X_2, X_6)$$
$$+ \text{Cov}(X_3, X_1) + \text{Cov}(X_3, X_2) + \text{Cov}(X_3, X_5) + \text{Cov}(X_3, X_6) \quad (3.15)$$
$$+ \text{Cov}(X_4, X_1) + \text{Cov}(X_4, X_2) + \text{Cov}(X_4, X_5) + \text{Cov}(X_4, X_6)$$

In a similar way,

$$\text{Cov}(Y_1, Y_3) = \text{Var}(X_1) + \text{Cov}(X_1, X_3) + \text{Cov}(X_1, X_5) + \text{Cov}(X_1, X_7)$$
$$+ \text{Cov}(X_2, X_1) + \text{Cov}(X_2, X_3) + \text{Cov}(X_2, X_5) + \text{Cov}(X_2, X_7)$$
$$+ \text{Cov}(X_3, X_1) + \text{Var}(X_3) + \text{Cov}(X_3, X_5) + \text{Cov}(X_3, X_7) \quad (3.16)$$
$$+ \text{Cov}(X_4, X_1) + \text{Cov}(X_4, X_3) + \text{Cov}(X_4, X_5) + \text{Cov}(X_4, X_7)$$

TABLE 3.16

Proportion of Favorable Response to Drugs

Drug	Mean	SD	n
A	0.6087	0.4934	46
B	0.6087	0.4934	46
C	0.3478	0.4815	46

Last

$$Cov(Y_2, Y_3) = Var(X_1) + Cov(X_1, X_3) + Cov(X_1, X_5) + Cov(X_1, X_7)$$
$$+ Cov(X_2, X_1) + Cov(X_2, X_3) + Cov(X_2, X_5) + Cov(X_2, X_7)$$
$$+ Cov(X_5, X_1) + Cov(X_5, X_3) + Var(X_5) + Cov(X_5, X_7)$$
$$+ Cov(X_6, X_1) + Cov(X_6, X_3) + Cov(X_6, X_5) + Cov(X_6, X_7)$$

(3.17)

In order to find the correlation, we need the variances of $Y_1, Y_2,$ and Y_3.

$$Var(Y_1) = Var(X_1) + Var(X_2) + Var(X_3) + Var(X_4)$$
$$+ 2Cov(X_1, X_2) + 2Cov(X_1, X_3) + 2Cov(X_1, X_4)$$
$$+ 2Cov(X_2, X_3) + 2Cov(X_2, X_4) + 2Cov(X_3, X_4)$$

(3.18)

$$Var(Y_2) = Var(X_1) + Var(X_2) + Var(X_5) + Var(X_6)$$
$$+ 2Cov(X_1, X_2) + 2Cov(X_1, X_5) + 2Cov(X_1, X_6)$$
$$+ 2Cov(X_2, X_5) + 2Cov(X_2, X_6) + 2Cov(X_5, X_6)$$

(3.19)

$$Var(Y_3) = Var(X_1) + Var(X_3) + Var(X_5) + Var(X_7)$$
$$+ 2Cov(X_1, X_3) + 2Cov(X_1, X_5) + 2Cov(X_1, X_7)$$
$$+ 2Cov(X_3, X_5) + 2Cov(X_3, X_7) + 2Cov(X_5, X_7)$$

(3.20)

Suppose we need to know the correlation between a favorable response to drug A and a favorable response to drug B, namely,

$$\rho(Y_1, Y_2) = \frac{Cov(Y_1, Y_2)}{\sqrt{VarY(Y_1)Var(Y_2)}}$$

(3.21)

where:
  $Cov(Y_1, Y_2)$ is given by (3.17)
  $Var(Y_1)$, the variance of Y_1, is given by (3.18)
  variance of Y_2 is given by (3.19)

Expressions similar to (3.21) will provide the correlations between Y_1 and Y_3 and Y_2 and Y_3. These three pair-wise correlations provide the investigator the associations between the favorable response F to the three drugs.

    Recall that the descriptive statistics consist of two parts: (1) the fraction who respond favorably to the three drugs and (2) the pair-wise correlations between $Y_1, Y_2,$ and Y_3.

3.10.6 Estimators and Estimates of Variances and Covariances between Y_1, Y_2, and Y_3

First, consider $\text{Cov}(Y_1, Y_2)$, given by (3.15), thus in order to estimate this moment, we need to estimate the components of the right-hand side of (3.15), namely:

1. $\text{Var}(X_i)$, $i = 1$ and 2
2. $\text{Cov}(X_i, X_j)$, for $i = 1, 2, 3,$ and 4 and $j = 1, 2, 5,$ and 6

where the corresponding estimators are:

$$n\left(\frac{Y_i}{n}\right)\left(\frac{1-Y_i}{n}\right), \text{for } i = 1 \text{ and } 2 \tag{3.22}$$

and

$$-n\left(\frac{Y_i}{n}\right)\left(\frac{Y_j}{n}\right), \text{for } i = 1, 2, 3, \text{and } 4 \text{ and } j = 1, 2, 5, \text{and } 6 \tag{3.23}$$

respectively

Based on the data in Table 3.15, the formula (3.15) for the $\text{Cov}(Y_1, Y_2)$, and the estimators given by (3.22) and (3.23), Table 3.17 provides the estimates for the various components of $\text{Cov}(Y_1, Y_2)$ of the drug study. The table provides

TABLE 3.17

Parameters and Estimates for $\text{Cov}(Y_1,Y_2)$ Formula (3.15)

Parameter	Estimate
$\text{Var}(X_1)$	5.2162
$\text{Var}(X_2)$	10.4344
$\text{Cov}(X_1, X_2)$	−0.2867
$\text{Cov}(X_1, X_3)$	−0.26033
$\text{Cov}(X_1, X_5)$	−0.2603
$\text{Cov}(X_1, X_6)$	−0.52156
$\text{Cov}(X_2, X_3)$	−0.6943
$\text{Cov}(X_2, X_4)$	−1.3902
$\text{Cov}(X_2, X_5)$	−0.6943
$\text{Cov}(X_2, X_6)$	−1.3902
$\text{Cov}(X_3, X_5)$	−0.0866
$\text{Cov}(X_3, X_6)$	−0.1734
$\text{Cov}(X_4, X_5)$	−0.17348
$\text{Cov}(X_4, X_6)$	−0.3473
$\text{Cov}(X_4, X_1)$	−0.52126

the necessary information to estimate the covariance between Y_1 and Y_2 given by formula (3.21).

What is your estimate of $\text{Cov}(Y_1, Y_2)$? This is left as an exercise.

Recall that Y_1 and Y_2 are the number of subjects who respond favorably to drugs A and B, respectively.

Referring to Table 3.16, the original data of responses to the three drugs, one can estimate the correlations of the binary responses between the three drugs with Kendall's taub (Table 3.18). See Agresti[11] for more information.

This shows a correlation estimated as 0.452 of the binary responses between drugs A and B, but the estimated correlation is negative at −0.163 between drugs B and C.

3.10.7 Bayesian Analysis for Drug Study

For the Bayesian analysis, we will take the actual observation (favorable and unfavorable) as the summary statistic for each individual. **BUGS CODE 3.2** assumes a Bernoulli distribution of the binary responses and a uniform prior for the probability of a favorable response. There are three parameters measuring the difference in the probabilities of a favorable response.

Let $a[i]$ be the responses of the ith subject to drug A and assume they have a Bernoulli distribution with parameter θ_a, and in a similar fashion suppose drug B has responses $b[i]$ that have a Bernoulli distribution with parameter θ_b. Last, suppose that $d[i]$, the responses to drug C, have a Bernoulli distribution with the probability of a favorable response θ_c, and suppose the prior distribution of the three probabilities is uniform (1,1). The main parameters of interest are

$$\text{diffab} = \theta_a - \theta_b \qquad (3.24)$$

the difference in the probability of a favorable response to drug A minus that to drug B.

$$\text{diffac} = \theta_a - \theta_c \qquad (3.25)$$

and

$$\text{diffbc} = \theta_b - \theta_c \qquad (3.26)$$

TABLE 3.18

Pair-Wise Correlations of Responses (Favorable and Unfavorable) between Drugs A, B, and C: Kendall's Taub

	A	B	C
A	1.000	0.452	−0.163
B		1.000	−0.163
C			1.000

BUGS CODE 3.2

```
Model;
{
for(i in 1:46){a[i]~dbern(thetaa)}
for(i in 1:46){b[i]~dbern(thetab)}
for (i in 1:46){d[i]~dbern(thetad)}
# prior distributions for thetaa,thetab, and thetad
thetaa~dbeta(1,1)
thetab~dbeta(1,1)
thetad~dbeta(1,1)
diffab<-thetaa-thetab
diffad<-thetaa-thetad
diffbd<-thetab-thetad

}
# data for drugs A,B, and C.
list(a = c(1,0,0,1,0,1,1,1,1,0,1,0,1,1,0,1,0,1,1,0,1,1,1,
0,1,0,1,0,0,1,1,1,1,0,0,1,1,1,1,0,1,0,0,1,1,0),
b = c(1,0,0,1,0,1,1,1,0,0,1,1,1,1,1,0,0,1,0,0,1,1,1,1,1,0
,0,0,1,1,0,1,1,1,1,1,1,0,0,1,0,0,1,1,1),

d = c(0,0,1,0,0,0,1,0,0,1,0,0,1,0,1,1,0,0,0,1,1,0,0,0,0,0
,0,1,0,0,1,1,0,1,0,0,0,1,0,1,0,0,1,0,1,0))
# initial values for the thetas
list(thetaa = .5,thetab = .5,thetad = .5)
```

BUGS CODE 3.2 follows closely the notation of Equations 3.24 through 3.26.
   The analysis is executed with 55,000 observations for the simulation, beginning with 5,000 observations and a refresh of 100; the results are reported in Table 3.19.
   This is an interesting outcome with a posterior mean of 0.6036 for the probability of a favorable response to drug A, compared to a posterior mean of 0.354 for drug C. Of course this is in agreement with the descriptive statistics of Table 3.16. A uniform prior density for the three probabilities of a favorable

TABLE 3.19

Bayesian Analysis for the Drug Study

Parameter	Mean	SD	Error	2½	Median	97½
diffab	0.000453	0.0988	0.000428	−0.194	0.000808	0.1926
diffac	0.2496	0.0974	0.000441	0.0556	0.2513	0.4359
Diffbc	0.2501	0.0978	0.000431	0.0548	0.2522	0.436
Thetaa (drug A)	0.6036	0.0696	0.000317	0.4634	0.6052	0.7351
Thetab (drug B)	0.6041	0.0697	0.000301	0.4637	0.6055	0.7352
Thetac (drug C)	0.354	0.0682	0.000307	0.2259	0.3525	0.4922

response ensures similar results given in Table 3.16. Note the small simulation error, which implies that the computed posterior mean is very close to the actual posterior mean.

3.11 Summary

This chapter introduces the basic concepts used in the analysis of repeated measures. It begins with an example involving a randomized study of epilepsy with one group, a placebo, and the other, where the patients receive progabide, a medication that was designed to alleviate the occurrence of epileptic seizures.

The chapter continues by introducing the notation that will be used in the remainder of the book. For example, the symbols for the observations, covariates, means, variances, covariances, and correlations are described and illustrated for the epilepsy study.

Of course, descriptive statistics are an integral part of any statistical analysis, and of special importance are the sample means, standard deviations, medians, and correlations. For the epilepsy study, the means, standard deviations, and median are computed for each time point and each group of the study. In addition, the pair-wise correlations between the five variables corresponding to the five time periods are computed, first for the placebo group and then for the progabide group.

Graphical techniques are employed to complement the descriptive statistics described in Chapter 2. For example, a scatter plot of the number of seizures versus the five time periods for each group complements the descriptive statistics for the mean and standard deviation at each time point for each group. A lowess curve reveals the trend for the two groups and demonstrates very little differences in the two, something that was implied by descriptive statistics. A matrix scatter plot of the five variables measuring number of seizures at each time point complements the estimated Pearson correlations.

Another graphical technique is the histogram, which helps to identify the distribution of the main response at each time point. This is important because the identification of the sample distribution of the response variable is essential for the proper choice of regression techniques that will be utilized to analyze the time profile of the various groups. For the epilepsy study, the distribution of the number of seizures at each time point appears as symmetric about its mean. Of course box plots and PP or QQ plots could also have been used to estimate the distribution of the repeated measures.

Three sources of variation are described for repeated measures and include the components: (1) between subject variation, (2) within subject biologic variation, and (3) measurement error. Each source of variation is defined and then illustrated for the epilepsy study. For example, to demonstrate interindividual variation, Figure 3.1 depicts the longitudinal responses for subjects 1 and 8.

It is shown how the use of prior information for the Bayesian analysis of repeated measures depends on previous studies similar to the one to be analyzed. Thus, it is essential that the statistician become an integral part of the study and be able to read the literature that includes the information that is necessary to set the prior information for the subsequent Bayesian analysis.

Summary statistics is one possible way to analyze a repeated measures study and consists of computing a summary statistics for each individual. This should be executed with caution because one is reducing the study to just one value of each subject, thus most likely losing valuable information about time–time variation in the response. For the epilepsy study, I chose summary statistics as the estimated least squares slope for each subject. That is, a linear regression of the number of seizures versus time was executed for each individual of the study. Assuming the slope are a random sample from the normal distribution for the placebo group, then for the treatment group, the mean slope of the placebo was to be compared to the average slope of the progabide group. See Davis[10] for additional information about summary statistics.

A Bayesian analysis is used to compare the slopes of the two groups. Assuming a normal population for the slopes of the placebo group, noninformative prior distributions were used for the normal population of placebo. That is, for the population mean μ_0 of the normal population, its prior distribution is assumed to be normal with a mean of zero and a precision of 0.0001 (= variance of 10,000). For the precision parameter τ_0 of the normal population, a gamma distribution with parameters 0.0001 and 0.0001 was assumed. This implies a prior mean of 1 for the precision. A similar prior analysis was assumed for the treatment group.

The main parameter of interest is the difference, diff = $\mu_0 - \mu_1$, where μ_1 is the average slope of the treatment group. A Bayesian analysis is executed with 55,000 observations for the simulation and the posterior distribution of difference determined with a 95% credible interval that included zero, implying very little difference in the mean slope between the two groups. Of course, this was also implied by the scatter plot and the descriptive statistics of the two groups.

The second major part of this chapter is the emphasis on the analysis of categorical repeated (clustered) data, and the basic ideas are presented. Basic ideas to be developed are descriptive statistics, graphical techniques, and a Bayesian analysis based on summary statistics, and the basic ideas are explained using an example involving the binary response to three drugs given to each of 46 subjects. This is an example of clustered data and is not formally speaking a longitudinal study. A subject responds in one of two ways to the administration of a drug, as favorably F or unfavorably U, and the experimental results are portrayed as the number of subjects who respond to the drugs in eight possible ways (profiles).

It is assumed that the number of subjects who respond in the eight categories follow a multinomial distribution, and this approach follows Davis[10] and will be used later to analyze repeated measures with categorical responses. Of interest is the number who respond favorably to each drug labeled as Y_1, Y_2, and Y_3, the number who respond favorably to drugs A, B, and C, respectively. It was shown how the variances, covariances, and correlations of Y_1, Y_2, and Y_3 can be estimated, which is useful for computing the descriptive statistics for the drug study.

A Bayesian analysis for the drug study is based on the binary response of each subject and is executed with **BUGS CODE 3.2** using 55,000 observations for the simulation. For a given drug, it is assumed that the binary response follows a Bernoulli distribution, where the probability of a favorable response is given, a priori, a uniform distribution. The posterior analysis estimates the probability of a favorable response for each of the three drugs and compares them, implying that there is very little difference between drugs A and B, but that the other two differences are indeed different. Of course, this is obvious from the descriptive statistics for the drug study.

Exercises

1. Using Table 3.1, verify the descriptive statistics of the epilepsy study given by Tables 3.3 and 3.4.

2. Using the information in Table 3.1, verify the estimated correlations of the epilepsy study given by Tables 3.5 and 3.6.

3. Verify Figure 3.1, a scatter plot for the data of Table 3.1 for the epilepsy study. This is a scatter plot of the number of seizures over five time periods comparing the two groups with a lowess curve.

4. Verify Figure 3.2, the matrix scatter plot of the number of seizures over five time periods for the placebo group.

5. Verify Figure 3.3, the matrix scatter plot of the number of seizures over five time periods for the progabide group.

6. Verify Figure 3.4, a histogram of the number of seizures for placebo at baseline.

7. Discuss Figure 3.5, which shows the variation between patients 1 and 8.

8. Refer to Figure 3.6 and discuss the measurement error of the two separate profiles of subjects 1 and 8.

9. Explain why the information from previous related experiments similar to the epilepsy study is needed in order to set the prior distributions for the Bayesian analysis.

10. For each individual, compute the least squares estimate of the slope using simple linear regression where the dependent variable is the number of seizures and the independent variable is time with five values.

11. Verify Figure 3.7, the histogram of the least squares slopes computed in problem 10.

12. Verify Table 3.8, the Bayesian posterior analysis for the epilepsy study. Use **BUGS CODE 3.1** with 55,000 observations for the simulation. Note that the least squares slopes for individuals of both groups appear in a list statement of **BUGS CODE 3.1**.

13. Based on Table 3.9, the information for the Alzheimer study, verify the scatter plot of Figure 3.8, the number of correctly remembered words at times 0, 1, 2, 4, and 6. Plot the linear regression line for each group. Is there a difference in the two groups? Explain why there is or there is not a difference.

14. Verify Table 3.11, which portrays the estimated Pearson correlation coefficients for the placebo group.

15. Verify Table 3.11, which portrays the estimated Pearson correlation coefficients for the lecithin group.

16. Using **BUGS CODE 3.1** and the information in Table 3.12, execute a Bayesian analysis with 55,000 observations for the simulation. Verify the posterior analysis of Table 3.13. Based on the difference in mu1−mu2, is there any difference in the mean slope between the two groups?

17. Verify Figure 3.8, the scatter plot of the number of correctly remembered words versus time at five time points. On the scatter plot, show the lowess curve for the two groups (placebo and treatment). Does the scatter plot show a difference in the two groups?

18. Verify Table 3.10, the descriptive statistics for the Alzheimer study.

19. Verify Table 3.12, the computed least squares slopes for each individual of the two groups.

20. Based on **BUGS CODE 3.1**, perform a Bayesian analysis with 55,000 observations for the simulation, and verify the analysis of Table 3.13. Note a list statement of **BUGS CODE 3.1** includes the data for the Alzheimer study.

21. In Table 3.21.1, there are the body weights in grams of guinea pigs at the end of weeks 1, 3, 4, 5, 6, and 7. The study is an investigation of diet supplement on growth, where at the beginning of week 1 the pigs were given a growth-inhibiting substance and at the start

TABLE 3.21.1

Weight in Grams of Guinea Pigs for Six Time Periods

	Time (weeks)					
Group	1	3	4	5	6	7
1	455	460	510	504	436	466
1	467	565	610	596	542	587
1	445	530	580	597	582	619
1	485	542	594	583	611	612
1	480	500	550	528	562	576
2	514	560	565	524	552	597
2	440	480	536	484	567	569
2	495	570	569	585	576	677
2	520	590	610	637	671	702
2	503	555	591	605	649	675
3	496	560	622	622	632	670
3	498	540	589	557	568	609
3	478	510	568	555	576	605
3	545	565	580	601	633	649
3	472	498	540	524	532	583

Source: Hand, D., and Crowder, M., *Practical Longitudinal Data Analysis*, Table A.12, Chapman & Hall, Boca Raton, FL, 1999.

of week 5 were given vitamin E. The three groups correspond to different doses (low, medium, high) of vitamin E. Thus, there are several research question, including

a. Does vitamin E have an effect on growth?

b. Is the effect different in the three groups?

c. Is there a difference in the overall profiles of the three?

Note this is not a randomized study.

d. Compute the descriptive statistics for the study, using the mean, standard deviation, and median at each time point by group. Your work should look similar to Table 3.21.2.

TABLE 3.21.2

Descriptive Statistics for Guinea Pig Study

Group	1	3	4	5	6	7
1	466.4	519.4	568.8	561.6	546.6	572
2	494.4	551	574.2	567.0	603.0	644.0
3	497.8	534.6	579.8	571.8	588.2	623.2

TABLE 3.21.3

Estimated Pearson Correlations for Group 1

	1	3	4	5	6	7
1	1	0.233	0.252	−0.049	0.426	0.244
3		1	0.998	0.933	0.726	0.826
4			1	0.939	0.760	0.849
5				1	0.727	0.850
6					1	0.965
7						1

e. Compute the Pearson correlation coefficient of the main response (body weight in grams) between the six time points by group. For the first group, your results should look similar to Table 3.21.3.

f. What is the histogram for the body weights at time 1 for the first group? Are the weights normally distributed?

22. What is the scatter plot of weight versus time for the three groups? Find the lowess curves for the three groups. Using the lowess curve, how do the three groups compare?

23. Compute the matrix scatter plot of the weight between the six time periods 1, 3, 4, 5, 6, and 7 weeks for each group separately. Explain how the scatter plot for the first group confirms the estimated correlation coefficients of Table 3.21.3. The scatter plot between time 3 and 4 weeks should appear to follow a straight line compared to the scatter plot between times 1 and 5 weeks. Explain why this is so.

24. What summary statistic for each guinea pig should be used in order to compare the three groups? Use the scatter plot of problem 22 to choose the summary statistic. Based on the scatter plot, I would fit a linear regression for all individuals at times 5, 6, and 7 weeks and compute the slope of the regression. It appears from the scatter plot of problem 22 that at time 5, the mean weight is approximately the same at this time and that the slope for group 2 (medium dose of vitamin E) is greater than that for the other two.

Describe the good points and bad points for the choice of slope as the summary statistic. Note that the baseline values for the three groups are not the same. I believe there is no obvious choice for a summary statistic.

25. For the drug study using Table 3.14, verify Table 3.15.

26. Assuming a multinomial distribution for the number of subjects that belong to the eight profiles of the drug study, what are the parameters of this multinomial distribution?

27. Show the variance of the number of subjects that belong to the ith profile of the drug study is given by (3.10).

28. Verify formula (3.11) for the covariance between the number of individuals who belong to categories i and j.

29. For the drug study, explain why $Y_1 = X_1 + X_2 + X_3 + X_4$ is the number of subjects who respond favorably to drug A.

30. Verify the descriptive statistics for the drug study given by Table 3.15.

31. Verify formula (3.15) for the covariance between Y_1 and Y_2.

32. Verify formulas (3.18) and (3.19) for the variances of Y_1 and Y_2, respectively.

33. Using formula (3.21), I computed a value of 0.6000 for the correlation between Y_1 and Y_2. Verify this value.

34. For the drug study, execute a Bayesian analysis using **BUGS CODE 3.2** with 55,000 observations for the simulation and verify the posterior analysis given by Table 3.19. What are the means of the posterior distributions for the probability of a favorable response to drugs A, B, and C. Based on the appropriate 95% credible interval, is there a difference in the probability of a favorable response?

    a. Between drugs A and B

    b. Between drugs A and C

    c. Between drugs B and C

35. Consider the following information about church attendance of elderly people of the Iowa 65+ Rural Health Study of Cornoni-Huntley et al.[12] The cohort was followed for six years, and every two years, subjects were asked about regular church attendance the previous two years with a simple yes or no answer. Of interest is the profile of church attendance for males and females and a comparison of attendance between the two (Table 3.35.4).

    Let X_i^F be the number of females who attend church on a regular basis for category i, and suppose these eight random variables follow a multinomial distribution with parameters $n = 1311$ and $(\theta_1^F, \theta_2^F, ..., \theta_8^F)$; thus, θ_i^F is the probability of a female belonging to category i, $i = 1,2,...,8$.

    Let X_i^M be the number of males who attend church on a regular basis for category i, and suppose the eight random variables follow a multinomial distribution with parameters $n = 662$ and $(\theta_1^M, \theta_2^M, ..., \theta_8^M)$; thus, θ_i^M is the probability of a male belonging to category i, $i = 1,2,...,8$.

    a. What is the obvious estimate that a female will attend church regularly for six years?

    b. What is the obvious estimate that a male will not attend church on a regular basis for the six-year period?

TABLE 3.35.4

Church Attendance Survey of the Iowa 65+ Rural Health Study

Category	Year 0	Year 3	Year 6	Count
Females				
1	Yes	Yes	Yes	904
2	Yes	Yes	No	88
3	Yes	No	Yes	25
4	Yes	No	No	51
5	No	Yes	Yes	33
6	No	Yes	No	22
7	No	No	Yes	30
8	No	No	No	158
Total				1311
Males				
1	Yes	Yes	Yes	391
2	Yes	Yes	No	36
3	Yes	No	Yes	12
4	Yes	No	No	26
5	No	Yes	Yes	15
6	No	Yes	No	21
7	No	No	Yes	18
8	No	No	No	143
Total				662

Source: Cornoni-Huntley, J. et al., *Established Populations for Epidemiologic Studies of the Elderly, Resource Data Book*, National Institutes of Health (NIH Publ. No. 86-2443), Bethesda, MD, 1986; Davis, C.S., *Statistical Methods for the Analysis of Repeated Measurements*, Table 7.6, Springer-Verlag, New York, p. 185, 2002.

c. What is the variance of the number of females who attend church on a regular basis for six years? That is, what is the variance of X_1^F? Use formula (3.10). What is your estimate of this variance based on the information in Table 3.35.4?

d. What is the covariance between X_1^M and X_8^M? Use formula (3.11). That is, what is the covariance between the number of males who answer yes to each of the three surveys and the number of males who respond no to each of the three surveys? What is your estimate of this covariance? Use Table 3.35.4.

36. For females, let

$$Y_1^F = X_1^F + X_2^F + X_3^F + X_4^F \tag{3.36.1}$$

What is Y_1^F? Is it the number of females who answer yes at the first survey at time 0? What is the actual value of Y_1^F? Refer to Table 3.35.4.

37. Suppose

$$Y_2^F = X_1^F + X_2^F + X_5^F + X_6^F \tag{3.37.2}$$

Then what is Y_2^F? Is it the number of females who answer yes to the second survey at time 3? What is the actual value of Y_2^F? See Table 3.35.4.

38. Let

$$Y_3^F = X_1^F + X_2^F + X_5^F + X_6^F \tag{3.38.3}$$

then what is Y_3^F?

39. Express the $\mathrm{Var}\left(Y_1^F\right)$ in terms of the variances and covariances of X_1^F, X_2^F, X_3^F, and X_4^F. See formula (3.18).

40. Express $\mathrm{Cov}\left(Y_1^F, Y_2^F\right)$ in terms of the variances and covariances of X_1^F, X_2^F, X_3^F, X_4^F, X_5^F, X_6^F, X_7^F, and X_8^F. See formula (3.15).

41. Suppose we want to compare the males with the females with regard to church attendance. Note that as the distribution of the X_i^F and X_i^M ($i = 1,2,...,8$) are multinomial and the random variables are independent according to gender, the marginal distribution of the Y_i^F and Y_i^M ($i = 1,2,3$) are independent binomial and independent according to gender.

To be more specific, Y_i^F is distributed binomial with parameters θ_i^F and n_f, where n_f is the number of females and θ_i^F is the probability of a female responded yes to survey i, where $i = 1,2,3$ corresponding to years 0, 3, and 6. In a similar fashion, Y_i^M is distributed binomial with parameters θ_i^M and n_m, where n_m is the number of males and θ_i^M is the probability of a male responded yes to survey i, where $i = 1,2,3$ corresponding to years 0, 3, and 6. Note we have observed values for the Y_i^F and Y_i^M and for n_f and n_m (see Table 3.35.4).

BUGS CODE 3.3 is the foundation for the Bayesian analysis of the church attendance study, and the analysis is executed with 55,000 observations, with 5,000 initial values and a refresh of 100. The primary focus is the three differences

$$d_i = \theta_i^F - \theta_i^M \tag{3.41.4}$$

for $i = 1,2,3$.

Note that the code closely follows the notation of the previous paragraph (see Table 3.41.5).

BUGS CODE 3.3

```
model;
{
        # binomial distributions for observed
                y1f~dbin(theta1f,nf)
                y2f~dbin(theta2f,nf)
                y3f~dbin(theta3f,nf)

                y1m~dbin(theta1m,nm)
                y2m~dbin(theta2m,nm)
                y3m~dbin(theta3m,nm)

        # differences between males and females
                d1<-theta1f-theta1m
                d2<-theta2f-theta2m
                d3<-theta3f-theta3m

        # prior distributions
                theta1f~dbeta(1,1)
                theta2f~dbeta(1,1)
                theta3f~dbeta(1,1)

                theta1m~dbeta(1,1)
                theta2m~dbeta(1,1)
                theta3m~dbeta(1,1)
}
# data for Cornoni-Huntley Study
list(y1f = 1068,y2f = 1047, y3f = 992,y1m = 465,y2m =
463,y3m = 436,nf = 1311,nm = 662)
# initial values
list(theta1f = .5,theta2f = .5,theta3f = .5,theta1m =
.5,theta2m = .5,theta3m = .5)
```

a. Is there a difference between females and males in the probability of a yes response at the first survey at time 0? Explain your answer.

b. What is the posterior median for the probability of a yes response for males at time 3 years (the second survey)?

c. Plot the posterior density of d_3, the difference between males and females in the probability of a yes response at the third survey.

d. What is the 95% credible interval for the posterior distribution of the difference between males and females in the probability of a yes response at the second survey?

TABLE 3.41.5

Bayesian Analysis for Church Attendance

Parameter	Mean	SD	Error	2½	Median	97½
d_1	0.1124	0.0208	0.0000925	0.0717	0.1123	0.1536
d_2	0.0994	0.021	0.0000955	0.0590	0.0992	0.1408
d_3	0.0981	0.0218	0.00001	0.0555	0.098	0.141
θ_1^F	0.8142	0.0107	0.0000475	0.7926	0.8143	0.8349
θ_1^M	0.7018	0.0177	0.0000793	0.6663	0.7021	0.7357
θ_2^F	0.7983	0.0110	0.0000465	0.776	0.7984	0.8194
θ_2^M	0.6988	0.0178	0.0000809	0.6634	0.699	0.7331
θ_3^F	0.7563	0.0118	0.0000579	0.7327	0.7565	0.7791
θ_3^M	0.6582	0.0187	0.0000831	0.6222	0.6583	0.6938

e. What is your overall assessment in the difference between males and females in their response to church attendance? Give a complete description.

f. The simulation error for estimating θ_3^F with the posterior mean is 0.0000579. Explain what this means.

42. Plot the proportion of yes responses to church attendance of males and females at times 0, 3, and 6 weeks. From this plot, explain any differences between males and females.

4

Linear Models for Repeated Measures and Bayesian Inference

4.1 Introduction

This chapter presents some introductory material, namely, some notation for the linear model, some preliminary information about modeling the mean profile of individuals over time and modeling the structure of the covariance matrix, reviewing some historical ways to analyze repeated measures, and last, a detailed presentation of the fundamentals of Bayesian inference. With regard to Bayesian inference, prior and posterior information is discussed as are Markov chain Monte Carlo (MCMC) techniques for the simulation of the posterior distribution of the model parameters, the parameters that model the mean profile and covariance structure of repeated measures.

The linear model approach to repeated measures is sufficiently general to encompass some of the historical approaches to repeated measures analysis, including the analysis of variance (ANOVA) and the multivariate ANOVA (MANOVA).

4.2 Notation for Linear Models

Consider a repeated measures study and suppose over n_i times, the response vector for individual i is

$$Y_i = \left(Y_{i1}, Y_{i2}, ..., Y_{in_i}\right)' \tag{4.1}$$

where:
   Y_{ij} is the response at time j, where $j = 1,2,...,n_i$

Suppose the jth occasion for subject i is indexed by t_{ij}, which allows for different occasions for different individuals, whereas the n_i allows for a different number of occasions for different individuals. There is a very special case, namely, that the occasions are the same for all subjects, and the number of occasions is also the same for all individuals. Of course it is often the case that the latter situation is not true, that is the occasions can differ as well as the number of occasions. In addition, it is often the case that there are missing observations due to dropout and other reasons. When the number of measurements is the same for all subjects, and there is no missing data, let $n_i = n$ for $i = 1,2,...,N$.

Of importance for repeated measures is the effect of various covariates on the main response, thus let the vector of p covariates be denoted by the p by 1 vector

$$X_{ij} = (X_{ij1}, X_{ij2}, ..., X_{ijp})'$$ (4.2)

for $i = 1,2,...,N$ and $j = 1,2,...,n_i$

In matrix form, let the covariate values for the ith individual be denoted

$$X_i = \begin{pmatrix} X_{i11}, X_{i12}, ..., X_{i1p} \\ X_{i21}, X_{i22}, ..., X_{i2p} \\ \vdots \\ X_{in_i1}, X_{in_i2}, ..., X_{in\,p} \end{pmatrix}$$ (4.3)

where first row is the 1 by p vector of values for the p covariates at the first occasion, while the last row is the vector of p covariate values at the last occasion n_i for individual i. Note that the jth occasion for subject i is the time t_{ij} and the set of occasions for one individual need not be the same as that for another individual and the number of occasions can also vary from individual to individual.

If

$$\beta = \begin{pmatrix} \beta_1 \\ \beta_2 \\ \vdots \\ \beta_p \end{pmatrix}$$ (4.4)

is a vector of unknown regression coefficients, then the linear repeated measures model for individual i is defined as

$$Y_i = X_i\beta + \varepsilon_i$$ (4.5)

where:

Y_i is the $n_i x1$ vector of observations for individual i

ε_i is a $n_i x1$ of residuals

Linear implies that the model is linear in the regression parameters $\beta_1, \beta_2, ..., \beta_p$. It is also assumed that the residuals of an expectation of zero, thus,

$$E(Y_i) = X_i\beta \tag{4.6}$$

where the expectation is taken over all hypothetical repetitions of the study.

Note another way to express the linear model (4.5) is as

$$Y_{ij} = \sum_{k=1}^{k=p} \beta_k X_{ijk} + \varepsilon_{ij} \tag{4.7}$$

or in turn, (4.7) can be expressed as

$$Y_{i1} = \beta_1 X_{i11} + \beta_2 X_{i12} + \cdots + \beta_p X_{i1p} + \varepsilon_{i1}$$

$$Y_{i2} = \beta_1 X_{i21} + \beta_2 X_{i22} + \cdots + \beta_p X_{i2p} + \varepsilon_{i2} \tag{4.8}$$

$$\vdots$$

$$Y_{in_i} = \beta_1 X_{in_i1} + \beta_2 X_{in_i2} + \cdots + \beta_p X_{in_ip} + \varepsilon_{in_i}$$

for $i = 1,2,...,N$. By convention, the first covariate X_{ij1}, the coefficient of β_1, is usually given the value 1 for all individuals.

It is important to know the distribution of the quantitative response Y for repeated measures. For the moment it is assumed that the distribution of Y at each time point is approximately normal with density

$$f(y) \propto \tau^{1/2} \exp\left(\frac{\tau}{2}\right)(y-\mu)^2 \tag{4.9}$$

where:

$y \in R$, mean μ, and positive precision τ are unknown parameters

If the density is plotted, one will observe the density is symmetric about the mean μ and that as τ increases, the density becomes more concentrated about the mean.

One would expect the mean and precision to vary over the n_i time points for individual i, and more generally it is assumed that the joint distribution of $Y_{i1}, Y_{i2}, ..., Y_{in}$ has a n_i-dimensional normal distribution with density

$$f(y) = (2\pi)^{-n_i/2} |\tau|^{1/2} \exp(y-\mu)'\tau\left(\frac{y-\mu}{2}\right) \tag{4.10}$$

where:

y is a $n_i \times 1$ vector of real numbers

the mean vector μ and precision matrix $\tau(n_i \times n_i)$ are unknown parameters

Note the similarity between the density (4.9) for the one-dimensional case and the density (4.10) for the multivariate case. The precision matrix is the inverse of the variance–covariance matrix, that is

$$\tau = \Omega^{-1} \tag{4.11}$$

where Ω is the $(n_i \times n_i)$ variance–covariance matrix of Y, that is

$$E(Y - \mu)(Y - \mu)' = \Omega \tag{4.12}$$

This is a symmetric matrix whose diagonal elements are the variances of the repeated measures at the n_i occasions of individual i and the off-diagonal elements are the covariances between the responses at the n_i time points. Of course, the mean vector μ and precision matrix τ can vary from individual to individual.

To demonstrate the various ideas presented in this section, consider the following example of Potthoff and Roy,[1] where the response is the distance from the pituitary to the pteryomaxillary fissure. The first column of Table 4.1 shows gender (1 = female, 2 = male) and the other four columns are the distances (in millimeters) measured at ages 8, 10, 12, and 14 years.

TABLE 4.1

Distances from the Pituitary to the Pteryomaxillary Fissure

Gender	8 Years	10 Years	12 Years	14 Years	Subject
1	21	20	21.5	23.0	1
1	21	21.5	24	25.5	2
1	20.5	24	24.5	26	3
1	23.5	24.5	25	26.5	4
1	21.5	23	22.5	23.5	5
1	20	21	21	22.5	6
1	21.5	22.5	23	25.0	7
1	23	23	23.5	24	8
1	20	21	22	21.5	9
1	16.5	19	19	19.5	10
1	24.5	25	28	28	11
2	26	25	29	31	12
2	21.5	22.5	23	26.5	13
2	23	22.5	24	27.5	14
2	25.5	27.5	26.5	27	15
2	20	23.5	22.5	26	16
2	24.5	25.5	27	28.5	17
2	22	22	24.5	26.5	18

(Continued)

TABLE 4.1 (*Continued*)

Distances from the Pituitary to the Pteryomaxillary Fissure

Gender	8 Years	10 Years	12 Years	14 Years	Subject
2	24	21.5	24.5	25.5	19
2	23	20.5	31	26	20
2	27.5	28	31	31.5	21
2	23	23	23.5	25	22
2	21.5	23.5	24	28	23
2	17	24.5	26	29.5	24
2	22.5	25.5	25.5	26	25
2	23	24.5	26	30	26
2	22	21.5	23.5	25	27

Source: Potthoff, R.F., and Roy, S.N., *Biometrika*, 51, 313–326, 1964; Hand, D., and Crowder, M., *Practical Longitudinal Data Analysis*, Table A.8, Chapman & Hall, Boca Raton, FL, 1996.

Our first objective is to demonstrate the notation introduced earlier. Consider subject 1, then the responses over the four occasions is given by the 4×1 vector

$$Y_1 = (21, 20, 21.5, 23.0)' \tag{4.13}$$

And note the subjects are divided by gender; thus, the vector of observations can be expressed as

$$Y_{ij} = \beta_1 X_{ij1} + \beta_2 X_{ij2} + \beta_3 X_{ij3} + \varepsilon_{ij} \tag{4.14}$$

where:

$X_{ij1} = 1$ for all i and j

$X_{ij2} = (t_j - 8)$ is the age for occasion j (minus the beginning age—8 years)

$X_{ij3} = (t_j - 8)$ group$_i$, where group$_i = 1$ for females and group$_i = 0$ for males

Thus, for males

$$E(Y_{ij} \mid X_{ij}) = \beta_1 + \beta_2(t_j - 8) \tag{4.15}$$

and β_1 is the average distance at age 8.

On the other hand, for females

$$E(Y_{ij} \mid X_{ij}) = \beta_1 + (\beta_2 + \beta_3)(t_j - 8) \tag{4.16}$$

Thus, when $t_j = 8$, β_1 is the average distance for females, the same as that for males, and one might want to use a different parameterization. One might want to account for different average baseline distance values. The model expressed by (4.15) and (4.16) assumes that time has a linear relationship with

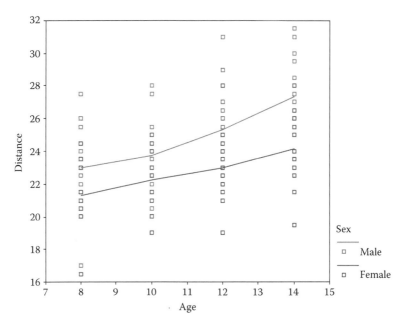

FIGURE 4.1
Scatterplot of distance from the pituitary to the pteryomaxillary fissure versus age by gender.

the average distance. Note that if $\beta_3 = 0$, then the two groups are similar in regard to the average distance over the four ages.

See Fitzmaurice, Laird, and Ware[3] for more information.

A scatter plot should reveal how realistic our assumptions are at this point; thus, consider Figure 4.1. Here, males appear to have a larger average distance (from the pituitary to the pteryomaxillary fissure) than females at baseline (age = 8), and the lowess plots show a linear relationship for females, but not for males. The plot shows a slight curvature for males compared to females; however, a linear relationship would not be misleading.

Our second goal is to determine the distribution of the main response: the distance from the pituitary to the pteryomaxillary fissure. This can be done in a number of ways: (1) the histogram, (2) a P-P or Q-Q plot, and (3) a box plot. I used all three methods, and a normal distribution is reasonable for the main response of this study. Of course, one should also compute the descriptive statistics, namely, for each group (males and females), compute the mean and standard deviation of the distance measure at each age group.

Table 4.2 corroborates the scatter plot of Figure 4.1 and implies that the distribution is more or less symmetric about the mean, thus assuming a normal distribution of the distance from the pituitary to the pteryomaxillary fissure is not unreasonable. It is also seen that the average value of the response at baseline is similar for males and females and so is the model specified by (4.15) and (4.16).

TABLE 4.2

Descriptive Statistics for Potthoff and Roy Study

Gender	Age	Mean	Median	SD
Female	8	21.18	21	2.124
Female	10	22.22	22.5	1.902
Female	12	23.09	23	2.364
Female	14	24.09	24	2.437
Male	8	22.87	23	2.452
Male	10	23.81	23.5	2.136
Male	12	25.71	25	2.651
Male	14	27.46	26.75	2.085

Source: Potthoff, R.F., and Roy, S.N., *Biometrika*, 51, 313–326, 1964.

4.3 Modeling the Mean

When modeling the mean response in a repeated measures study, one needs to refer to scatter plot of the main response versus time as well as the descriptive statistics of the outcome. For example, for the Potthoff and Roy[1] study, what does Figure 4.1 imply about the mean response over the four time periods? In particular, is the model (4.14) a reasonable tentative model to consider for representing the mean response? Of course, one should also refer to the Table 4.2 of descriptive statistics in choosing a model for the mean response. Recall that Equation 4.15 is our tentative choice to model the mean response for females, while Equation 4.16 gives the mean response for males. The model shows that at $t_j = 8$, the mean response is β_1 at baseline for both groups, that the average response per year increases at a rate of β_2 from ages 8 to 14 and at a rate of $\beta_2 + \beta_3$ for males, thus when β_3 equals to 0, the mean response is the same for both groups. The latter situation will be considered when making Bayesian inferences with the model.

Consider another representation of the mean response as follows: For females, consider

$$E(Y_{ij} \mid X_{ij}) = \beta_1 + \beta_2 t_j \tag{4.17}$$

and for males

$$E(Y_{ij} \mid X_{ij}) = \beta_3 + (\beta_2 + \beta_4)t_j \tag{4.18}$$

and the mean response is the same for both males and females when

$$\beta_1 = \beta_3 \text{ and } \beta_4 = 0$$

Thus, different parameterization can lead to model that investigate the research question from different viewpoints.

4.4 Modeling the Covariance Matrix

When using regression techniques to model repeated measures, one must consider the covariance of the repeated measures between the various times. This is because the various estimation techniques used for estimating the mean are so-called weighted estimators, that is, the mean response estimators depend on the estimated covariance matrix.

Consider the Potthoff and Roy[1] study; how does one chose a covariance structure for the model? One is guided by the choice of a correlation structure by basing the choice on the estimated correlations that appear in Table 4.3.

Thus, for females, the estimated Pearson correlation of the distance (from the pituitary to the pteryomaxillary fissure) is 0.830 between ages 8 and 10, is 0.862 between ages 8 and 12, and is 0.841 between ages 8 and 14.

For males, the estimated correlation of distance (from the pituitary to the pteryomaxillary fissure) is 0.437 between ages 8 and 10, is 0.558 between ages 8 and 12, and is 0.315 between ages 8 and 14 (Table 4.4). One notices the relative large correlation between the various ages for females compared to males. These estimated correlations will play an important rule when using Bayesian methods to estimate the mean response. Based on these estimated correlations, what structure should be placed on the covariance matrix?

Scatter plots of the distance from the pituitary to the pteryomaxillary fissure versus age reveal the association between these two measurements and are presented in Figure 4.2 for females and Figure 4.3 for males.

TABLE 4.3

Estimated Pearson Correlations of Distance between Ages for Females

Ages	8	10	12	14
8	1	0.830	0.862	0.841
10		1	0.895	0.879
12			1	0.948
14				1

TABLE 4.4

Estimated Pearson Correlations of Distance between Ages for Males

Ages	8	10	12	14
8	1	0.437	0.558	0.315
10		1	0.387	0.631
12			1	0.586
14				1

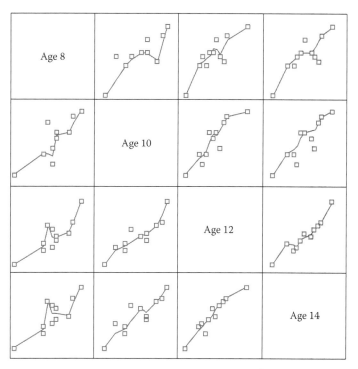

FIGURE 4.2
Matrix plot of distance versus age for females at four ages.

One should compare Figure 4.3 with Table 4.4 to see if each corroborates the other. One should also compare Figure 4.2 with Table 4.3 to see if one corroborates the other. From Figures 4.3 and 4.4, it appears to me as if the association between distance and age is stronger for females compared to males. Of course this is also implied by Tables 4.3 and 4.4.

4.5 Historical Approaches

This section will introduce two historical approaches to the analysis of repeated measures: (1) the univariate ANOVA and (2) MANOVA.

4.5.1 Univariate Analysis of Variance

I will first present the non-Bayesian approach for using ANOVA to analyze repeated measures.

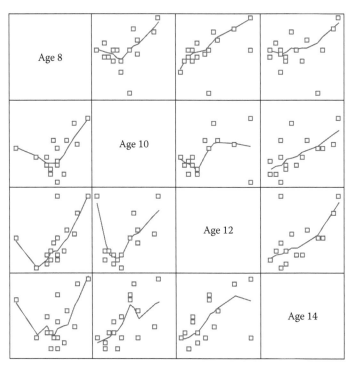

FIGURE 4.3
Matrix plot of distance versus age for males at four ages.

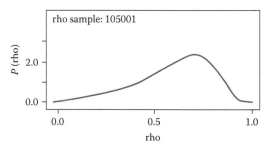

FIGURE 4.4
Posterior density of ρ.

In this approach, the various time periods serve as groups, and inferences are concerned with comparing those groups, thus the less complicated or involved approach is the one-way ANOVA, where the between and within sum of squares are computed. For the case where there are N individuals and n time points, and the time points are the same for all individuals, the model is

$$y_{ij} = \mu + \theta_j + \varepsilon_{ij} \tag{4.19}$$

where:

y_{ij} is the response for individual j at occasion j, where $i = 1,2,...,N$ and $j = 1,2,...,n$

μ is the overall mean of the responses

θ_j is the effect on the response at time j

ε_{ij} is the error (the difference between the response and its mean value)

The analysis consists of testing for differences in the mean values of the response among n groups, followed by multiple comparisons testing for pair-wise differences in the mean values of the main response.

Note the model does not take into account explicitly the variation between individuals per se. This will be considered later in the next model to be considered for repeated measures.

Consider the Potthoff and Roy[1] study where the main response is the distance from the pituitary to the pteryomaxillary fissure for males and females reported in Table 4.1. The one-way ANOVA for females is given by Table 4.5.

The ANOVA implies that there is a significant difference in the mean distance between the four ages (8, 10, 12, 14 years), thus a multiple comparison procedure is used to perform pair-wise differences between the four ages. Using the Scheffe multiple comparison procedure, see Miller (p. 144),[4] it can be shown that the average mean distance at time 8 years is significantly different than the mean distance at 14 years. This is the only significant difference detected by Scheffe. This in turn implies there are two sets of homogeneous mean differences, namely, those at 8, 10, and 12 years and those at 10, 12, and 14 years.

As for males, the one-way ANOVA is given by Table 4.6.

TABLE 4.5

ANOVA for Potthoff and Roy Study for Females

Source	Sum of Squares	df	Mean Square	Sig.	F
Between ages	50.563	3	16.884	0.026	3.435
Within groups	196.636	40	4.916		
Total	247.290	43			

TABLE 4.6

ANOVA for Potthoff and Roy Study for Males

Source	Sum of Squares	df	Mean Square	Sig.	F
Between ages	200.531	3	66.844	0.000	12.175
Within groups	329.406	60	5.49		
Total	529.938	63			

As with females, there is a significant difference in the average distance between the four time periods (8, 10, 12, and 14 years) for males, and the Scheffe multiple comparison procedure (with alpha = 0.05) detects the following differences: The average distance at age 8 is significantly different than the average distance at times 12 and 14 years. The average distance at time 10 years is significantly different than that at 14 years, and one can conclude the following homogenous average distances: (1) at 8 and 10 years, (2) at 10 and 12 years, and (3) at 12 and 14 years.

The one-way ANOVA is not the only way to analyze this study; thus we will consider a more appropriate model:

$$y_{ijk} = \mu + a_i + b_j + \varepsilon_{ijk} \tag{4.20}$$

where:

y_{ijk} is the distance for subject i at time j for group k, where $j = 1,2,3,4$, and $k = 1,2$

μ is the overall mean

a_i is the effect of individual i

b_j is the effect of age j

ε_{ijk} is the error

The so-called repeated measures ANOVA is presented as follows. I first used a model with compound symmetry for the covariance of distance between the four time periods and included the age by sex interaction, where sex and age are considered fixed effects. The ANOVA numbers appear as in Table 4.7.

As the sex by age interaction has a p value of 0.062, one is inclined to believe that there is no difference in the average distance between males and females. However, one should proceed with caution because the p value is "small" but larger than the nominal 0.05.

Because of the assumption of compound symmetry for the covariance matrix, the estimated common correlation between the four ages is 0.6292 (0.0888), and Table 4.8 provides the estimated variances.

Additional information about the analysis is provided by the estimated marginal means of distance at the four ages given in Table 4.9.

TABLE 4.7

AVOVA Repeated Measures for Potthoff and Roy Study

Source of Variation	Degrees of Freedom	F	p value
Intercept	1	3911.796	0.000
Age	3	36.712	0.000
Sex	1	24.441	0.005
Sex × age	3	55.952	0.062

Source: Potthoff, R.F., and Roy, S.N., *Biometrika*, 51, 313–326, 1964.

TABLE 4.8

Estimated Variance of Distance at the Four Ages

Age	Variance Estimate	Standard Error
8	5.670	1.618
10	4.221	1.1900
12	6.314	1.762
14	4.835	1.345

TABLE 4.9

Estimated Means of Distance at the Four Ages

Age	Variance Estimate	Standard Error
8	22.628	0.466
10	23.020	0.402
12	24.405	0.492
14	25.786	0.431

Based on the estimated means and standard errors, one can show all pair-wise differences are significant at the 0.05 level.

The preceding analyses are non-Bayesian. What is a Bayesian analysis? Consider the model for females as

$$y_{ij} = \theta + \alpha_i + \beta_j + \varepsilon_{ij} \tag{4.21}$$

where:
θ is the overall mean of the response
α_i is the effect of the ith individual
β_j is the effect of the jth time period
ε_{ij} is the error

In addition, the $\alpha_i \sim n(0, \tau_\alpha)$, the $\beta_j \sim n(0, \tau_\beta)$, and the $\varepsilon_{ij} \sim n(0, \tau)$ and all random variables are independent. This is a so-called random model, where the effects are given normal distributions, which induces the intraclass correlation

$$\rho = \frac{\sigma_\alpha^2}{\sigma_\alpha^2 + \sigma_\beta^2 + \sigma^2} \tag{4.22}$$

which is the common correlation of the response between the various time periods. Note that

$$\sigma_\alpha^2 = \frac{1}{\tau_\alpha} \tag{4.23}$$

$$\sigma_\beta^2 = \frac{1}{\tau_\beta} \tag{4.24}$$

and

$$\sigma^2 = \frac{1}{\tau} \tag{4.25}$$

are the corresponding variance components, namely the reciprocals of the precision components.

BUGS CODE 4.1 will execute the Bayesian analysis for the Potthoff and Roy[1] study. The program statements are very similar to formulas (4.21) through (4.25), and the main parameter is the intraclass correlation coefficient ρ of the distance response between the four ages 8, 10, 12, and 14 years for females. Prior information for the parameters is noninformative, namely:

$$\theta \sim \text{normal}(0, 0.0001)$$

$$\alpha_i \sim \text{normal}(0, \tau_\alpha)$$

$$\beta_j \sim \text{normal}(0, \tau_\beta)$$

$$\tau_\alpha \sim \text{gamma}(0.0001, 0.0001) \tag{4.26}$$

$$\tau_\beta \sim \text{gamma}(0.0001, 0.0001)$$

and

$$\tau \sim \text{gamma}(0.0001, 0.0001)$$

For θ, the prior distribution is normal with mean 0 and precision 0.0001 and variance 10,000. On the other hand, the prior distribution for τ_α is gamma with parameters 0.0001 and 0.0001, which has a mean of 1 and a variance of 10,000. The large variances imply noninformative information for those parameters. For additional information about Bayesian inference using compound symmetry for the covariance matrix, see Broemeling (p. 204).[5]

Note the list statement contains the distance measurements for females found in Table 4.10.

Based on Table 4.10 and **BUGS CODE 4.1**, a Bayesian analysis is executed with 100,000 observations for the simulation, initially with 5,000 observations and a refresh of 100.

Our main interest is in the correlation ρ, the intraclass correlation coefficient, and one sees its posterior density has a mean of 0.5985 (0.1894) and median of 0.6313, a skewness confirmed by the plot in Figure 4.4. Recall that the conventional analysis, assuming compound symmetry, gives, via maximum likelihood, an estimate of 0.6292 (0.0888). Thus, the conventional and Bayesian estimates are quite similar. A use of noninformative priors (4.26) explains why the two approaches give similar estimates for ρ.

TABLE 4.10

Bayesian Analysis of the Potthoff and Roy Study for Females: Compound Symmetry

Parameter	Mean	SD	Error	2½	Median	97½
ρ	0.5985	0.1894	0.00323	0.1542	0.6313	0.879
σ^2	0.6973	0.1944	0.000856	0.4152	0.6651	1.163
σ_α^2	5.318	3.184	0.02189	1.964	4.524	13.38
σ_β^2	4.166	19.74	0.2411	0.4167	1.843	21.04
θ	22.57	1.131	0.04982	20.27	22.6	24.74

Source: Potthoff, R.F., and Roy, S.N., *Biometrika*, 51, 313–326, 1964.

BUGS CODE 4.1

```
model;
{
for(i in 1:n){for(j in 1:p){y[i,j]~dnorm(mu[i,j],tau)}}
for(i in 1:n){for(j in 1:p)
{mu[i,j]<- theta+alpha[i]+beta[j]}}

for(i in 1:n){alpha[i]~dnorm(0,tau.alpha)}
for(j in 1:p){beta[j]~dnorm(0,tau.beta)}
for(i in 1:n){for(j in 1:p){z[i,j]~dnorm(mu[i,j],tau)}}
# prior distributions
        theta~dnorm(0,.0001)
tau.alpha~dgamma(.0001,.0001)
tau.beta~dgamma(.0001,.0001)
tau~dgamma(.0001,.0001)
sigma.alpha<-1/tau.alpha
sigma.beta<-1/tau.beta
sigma<-1/tau

rho<-sigma.alpha/(sigma.alpha+sigma.beta+sigma)
}
# example with females

# Potthoff and Roy[1]
list(n = 11, p = 4, y = structure(.Data =
c(21,20,21.5,23,
  21,21.5,24,25.5,
  20.5,24,24.5,26,
  23.5,24.5,25,26.5,
  21.5,23,22.5,23.5,
  20,21,21,22.5,
  21.5,22.5,23,25,
  23,23,23.5,24,
  20,21,22,21.5,
```

```
   16.5,19,19,19.5,
   24.5,25,28,28),.Dim = c(11,4)))

# initial values
list(alpha = c(0,0,0,0,0,0,0,0,0,0,0),
beta = c(0,0,0,0), tau.alpha = 1, tau.beta = 1, tau = 1,
theta = 0)
```

4.5.2 Multivariate Analysis of Variance

Recall formulas (4.10) and (4.11) of the density of a multivariate normal, where one would expect the mean and precision to vary over the n_i time points for individual i. It is assumed that the joint distribution of $Y_{i1}, Y_{i2}, ..., Y_{in}$ has a n_i-dimensional normal distribution with density

$$f(y) = (2\pi)^{-n_i/2} |\tau|^{1/2} \exp(y-\mu)' \tau(y-\mu)/2 \qquad (4.10)$$

where y is a $n_i \times 1$ vector of real numbers with mean vector μ and precision matrix τ $(n_i \times n_i)$ are unknown parameters. Note the similarity between the density (4.9) for the one-dimensional case and (4.10) the multivariate case. The precision matrix is the inverse of the variance–covariance matrix, that is

$$\tau = \Omega^{-1} \qquad (4.11)$$

where Ω is the $(n_i \times n_i)$ variance–covariance matrix of Y, that is

$$E(Y-\mu)(Y-\mu)' = \Omega \qquad (4.12)$$

This is a symmetric matrix whose diagonal elements are the variances of the repeated measures at the n_i occasions of individual i and the off-diagonal elements are the covariances between the responses at the n_i time points. Of course, the mean vector μ and precision matrix τ can vary from individual to individual.

The MANOVA assumes a multivariate normal distribution for the n_i repeated measures of individual i with an unknown covariance matrix Ω and a mean vector that depends on within and between subject factors. For example, for the Potthoff and Roy[1] distance (from the pituitary to the pteryomaxillary fissure) study, there is one within subject factor, namely, the age at which the distance is measured and one between subject factor, namely, the gender of a patient.

An SPSS statistics for an MANOVA for the distance study revealed that there is no significant difference between males and females, thus a repeated measures study is conducted to examine the mean distance over the four ages. The analysis assumes an unstructured covariance matrix, and recall the estimated covariance matrixes reported in Tables 4.3 and 4.4. When using the univariate ANOVA, a structured compound symmetry was assumed for the covariance matrix of the distance measured over the four ages.

TABLE 4.11

P Values Comparing Mean Distance between Ages for Females

Age	8	10	12	14
8		0.094	0.002	0.000
10			0.142	0.002
12				0.010

TABLE 4.12

P Values Comparing Mean Distance between Ages for Males

Age	8	10	12	14
8		0.879	0.002	0.000
10			0.074	0.000
12				0.039

The MANOVA computes a pair-wise comparison of the average distance between the four ages for both sexes (see Tables 4.11 and 4.12).

The *P* values are computed using a Bonferonni (see Miller, p. 75)[4] correction for multiple comparisons, and one sees that for females, there is a significant difference between the average distance at age 8 and that of the other three ages (10, 12, and 14 years), and that the average distance at age 10 is significantly different than that at age 14. Last, for females, the average distance at age 12 is significantly different than that at age 14. *P* values <0.05 are designated as significant.

For males, the average distance at age 8 is significantly different than that at ages 12 and 14, while the average distance at age 10 is significantly different than that at age 14, and last, for age 12, the average age is significantly different than that at age 14. The reader should refer to Table 4.2 and Figure 4.1 when interpreting Tables 4.11 and 4.12.

The Bayesian analysis uses the model where for individual i, the observations $Y_{i1}, Y_{i2}, ..., Y_{in_i}$ has a n_i dimensional normal distribution with density (see Equation 4.10), where y_i is a n_i by 1 vector of repeated measures $y_{i1}, y_{i2}, ..., y_{in}$ of real numbers, μ is the n_i by 1 mean vector and the variance–covariance matrix of Y_i is τ^{-1}, and τ is referred to as the precision matrix. For the Bayesian analysis, the precision matrix is assumed to have a Wishart distribution with parameters R and 4 for the Potthoff and Roy[1] study.

The density of τ is Wishart with parameters R and k, namely

$$f(\tau) \propto |R|^{k/2} |\tau|^{(k-p-1)/2} \exp\left(\frac{-1}{2}\right) Tr(R\tau) \qquad (4.27)$$

where:

τ is a pth order positive definite symmetric matrix

k is the number of individuals

p is the number of time points

Tr is the trace operation of the matrix $R\tau$

For this study and for males, one has a random sample of 16 males from a four-dimensional multivariate normal with mean vector and 4 by 4 precision matrix τ. The mean vector for each male is assumed to be linear, namely

$$\mu_{ij} = \beta_0 + \beta_1 age_j \qquad (4.28)$$

for $i = 1,2,\ldots,16$ and $j = 1,2,3,4$. Note that the mean vector (4.27) for individual i is the same for all i. **BUGS CODE 4.2** closely follows the notation used in formulas (4.10) and (4.12), and the four ages are 8, 10, 12, and 14 years.

BUGS CODE 4.2

```
        model
        {

# prior distributions of the beta coefficients

              beta0 ~ dnorm(0.0, 0.001)
              beta1 ~ dnorm(0.0, 0.001)
              for (i in 1:N) {
                      Y[i, 1:M] ~ dmnorm(mu[], Omega[,])
                      Z[i, 1:M] ~ dmnorm(mu[], Omega[,])

              }
              for(j in 1:M) {
                      mu[j] <- beta0 + beta1 * age[j]
              }

# prior distribution of the precision matrix Omega

              Omega[1 : M, 1 : M] ~ dwish(R[,],4)
              Sigma[1 : M, 1 : M] <- inverse(Omega[,])

        }

# Potthoff and Roy[1]
list(M = 4, N = 16, Y = structure(
        .Data = c(26,25,29,31,
                  21.5,22.5,23,26.5,
                  23,22.5,24,27.5,
                  25.5,27.5,26.5,27,
                  20,23.5,22.5,26,
                  24.5,25.5,27,28.5,
                  22,22,24.5,26.5,
                  24,21.5,24.5,25.5,
                  23,20.5,31,26,
                  27.5,28,31,31.5,
```

```
                23,23,23.5,25,
                21.5,23.5,24,28,
                17,24.5,26,29.5,
                22.5,25.5,25.5,26,
                23,24.5,26,30,
                22,21.5,23.5,25),.Dim = c(16, 4)),
     age = c(8.0, 10, 12, 14),
     R = structure(
                .Data = c(1,0,0,0,0,1,0,0,0,0,1,0,0,0,
                0,1),.
                Dim = c(4, 4)))

# Initial Values
list(beta0 = 40, beta1 = 1)
```

Based on Table 4.1 for males and **BUGS CODE 4.2**, the Bayesian analysis is executed with 55,000 observations for the MCMC simulation with a beginning set of 5,000 observations and a refresh of 100; the posterior analysis is reported in Table 4.13. Note that the prior distribution of the precision matrix is Wishart where the parameter matrix R is the 4×4 identity matrix, and the prior distributions of the two regression coefficients are normal with mean 0 and precision 0.0001 (or variance 10,000). Using 4 as the second parameter of the Wishart implies noninformative prior information for the covariance matrix.

Based on Table 4.11, the profile for the mean response (the distance from the pituitary to the pteryomaxillary fissure) is estimated as

$$\tilde{\mu}_{ij} = 15.83 + 0.834 \times \text{age} \tag{4.29}$$

TABLE 4.13

Bayesian Analysis for Males of Potthoff and Roy Study

Parameter	Mean	SD	Error	2½	Median	97½
Sigma(1,1)	6.561	2.668	0.01414	2.074	3.394	6.076
Sigma(1,2)	2.411	1.75	0.00853	−0.1976	2.133	6.66
Sigma(1,3)	3.766	2.176	0.01053	0.6713	3.385	9.149
Sigma(1,4)	1.722	1.631	0.00836	−0.9461	1.514	5.533
Sigma(2,2)	4.989	2.022	0.00972	2.43	4.556	10.1
Sigma(2,3)	2.457	1.83	0.00900	−0.2831	2.165	6.92
Sigma(2,4)	3.01	1.624	0.00765	0.8093	2.695	7.083
Sigma(3,3)	7.342	2.902	0.01489	3.628	6.728	14.56
Sigma(3,4)	3.506	1.934	0.00957	0.8943	3.131	8.299
Sigma(4,4)	4.738	1.937	0.00581	2.299	4.319	9.645
Beta0	15.83	1.214	0.02078	13.45	15.81	18.26
Beta1	0.834	0.1012	0.001726	0.6314	0.8348	1.032

Source: Potthoff, R.F., and Roy, S.N., *Biometrika*, 51, 313–326, 1964.

thus, at age 8 at baseline, the average distance is estimated as 15.83 + 0.834 × (8) = 22.502, which agrees with the value in 22.87 of Table 4.2. At age 14, the estimated value is 15.82 + 0.8346 × (14) = 27.49 compared to the value of 27.46 of Table 4.2, thus overall the simple linear regression model provides accurate estimates of the mean distance profile for males (see Figure 4.1).

With regard to the covariance matrix, these results provide estimates of the matrix. For example, sigma(4,4) denotes the variance of the distance at age 14 at 4.738 via the posterior mean with a standard deviation of 1.93 and 95% credible interval of (2.299, 9.645). On the other hand, the covariance of the distance between ages 12 and 14 is estimated at 3.506 with a standard deviation of 1.93 and 95% credible interval of (0.8843, 8.299). Of course these values can be used to estimate the correlation matrix. For example, what is the estimate of the correlation of the distance response between 8 and 10 years, thus the estimated correlation between ages 8 and 10 years is

$$\tilde{\rho}_{12} = \frac{2.411}{\sqrt{(6.591 \times 4.8189)}} = 0.4278 \tag{4.30}$$

One should compute the other values to see if there is any pattern in the correlation structure.

The posterior density of the covariance of the distance between ages 8 and 10 is depicted in Figure 4.5.

Table 4.14 lists the observed and predicted distance values, where the predicted values are computed from (4.13), the linear regression for the Potthoff and Roy study. Note that the predicted values are the values in the Z vector of **BUGS CODE 4.2**.

Based on Table 4.14, Figure 4.6 is a graph of the predicted versus observed distance values of Potthoff and Roy[1] with the simple linear regression line imposed. The latter shows a linear relation between the two but also depicts the variation of both values for each of the four ages. It appears that the range of observations is the same for each age.

How well does the model fit the data? It is interesting that the Pearson correlation is estimated as 0.586.

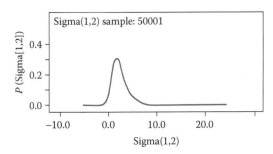

FIGURE 4.5
Posterior density of the covariance between ages 8 and 10.

TABLE 4.14

Observed and Predicted Distance Values of Potthoff and Roy Study for Males

				Subject
26,22.5	25,24.18	29,25.81	31,27.49	1
21.5,22.49	22.5,24.16	23,25.81	26.5,27.5	2
23,22.51	22.5,24.18	24,25.85	27.5,27,51	3
25.5,22.49	27.5,24.17	26.5,25.85	27,27.51	4
20,22.49	23.5,24.17	22.5,25.84	26,27.51	5
24.5,22.49	25.5,24.15	27,25.84	28.5,27.48	6
22,22.48	22,24.16	24.5,25.81	26.5,27.5	7
24,22.5	21.5,24.17	24.5,25.83	25.5,27.5	8
23,22.48	20.5,24.17	31,25.82	26,27.5	9
27.5,22.5	28,24.16	31,25.83	31.5,27.5	10
23,22.5	23,24.17	23.5,25.84	25,27.5	11
21.5,22.51	23.5,24.18	24,25.84	28,27.51	12
17,22.48	24.5,24.15	26,25.82	29.5,27.49	13
22.5,22.48	25.5,24.15	25.5,25.82	26,27.48	14
23,22.51	24.5,24.18	26,25.83	30,27.5	15
22,22.51	21.5,24.18	23.5,25.83	25,27.51	16

Source: Potthoff, R.F., and Roy, S.N., *Biometrika*, 51, 313–326, 1964.

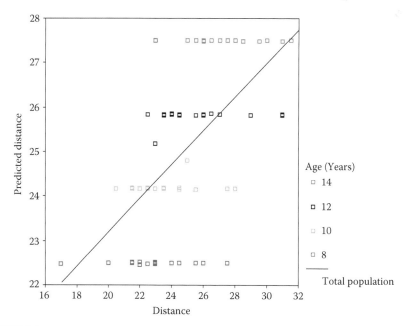

FIGURE 4.6

Predicted versus observed distance.

4.6 Bayesian Inference

The Bayesian approach to inference is briefly described in the upcoming sections and consists of the following concepts: (1) prior information, (2) sample information and the likelihood, and (3) posterior inferences. For a general introduction to Bayesian inference, see Leonard and Hsu.[6] Recall the review of Bayesian inference presented in Chapter 1.

4.6.1 Prior Information

With the Bayesian approach, prior information is taken into account by the prior distribution of the parameters of the model. Suppose one is conducting a study where the study conclusions are based on a statistical model. Where does the prior information come from for the Bayesian analysis? It comes from previous experimental studies related to the "present" study. If the present study is essentially a replication of past studies, the prior distribution is the likelihood function for the parameters of the model. If a previous study is a replication of the present study, the statistical model for the past study is the same as that for the present study. However, it is not likely that previous related experiments will be replications of the present. This implies that the prior distribution for the parameters of the present experiment needs to be chosen with care. In the extreme case, where one cannot base the present prior on previous studies, one may specify so-called noninformative priors for the parameters.

A Bayesian statistician collaborating on a study needs to be deeply involved in the investigation, and this involves reading the literature of previous related studies. In this way, a prior for the "present" study can be determined; however, it is always difficult to minimize the subjectivity for the choice of the prior.

4.6.2 The Model

Specifying the statistical model will determine the likelihood function, which is combined with the prior distribution to give the posterior distribution of the parameters. For example, with regard to the Potthoff and Roy[1] study, the model is postulated as Equation 4.17 for females and Equation 4.18 for males, where the times $t_j = 8, 10, 12$, and 14 years.

The model accounts for possibly different intercepts β_1 for females and β_3 for males. As for the slope, they are β_2 for females and $\beta_2 + \beta_4$ for males. Recall the slope measures the rate of change in distance per year for both groups. The model formulation induces the likelihood function. For example, for the Potthoff and Roy[1] study, the likelihood function depends on the assumptions made about the conditional distribution of the repeated response given the covariates and unknown parameters. The assumption made was that the joint distribution of the repeated responses is multivariate normal with mean vector given by the means (4.17) and (4.18) and unknown covariance matrix. Actually, there are two likelihood functions, one for the females with mean (4.17) and one for the males with mean vector (4.18). I will assume that the two populations, females and males, have different correlation matrices.

Thus, for female i, denote the likelihood function by $l_1(\beta_1, \beta_2, \Omega_1 | y_i^f, x_i^f)$, where β_1 and β_2 are the intercept and slope for the females given by (4.17) and Ω_1 is the covariance matrix. And for males, let $l_2(\beta_3, \beta_4, \Omega_1 | y_i^m, x_i^m)$ be the corresponding likelihood function. Note that y_i is the vector of repeated measures for individual i, and x_i the corresponding matrix of p of covariate values.

4.6.3 Posterior Inferences

The model formulation induces the likelihood function, which is combined with the prior information for the parameters, thereby inducing the posterior distribution of the parameters. Consider the likelihood function for the Potthoff and Roy[1] study, which is the product

$$L(\beta_1, \beta_2, \beta_3, \beta_4, \Omega_1, \Omega_2 | data) = \prod_{i=1}^{i=N_1} l_1(\beta_1, \beta_2, \Omega_1 | y_i^f, x_i^f) \prod_{j-1}^{i=N_2} l_2(\beta_3, \beta_4, \Omega_1 | y_i^m, x_i^m)$$

(4.31)

where:
    β_i, $i = 1,2,3,4$, and Ω_j, $j = 1$ and 2 are unknown regression parameters and variance–covariance matrices

Note that it is assumed that the repeated measures for males are independent of those for the females, thus the product in (4.19), and that the observations of one individual is independent to that of another.

To conduct a Bayesian analysis, one must specify a prior distribution for the unknown parameters β_i, for $i = 1,2,3,4$, and Ω_j, for $j = 1$ and 2. In the following Bayesian analysis, it will be assumed that the prior density of the four regression coefficient is normal with a mean of 0 and a precision of 0.0001 (variance of 10,000), and the prior distribution of the variance–covariance matrices is a Wishart with the identity matrix for the prior.

Using **BUGS CODE 4.3**, the analysis is carried out with 55,000 observations for the simulation, with 5,000 initial observations and a refresh of 100.

The posterior analysis is reported in Table 4.15.

Recall that the model for females was a linear regression with intercept β_1 and slope β_2, while that for males is a linear regression with intercept β_3 and slope $\beta_2 + \beta_4$. Of primary importance is the difference $d13 = \beta_1 - \beta_3$ with 95% credible interval (−1.264, 4.35), which implies there is little difference in the two intercepts. In addition, as the 95% credible interval for β_4 does not include 0, the implication is that the slope for males is indeed $\beta_2 + \beta_4$. Upon inspection of Table 4.15, one sees that the five posterior distributions are symmetric about the mean and that the simulation errors are quite small.

BUGS CODE 4.3

```
model
{
        beta1 ~ dnorm(0.0, 0.001)
        beta2 ~ dnorm(0.0, 0.001)
        beta3 ~ dnorm(0.0,.0001)
        beta4 ~ dnorm(0.0,.0001)

        for (i in 1:N2) {
                Y[i, 1:M1] ~ dmnorm(mu2[], Omega2[,])
        }
        for(j in 1:M2) {
                mu2[j] <- beta3 + (beta2+beta4)* age[j]
        }
        Omega2[1 : M2, 1 : M2] ~ dwish(R[,], 4)
        Sigma2[1 : M2, 1 : M2] <- inverse(Omega2[,])

    for (i in 1:N1) {
                Z[i, 1:M1] ~ dmnorm(mu1[], Omega1[,])
        }
        for(j in 1:M1) {
                mu1[j] <- beta1 + beta2* age[j]
        }
        Omega1[1 : M1, 1 : M1] ~ dwish(R[,], 4)
        Sigma1[1 : M2, 1 : M2] <- inverse(Omega1[,])

   d13<-beta1-beta3


        }


# Potthoff and Roy
list(M2 = 4, N2 = 16, M1 = 4,N1 = 11,

Z = structure(
       .Data = c(21,20,21.5,23,
                 21,21.5,24,25.5,
                 20.5,24,24.5,26,
                 23.5,24.5,25,26.5,
                 21.5,23,22.5,23.5,
                 20,21,21,22.5,
                 21.5,22.5,23,25,
                 23,23,23.5,24,
                 20,21,22,21.5,
                 16.5, 19,19,19.5,
                 24.5,25,28,28
                 ),.Dim = c(11, 4)),
```

```
Y = structure(
            .Data = c(26,25,29,31,
                      21.5,22.5,23,26.5,
                      23,22.5,24,27.5,
                      25.5,27.5,26.5,27,
                      20,23.5,22.5,26,
                      24.5,25.5,27,28.5,
                      22,22,24.5,26.5,
                      24,21.5,24.5,25.5,
                      23,20.5,31,26,
                      27.5,28,31,31.5,
                      23,23,23.5,25,
                      21.5,23.5,24,28,
                      17,24.5,26,29.5,
                      22.5,25.5,25.5,26,
                      23,24.5,26,30,
                      22,21.5,23.5,25),.Dim = c(16, 4)),
      age = c(8.0, 10, 12, 14),
      R = structure(
                .Data = c(1, 0, 0, 0, 0, 1, 0, 0, 0,
                          0, 1, 0, 0, 0, 0, 1),.
                Dim = c(4, 4)))

# Initial Values
list(beta1 = 0, beta2 = 0, beta3 = 0,beta4 = 0)
```

TABLE 4.15

Posterior Analysis for Potthoff and Roy Study

Parameter	Mean	SD	Error	2½	Median	97½
β_1	17.4	0.7424	0.00855	15.89	17.41	18.86
β_2	0.4831	0.06617	0.00096	0.3524	0.4828	0.6147
β_3	15.83	1.2	0.0215	13.45	15.82	18.22
β_4	0.3506	0.1203	0.00229	0.1111	0.3502	0.5866
$d13$	1.54	1.416	0.02449	−1.264	1.53	4.35

Based on this analysis, the model is revised as follows: For females, the mean profile is

$$E(Y_{ij} \mid X_{ij}) = \beta + \beta_2 t_j \qquad (4.32)$$

while for males, the mean profile is

$$E(Y_{ij} \mid X_{ij}) = \beta + (\beta_2 + \beta_4) t_j \qquad (4.33)$$

Thus, the males and females share a common intercept but have different slopes with a larger slope for the males. Based on the revised model described by (4.31) and (4.32), a Bayesian analysis is executed with 55,000 observations for the MCMC simulation, with 5000 initial observations and a refresh of 100. The code is given by **BUGS CODE 4.4**.

For females, Bayesian inference shows a posterior mean for the common (with males) intercept β is 16.97 and a slope estimated as 0.51, with the posterior mean, while for males, the slope delta $= \beta_2 + \beta_4$ is estimated as 0.7458, thus the average distance for males is increasing at a higher rate compared to females (Table 4.16).

```
                          BUGS CODE 4.4
        model
        {
                beta  ~ dnorm(0.0, 0.001)
                beta2 ~ dnorm(0.0, 0.001)

        beta4 ~ dnorm(0.0,.0001)
                for (i in 1:N2) {
                        Y[i, 1:M2] ~ dmnorm(mu2[], Omega2[,])
                }
                for(j in 1:M2) {
                        mu2[j] <- beta + (beta2+beta4)* age[j]


                }
                Omega2[1 : M2, 1 : M2] ~ dwish(R[,], 4)
                Sigma2[1 : M2, 1 : M2] <- inverse(Omega2[,])

        for (i in 1:N1) {
                        Z[i, 1:M1] ~ dmnorm(mu1[], Omega1[,])
                }
                for(j in 1:M1) {
                        mu1[j] <- beta + beta2* age[j]
                }
                Omega1[1 : M1, 1 : M1] ~ dwish(R[,],4)
                Sigma1[1 : M2, 1 : M2] <- inverse(Omega1[,])

    delta<-beta2+beta4
        }

# Potthoff and Roy
list(M2 = 4, N2 = 16, M1 = 4,N1 = 11,
# for males
Z = structure(
        .Data = c(21,20,21.5,23,
```

```
                    21,21.5,24,25.5,
                    20.5,24,24.5,26,
                    23.5,24.5,25,26.5,
                    21.5,23,22.5,23.5,
                    20,21,21,22.5,
                    21.5,22.5,23,25,
                    23,23,23.5,24,
                    20,21,22,21.5,
                    16.5, 19,19,19.5,
                    24.5,25,28,28
                    ),.Dim = c(11, 4)),

# for females
Y = structure(
        .Data = c(26,25,29,31,
                    21.5,22.5,23,26.5,
                    23,22.5,24,27.5,
                    25.5,27.5,26.5,27,
                    20,23.5,22.5,26,
                    24.5,25.5,27,28.5,
                    22,22,24.5,26.5,
                    24,21.5,24.5,25.5,
                    23,20.5,31,26,
                    27.5,28,31,31.5,
                    23,23,23.5,25,
                    21.5,23.5,24,28,
                    17,24.5,26,29.5,
                    22.5,25.5,25.5,26,
                    23,24.5,26,30,
                    22,21.5,23.5,25),.Dim = c(16, 4)),
        age = c(8.0, 10, 12, 14),
        R = structure(
                    .Data = c(1, 0, 0, 0, 0, 1, 0, 0, 0,
                              0, 1, 0, 0, 0, 0, 1),.
                    Dim = c(4, 4)))

# Initial Values
list(beta = 0, beta2 = 0,beta4 = 0)
```

TABLE 4.16

Bayesian Analysis for Potthoff and Roy Study: Revised Model

Parameter	Mean	SD	Error	2½	Median	97½
β	16.97	0.6341	0.00735	15.67	16.98	18.17
β_2	0.5101	0.06101	0.000742	0.3932	0.5084	0.6345
β_4	0.2357	0.061	0.000584	0.1162	0.2357	0.355
Delta	0.7458	0.062	0.000620	0.6245	0.7454	0.8683

4.7 Another Example

4.7.1 Introduction

The previous ideas and concepts involving repeated measures for continuous observations will be reviewed by analyzing another example. Recall that the previous ideas and concepts include (1) notation for linear models, (2) modeling the mean response, (3) modeling the correlation, (4) historical approaches to repeated measures such as univariate ANOVA and MANOVA, and (5) Bayesian inference with a review of prior information, sample information, and posterior inferences.

4.7.2 Example

Consider the following study by Crowder and Hand[7] about body weight in grams of guinea pigs, where there are three groups corresponding to three doses (low, medium, and high) of vitamin E (Table 4.17). All were given a growth-inhibiting substance at time 1 and the weights were observed at weeks 1, 3, 4, 5, 6, and 7. This example is somewhat more complex than the other in that there are three groups that will be compared. After graphing the data, a linear model for three groups will be proposed.

TABLE 4.17

Weight of the Three Groups of Guinea Pigs

Group	1	3	4	5	6	7
1	455	460	510	504	436	466
1	467	565	610	596	542	587
1	445	530	580	597	582	619
1	485	542	594	583	611	612
1	480	500	550	528	562	576
2	514	560	565	524	552	597
2	440	480	536	484	567	569
2	495	570	569	585	576	677
2	520	590	610	637	671	702
2	503	555	591	605	649	675
3	496	560	622	622	632	670
3	498	540	589	557	568	609
3	478	510	568	555	576	605
3	545	565	580	601	633	649
3	472	498	540	524	532	583

Source: Crowder, M.J., and Hand, D.J., *Analysis of Repeated Measures*, Chapman & Hall, New York, 1990.

4.7.3 Scatter Plots for Crowder and Hand

Figure 4.7 portrays the body weight versus time for three groups of guinea pigs, and the lowess plots delineate the trend of the three groups. What model would you propose for the mean profile of the three groups and how would you express it with a linear model?

The plot reveals that there is a linear association between weight and weeks, where for groups 2 and 3, the linear associations are similar, while for group 1 the intercept at 0 appears smaller compared to the other two, and the slope appears to be slightly smaller compared to the other two.

4.7.4 Descriptive Statistics

Let us compute the descriptive statistics for the Crowder and Hand[7] study based on Table 4.18.

4.7.5 Modeling the Mean Profile

Based on the descriptive statistics depicted in Table 4.16 and scatter plot Figure 4.6, a model for the average response over time is specified by three equations:

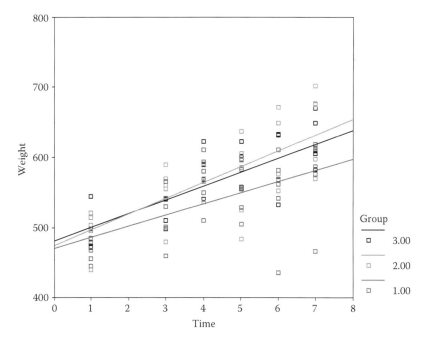

FIGURE 4.7
Weight in grams of three groups of guinea pigs by week.

TABLE 4.18

Descriptive Statistics for Guinea Pig Weights

Group	Time	Mean	SD
1	1	466.40	16.72
1	3	519.40	40.64
1	4	568.80	39.58
1	5	561.60	42.84
1	6	546.60	66.87
1	7	572.00	61.81
2	1	494.40	31.91
2	3	551.00	41.89
2	4	574.20	27.99
2	5	567.00	62.06
2	6	603.00	53.30
2	7	644.00	57.54
3	1	497.80	28.67
3	3	534.60	29.76
3	4	579.80	29.95
3	5	571.80	39.23
3	6	588.20	43.70
3	7	623.20	35.37

For group 1

$$E\left[y_{ij}^{(1)}\right] = \beta_1 + \beta_2 t_j \tag{4.34}$$

and for group 2,

$$E\left[y_{ij}^{(2)}\right] = \beta_3 + \beta_4 t_j \tag{4.35}$$

and finally for group 3,

$$E\left[y_{ij}^{(3)}\right] = \beta_5 + \beta_6 t_j \tag{4.36}$$

where:

E denotes expectation over all replications of the study

Note for group 1, the linear association between the weight $y_{ij}^{(1)}$ of subject i at time t_j is a linear regression with intercept β_1 and slope β_2. Thus, for group 2, the intercept is β_3 and the slope is β_4, while for group 3, the intercept is β_5 and the slope is β_6.

4.7.6 Correlation Structure

One should always compute estimates of the pair-wise correlations between the weeks.

TABLE 4.19

Correlations of Guinea Pig Weights

Time	1	3	4	5	6	7
1	1	0.714	0.460	0.510	0.589	0.619
3		1	0.847	0.830	0.718	0.843
4			1	0.881	0.749	0.761
5				1	0.750	0.807
6					1	0.913
7						1

Table 4.19 reports the estimated Pearson pair-wise correlations between the six weeks. For example, the estimated correlation between weeks 3 and 4 is 0.847 and is estimated as 0.881 between weeks 4 and 5. Do the estimated correlations suggest any structure covariance matrix? In the following chapters, various structures for the covariance matrix will be assumed when conducting a Bayesian analysis.

The scatter plots of Figure 4.8 verify, to some extent, the estimated Pearson correlations appearing in Table 4.19. For example, the plot of week 3 versus

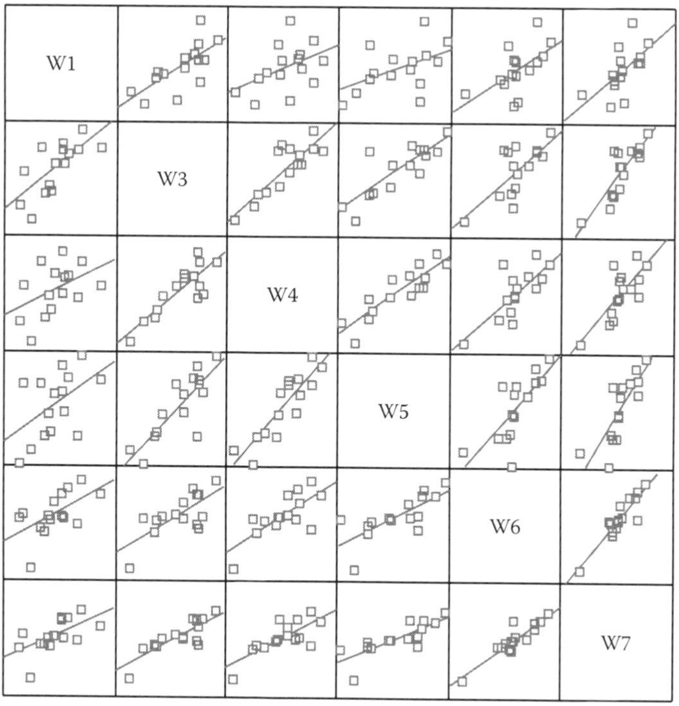

FIGURE 4.8
Matrix plots of guinea pig weights.

week 4 looks very strong and corresponds to an estimated correlation of 0.847 compared to the plot of week 1 versus week 5, which looks quite scattered and corresponds to an estimated correlation of 0.510.

4.7.7 Bayesian Analysis

The Bayesian analysis is based on the model specification given in formulas (4.34) through (4.36), and the notation of **BUGS CODE 4.5** closely follows the symbols of those formulas. Recall that distinct linear regressions are assumed for the three groups, with different intercepts and slopes. In addition, a multivariate normal distribution is assumed for the guinea pig weights with a linear regression for each group and a common precision matrix for the correlations between the six weeks. The precision matrix is given a Wishart distribution with the identity matrix for its parameter, and the precision matrix is later inverted to give an estimate of the 6 by 6 variance–covariance matrix. Note that the list statement for the data contains the weight in grams of the guinea pigs denoted by Y1 for group 1, Y2 for group 2, and Y3 for group 3.

4.7.7.1 Mean Profile

The mean profile for the three groups is assumed to be linear over weeks, while the common covariance matrix is assumed to be unstructured. Note that separate means and slopes are assumed for the three linear regressions and the three groups are assumed to have the same covariance matrix.

Based on **BUGS CODE 4.5**, a Bayesian analysis is executed with 55,000 observations for the simulation, with 5,000 starting values and a refresh of 100.

```
                          BUGS CODE 4.5

model;
        {
                beta1 ~ dnorm(0.0,.0001)
                beta2 ~ dnorm(0.0,.0001)
                beta3 ~ dnorm(0.0,.0001)
                beta4 ~ dnorm(0.0,.0001)
                beta5 ~ dnorm(0.0,.0001)
                beta6 ~ dnorm(0.0,.0001)

                for (i in 1:N) {
                        Y1[i, 1:M] ~ dmnorm(mu1[], Omega[,])
                }
                for(j in 1:M) {
                mu1[j] <- beta1 + beta2* age[j]


                }
```

```
            Omega[1 : M, 1 : M] ~ dwish(R[,], 6)
            Sigma[1 : M, 1 : M] <- inverse(Omega[,])
    for (i in 1:N) {
                Y2[i, 1:M] ~ dmnorm(mu2[], Omega[,])
            }
            for(j in 1:M) {
                mu2[j] <- beta3+ beta4*age[j]
            }

            for (i in 1:N) {
                Y3[i, 1:M] ~ dmnorm(mu3[], Omega[,])
            }
            for(j in 1:M) {
                mu3[j] <- beta5+beta6* age[j]

            }
            d13<-beta1-beta3
            d35<-beta3-beta5
            d15<-beta1-beta5
    }


# guinea pigs

list(M = 6, N = 5, age = c(1,3,4,5,6,7),
Y1 = structure(.Data = c(455,460,510,504,436,466,
                         467,565,610,596,542,587,
                         445,530,580,597,582,619,
                         485,542,594,583,611,612,
                         480,500,550,528,562,576),.Dim =
                         c(5, 6)),

Y2 = structure(.Data = c(514,560,565,524,552,597,
                         440,480,536,484,567,569,
                         495,570,569,585,576,677,
                         520,590,610,637,671,702,
                         503,555,591,605,649,675),.Dim =
                         c(5, 6)),

    R = structure(
                .Data = c(1,0,0,0,0,0,0,0,1,0,0,0,0,0,0,
                          1,0,0,0,0,0,0,1,0,0,0,0,0,0,0,1,
                          0,0,0,0,0,0,1),.Dim = c(6,6)),

Y3 = structure(.Data = c(496,560,622,622,632,670,
                         498,540,589,557,568,609,
                         478, 510,568,555,576,605,
```

```
                        545, 565, 580,601,633,649,
                        472,498,540,524,532,583),.Dim =
                        c(5,6)))

# Initial Values
list(beta1 = 1, beta2 = 0, beta3 = 0,beta4 = 0,beta5 = 0,
beta6 = 0)
```

TABLE 4.20

Posterior Analysis from Crowder and Hand: Guinea Pig Weights; Mean Profile of Three Groups

Parameter	Mean	SD	Error	2½	Median	97½
β_1	470.6	21.3	0.5768	423.0	472.6	506.6
β_2	24.93	5.732	0.1466	14.71	24.56	37.36
β_3	462.5	16.17	0.3434	426.7	463.7	590.6
β_4	31.0	4.42	0.0876	22.88	30.78	40.49
β_5	487.0	18.18	0.5	446.3	489	516
β_6	32.82	4.831	0.1235	24.43	32.46	43.44
$d13$	8.101	24.41	0.5046	−43.09	8.818	54.22
$d35$	−16.48	18.34	0.2934	−53.48	−16.24	19.23
$d15$	−24.58	18.66	0.342	−59.3	−25.22	14.47

Based on the posterior analysis provided by Table 4.20, and the 95% credible intervals for $d13$, $d35$, and $d15$, one is inclined to believe that the three groups have the same intercept, thus the mean profile for weight will be revised to the following:

For group 1

$$E\left[y_{ij}^{(1)}\right] = \beta + \beta_2 t_j \tag{4.37}$$

and for group 2

$$E\left[y_{ij}^{(2)}\right] = \beta + \beta_4 t_j \tag{4.38}$$

and finally for group 3

$$E\left[y_{ij}^{(3)}\right] = \beta + \beta_6 t_j \tag{4.39}$$

BUGS CODE 4.6 closely follows the notation for the model specified by formulas (4.36) through (4.38). Note it is assumed that the repeated measures over the six weeks have the same correlation for the three groups. That is to say, the repeated measures of the three groups have a common correlation matrix.

BUGS CODE 4.6

```
     model;
     {

             beta2 ~ dnorm(0.0,.0001)
beta4 ~ dnorm(0.0,.0001)
     beta ~ dnorm(0.0,.0001)
beta6 ~ dnorm(0.0,.0001)

# for group 1
             for (i in 1:N) {
                     Y1[i, 1:M] ~ dmnorm(mu1[], Omega[,])
             }
             for(j in 1:M) {
             mu1[j] <- beta + beta2* age[j]


             }
             Omega[1 : M, 1 : M] ~ dwish(R[,], 6)
             Sigma[1 : M, 1 : M] <- inverse(Omega[,])

# for group 2
     for (i in 1:N) {
                     Y2[i, 1:M] ~ dmnorm(mu2[], Omega[,])
             }
             for(j in 1:M) {
                     mu2[j] <- beta+ beta4*age[j]
             }

# for group 3
             for (i in 1:N) {
                     Y3[i, 1:M] ~ dmnorm(mu3[], Omega[,])
             }
             for(j in 1:M) {
                     mu3[j] <- beta+beta6* age[j]


             }

             # difference in slopes

             d24<-beta2-beta4
             d26<-beta2-beta6
             d46<-beta4-beta6


     }
```

```
# guinea pig weigh information


list(M = 6, N = 5, age = c(1,3,4,5,6,7),
# group 1
Y1 = structure(.Data = c(455,460,510,504,436,466,
                         467,565,610,596,542,587,
                         445,530,580,597,582,619,
                         485,542,594,583,611,612,
                         480,500,550,528,562,576),.Dim
                       = c(5, 6)),
# group 2
Y2 = structure(.Data = c(514,560,565,524,552,597,
                         440,480,536,484,567,569,
                         495,570,569,585,576,677,
                         520,590,610,637,671,702,
                         503,555,591,605,649,675),.Dim
                       = c(5, 6)),
            # R is the parameter of the Wishart
distribution for the #correlation of the observations

            R = structure(
                    .Data = c(1,0,0,0,0,0,0,1,0,0,0,
                    0,0,0,1,0,0,0,0,0,0,0,1,0,0,0,0,0,
                    0,1,0,0,0,0,0,0,1),.Dim =
                    c(6,6)),
# group 3
Y3 = structure(.Data = c(496,560,622,622,632,670,
                         498,540,589,557,568,609,
                         478,510,568,555,576,605,
                         545,565,580,601,633,649,
                         472,498,540,524,532,583),.Dim =
                         c(5,6)))


# Initial Values

list(beta = 0, beta2 = 0, beta4 = 0,beta6 = 0)
```

As before, it is assumed that the repeated measures follow a multivariate normal distribution for the six weeks with a common correlation matrix for the three groups. It is assumed that the variance–covariance matrix is not structured.

Based on **BUGS CODE 4.6**, the analysis is executed with 55,000 observations for the simulation, with 5,000 starting values and a refresh of 100, and the results are presented in Table 4.21.

Inspecting Table 4.21, one sees that the common intercept is estimated as 482.5 g using the posterior mean and that the posterior distribution of all parameters appear to be symmetric about the posterior mean. Based on the

TABLE 4.21

Posterior Analysis for Three Groups of Guinea Pigs with the Same Intercept

Parameter	Mean	SD	Error	2½	Median	97½
β	482.5	10.71	0.2158	460.2	482.8	502.9
β_2	22.37	4.071	0.0706	14.47	22.3	30.55
β_4	26.69	3.553	0.0602	19.55	26.7	33.76
β_6	33.83	3.783	0.06607	26.66	33.71	41.59
$d24$	−4.319	4.325	0.05103	−12.83	−4.331	4.34
$d26$	−11.46	3.215	0.02789	−17.79	−11.47	−5.085
$d46$	−7.14	3.591	0.03766	−14.27	−7.13	−0.06084

95% credible intervals, groups 1 and 2 have different slopes; however, groups 1 and 3 have the same slope, and groups 2 and 3 have the same slope. What are the implications for comparing the three groups?

4.7.7.2 Structure of Covariance Matrix

The Bayesian analysis continues with Bayesian inferences for the variance-covariance matrix of the weight of guinea pigs over the weeks 1, 3, 4, 5, 6, and 7, that is, the covariance between the response at time week i and week j, where $i, j = 1, 3, 4, 5, 6$, and 7. Referring to Table 4.22, the posterior mean of the covariance between weeks 1 and 3 is 1001, for other variances and covariances. For example, the estimated variance of the weight at time week 1 is 921.6, while the estimated variance at week 3 is 1635. Using the posterior means, I would estimate the correlation between weeks 1 and 3 as

$$\tilde{\rho}_{13} = \frac{1479}{\sqrt{(1428)(2572)}} = \frac{1479}{1916.4592} = 0.7717 \tag{4.40}$$

Additionally, referring to Table 4.21, the correlation between weeks 3 and 7, would be estimated as

$$\tilde{\rho}_{37} = \frac{4122}{\sqrt{(2572)(7996)}} = \frac{4122}{4534} = 0.9091 \tag{4.41}$$

It should be pointed out that these two estimates are based on the posterior mean of the relevant posterior distributions. However, note that the posterior distribution of the covariances are highly skewed, thus if the estimated correlations are based on the posterior medians, how are the estimated correlations affected?

Thus, based on the medians of the posterior distributions, consider again the correlation of the weight between weeks 1 and 3, then

$$\overset{*}{\rho}_{13} = \frac{1300}{\sqrt{(1251)(2319)}} = \frac{1300}{1703} = 0.7633 \tag{4.42}$$

TABLE 4.22

Posterior Distribution of the Covariance Matrix

Parameter	Mean	SD	Error	2½	Median	97½
Sigma(1,1)	1428	758.4	8.757	522	1251	3354
Sigma(1,2)	1479	843.1	7.982	424.4	1300	3596
Sigma(1,3)	770.4	613.4	5.407	−64.39	653.3	2287
Sigma(1,4)	1880	1183	11.06	318.6	1664	4828
Sigma(1,5)	2469	1462	13.66	540.6	2187	6095
Sigma(1,6)	2314	1398	12.47	481.9	2042	5762
Sigma(2,2)	2572	1196	8.448	1078	2319	5596
Sigma(2,3)	1833	943.4	7.059	669.4	1630	4206
Sigma(2,4)	3490	1733	12.16	1341	3122	7838
Sigma(2,5)	4001	2071	15.58	1408	3562	9207
Sigma(2,6)	4122	2090	16.4	1532	3671	9369
Sigma(3,3)	1677	812	5.934	693.4	1492	3739
Sigma(3,4)	2782	1449	11.25	987.1	2461	6442
Sigma(3,5)	3058	1738	14.78	882.6	2682	7421
Sigma(3,6)	3026	1726	15.26	919.2	2643	7390
Sigma(4,4)	5803	2745	20.94	2393	5196	12810
Sigma(4,5)	6423	3262	27.64	2373	5705	14710
Sigma(4,6)	6326	3270	29.9	2322	5594	14640
Sigma(5,5)	8649	4187	37.74	3439	7727	19330
Sigma(5,6)	8010	4130	40.81	2925	7084	18500
Sigma(6,6)	7966	4188	43.82	2920	7039	18690

Comparing (4.40) with (4.42), it appears that there is very little difference (Table 4.22).

One should consult Table 4.19, the estimated Pearson correlation coefficients, and compare the entries to the correlations implied by Table 4.19, the estimates based on the Wishart distribution. Based on Table 4.19, is any structure implied for the variance–covariance matrix? It appears to me that approximately the correlations are the same, implying a compound symmetry for the correlation.

4.7.7.3 Analysis Based on Compound Symmetry for Covariance

In order to account for the compound symmetry structure of the covariance matrix, let

$$y_{ijk} = \theta + a_i + b_j + c_k + e_{ijk} \tag{4.43}$$

where y_{ijk} is the weight in grams for group i, for guinea pig j, and for time k. Suppose θ is a constant, a_i are independent normally distributed random

TABLE 4.23

Estimates of Variance Components: Guinea Pig Weights

Parameter	ML Estimate	Variance of Estimate
σ_a^2 Group	16.262	109859.85
σ_b^2 Pig	1363.613	3576853.62
σ_c^2 Time (weeks)	1860.426	1445151.1
σ_e^2 Error	604.505	10440.743

variables with mean 0 and precision τ_a, b_j are independent normally distributed variables with mean 0 and precision τ_b, c_k are independent and normally distributed variables with mean 0 and precision τ, and e_{ijk} are normally distributed variables with mean 0 and precision τ_e. In addition, all random variables are mutually independent.

It can be shown the common correlation between observations at times k and k' is

$$\rho = \frac{(\sigma_a^2 + \sigma_b^2)}{(\sigma_a^2 + \sigma_b^2 + \sigma_c^2 + \sigma_e^2)} \tag{4.44}$$

which is often referred to as the interclass correlation coefficient.

The analysis is conducted using a non-Bayesian conventional approach. The analysis was conducted using SPSS, which gives the maximum likelihood estimate of the variance components. Also provided are the estimated variances of the maximum likelihood estimates of the variance components. Based on the estimates for variance components provided by Table 4.23, and formula (4.44), the maximum likelihood estimate of the intra class correlation coefficient is

$$\tilde{\rho} = \frac{(16.62 + 1363.613)}{(16.62 + 1363.613 + 1860.426 + 604.505)(4.45)} = 0.3588$$

Thus, 0.3588 is the estimate of the common correlation of the weights between the six time points. That is, the correlation between weeks 1 and 3 is estimated as 0.3588, as it is between weeks 3 and 4.

4.8 Summary and Conclusions

The chapter begins with the objectives and states that the subject matter to be stressed is that the analysis for repeated measures will be based on the linear model, that is to say, on regression techniques that take into account the correlation between observations of the same individual and the possibility of different variances for the observations taken at distinct occasions. Notation for the linear model is explained where the vector of observations for individual i is expressed as a linear function of the covariates and unknown regression parameters.

In order to illustrate the notation for the linear model and the subsequent analysis, the Potthoff and Roy[1] example of two groups (males and females) where the main response is the distance from the pituitary to the pteryomaxillary fissure of a person is used. Such a measure is important in dental science in order to follow the development of adolescents over time. Initially, a linear response of distance versus time is tentatively entertained for both groups, where the intercepts of the two groups are distinct and corresponding two slopes are not assumed to be the same. In order to estimate the mean profile of the two groups, it is important to know with some confidence the structure of the correlation matrix of the distance (from the pituitary to the pteryomaxillary fissure), that is, the correlation between the four time points of 8, 10, 12, and 14 years of an adolescent. To do this, the Pearson correlations are estimated, followed by a matrix plot of the pair-wise distances over the four time points. The matrix plot should corroborate the six estimated Pearson correlations.

Next, a review of historical approaches to repeated measures is presented. The univariate ANOVA is described, and using the female data from Potthoff and Roy,[1] the conventional analysis is presented with three sources of variation, namely, the between subjects, the between times, and the error. See Table 4.4 for females and Table 4.5 for males. The analysis revealed a significant difference in the distance from the pituitary to the pteryomaxillary fissure, between the four time points 8, 10, 12, and 14 years. This was followed by a multiple comparison procedure that compared the average distance pair-wise between the four years 8, 10, 12, and 14.

Assuming compound symmetry for the 4 by 4 covariance matrix, the results of a Bayesian analysis based on **BUGS CODE 4.1** is portrayed in Table 4.10, and the posterior mean estimated the intraclass correlation coefficient as 0.587, that is, the correlation of the distance response between any two years (8, 10, 12, and 14 years) is estimated as 0.587.

We review the Basic components of a Bayesian analysis, the prior information, the sample information, and the posterior analysis. With reference to the Potthoff and Roy study, it is important to review the references in that article to previous studies, so that one is able to place realistic priors on the parameters (the regression parameters and correlation parameters) of the model. A Bayesian analysis for the Potthoff and Roy study is performed via **BUGS CODE 4.3**, where the mean profile of the two groups are linear regressions with separate intercept and slopes and the variance covariance matrix of the observations is not the same for the two groups, females and males. It is shown that the intercepts of the two groups are not the same, thus the mean profile is revised to include a common intercept and the analysis is executed with **BUGS CODE 4.4**.

The chapter continues with a final example, based on a study of Crowder and Hand,[7] with three groups of guinea pigs where the weight is measured at six times. For the analysis, the mean profile is assumed to be different linear regressions with distinct intercepts and slopes but a common covariance matrix across the three groups. The three groups correspond to three different vitamin E supplements and is of interest to compare the three groups.

The analysis begins with a scatter plot of weight versus time for the three groups, and the association between weight and weeks appears to be linear for each group. This is followed by Table 4.18, which contains the descriptive statistics for the study and complements the scatter plot. Tentatively, the mean profile is modeled as a separate linear regression with distinct intercepts and slopes for the three groups.

Next, the correlations of the weight between the six weeks are computed assuming a common correlation structure for the three groups of guinea pigs.

A Bayesian analysis for the study based on **BUGS CODE 4.5** is performed and the posterior analysis is reported in Table 4.20, and the analysis implies that there is a common intercept, thus, accordingly, a revised model for the mean profile of the three groups is put forward. The Bayesian analysis assumes a common correlation matric for the three groups, and the precision matrix of the multivariate normal observations is given a noninformative prior.

Based on **BUGS CODE 4.6**, the revised analysis is reported in Table 4.21, which lists the posterior analysis for the common intercept and the three slopes. Also reported is posterior distribution of the inverse of the precision matrix, the variance covariance matrix of the multivariate normal observations (of the weights for each guinea pig) in Table 4.21.

The chapter concludes with a conventional analysis. Using SPSS, the variance components are estimated with maximum likelihood, and the results appear in Table 4.22. The estimated variances of the maximum likelihood estimates of the variance components are also reported.

There are several references that are relevant to the ideas presented in this chapter. For example, with regard to the ANOVA for repeated measures, Hand and Crowder[2] develop the methodology, and the book contains many data sets that will be used in this book. For a Bayesian approach using the multivariate normal distribution, the book by Box and Tiao[8] presents an in-depth view of the subject and is one of the best introductions to Bayesian inference. Broemeling[9] is another Bayesian reference that deals with the linear model, including the multivariate linear model used in this chapter to model repeated measures. Finally, for related topics to repeated measures using weighting methods in regression, see Carroll and Ruppert,[10] and related to this reference, but with a nonlinear flavor for repeated measures, is the book by Davidian and Giltinan.[11] For a more general treatment of repeated measures emphasizing clustered data, see Aerts et al.[12]

Exercises

1. Describe the model given by (4.15) and (4.16) for the Potthoff and Roy[1] study.

2. Does the scatter plot of Figure 4.1 reveal a linear or quadratic association between distance and years? Explain your answer.

3. Do the descriptive statistics of Table 4.2 corroborate the scatter plot of the two groups given by Figure 4.1?

4. The mean profile of the two groups, females and males, of the Potthoff and Roy study is defined by (4.15) and (4.16). The mean profile is also defined by (4.17) and (4.18). Show the two model representations are essentially the same.

5. Based on (4.15) and (4.16), how would you describe a mean profile for the two groups with a common intercept?

6. Using information from Table 4.1, validate the correlations given for females by Table 4.3 and for males by Table 4.4. Explain how you would calculate the estimated correlations.

7. Refer to Figure 4.2, the matrix plot of the distance versus age for females. Does the plot corroborate Table 4.3?

8. Does the matrix plot of Figure 4.2 for males corroborate the estimated correlations of Table 4.4?

9. Refer to sections 4.5.1 and 4.5.2, the univariate ANOVA and MANOVA for repeated measures. Explain the difference in the two approaches.

10. Verify Table 4.5, the one-way ANOVA for females of the Potthoff and Roy study.[1]

11. Verify Table 4.6, the one-way ANOVA for males.

12. Verify Table 4.7, the repeated measures of ANOVA for both groups of Potthoff and Roy.[1]

13. Describe and explain the random effects model defined by (4.21).

14. Using **BUGS CODE 4.1** and 55,000 observations for the simulation, verify Table 4.10, the posterior analysis for the variance components for females.

15. Based on Table 4.10, is the posterior distribution of ρ symmetric about the posterior mean? Explain why or why not.

16. What does the posterior distribution of σ_β^2 symmetric about its posterior mean? Refer to Table 4.10 and explain your answer.

17. Explain how the posterior density for ρ, portrayed in Figure 4.4, corroborates the posterior distribution of ρ given in Table 4.10.

18. Refer to (4.27) the posterior density of a Wishart distribution of the precision matrix τ of the multivariate normal distribution defined by (4.10). What is the variance–covariance matrix of the multivariate normal distribution?

19. Using **BUGS CODE 4.2** with 55,000 observations for the simulation and 5000 for the starting values, verify the posterior analysis of Table 4.13 for males of Potthoff and Roy.[1]

    a. What is your estimate of the intercept?

    b. What is your estimate of the slope?

    c. What is your estimate of the variance of the distance at time 1 (8 years)?

20. Based on 55,000 observations for the simulation, execute a Bayesian analysis that generates future values for the Potthoff and Roy study. The code in **BUGS CODE 4.2** that generates the future values in Table 4.14 is Z[I,1:M]~dmnorm(m[], omega[,]). Verify Table 4.14.

21. Verify the observed and predicted distance values of Potthoff and Roy, thus verify Figure 4.6.

22. Refer to section 4.6.1 about prior information for a Bayesian analysis. Read Potthoff and Roy[1] and the references. Based on the previous studies given in the references, did you find information useful for determining prior information for the Bayesian analysis? If so, please elaborate.

23. Refer to the mean profile given by (4.17) and (4.18).

    a. Under what conditions are the two intercepts the same?

    b. What does $\beta_4 = 0$ imply about the two slopes?

24. Based on (4.17) and (4.18), perform a Bayesian analysis using **BUGS CODE 4.3** and verify Table 4.15.

    a. Is the posterior distribution of β_1 symmetric about its mean of 17.4?

    b. $d13$ is the parameter that measures the difference in the two intercepts. Based on its posterior distribution, do the two groups (females and males) have a common intercept? Explain your answer.

25. Verify Figure 4.7, a scatter plot of the weight versus week for the three groups of guinea pigs in the Crowder and Hand study. Use the information in Table 4.17.

26. Based on Figure 4.7, describe the mean profile (trends) of the three groups. Is the association linear?

27. Do the descriptive statistics of Table 4.18 corroborate the scatter plot of Figure 4.7? Explain your answer.

28. Equations 4.34 through 4.36 define the mean profile of the three groups. Interpret the parameters of the three equations.

a. What are the three intercepts?

b. What are the three slopes?

c. What is true about the parameter values in order for the three groups to share a common intercept?

29. Verify Table 4.19, the computed correlation coefficients of the weights observed at the six time points. Is any structure implied by the values of Table 4.19?

30. Do the matrix plots of Figure 4.8 corroborate the correlations appearing in Table 4.15?

31. Equations 4.34 through 4.36 define the mean profile of three groups of the Crowder and Hand[7] study and appear in **BUGS CODE 4.5**. Perform a Bayesian analysis with 55,000 observations for the simulation and 5,000 beginning observations. Is there a common intercept for the three groups of guinea pigs?

32. Assuming a common intercept for the three groups, the mean profile for the three groups is described by Equations 4.37 through 4.39. What parameters represent the slopes of the three groups?

33. Based on **BUGS CODE 4.6**, generate 55,000 observations for the Bayesian analysis reported in Table 4.21.

a. What is your estimate of the common intercept?

b. Are the slope for groups 1 and 2 the same? Explain your answer.

34. Describe the random effects model defined by (4.43). Why does the model imply compound symmetry for the correlation matrix?

Verify Table 4.23 using a conventional maximum likelihood approach to the estimation of the four variance components.

5

Estimating the Mean Profile
of Repeated Measures

5.1 Introduction

In this chapter, parametric techniques that model the mean of the repeated measures over time are developed, and there are many ways to estimate the mean profile for repeated measures, which were introduced in Chapter 4. First and foremost is computing the descriptive statistics (mean, median, mode, and standard deviation) of the response at each time point, which should be replicated for each covariate (e.g., treatment group and other values such as age). In addition to the descriptive statistics, a scatter plot with a lowess curve should be conducted, and the plot reconciled with the descriptive statistics, that is, the plot should corroborate the descriptive statistics. This will entail a give-and-take process, whereby some plots (e.g., varying the type of plot such as fitting linear, quadratic, and lowess curves) will agree more and some will agree less with the descriptive statistics. Thus, one's choice of a parametric model (linear, quadratic, and linear shift point) is to a large extent subjective. That is, before one chooses a particular parametric model, one will have to determine the degree of agreement and various scatter plots of the data.

The approach taken in the chapter to some extent follows that of Fitzmaurice, Laird, and Ware,[1] where a continuous response (following a normal distribution) is assumed and where the estimation of the parameters is based on restricted maximum likelihood. Instead of maximum likelihood, the Bayesian approach is followed, where prior information is required for the distribution of the parameters of the parametric model. Fitzmaurice, Laird, and Ware use the data from the Vlagtwedde–Vlaardingen study of Van der Lende et al.[2] to illustrate the conventional approach. With the Bayesian approach, the same repeated measures example will be performed and the results will be contrasted with those of the conventional analysis. In order to determine the prior distribution for the Bayesian method, one will have to study the references in Van der Lende et al.[2]

Fitting a parametric model for the mean profile of a repeated measures study has certain advantages, in that one would expect the profile to be well behaved, that is, not to be very complex over the time period of the study. That is, one can most likely expect the mean response to follow a linear, a quadratic, or a linear-type shift point model (linear spline). Also, in many situations, the response is not measured at the same number of time points for all individuals, and parametric models have certain advantages in such circumstances.

This chapter continues with a description of three parametric models (polynomials): a linear representation, a quadratic representation, and a linear shift point model. Using Van der Lende et al.[2] as an example, it will be shown that one group (smokers) follows a linear representation, while another group (nonsmokers) follows a linear shift point model. The Bayesian estimates of the model parameters will be compared with the restricted maximum likelihood of Van der Lende et al.[2]

The mean profile will be represented using the notation based on the linear models (see Section 4.2), thus demonstrating the versatility of the linear model representation. Finally, the chapter concludes with several more examples that will demonstrate the use of polynomials to model the mean profile of a repeated measures study.

5.2 Polynomials for Fitting the Mean Profile

Possible candidates for the mean profile will be considered in this section, and they include (1) a linear association for both groups of subjects, (2) a quadratic fit for the mean profiles, and (3) a shift point model with one shift point and a linear regression for both segments.

5.2.1 Linear Regression for Two Groups

The first case to be considered is appropriate for a two-group repeated measures, where a linear regression is assumed for the two mean profiles. Let i denote the ith subject and t_{ij} the time of observation for subject i at occasion j, and suppose the expected value for Y_{ij}, the observation for subject i at occasion j, is

$$E(Y_{ij}) = \beta_1 + \beta_2 t_{ij} + \beta_3 g_i + \beta_4 t_{ij} g_i \qquad (5.1)$$

where E denotes expectation with respect to hypothetical replications of the study, $g_i = 1$ when subject i is a member of the treatment group, and $g_i = 0$ when subject i is a member of the control group. Thus, for the treatment group

$$E(Y_{ij}) = \beta_1 + \beta_3 + (\beta_2 + \beta_4) t_{ij} \qquad (5.2)$$

and for the control group

$$E\left(Y_{ij}\right) = \beta_1 + \beta_2 t_{ij} \tag{5.3}$$

therefore, when $\beta_3 = \beta_4 = 0$, the two groups share a common mean profile given by (5.3).

In order to explain the Bayesian method for estimating the mean profile of two groups, consider the example described by Van der Lende et al.,[2] which is a study conducted in two regions of Holland, a rural area of Vlagtwedde and more urban industrial area of Vlaardingen. All subjects were followed over time to determine data concerning risk factors for chronic obstructive lung disease. Participants were questioned every 3 years for up to 21 years. The questionnaire enquired about information of respiratory symptoms and smoking status. The main repeated measure was forced expiratory volume, FEV1, which is measured by spirometry. Once every 3 years for the first 15 years and once at year 19, FEV1 is measured.

The study will show that a linear association between FEV1 and time for smokers and separately for nonsmokers is appropriate, where the mean profile is modeled by (5.1). To be more specific, the sample included in the example consists of 133 subjects aged at least 36 years and whose smoking status did not change during the course of the study; thus, smoking status is not a time-dependent covariate.

My approach will be to use the descriptive statistics and scatter plot to determine a linear association between FEV1 and time of survey for smokers and nonsmokers separately. First, consider the scatter plot of FEV1 versus time given by Figure 5.1. A lowess curve is depicted for smokers and non-smokers and reveals that a linear association is not an unreasonable assumption for the association. It should be noted that the number to time points at the times (years from baseline) 0, 3, 6, 9, 12, 15, and 19 is not the same for all individuals. See Table 5.1 for the imbalance for the various time points for nonsmokers and smokers.

The number of observations vary from a low of 23 at time 0 to a maximum of 30 observed at time 9, suggesting the imbalance is not too serious.

With regard to smokers, the fewest number of observations occurs at time 15 with 73 measurements to a high of 95 occurring at time 3 years, suggesting the imbalance is somewhat more serious for smokers.

Consider the descriptive statistics given by Table 5.2 for nonsmokers and smokers.

The descriptive statistics should be compared to the scatter plot of Figure 5.1. I notice that the distribution of FEV1 appears to be symmetric at each time point for each group and that the mean at each time point is reflected with the lowess curves of Figure 5.1. Considering the mean FEV1 value of 3.09 at time 6 years for smokers, can you identify the corresponding point on Figure 5.1?

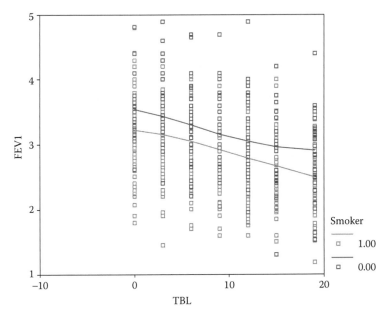

FIGURE 5.1
Association between FEV1 and time since baseline.

TABLE 5.1

Number of Observations at Each Time
Point for Nonsmokers and Smokers

Time Since Baseline	N
Nonsmokers	
0	23
3	27
6	28
9	30
12	29
15	24
19	28
Smokers	
0	85
3	95
6	89
9	85
12	81
15	73
19	74

TABLE 5.2

Descriptive Statistics for FEV1 for Nonsmokers and Smokers

Time	Mean	Median	SD
Nonsmokers			
0	3.519	3.4	0.4856
3	3.577	3.45	0.6277
6	3.023	3.24	0.622
9	3.171	3.175	0.581
12	3.136	3.1	0.613
15	2.87	2.95	0.609
19	2.90	3.09	0.646
Smokers			
0	3.229	3.3	0.5883
3	3.119	3.1	0.548
6	3.09	3.2	0.564
9	2.87	2.95	0.562
12	2.797	2.85	0.544
15	2.681	2.7	0.513
19	2.497	2.465	0.504

Suppose we represent the observations FEV1 with the regression model given by Equation 5.1 and suppose that when $g_i = 0$, then subject i is a non-smoker, but when $g_i = 1$, subject i is a smoker. Therefore, the expectation (5.1) reduces to (5.2) for nonsmokers and (5.3) for smokers.

In the Bayesian analysis, we want to know if $\beta_3 = \beta_4 = 0$, because if this is indeed the case, the two groups share a common mean profile given by (5.2).

The Bayesian analysis is executed with **BUGS CODE 5.1** using 55,000 observations for the simulation and a refresh of 100. Note that the code closely follows the notation for the mean profile expressed by (5.2) and (5.3). Also, the prior distributions for the regression coefficients are noninformative, and the prior distributions for the precision matrix of the multivariate normal are also noninformative. It is also important to know that an unstructured 7 by 7 precision matrix is assumed for the correlation between the seven possible time points for the FEV1 measurement. In addition, it is assumed that the two populations, smokers and nonsmokers, share a common precision matrix labeled as Omega in the code. Chapter 6 will describe various structures for the precision matrix of the repeated measures.

The results for the posterior analysis are provided by Table 5.3.

Therefore, the intercept for nonsmokers is estimated (via the posterior mean) as 3.523, the slope is estimated as −0.03377, and the 95% credible interval for the slope is (−0.04021, −0.027365). On the other hand, for smokers, the

BUGS CODE 5.1

```
model;
{

# prior distribution for the regression coefficients.
            beta1 ~ dnorm(0.0, 0.001)
            beta2 ~ dnorm(0.0, 0.001)
            beta3 ~ dnorm(0.0,.0001)
            beta4 ~ dnorm(0.0,.0001)

for(i in 1:N1){Y1[i,1:M1]~dmnorm(mu1[],Omega[,])}
for (j in 1:M1){mu1[j]<-beta1+beta2*age[j]}
# non-informative precision matrix
Omega[1:M1,1:M1]~dwish(R[,],7)
Sigma[1:M1,1:M1]<-inverse(Omega[,])

for(i in 1:N2){Y2[i,1:M2]~dmnorm(mu2[],Omega[,])}
for (j in 1:M2){mu2[j]<-beta3+beta4*age[j]}
d13<-beta1-beta3
d23<-beta2-beta4
}

# non smokers
list(N1 = 32,N2 = 100,M1 = 7,M2 = 7,Y1 = structure(.Data
= c(3.4,3.4,3.5,3.2,NA,3.0,2.4,
3.3,3.8,3.5,3.0,3.1,NA,NA,
NA,2.4,2.4,2.4,NA,2.2,2.0,
4.4,4.4,4.1,4.1,4.0,NA,3.6,
3.2,3.4,2.7,3.0,3.1,2.6,1.6,
3.1,NA,NA,2.4,2.5,2.5,2.1,
3.7,3.8,3.9,3.8,3.8,3.5,3.2,
3.8,3.2,3.3,3.2,3.1,3.0,NA,
NA,4.6,4.7,4.7,4.9,NA,4.4,
3.2,NA,2.7,2.8,2.7,2.5,2.5,
3.6,3.7,3.4,3.5,3.3,3.2,3.4,
3.6,4.0,NA,3.6,3.8,NA,3.4,
4.8,4.6,4.7,4.0,3.9,4.2,3.3,
2.6,2.4,2.5,2.6,2.6,2.4,2.3,
NA,NA,3.6,2.7,3.0,3.1,3.2,
NA,NA,2.2,2.1,2.0,1.3,1.5,
3.7,3.3,3.0,3.2,NA,2.9,NA,
3.4,3.3,3.1,3.0,3.0,NA,2.8,
4.1,4.1,3.7,3.8,3.7,3.7,3.6,
NA,2.9,2.8,2.8,2.6,2.6,2.8,
4.1,4.3,3.5,NA,3.3,NA,3.5,
3.2,3.0,2.9,3.0,2.8,2.5,2.6,
NA,3.2,3.2,3.4,3.1,NA,3.2,
```

```
3.5,3.6,3.5,3.2,3.0,3.0,3.2,
3.2,3.5,3.4,3.4,3.1,3.0,3.0,
NA,4.9,3.9,3.7,4.0,3.8,3.6,
3.3,3.3,3.2,3.3,2.9,3.2,3.1,
3.1,3.3,3.0,NA,2.6,NA,2.5,
NA,NA,3.7,3.6,3.4,3.4,3.1,
NA,NA,2.7,2.5,2.3,2.2,2.5,
NA,3.2,3.0,2.8,2.8,2.6,2.4,
3.7,3.7,NA,2.7,3.3,3.0,3.1),.Dim = c(32,7)),
# smokers
Y2 = structure(.Data = c(3.1,3.2,3.5,3.0,2.9,NA,NA,
3.6,3.5,3.5,3.1,NA,2.8,2.7,
NA,2.7,2.9,2.7,2.7,2.5,NA,
3.4,3.3,2.9,2.3,2.5,2.4,NA,
NA,2.5,2.5,2.1,2.4,NA,2.3,
3.9,4.0,4.1,3.8,4.0,NA,NA,
2.7,NA,3.3,2.4,2.2,2.2,2.3,
3.0,2.9,3.0,2.8,2.8,2.7,2.8,
2.8,2.7,2.1,1.9,1.8,NA,1.7,
3.9,3.8,3.5,3.3,3.5,3.2,3.3,
3.2,NA,3.0,2.8,2.9,2.5,2.2,
3.3,3.4,3.4,3.1,3.5,3.0,2.8,
NA,3.1,3.4,3.1,3.2,NA,2.6,
3.2,3.2,3.3,3.1,3.2,2.8,NA,
4.0,3.6,3.7,3.7,3.2,3.0,NA,
3.3,3.6,3.3,3.3,3.0,2.9,NA,
NA,3.0,3.2,3.0,2.9,2.4,NA,
2.8,2.5,2.6,2.5,2.4,2.3,2.2,
3.6,3.9,3.7,NA,3.3,3.1,NA,
3.6,3.4,3.4,NA,3.0,NA,2.7,
NA,2.0,1.8,1.7,1.8,1.6,1.5,
3.6,3.4,3.2,3.3,3.3,NA,2.6,
2.6,3.6,NA,2.9,3.2,3.1,2.2,
3.3,3.0,2.9,3.0,2.9,2.8,2.4,
3.6,3.6,3.4,NA,NA,2.8,2.8,
1.8,1.5,1.7,NA,1.8,1.5,1.2,
3.6,3.8,3.2,3.1,3.2,3.2,2.8,
4.2,3.8,3.5,3.8,3.5,3.4,3.3,
NA,3.9,4.0,NA,3.7,3.3,3.4,
3.7,NA,4.1,2.7,2.9,2.9,2.5,
2.6,2.6,2.7,2.6,2.6,2.4,2.5,
3.5,3.6,3.1,3.0,2.9,3.1,2.7,
2.2,2.3,2.3,2.1,NA,NA,1.5,
NA,3.5,3.3,3.8,3.6,3.9,3.4,
NA,3.1,3.3,3.1,NA,2.8,2.7,
3.7,3.7,3.7,3.3,3.1,NA,NA,
NA,2.4,2.3,2.1,1.9,2.2,2.0,
```

```
4.4,4.2,4.1,4.1,NA,4.0,NA,
2.9,2.7,2.8,NA,2.5,NA,2.1,
3.1,2.6,2.7,2.5,2.4,2.1,2.3,
2.5,2.4,2.0,NA,NA,2.1,2.3,
3.3,3.1,3.0,3.2,NA,3.3,2.9,
2.8,2.7,2.7,2.4,NA,2.1,1.8,
2.7,2.4,3.0,1.9,2.6,2.5,2.4,
3.4,3.4,3.4,3.1,2.9,NA,2.5,
3.7,3.4,3.3,3.2,2.7,NA,2.8,
NA,2.5,2.4,2.6,2.0,NA,2.0,
3.6,3.4,3.5,3.5,3.6,3.3,3.2,
2.6,2.7,2.7,2.7,NA,NA,2.4,
3.4,3.5,3.7,3.4,3.6,NA,3.0,
3.4,3.1,3.0,2.9,NA,2.6,2.4,
3.6,3.3,NA,3.2,3.0,2.7,NA,
4.1,3.6,3.8,3.9,3.6,3.5,NA,
3.7,3.6,3.3,3.1,3.3,3.1,NA,
4.8,3.8,3.8,3.5,3.6,NA,3.3,
3.7,4.2,3.9,3.3,3.2,3.5,3.6,
2.8,2.9,NA,NA,2.7,2.4,2.5,
3.4,3.4,3.3,2.7,3.0,3.1,3.1,
3.0,3.1,2.8,1.7,2.9,2.7,2.2,
3.0,2.5,NA,2.1,2.3,NA,1.9,
3.5,3.4,2.9,3.1,NA,2.8,NA,
2.9,2.5,NA,2.4,2.7,2.3,2.1,
NA,3.5,NA,3.0,3.0,3.2,3.2,
2.3,2.6,2.6,NA,2.4,2.3,2.2,
3.5,3.6,3.4,3.3,NA,2.8,2.8,
3.4,3.3,NA,NA,2.8,2.7,2.6,
3.5,3.1,2.9,2.6,2.6,2.3,NA,
3.8,3.7,3.6,3.4,3.2,3.0,2.8,
3.9,3.1,3.6,3.4,3.3,NA,3.0,
3.5,3.8,4.0,3.8,NA,3.2,3.0,
3.4,3.0,3.1,3.1,NA,2.8,NA,
2.4,2.8,2.7,2.1,NA,2.0,1.9,
1.9,1.9,1.8,1.6,1.6,NA,NA,
3.0,2.8,2.5,NA,NA,2.2,2.2,
2.7,3.0,NA,NA,2.3,1.9,1.6,
2.8,3.4,NA,3.0,2.5,2.5,NA,
NA,3.1,NA,2.8,2.5,2.0,2.7,
NA,4.4,3.5,3.6,NA,3.4,3.2,
4.1,3.7,3.9,2.8,3.4,3.0,3.2,
2.1,2.2,2.4,2.3,2.2,NA,NA,
3.2,2.9,2.8,2.7,2.6,2.4,2.3,
3.5,3.5,3.9,3.3,NA,2.6,2.5,
3.2,3.7,3.4,3.4,3.5,NA,NA,
3.5,3.5,3.2,3.0,3.1,2.8,2.6,
```

```
4.3,NA,4.1,4.1,3.8,3.7,3.5,
2.9,3.1,2.7,2.8,NA,2.2,2.2,
2.3,2.2,2.0,2.0,1.9,NA,NA,
3.8,3.4,3.3,3.1,3.1,NA,NA,
4.0,3.3,3.5,NA,3.3,2.8,2.7,
2.8,2.9,2.8,2.4,2.5,2.4,2.3,
3.0,NA,3.2,2.5,2.5,2.7,NA,
NA,2.6,3.0,2.4,2.5,2.1,NA,
3.0,3.1,2.9,2.8,2.5,2.4,2.1,
3.0,2.7,NA,2.4,2.2,2.2,2.3,
2.4,2.3,2.3,2.0,1.9,NA,NA,
NA,2.6,2.5,NA,2.1,2.2,1.7,
NA,3.1,2.9,NA,2.4,2.7,2.2,
3.7,3.5,3.3,2.9,3.1,NA,2.6,
2.9,3.1,2.9,2.8,2.6,NA,NA,
3.0,3.0,2.7,2.5,2.1,2.0,2.2),.Dim = c(100,7)),
age = c(0,3,6,9,12,15,19),
R = structure(.Data = c(1,0,0,0,0,0,0,0,1,0,0,0,0,0,0,0,1,
0,0,0,0,0,0,0,1,0,0,0,0,0,0,0,1,0,0,0,0,0,0,0,
1,0,0,0,0,0,0,0,1),.Dim = c(7,7)))
# initial values
list(beta1 = 0,beta2 = 0,beta3 = 0,beta4 = 0)
```

TABLE 5.3

Posterior Analysis for Linear–Linear Mean Profile

Parameter	Mean	SD	Error	2½	Median	97½
β_1	3.523	0.1007	0.000576	3.326	3.524	3.721
β_2	−0.033770	0.00327	0.0000201	−0.04021	−0.03376	−0.027365
β_3	3.263	0.05863	0.000315	3.148	3.263	3.379
β_4	−0.03828	0.001924	0.0000127	−0.04206	−0.03829	−0.034485
d_{13}	0.2606	0.1157	0.000601	0.0344	0.2607	0.4879
d_{24}	0.00451	0.00378	0.0000212	−0.002943	0.00454	0.011895

intercept is estimated as 3.263, the slope as −0.03828, and the 95% credible interval for the slope as (−0.04206, 0.034485). It is interesting to note that the posterior distributions appear to be symmetric about the posterior mean for both the intercept and the slope. Simulation errors are very small, indicating the posterior mean is very close to the "true" posterior mean for each of the four parameters. Of special interest are the questions: (1) are the intercepts of the two groups the same and (2) are the two slopes different?

The 95% credible interval for d_{13}, the difference in the intercepts is (0.0344, 0.2607), while that for the slopes is (−0.002943, 0.011895). Referring to Figure 5.1, it does appear that the intercepts are not the same, but the slopes appear to be the same.

This example was also examined by Fitzmaurice, Laird, and Ware,[1] and the Bayesian estimates (based on the posterior mean) for the intercepts and slopes are almost the same as their maximum-likelihood estimates. See page 155 of the second edition, which reports 3.057 and −0.033 for the intercept and slope, respectively, of nonsmokers and 3.245 and −0.038, respectively, for smokers. The two sets estimated are almost the same because noninformative priors are used in the Bayesian analysis.

Using polynomials for this example is an advantage because the number of time points at which the FEV1 is measured is not the same for all subjects. The possible points at which measurements can be made are at 0, 3, 6, 9, 12, 15, and 19 years; thus the maximum number of observations per subject is 7, but there are many subjects with less than 7 observations. Having an unequal number of observations from subject to subject makes it difficult to use descriptive statistics and t test to estimate the mean profile for this study. When there are missing observations, one must investigate the implications for biased estimates of the mean profile, which is a special topic to be considered in Chapters 6 through 10.

Based on the Bayesian analysis, it appears that a plausible model, one that includes a common slope for smokers and nonsmokers; thus, let the expectation for nonsmokers be

$$E\left(Y_{ij}\right) = \beta_1 + \beta t_{ij} \tag{5.4}$$

and for smokers, consider

$$E\left(Y_{ij}\right) = \beta_3 + \beta t_{ij} \tag{5.5}$$

The Bayesian analysis is repeated using a common slope for smokers and nonsmokers. Based on a slight variation of **BUGS CODE 5.1** with 65,000 observations for the simulation, initially with 5000 observations to start and a refresh of 100, the posterior analysis is given by Table 5.4.

Therefore, the intercept for nonsmokers is estimated (posterior mean) as 3.561 and as 3.251 for smokers, and in addition, the common slope is estimated as −0.037110. See Figure 5.2 for a plot of the posterior density of the common

TABLE 5.4

Posterior Analysis for Smokers and Nonsmokers

Parameter	Mean	SD	Error	2½	Median	97½
β	−0.037110	0.00167	0.00000947	−0.04035	−0.03711	−0.033825
β_1	3.561	0.09529	0.0004483	3.374	3.561	3.747
β_3	3.251	0.05765	0.0002907	3.137	3.251	3.363
d_{13}	0.3108	0.1075	0.0005101	0.09971	0.3109	0.521

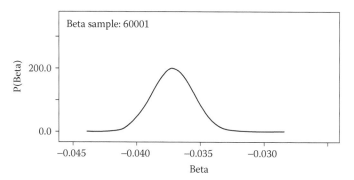

FIGURE 5.2
Posterior density of the common slope β.

slope β and note the symmetry about the posterior mean of −0.037110. The 95% credible interval for $d13 = \beta_1 - \beta_3$ implies that the intercepts are indeed not the same; thus, the final model for this study is that the two groups have the same slope but different intercepts.

5.2.2 Quadratic Mean Profile

In the previous section, a linear relation between the main response and time was assumed for the mean profile of two groups of subjects. From experience one would expect the mean profile to not be too complex and that simple polynomials or linear shift point models should be able to indeed model the mean profile. For our next case, a quadratic will be proposed to model the mean profile of two groups. A good example of this is the study selected from Table 5.2 of Crowder and Hand (p. 79),[3] where the hematocrit of hip-replacement patients is measured on four occasions. Also measured is the age of each subject at the beginning of the study, and there are two groups, male and female patients. Hematocrit is a measure of how much volume the red blood cells take up in the blood and is measured as a percentage; thus, an observed value of 33 means red blood cells comprise 33% by volume of the blood. For our study, one follows the hematocrit of hip-replacement patients in order to see if they are becoming anemic. The data for Crowder and Hand is found in Table 5.5, where the four times are denoted by t1, t2, t3, and t4, and missing values are denoted by periods.

Figure 5.3 demonstrates that a quadratic mean response is a possibility for the two groups. The expectation for the mean profile of this study is described by

$$E\left(Y_{ij}\right) = \beta_1 + \beta_2 t_{ij} + \beta_3 t_{ij}^2 + \beta_4 g_i + \beta_5 t_{ij} g_i + \beta_6 t_{ij}^2 g_i \qquad (5.6)$$

where:
Y_{ij} is the hematocrit of subject i at time t_{ij}
$g_i = 0$ if subject i is a male, otherwise $g_i = 1$

TABLE 5.5

Crowder and Hand Hip-Replacement Study

Subject	Age	t1	t2	t3	t4
1.00	66.00	47.1	31.1	.	32.8
2.00	70.00	44.1	31.5	.	37.0
3.00	44.00	39.7	33.7	.	24.5
4.00	70.00	43.3	18.4	.	36.6
5.00	74.00	37.4	32.3	.	29.1
6.00	65.00	45.7	35.5	.	39.8
7.00	54.00	44.9	34.1	.	32.1
8.00	63.00	42.9	32.1	.	.
9.00	71.00	46.1	28.8	.	37.8
10.00	68.00	42.1	34.4	34.0	36.1
11.00	69.00	38.3	29.4	32.9	30.5
12.00	64.00	43.0	33.7	34.1	36.7
13.00	70.00	37.8	26.6	26.7	30.6
14.00	60.00	37.3	26.5	.	38.5
15.00	52.00	.	28.0	.	33.9
16.00	52.00	27.0	32.5	.	32.0
17.00	75.00	38.4	32.3	.	37.9
18.00	72.00	38.8	32.6	.	26.9
19.00	54.00	44.7	32.2	.	34.2
20.00	71.00	38.0	27.1	.	37.9
21.00	58.00	34.0	23.2	.	26.0
22.00	77.00	44.8	37.2	.	29.7
23.00	66.00	46.0	29.1	.	26.7
24.00	53.00	41.9	32.0	37.1	37.6
25.00	74.00	38.0	31.7	38.4	35.7
26.00	78.00	42.2	34.0	32.9	33.3
27.00	74.00	39.7	33.5	26.6	32.7
28.00	79.00	37.5	28.2	28.8	30.3
29.00	71.00	34.6	31.0	30.1	28.7
30.00	68.00	35.5	24.7	28.1	29.8

Source: Crowder, M.J., and Hand, D.J., *Analysis of Repeated Measures*, Table 5.2, Chapman & Hall, London, 1990.

Therefore, the expectation (5.6) reduces to

$$E\left(Y_{ij}\right) = \beta_1 + \beta_2 t_{ij} + \beta_3 t_{ij}^2 \qquad (5.7)$$

for males and reduces to

$$E\left(Y_{ij}\right) = \beta_1 + \beta_4 + \left(\beta_2 + \beta_5\right)t_{ij} + \left(\beta_3 + \beta_6\right)t_{ij}^2 \qquad (5.8)$$

for females.

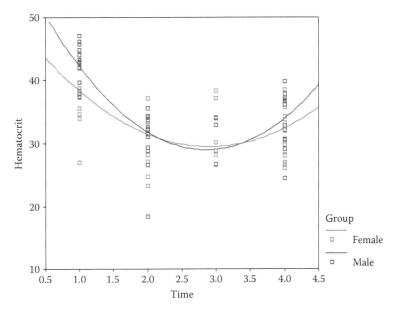

FIGURE 5.3
Hematocrit of male and female hip-replacement patients.

Note that there are 13 male and 17 female hip-replacement patients in the Crowder and Hand (p. 79)[3] study. Note that not all the patients have a complete set of four measurements. Indeed some have four, some three, and some two, but all have at least two repeated values.

Note the quadratic fit to the two groups, where group 1 is males and group 2 is females. From this graph, one could conclude that the two groups have the same quadratic main profile. One should be careful in extrapolating the quadratic response outside of the time range of four occasions.

A Bayesian analysis is performed with **BUGS CODE 5.2**.

The Bayesian analysis is executed with 65,000 observations for the simulation, with 5000 starting values and a refresh of 100. Note that the Bayesian analysis assumes an unstructured covariance matrix for the observations at the four time points (Table 5.6).

When making Bayesian inference about the quadratic response for the mean profile of the two groups, one should refer to Equations 5.7 and 5.8. One should notice that the 95% credible interval for β_4, β_5, and β_6 suggests that males and females share the same quadratic response, namely that given by Equation 5.5. These credible intervals all include the number 0; thus, if one replaces these parameters by 0, one would arrive at

$$E(Y_{ij}) = \beta_1 + \beta_2 t_{ij} + \beta_3 t_{ij}^2 \tag{5.9}$$

for the common mean response (hematocrit) for males and females. It also implies that one should repeat the Bayesian analysis based on (5.7) for the

BUGS CODE 5.2

```
model;
{
    beta1 ~ dnorm(0.0, 0.001)
    beta2 ~ dnorm(0.0, 0.001)
  beta3 ~ dnorm(0.0,.0001)
  beta4 ~ dnorm(0.0,.0001)
  beta5 ~ dnorm(0.0,.0001)
  beta6 ~ dnorm(0.0,.0001)
# Y1 is hematocrit for males
for(i in 1:N1){Y1[i,1:M1]~dmnorm(mu1[],Omega[,])}
for (j in 1:M1){mu1[j]<-beta1+beta2*age[j]+beta3*age[j]
*age[j]}
Omega[1:M1,1:M1]~dwish(R[,],4)
Sigma[1:M1,1:M1]<-inverse(Omega[,])

# Y2 is hematocrit for females
for(i in 1:N2){Y2[i,1:M2]~dmnorm(mu2[],Omega[,])}
for (j in 1:M2){mu2[j]<-beta1+beta4+(beta2+beta5)*age[j]+
(beta3+beta6)*age[j]*age[j]}

}

# Males

list(N1 = 13,N2 = 17,M1 = 4,M2 = 4,
Y1 = structure(.Data =
c(47.1,31.1,NA,32.8,
44.1,31.5,NA,37.0,
39.7,33.7,NA,24.5,
43.3,18.4,NA,36.6,
37.4,32.3,NA,29.1,
45.7,35.5,NA,39.8,
44.9,34.1,NA,32.1,
42.9,32.1,NA,NA,
46.1,28.8,NA,37.8,
42.1,34.4,34.0,36.1,
38.3,29.4,32.9,30.5,
43.0,33.7,34.1,36.7,
37.8,26.6,26.7,30.6),.Dim = c(13,4)),
# females
Y2 = structure(.Data =
c(37.3,26.5,NA,38.5,
NA,28.0,NA,33.9,
27.0,32.5,NA,32.0,
38.4,32.3,NA,37.9,
38.8,32.6,NA,26.9,
```

```
44.7,32.2,NA,34.2,
38.0,27.1,NA,37.9,
34.0,23.2,NA,26.0,
44.8,37.2,NA,29.7,
46.0,29.1,NA,26.7,
41.9,32.0,37.1,37.6,
38.0,31.7,38.4,35.7,
42.2,34.0,32.9,33.3,
39.7,33.5,26.6,32.7,
37.5,28.2,28.8,30.3,
34.6,31.0,30.1,28.7,
35.5,24.7,28.1,29.8),.Dim = c(17,4)),
age = c(1,2,3,4),
R = structure(.Data = c(1,0,0,0,
                        0,1,0,0,
                        0,0,1,0,
                        0,0,0,1),.Dim = c(4,4)))
# initial values
list(beta1 = 0,beta2 = 0,beta3 = 0,beta4 = 0,beta5 =
0,beta6 = 0)
```

TABLE 5.6

Posterior Analysis for Hip-Replacement Study

Parameter	Mean	SD	Error	2½	Median	97½
β_1	61.17	3.1	0.178	54.65	61.36	66.7
β_2	−22.62	2.901	0.1737	−27.78	−22.81	−16.54
β_3	3.892	0.5597	0.03345	2.735	3.927	4.888
β_4	−8.559	4.084	0.2361	−15.8	−8.791	0.5215
β_5	6.402	3.769	0.2267	−2.198	6.676	12.8
β_6	−1.156	0.7173	0.04317	−2.368	−1.205	0.4907

two groups. Will the repeat analysis give the same posterior analysis for β_1, β_2, and β_3?

A slight revision of **BUGS CODE 5.2** allows one to perform a Bayesian analysis with a common quadratic mean profile for males and females, giving the posterior analysis portrayed in Table 5.7.

The intercept is estimated as 56.69%, with the posterior mean, and the linear effect by −19.36, and finally the quadratic effect by 3.308. The estimated quadratic mean profile is presented in Figure 5.4.

Assuming no difference between males and females, Table 5.8 gives the descriptive statistics for the hip-replacement study.

Thus, the mean hematocrit at time 1 is 40.37%, with a low of 30.66% at time 2, to a mean value of 32.74% at time 4. One should compare Table 5.8 with Figure 5.3. Do the descriptive statistics appear compatible with Figure 5.3?

TABLE 5.7

Posterior Distribution for the Common Mean Profile of Males and Females:
Hip-Replacement Study

Parameter	Mean	SD	Error	2½	Median	97½
β_1	56.69	2.645	0.1503	51.55	56.85	61.59
β_2	−19.36	2.459	0.1442	−23.89	−19.48	−14.7
β_3	3.308	0.4729	0.02773	2.417	3.326	4.193

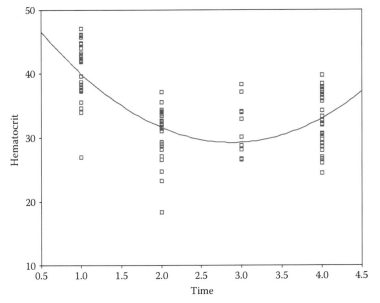

FIGURE 5.4
The common quadratic mean profile for hematocrit at four time points.

TABLE 5.8

Descriptive Statistics for Hip-Replacement Study

Time	Mean	SD	Median	N
1	40.372	4.527	39.7	29
2	30.669	4.018	32.0	29
3	31.791	4.035	32.9	11
4	32.741	4.338	32.7	29

TABLE 5.9

Posterior Analysis for Estimated Covariance Matrix

Parameter	Mean	SD	Error	2½	Median	97½
σ_1^2	21.2	5.946	0.0402	12.52	20.23	35.56
σ_{12}	4.67	3.835	0.0265	−2.136	4.373	13.23
σ_{13}	−13.31	10.82	0.2765	−37.28	−12.45	5.678
σ_{14}	5.714	4.087	0.0283	−1.309	5.308	15.03
σ_2^2	16.84	4.792	0.0456	9.919	16.05	28.39
σ_{23}	−0.8894	8.167	0.1315	−17.57	−0.8274	15.25
σ_{24}	−0.007903	3.529	0.0345	−7.075	0.00863	7.074
σ_3^2	68.14	32.57	0.7066	25.77	61.58	148.2
σ_{34}	24.98	10.27	0.119	9.073	23.57	48.99
σ_4^2	19.15	5.483	0.03676	11.22	18.22	32.59

Before concluding this section, it should be noted that one is assuming the main response is approximately normally distributed, a fact that can be verified with a PP plot of the data.

Last, remember that an unstructured covariance matrix is assumed for the repeated measures of hip-replacement example. Recall that, for this case, there were two situations for the mean profile of the two groups: (1) two quadratic curves with different coefficients and (2) one quadratic curve that is common to both groups. Assuming the latter case, the 4 by 4 variance–covariance matrix of the repeated measure at four times 1, 2, 3, and 4 is estimated with the entries given in Table 5.9.

Can you detect any pattern in this covariance matrix? Looking at the four variances, I cannot find any trend. Based on the estimated covariance matrix, one can easily compute estimates of the corresponding correlation matrix. For example, consider the correlation of hematocrit between time 1 and time 2, namely

$$\rho_{12} = \frac{\sigma_{12}}{\sqrt{\sigma_{11}\sigma_{22}}} \tag{5.10}$$

which is estimated by

$$\tilde{\rho}_{12} = \frac{4.67}{\sqrt{(21.2)(16.84)}} = 0.2471$$

5.2.3 Linear–Linear with One Join Point

An interesting study using a spline with one join point is analyzed by Fitzmarurice, Laird, and Ware[1] and is based on blood levels from lead-exposed children randomized study with placebo and a treatment arm using succimer, a chelating agent. See Rhoads[4] for additional information. It is hoped that succimer will decrease the blood levels of lead in children, where the lead level in µg/dL is measured at baseline and at weeks 0, 1, 4, and 6, thus, it is of interest to compare the two groups. The comparison will be based on fitting a linear association for the placebo group over all time points and fitting a spline with one known join point at week 1. That is, a linear association between blood level and time will be assumed for weeks 0 and 1 and another linear association over times 1, 4, and 6 weeks. Such a mean profile is implied by the scatter plot of Figure 5.5 where the mean profile of the two groups is delineated by a lowess plot. Group A is the one using the chelating agent succimer and P denotes placebo. Note that there are 200 children in each group.

Consider the following model for the mean profiles:

1. The placebo group

$$Y_{ij} = \beta_1 + \beta_2 t_{ij} \tag{5.11}$$

where:
Y_{ij} is the lead level for subject i at time $j = 0,1,4,6$

2. The succimer group
   a. For the second segment with time points 1, 4, and 6 weeks.

$$Y_{ij} = \beta_3 + \beta_4 t_{ij} \tag{5.12}$$

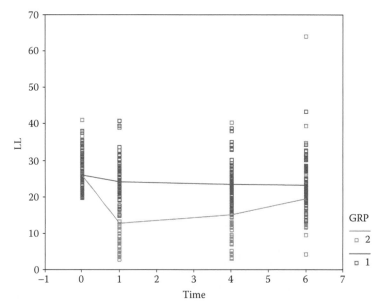

FIGURE 5.5
Lead levels in children with two groups, placebo and succimer.

where:

Y_{ij} is the lead level for subject i at time $j = 1,4,6$

b. For the first segment with time points 0 and 1 week.

$$Y_{ij} = \beta_5 + \beta_6 t_{ij} \qquad (5.13)$$

where:

Y_{ij} is the lead level for subject at time $j = 0,1$

The scatter plot of Figure 5.5 uses the time points 0, 1, 4, and 6 weeks; thus, the baseline measurement is at 0. From the plot, it does indeed seem reasonable to use a linear regression for the placebo but use a spline with two linear segments and a join point at time 1. It is also apparent that succimer does indeed have an effect reducing the lead level at time 1 but increasing slightly over time 1, 4, and 6. Table 5.10 presents the descriptive statistics of placebo and succimer group and shows that the baseline levels of both groups are the same.

One sees at time 1, the mean lead level for placebo is 24.66 µg/dL compared to 13.52 for the succimer group, demonstrating that at least it temporarily has a efficacious effect on the children.

Using the information from the succimer study, see Rhoads.[4] **BUGS CODE 5.3** will execute the Bayesian analysis and the code closely follows the formulas for the mean profile given by (5.11), (5.12), and (5.13). The analysis will focus on the six beta parameters that define the mean profile of the two groups. For example, β_1 and β_2 are the intercept and slope for the placebo group, while β_5 and β_6 are the intercept and slope for the first linear segment over time points 0 and 1, where 1 is the join point of the two segment mean profile. For the simulation, 65,000 observations are used with 5000 initial observations and a refresh of 100.

TABLE 5.10

Descriptive Statistics for Placebo Group and Succimer Group

Time	Mean	SD	Median
Placebo Group			
0	26.27	5.02	26.25
1	24.66	5.41	24.10
4	24.07	5.75	22.45
6	23.64	5.63	22.35
Succimer Group			
0	26.54	5.02	26.20
1	13.52	7.67	12.25
4	15.51	7.85	15.35
6	20.76	9.24	18.85

An unstructured covariance matrix is assumed for the multivariate normal distribution over the four time points, and the other parameters of interest are the precision and covariance matrices for the repeated measures over the time points.

When interpreting the posterior analysis, remember that baseline is time 0.

BUGS CODE 5.3

```
model;
{
    beta1 ~ dnorm(0.0, 0.001)
    beta2 ~ dnorm(0.0, 0.001)
  beta3 ~ dnorm(0.0,.0001)
  beta4 ~ dnorm(0.0,.0001)
  beta5 ~ dnorm(0.0,.0001)
  beta6 ~ dnorm(0.0,.0001)

for(i in 1:N1){YP[i,1:M1]~dmnorm(mu1[],Omega1[,])}
for (j in 1:M1){mu1[j]<-beta1+beta2*time1[j]}

Omega1[1:M1,1:M1]~dwish(R1[,],4)
Sigma1[1:M1,1:M1]<-inverse(Omega1[,])

for(i in 1:N2){YS146[i,1:M2]~dmnorm(mu2[],Omega2[,])}
for (j in 1:M2){mu2[j]<-beta3+beta4*time2[j]}

Omega2[1:M2,1:M2]~dwish(R2[,],3)
Sigma2[1:M2,1:M2]<-inverse(Omega2[,])

for(i in 1:N3){YS01[i,1:M3]~dmnorm(mu3[],Omega3[,])}
for (j in 1:M3){mu3[j]<-beta5+beta6*time3[j]}

Omega3[1:M3,1:M3]~dwish(R3[,],2)
Sigma3[1:M3,1:M3]<-inverse(Omega3[,])
}


list(N1 = 50,N2 = 50,N3 = 50,M1 = 4,M2 = 3,M3 = 2,time1
= c(0,1,4,6),
time2 = c(1,4,6),time3 = c(0,1),

# regression for placebo
YP = structure(.Data =

c(30.8,26.9,25.8,23.8,
24.7,24.5,22.0,22.5,
28.6,20.8,19.2,18.4,
```

```
33.7,31.6,28.5,25.1,19.7,14.9,15.3,14.7,31.1,31.2,29.2,30.1,
19.8,17.5,20.5,27.5,21.4,26.3,19.5,19.0,21.1,20.3,18.4,20.8,
20.6,23.9,19.0,17.0,24.0,16.7,21.7,20.3,37.6,33.7,34.4,31.4,
31.9,27.9,27.3,34.2,26.2,26.8,25.3,24.8,20.5,21.1,17.4,21.1,
33.3,26.2,34.0,28.2,27.9,21.6,23.6,27.7,24.7,21.2,22.9,21.9,
28.8,26.4,23.8,22.0,32.0,30.2,30.2,27.5,21.8,19.3,16.4,17.6,
24.9,20.9,22.2,19.8,19.8,18.9,18.9,15.5,35.4,30.4,26.5,28.1,
25.3,23.9,22.2,27.2,20.3,21.0,16.7,13.5,20.4,17.2,15.9,17.7,
24.1,20.1,17.9,18.7,28.5,32.6,27.5,22.8,26.6,22.4,21.8,21.0,
20.5,17.5,19.6,18.4,25.2,25.1,23.4,22.2,34.7,39.5,38.6,43.3,
30.3,29.4,33.1,28.4,26.6,25.3,25.1,27.9,20.7,19.3,21.9,21.8,
28.9,28.9,32.8,31.8,27.2,28.5,35.0,30.5,22.4,22.0,19.1,18.7,
32.5,25.1,27.8,27.3,24.9,23.6,21.2,21.1,24.6,25.0,21.7,23.9,
23.1,20.9,21.7,19.9,25.8,21.9,23.6,24.8,30.0,27.6,24.0,23.7,
20.0,22.7,21.2,20.5,38.1,40.8,38.0,32.7,25.1,28.1,27.5,24.8,
22.1,21.1,21.5,20.6,25.4,24.3,22.7,20.1),.Dim = c(50,4)),
# Succimer segment for times 1,4,6.

YS257 = structure(.Data =

c(14.8,19.5,21.0,23.0,19.1,23.2,2.8,3.2,9.4,5.4,4.5,11.9,23.1,
24.6,30.9,6.3,18.5,16.3,25.5,26.3,30.3,15.8,22.9,25.9,15.8,
23.7,23.4,6.5,7.1,16.0,12.0,16.8,19.2,4.2,4.0,16.2,11.5,9.5,
14.5,3.9,12.8,12.7,21.4,21.0,22.4,13.2,14.6,11.6,17.5,21.0,24.2,
16.4,11.6,16.6,14.9,14.5,63.9,6.4,5.1,15.1,20.4,19.3,23.8,10.6,
9.0,16.0,17.5,17.4,18.6,10.0,15.6,15.2,14.9,18.1,21.3,39.0,28.8,
34.7,5.1,8.2,23.6,4.0,4.2,11.7,24.3,18.4,27.8,23.3,40.4,39.3,10.7,
12.6,21.2,19.0,16.3,18.6,9.2,8.3,18.4,15.3,24.6,32.4,10.6,14.4,
18.7,5.6,7.3,12.3,21.0,8.6,24.6,12.5,16.7,22.2,11.6,13.0,23.1,7.9,
12.4,18.9,16.8,15.1,18.8,3.5,3.0,11.5,28.2,27.0,25.5,7.1,17.2,
18.7,10.8,19.8,22.2,3.9,7.0,17.8,15.1,10.9,27.1,22.1,25.3,
4.1,7.6,10.8,13.0,8.1,25.7,12.3),.Dim = c(50,3)),

# succimer segment from 0,1
YS12 = structure(.Data =

c(26.5,14.8,25.8,23.0,20.4,2.8,20.4,5.4,24.8,23.1,27.9,6.3,
35.3,25.5,28.6,15.8,29.6,15.8,
21.5,6.5,21.8,12.0,23.0,4.2,22.2,11.5,25.0,3.9,26.0,21.4,
19.7,13.2,29.6,17.5,24.4,16.4,33.7,14.9,26.7,6.4,26.8,
20.4,20.2,10.6,20.2,17.5,24.5,10.0,27.1,14.9,34.7,39.0,24.5,
5.1,
27.7,4.0,24.3,24.3,36.6,23.3,34.0,10.7,32.6,19.0,29.2,9.2,
26.4,15.3,21.8,10.6,21.1,5.6,
22.1,21.0,28.9,12.5,19.8,11.6,23.5,7.9,29.1,16.8,30.3,3.5,
30.6,28.2,22.4,7.1,31.2,10.8,
```

```
31.4,3.9,41.1,15.1,29.4,22.1,21.9,7.6,20.7,8.1),.Dim =
c(50,2)),

R1 = structure(.Data = c(1,0,0,0,
                         0,1,0,0,
                         0,0,1,0,
                         0,0,0,1),.Dim = c(4,4)),

R2 = structure(.Data = c(1,0,0,
                         0,1,0,
                         0,0,1),.Dim = c(3,3)),

R3 = structure(.Data = c(1,0,
                         0,1),.Dim = c(2,2)))

# initial values

list(beta1 = 0,beta2 = 0,beta3 = 0,beta4 = 0,beta5
= 0,beta6 = 0)
```

TABLE 5.11

Posterior Distribution for Succimer Study

Parameter	Mean	SD	Error	2½	Median	97½
β_1	26.05	0.7426	0.003472	24.59	26.05	27.50
β_2	−0.3969	0.0895	0.000426	−0.574	−0.3964	−0.222
β_3	12.37	1.259	0.008081	9.9	12.36	14.84
β_4	1.11	0.2123	0.001319	0.6908	1.111	1.527
β_5	26.54	0.7163	0.003261	25.12	26.54	27.94
β_6	−13.02	1.046	0.005068	−15.07	−13.02	−10.95

The Bayesian analysis is portrayed in Table 5.11.

Thus, for the placebo group, the intercept is estimated as 26.05 with a slope of −0.3969. For the succimer group, the first linear segment over times 0 and 1 has an estimated intercept of 26.54 and an estimated slope of −13.02. For the linear segment over time points 1, 4, and 6, the estimated slope is 1.11 and the estimated intercept is 12.37. One should refer to Table 5.10 when interpreting the Bayesian analysis of Table 5.11. Also, all the posterior distributions appear to be symmetric about the posterior mean, and Markov chain Monte Carlo (MCMC) errors appear to be sufficiently small.

Last, the covariance matrix is estimated by the Sigma1 matrix of the Bayesian analysis. See **BUGS CODE 5.3**, where Sigma1 is the inverse of the precision matrix Omega1, the precision matrix for the lead level between

TABLE 5.12

Posterior Means of Variance–Covariance Matrix
at Four Time Points for Placebo Group

Time	0	1	4	6
0	25.9	23.06	24.69	21.92
1		31.39	27.98	23.95
4			33.88	28.88
6				32.55

the four times of the placebo group. The prior distribution for Omega1 is uninformative, and the posterior distribution of the entries of the variance–covariance matrix appears in Table 5.12. Also, the variances and covariances are estimated by the posterior mean.

Therefore, the variance of the lead level at time 0 for the placebo group is estimated by 25.9, while the covariance between times 0 and 1 is estimated as 23.06. Note that the corresponding correlations can be estimated as follows. Consider the correlation of the lead levels between times 0 and 1, namely

$$\rho_{01} = \frac{\sigma_{01}}{\sqrt{\sigma_{00}\sigma_{11}}} \tag{5.14}$$

which is estimated by

$$\frac{23.06}{\sqrt{(25.99)(31.39)}} = 0.8074 \tag{5.15}$$

All the relevant correlations can be estimated in a similar way.

5.3 Modeling the Mean Profile for Discrete Observations

Suppose we consider a repeated measure study with discrete observations, where the most elementary case is when the observations are binary (e.g., yes or no responses). For determining the mean profile, it is of interest to plot the percent of yes responses at each time point. For the case of two or more groups, the mean profile of percents can be compared between the two groups. In the case of more than two responses, the mean (of the several discrete observations) response can be used to delineate the profile of a group of subjects.

First to be considered is the binary case such as the repeated measures study of church attendance by Cornoni-Huntley et al.,[5] where the data are portrayed in Table 5.13. This is part of the Iowa 65+ Rural Health Study,

TABLE 5.13

Church Attendance for Iowa Rural Health Study

Gender	Year 0	Year 3	Year 6	Count
Female	Yes	Yes	Yes	904
	Yes	Yes	No	88
	Yes	No	Yes	25
	Yes	No	No	51
	No	Yes	Yes	33
	No	Yes	No	22
	No	No	Yes	30
	No	No	No	158
Total				1311
Male	Yes	Yes	Yes	391
	Yes	Yes	No	36
	Yes	No	Yes	12
	Yes	No	No	26
	No	Yes	Yes	15
	No	Yes	No	21
	No	No	Yes	18
	No	No	No	143
Total				662

Source: Cornoni-Huntley, J., Brock, D.B., Ostfeld, A., Taylor, J.O., and Wallace, R.B., *Established Populations for Epidemiologic Studies of the Elderly, Resource Data Book*, National Institutes of Health (NIH Publ. No. 86-2443), Bethesda, MD, 1986; Davis, C.S., *Statistical Methods for the Analysis of Repeated Measurements*, Springer-Verlag, New York, p. 197, 2002.

which followed a cohort of elderly males and females over a six-year period. At each time point, the subject answered yes or no regarding their regular church attendance. This study was analyzed by Davis,[6] and the data appears in his book as Table 7.6.

There are a total of 1973 members of the cohort, of which 1311 are female and 662 are male. Thus, for females, the fraction who responded yes at baseline is 0.8146 compared to 0.7024 for males. This is summarized in Table 5.14, where the number in parentheses is the estimated standard deviation of the estimated fraction of attendance.

TABLE 5.14

Proportion of People with Regular Church Attendance for Males and Females

Gender	Year 0	Year 3	Year 6
Females	0.8146 (0.0107)	0.7986 (0.0110)	0.7566 (0.0118)
Males	0.7024 (0.0177)	0.6993 (0.0178)	0.6586 (0.0184)

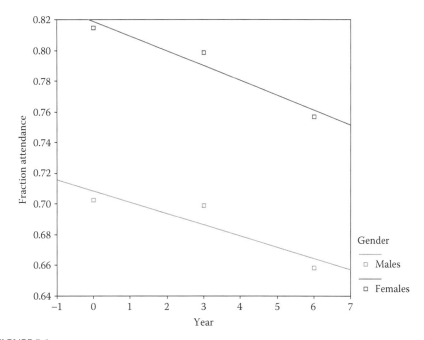

FIGURE 5.6
Fraction of church attendance for females and males.

Thus, it appears that for both groups, the fraction attending church on a regular basis is in decline, and that females have a higher fraction of attendance for each of the three years. Of course, of interest is the difference in the rate of decline for males compared to females. See Figure 5.6, which suggests that the rate of decline in church attendance is similar for males and females.

What is the Bayesian analysis for this binary study? One approach is to assume the attendance pattern for males and females follow a multinomial distribution, then one can employ the Dirichlet distribution as the posterior distribution of the attendance fraction separately for males and females. Of course one must assume a prior distribution for the fraction, and initially an uninformative prior will be adopted.

If a uniform prior is adopted for the female fractions $\theta = \theta_1, \theta_2, \theta_3, \theta_4, \theta_5, \theta_6, \theta_7, \theta_8$), then it can be shown that the posterior distribution of the θ_i, $i = 1,2,...,8$, is Dirichlet with parameter vector (905, 89, 26, 52, 34, 23, 31, 159).

In a similar way, let $\phi = (\phi_1, \phi_2, \phi_3, \phi_4, \phi_5, \phi_6, \phi_7, \phi_8)$ be the vector of fractions for males, then assuming a uniform prior the posterior distribution of the ϕ_i, $i = 1,2,...,8$, is Dirichlet with parameter vector (392, 37, 13, 27, 16, 22, 19, 144).

For additional information about using the Dirichlet for Bayesian inference, see DeGroot (p. 40)[7] and Broemeling.[8]

Note that there are eight patterns of yes or no responses for females; thus, let m_i be the number in category i (= 1,2,...,8) of Table 5.13; then assuming these eight frequencies follow a multinomial distribution with probability mass function

$$f\left(m_1,m_2,m_3,m_4,m_5,m_6,m_7,m_8\big|\theta\right) \propto \prod_{i=1}^{i=8} \theta_i^{m_i} \tag{5.16}$$

where $0 < \theta_i < 1$, $i = 1,2,...,8$, and $\sum_{i=1}^{i=8}\theta_i = 1$. Assume the uniform density for the prior distribution of the θ_i, namely, $g(\theta) = 1$, where $0 < \theta_i < 1$, $i = 1,2,...,8$, and $\sum_{i=1}^{i=8}\theta_i = 1$. Then applying Bayes's theorem one may show that the posterior distribution of the θ_i, $i = 1,2,...,8$, is Dirichlet with parameter vector $(m_1 +1, m_2 +1, m_3 +1, m_4 +1, m_5 +1, m_6 +1, m_7 +1, m_8 +1)$.

It is easy to show that the posterior mean of θ_i is

$$E\left(\theta_i\big|\text{data}\right) = \frac{m_i +1}{m+8}$$

where

$$m = \sum_{i=1}^{i=8} m_i, \quad \text{for } i = 1,2,...,8$$

If one adopts the improper and uninformative prior

$$g(\theta) = \prod_{i=1}^{i=8} \theta_i^{-1} \tag{5.17}$$

where $0 < \theta_i < 1$, $i = 1,2,...,8$, and $\sum_{i=1}^{i=8}\theta_i = 1$, then one may show the posterior distribution of the θ_i is Dirichlet with parameter vector $(m_1,m_2,m_3,m_4,m_5,m_6, m_7, m_8)$.

Thus, the posterior mean of θ_i is

$$E\left(\theta_i\big|\text{data}\right) = \frac{m_i}{m} \tag{5.18}$$

and

$$m = \sum_{i=1}^{i=8} m_i \tag{5.19}$$

In addition, the variance of the posterior distribution of θ_i is

$$\text{Var}\left(\theta_i\right) = \frac{m_i\left(m-m_i\right)}{m^2\left(m+1\right)} \tag{5.20}$$

where

$$m = \sum_{i=1}^{i=8} m_i \tag{5.21}$$

Also, the posterior covariance between θ_i and θ_j is

$$\text{Cov}\left(\theta_i, \theta_j\right) = -\frac{m_i m_j}{m^2(m+1)} \tag{5.22}$$

Note that $m = 1311$ is the total number of females in the cohort and m_i is the number of subjects in the ith category (see Table 5.13). Of course similar expressions for the males are given as follows.

Note that there are eight patterns of yes or no responses for males; thus, let n_i be the number in category i (= 1,2,...,8) of Table 5.13; then assuming these eight frequencies follow a multinomial distribution with probability mass function

$$f\left(n_1, n_2, n_3, n_4, n_5, n_6, n_7, n_8 \mid \phi\right) \propto \prod_{i=1}^{i=8} \phi_i^{n_i} \tag{5.23}$$

where $0 < \phi_i < 1$, $i = 1,2,...,8$, and $\sum_{i=1}^{i=8} \phi_i = 1$. Assume the uniform density for the prior distribution of the ϕ_i, namely, $g(\phi) = 1$, where $0 < \phi_i < 1$, $i = 1,2,...,8$, and $\sum_{i=1}^{i=8} \phi_i = 1$, then applying Bayes' theorem, one may show that the posterior distribution of the ϕ_i, $i = 1,2,...,8$ is Dirichlet with parameter vector $\left(n_1 + 1, n_2 + 1, n_3 + 1, n_4 + 1, n_5 + 1, n_6, +1, n_7 + 1, n_8 + 1\right)$.

It is easy to show that the posterior mean of ϕ_i is

$$E\left(\phi_i \mid \text{data}\right) = \frac{n_i + 1}{n + 8} \tag{5.24}$$

where:

$$n = \sum_{i=1}^{i=8} n_i, \quad \text{for } i = 1, 2, \dots, 8$$

If one adopts the improper and uninformative prior

$$g(\phi) = \prod_{i=1}^{i=8} \phi_i^{-1} \tag{5.25}$$

where $0 < \phi_i < 1$, $i = 1,2,...,8$, and $\sum_{i=1}^{i=8} \phi_i = 1$, then one may show the posterior distribution of the ϕ_i is Dirichlet with parameter vector $\left(n_1, n_2, n_3, n_4, n_5, n_6, n_7, n_8\right)$.

Thus, the posterior mean of ϕ_i is

$$E\left(\phi_i \mid \text{data}\right) = \frac{n_i}{n} \tag{5.26}$$

and

$$n = \sum_{i=1}^{i=8} n_i \tag{5.27}$$

In addition, the variance of the posterior distribution of ϕ_i is

$$\text{Var}(\phi_i) = \frac{n_i(n-n_i)}{n^2(n+1)} \tag{5.28}$$

where

$$n = \sum_{i=1}^{i=8} n_i \tag{5.29}$$

Also, the posterior covariance between ϕ_i and ϕ_j is

$$\text{Cov}(\phi_i,\phi_j) = -\frac{n_i n_j}{n^2(n+1)} \tag{5.30}$$

Note that $n = 662$ is the total number of males in the cohort and n_i is the number of subjects in the ith category (see Table 5.13).

Our main focus is to determine the mean, variance, and appropriate covariances of the three entities

$$\eta_0 = \theta_1 + \theta_2 + \theta_3 + \theta_4 \tag{5.31}$$

where:

η_0 is the proportion of females who answer yes to regular church attendance at time 0

$$\eta_3 = \theta_1 + \theta_2 + \theta_5 + \theta_6 \tag{5.32}$$

where:

η_3 is the probability of a subject answering yes at time 3

and

$$\eta_6 = \theta_1 + \theta_3 + \theta_5 + \theta_7 \tag{5.33}$$

where:

η_6 is the probability of a woman answering yes at year 6 to regular church attendance the previous three years

For this part of the presentation, the uninformative prior (5.17) will be assumed jointly for the θ_i, $i = 1,2,\dots,8$.

Note that the posterior mean of η_0 is

$$E(\eta_0|\text{data}) = \frac{1}{m} \sum_{i=1}^{i=4} m_i \tag{5.34}$$

$$= \left(\frac{1}{1311}\right)(1068)$$

$$E\left(\eta_3 | \text{data}\right) = \frac{1}{m}\left(m_1 + m_2 + m_5 + m_6\right)$$

$$= \left(\frac{1}{1311}\right)\left(904 + 88 + 33 + 22\right) \tag{5.35}$$

$$= 0.7986$$

and finally for the time point 6

$$E\left(\eta_6 | \text{data}\right) = 0.7566 \tag{5.36}$$

Thus, for females, the posterior probability that the answer is yes at baseline is estimated as 0.8146 via the posterior mean.

Also of interest is the correlation between the η_i, $i = 0,3,6$. Consider

$$\begin{aligned}
\text{Cov}\left(\eta_0, \eta_3 \mid \text{data}\right) &= \text{Cov}\left(\theta_1 + \theta_2 + \theta_3 + \theta_4, \theta_1 + \theta_2 + \theta_5 + \theta_6\right)\text{Cov}\left(\theta_1, \theta_1\right) \\
&+ \text{Cov}\left(\theta_1, \theta_2\right) + \text{Cov}\left(\theta_1, \theta_5\right) + \text{Cov}\left(\theta_1, \theta_6\right) \\
&+ \text{Cov}\left(\theta_1, \theta_2\right) + \text{Cov}\left(\theta_2, \theta_2\right) + \text{Cov}\left(\theta_2, \theta_5\right) \\
&+ \text{Cov}\left(\theta_2, \theta_6\right) + \text{Cov}\left(\theta_3, \theta_1\right) + \text{Cov}\left(\theta_3, \theta_2\right) \\
&+ \text{Cov}\left(\theta_3, \theta_5\right) + \text{Cov}\left(\theta_3, \theta_6\right) + \text{Cov}\left(\theta_4, \theta_1\right) \\
&+ \text{Cov}\left(\theta_4, \theta_2\right) + \text{Cov}\left(\theta_4, \theta_5\right) + \text{Cov}\left(\theta_4, \theta_6\right)
\end{aligned} \tag{5.37}$$

This represents the covariance between the proportion of respondents who answer yes at baseline and the proportion of respondents who answer yes at time point 3. The covariances on the right-hand side will be estimated via formula (5.22). These formulas allow one to do a complete Bayesian analysis without using simulation techniques.

On the other hand, one may want to employ a Bayesian simulation procedure; thus, in lieu of using formulas (5.26) and (5.37) for the posterior analysis, one may employ **BUGS CODE 5.4** with statements that closely follow the notation of those formulas.

A Bayesian analysis is performed with 65,000 observations for the simulation, beginning with 5000 observations and a refresh of 100 (Table 5.14).

As check on the computations, the sum of the thetas should be one.

For females, the estimated proportion (via the posterior mean) who respond yes to church attendance at baseline is 0.8146, a standard deviation of 0.0106, and 95% credible interval (0.7934, 0.8349) compared to males who have a posterior mean of 0.7024, a standard deviation of 0.0176, and 95% credible interval (0.6673, 0.7365). This pattern continues where at time 3, the estimated proportion of yes is 0.7986 for females compared to 0.6994 for males.

BUGS CODE 5.4

```
model;
{
g1~dgamma(m1,2)
g2~dgamma(m2,2)
g3~dgamma(m3,2)
g4~dgamma(m4,2)
g5~dgamma(m5,2)
g6~dgamma(m6,2)
g7~dgamma(m7,2)
g8~dgamma(m8,2)

sg<-g1+g2+g3+g4+g5+g6+g7+g8

# probabilities of female categories
theta1<-g1/sg
theta2<-g2/sg
theta3<-g3/sg
theta4<-g4/sg
theta5<-g5/sg
theta6<-g6/sg
theta7<-g7/sg
theta8<-g8/sg

eta0<-theta1+theta2+theta3+theta4

eta3<-theta1+theta2+theta5+theta6

eta6<-theta1+theta3+theta5+theta7

h1~dgamma(n1,2)
h2~dgamma(n2,2)
h3~dgamma(n3,2)
h4~dgamma(n4,2)
h5~dgamma(n5,2)
h6~dgamma(n6,2)
h7~dgamma(n7,2)
h8~dgamma(n8,2)

sh<-h1+h2+h3+h4+h5+h6+h7+h8
# probability of male categories
phi1<-h1/sh
phi2<-h2/sh
phi3<-h3/sh
phi4<-h4/sh
phi5<-h5/sh
phi6<-h6/sh
```

```
phi7<-h7/sh
phi8<-h8/sh
# proportion who answer yes at time 0
delta0<-phi1+phi2+phi3+phi4
delta3<-phi1+phi2+phi5+phi6
delta6<-phi1+phi3+phi5+phi7


}

# assumes uninformative prior
list(m1 = 904,m2 = 88,m3 = 25,m4 = 51,m5 = 33,m6 = 22,m7
= 30,m8 = 158,
n1 = 391,n2 = 36,n3 = 12,n4 = 26,n5 = 15,n6 = 21,n7 =
18,n8 = 143)
```

TABLE 5.15

Posterior Analysis for Church Attendance

Parameter	Mean	SD	Error	2½	Median	97½
ζ_0	0.7024	0.0176	0.00000176	0.6673	0.7026	0.7365
ζ_3	0.6994	0.0177	0.00000177	0.6641	0.6995	0.7336
ζ_6	0.6578	0.0184	0.00000184	0.6221	0.6589	0.6941
η_0	0.8146	0.0106	0.00000106	0.7934	0.8148	0.8349
η_3	0.7986	0.0110	0.00000439	0.7766	0.7988	0.8189
η_6	0.7566	0.0118	0.00000990	0.733	0.7568	0.7795
θ_1	0.6895	0.0128	0.00005099	0.664	0.6896	0.7141
θ_2	0.0671	0.0069	0.0000279	0.0542	0.0669	0.0811
θ_3	0.0190	0.0037	0.0000157	0.0124	0.0188	0.0271
θ_4	0.0389	0.0053	0.0000209	0.0291	0.0386	0.0499
θ_5	0.0251	0.0043	0.0000165	0.0174	0.0249	0.0342
θ_6	0.0167	0.0034	0.0000142	0.0105	0.0165	0.0244
θ_7	0.0228	0.0041	0.0000161	0.0158	0.0226	0.0316
θ_8	0.1205	0.0089	0.0000398	0.1035	0.1203	0.1386

The entries of Table 5.15 should be compared to those of Table 5.14. One will see the corresponding entries are almost identical. Why?

Also of interest is the correlation $\text{Cov}(\eta_0, \eta_3 \mid \text{data})$ between the proportion of females who answer yes at time 0 and the proportion of females who answer yes at time 3. This can be computed with the formulas (5.37) or with **BUGS CODE 5.4**. Using the inference menu of WinBUGS, one can verify that $\text{Cov}(\eta_0, \eta_3 \mid \text{data})$ is estimated as 0.6802. Table 5.16 provides additional information about the correlations.

Thus, for males, the correlation between the proportion who respond yes at time 3 and the proportion who respond yes at time 6 is estimated as 0.7019.

TABLE 5.16

Correlations for Females and Males

Female

	η_0	η_3	η_6
η_0	1	0.6802	0.5545
η_3			0.6444
η_6			1

Male

	ζ_0	ζ_3	ζ_6
ζ_0	1	0.7346	0.6739
ζ_3			0.7019
ζ_6			1

5.4 Examples

The remainder of the chapter will present additional examples of modeling the mean profile of continuous and discrete responses. Of special interest is demonstrating the use of polynomials and splines (shift point models) to represent the then mean profile of continuous responses. Determining the mean profile rests on graphical techniques such as scatter plots. Descriptive statistics are also useful in aiding one in modeling the mean profile. Graphical techniques such as histograms are used to justify the approximate normal distribution of continuous responses, while for discrete observations multinomial distribution of various categories identified over the time points are sufficient for modeling the mean profile.

5.4.1 Plasma Inorganic Phosphate

Table 5.17 displays the inorganic phosphate measurements for 13 control and 20 obese patients at hours 0, 0.5, 1, 1.5, 2, and 3 after an oral glucose challenge. See Zerbe[9] for additional details.

Figure 5.7 reveals the association between inorganic phosphate and time for the two groups.

Figure 5.7 provides preliminary evidence of the mean profile of the two groups. The control subjects follow a linear association from baseline to hour 1, another linear regression from hour 1 to hour 1.5, still another linear association from hour 1.5 to hour 2, and last between hours 2 and 3 after a glucose challenge. Thus, for the control group, there are four linear segments with join points 0, 1, 1.5, 2, and 3 hours.

For the obese group, the plot suggests two linear segments, the first from baseline to time 2 and the other from hour 2 to hour 3.

TABLE 5.17

Inorganic Phosphate of Control Subjects and Obese Subjects

t0	t0.5	t1	t1.5	t2	t3
Control Subjects					
4.30	3.30	3.00	2.60	2.20	2.50
3.70	2.60	2.60	1.90	2.90	3.20
4.00	4.10	3.10	2.30	2.90	3.10
3.60	3.00	2.20	2.80	2.90	3.90
4.10	3.80	2.10	3.00	3.60	3.40
3.80	2.20	2.00	2.60	3.80	3.60
3.80	3.00	2.40	2.50	3.10	3.40
4.40	3.90	2.80	2.10	3.60	3.80
5.00	4.00	3.40	3.40	3.30	3.60
3.70	3.10	2.90	2.20	1.50	2.30
3.70	2.60	2.60	2.30	2.90	2.20
4.40	3.70	3.70	3.20	3.70	4.30
4.70	3.10	3.20	3.30	3.20	4.20
Obese Subjects					
4.30	3.30	3.00	2.60	2.20	2.50
5.00	4.90	4.10	3.70	3.70	4.10
4.60	4.40	3.90	3.90	3.70	4.20
4.30	3.90	3.10	3.10	3.10	3.10
3.10	3.10	3.30	2.60	2.60	1.90
4.80	5.00	2.90	2.80	2.20	3.10
3.70	3.10	3.30	2.80	2.90	3.60
5.40	4.70	3.90	4.10	2.80	3.70
3.00	2.50	2.30	2.20	2.10	2.60
4.90	5.00	4.10	3.70	3.70	4.10
4.80	4.30	4.70	4.60	4.70	3.70
4.40	4.20	4.20	3.40	3.50	3.40
4.90	4.30	4.00	4.00	3.30	4.10
5.10	4.10	4.60	4.10	3.40	4.20
4.80	4.60	4.60	4.40	4.10	4.00
4.20	3.50	3.80	3.60	3.30	3.10
6.60	6.10	5.20	4.10	4.30	3.80
3.60	3.40	3.10	2.80	2.10	2.40
4.50	4.00	3.70	3.30	2.40	2.30
4.60	4.40	3.80	3.80	3.80	3.60

Source: Davis, C.S., *Statistical Methods for the Analysis of Repeated Measurements*, Springer-Verlag, New York, p. 68, 2002.

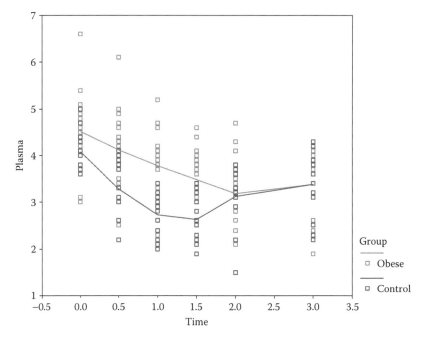

FIGURE 5.7
Plasma inorganic phosphate for controls and obese subjects.

Of course, the descriptive statistics should corroborate the scatter plot of the repeated measures of Figure 5.7. Comparing the mean to the median, the distribution of the inorganic phosphate values appear to be symmetric about the mean (Table 5.18).

I computed the histogram of the response separately for each group and found that the observations appear to be approximately normally distributed; thus, when estimating the mean profile, I will assume the endpoint is normally distributed.

The following equations define the mean profile of two groups, and it appears that this example is the most complex of the spline-type mean profiles considered so far.

For the control group, let y_{ij} be the value of plasma inorganic phosphate for subject i at hour j and suppose

$$E(y_{ij}) = \beta_{11} + \beta_{12}j \quad j = 0, 0.5, 1$$

$$E(y_{ij}) = \beta_{13} + \beta_{14}j \quad j = 1, 1.5$$

$$E(y_{ij}) = \beta_{15} + \beta_{16}j \quad j = 1.5, 2$$

$$E(y_{ij}) = \beta_{17} + \beta_{18}j \quad j = 2, 3$$

(5.38)

TABLE 5.18

Descriptive Statistics of Plasma Inorganic Phosphate

Time	Mean	SD	Median	Group
0	4.092	0.4387	4.0	Control
0.5	3.277	0.5988	3.3	
1	2.723	0.4475	2.8	
1.5	2.631	0.4803	2.6	
2	3.046	0.6385	3.1	
3	3.346	0.6765	3.4	
0	4.530	0.8033	4.6	Obese
0.5	4.140	0.8338	4.25	
1	3.780	0.7120	3.85	
1.5	3.480	0.6756	3.65	
2	3.195	0.7681	3.30	
3	3.375	0.7144	3.60	

where β_{1k}, $k = 1, 2, 3, 4, 5, 6, 7$, and 8, are unknown parameters to be estimated with Bayesian techniques.

For the obese group, let

$$E\left(y_{ij}\right) = \beta_{21} + \beta_{22}j \quad j = 1, 0.5, 1, 1.5, 2$$

$$E\left(y_{ij}\right) = \beta_{23} + \beta_{24}j \quad j = 2, 3$$

(5.39)

The Bayesian analysis is based on **BUGS CODE 5.5**, and the program statements closely follow the notation of formulas (5.38) and (5.39). I used 110,000 observations for the MCMC simulation with 5000 starting values and a refresh of 100.

The Bayesian analysis focuses on estimation of the coefficients of the various linear regressions for the various segments of the two splines, one for the control group and the other for the obese group (Table 5.19).

For the first segment of the spline for the control group, the intercept is estimated as 4.066 and the slope is estimated as −1.338. On the other hand, the intercept for the first segment for the obese group is estimated as 4.483 and the slope is estimated as −0.6664. Note for the second segment of the obese group, the slope estimated as 0.1816 has a credible interval (−0.1039, 0.47), which includes 0. Is the slope actually 0 for the second segment of the obese group? From Figure 5.7, note that the two last segments for both groups appear to have slopes that are almost parallel, and both estimates have 95% credible intervals that include 0. The posterior means are computed assuming a general covariance structure for the observations between the six time points.

BUGS CODE 5.5

```
    model;
    {
            beta11 ~ dnorm(0.0, 0.001)
            beta12 ~ dnorm(0.0, 0.001)
beta13 ~ dnorm(0.0,.0001)
beta14 ~ dnorm(0.0,.0001)
beta15 ~ dnorm(0.0,.0001)
beta16 ~ dnorm(0.0,.0001)
beta17 ~ dnorm(0.0,.0001)
beta18 ~ dnorm(0.0,.0001)
beta21 ~ dnorm(0.0,.0001)
beta22 ~ dnorm(0.0,.0001)
beta23 ~ dnorm(0.0,.0001)
beta24 ~ dnorm(0.0,.0001)

for(i in 1:N1){Y1[i,1:M1]~dmnorm(mu1[],Omega1[,])}
for(j in 1:M1){mu1[j]<-beta11+beta12*time1[j]}

Omega1[1:M1,1:M1]~dwish(R1[,],3)
Sigma1[1:M1,1:M1]<-inverse(Omega1[,])

for(i in 1:N2){Y2[i,1:M2]~dmnorm(mu2[],Omega2[,])}
for(j in 1:M2){mu2[j]<-beta13+beta14*time2[j]}

Omega2[1:M2,1:M2]~dwish(R2[,],3)
Sigma2[1:M2,1:M2]<-inverse(Omega2[,])

for(i in 1:N3){Y3[i,1:M3]~dmnorm(mu3[],Omega3[,])}
for(j in 1:M3){mu3[j]<-beta15+beta16*time3[j]}

Omega3[1:M3,1:M3]~dwish(R3[,],2)
Sigma3[1:M3,1:M3]<-inverse(Omega3[,])

for(i in 1:N4){Y4[i,1:M4]~dmnorm(mu4[],Omega4[,])}
for(j in 1:M4){mu4[j]<-beta17+beta18*time4[j]}

Omega4[1:M4,1:M4]~dwish(R4[,],2)
Sigma4[1:M4,1:M4]<-inverse(Omega4[,])

for(i in 1:N5){Y5[i,1:M5]~dmnorm(mu5[],Omega5[,])}
for(j in 1:M5){mu5[j]<-beta21+beta22*time5[j]}

Omega5[1:M5,1:M5]~dwish(R5[,],5)
Sigma5[1:M5,1:M5]<-inverse(Omega5[,])
```

```
for(i in 1:N6){Y6[i,1:M6]~dmnorm(mu6[],Omega6[,])}
for(j in 1:M6){mu6[j]<-beta23+beta24*time6[j]}

Omega6[1:M6,1:M6]~dwish(R6[,],2)
Sigma6[1:M6,1:M6]<-inverse(Omega6[,])


}

list(N1 = 13,N2 = 13,N3 = 13,N4 = 13,N5 = 20,N6 = 20,
M1 = 3,M2 = 2,M3 = 2,M4 = 2,M5 = 5,M6 = 2,

time1 = c(0,.5,1),
time2 = c(1,1.5),
time3 = c(1.5,2),
time4 = c(2,3),

time5 = c(0,.5,1,1.5,2),
time6 = c(2,3),

Y1 = structure(.Data =

c(4.30,        3.30,  3.00,
3.70,  2.60,  2.60,
4.00,  4.10,  3.10,
3.60,  3.00,  2.20,
4.10,  3.80,  2.10,
3.80,  2.20,  2.00,
3.80,  3.00,  2.40,
4.40,  3.90,  2.80,
5.00,  4.00,  3.40,
3.70,  3.10,  2.90,
3.70,  2.60,  2.60,
4.40,  3.70,  3.70,
4.70,  3.10,  3.20),.Dim = c(13,3)),

Y2 = structure(.Data =

c(
3.00,  2.60,
2.60,  1.90,
3.10,  2.30,
2.20,  2.80,
2.10,  3.00,
2.00,  2.60,
2.40,  2.50,
2.80,  2.10,
```

```
3.40,   3.40,
2.90,   2.20,
2.60,   2.30,
3.70,   3.20,
3.20,   3.30)
,.Dim = c(13,2)),

Y3 = structure(.Data =

c(

2.60,   2.20,
1.90,   2.90,
2.30,   2.90,
2.80,   2.90,
3.00,   3.60,
2.60,   3.80,
2.50,   3.10,
2.10,   3.60,
3.40,   3.30,
2.20,   1.50,
2.30,   2.90,
3.20,   3.70,
3.30,   3.20
),.Dim = c(13,2)),

Y4 = structure(.Data =

c(2.20,        2.50,
2.90,   3.20,
2.90,   3.10,
2.90,   3.90,
3.60,   3.40,
3.80,   3.60,
3.10,   3.40,
3.60,   3.80,
3.30,   3.60,
1.50,   2.30,
2.90,   2.20,
3.70,   4.30,
3.20,   4.20
),.Dim = c(13,2)),

Y5 = structure(.Data =

c(4.30,        3.30,   3.00,   2.60,   2.20,
5.00,   4.90,   4.10,   3.70,   3.70,
```

```
4.60,   4.40,   3.90,   3.90,   3.70,
4.30,   3.90,   3.10,   3.10,   3.10,
3.10,   3.10,   3.30,   2.60,   2.60,
4.80,   5.00,   2.90,   2.80,   2.20,
3.70,   3.10,   3.30,   2.80,   2.90,
5.40,   4.70,   3.90,   4.10,   2.80,
3.00,   2.50,   2.30,   2.20,   2.10,
4.90,   5.00,   4.10,   3.70,   3.70,
4.80,   4.30,   4.70,   4.60,   4.70,
4.40,   4.20,   4.20,   3.40,   3.50,
4.90,   4.30,   4.00,   4.00,   3.30,
5.10,   4.10,   4.60,   4.10,   3.40,
4.80,   4.60,   4.60,   4.40,   4.10,
4.20,   3.50,   3.80,   3.60,   3.30,
6.60,   6.10,   5.20,   4.10,   4.30,
3.60,   3.40,   3.10,   2.80,   2.10,
4.50,   4.00,   3.70,   3.30,   2.40,
4.60,   4.40,   3.80,   3.80,   3.80)
,.Dim = c(20,5)),

Y6 = structure(.Data =

c(2.20,        2.50,
3.70,   4.10,
3.70,   4.20,
3.10,   3.10,
2.60,   1.90,
2.20,   3.10,
2.90,   3.60,
2.80,   3.70,
2.10,   2.60,
3.70,   4.10,
4.70,   3.70,
3.50,   3.40,
3.30,   4.10,
3.40,   4.20,
4.10,   4.00,
3.30,   3.10,
4.30,   3.80,
2.10,   2.40,
2.40,   2.30,
3.80,   3.60

),.Dim = c(20,2)),

R1 = structure(.Data = c(1,0,0,
                         0,1,0,
```

```
                            0,0,1
                            ),.Dim = c(3,3)),

    R2 = structure(.Data = c(1,0,
                             0,1),.Dim = c(2,2)),


    R3 = structure(.Data = c(1,0,
                             0,1),.Dim = c(2,2)),


    R4 = structure(.Data = c(1,0,
                             0,1),.Dim = c(2,2)),


    R5 = structure(.Data = c(1,0,0,0,0,
                             0,1,0,0,0,
                             0,0,1,0,0,
                             0,0,0,1,0,
                             0,0,0,0,1),.Dim = c(5,5)),

    R6 = structure(.Data = c(1,0,
                             0,1),.Dim = c(2,2)))
    # initial values

    list(beta11 = 0,beta12 = 0,beta13 = 0,beta14 = 0,beta15 =
    0,beta16 = 0,
    beta17 = 0, beta18 = 0, beta21 = 0,beta22 = 0,beta23 =
    0,beta24 = 0)
```

TABLE 5.19

Posterior Analysis for the Plasma Inorganic Phosphate Study

Parameter	Mean	SD	Error	2½	Median	97½
β_{11}	4.066	0.149	0.000687	3.77	4.066	4.363
β_{12}	−1.338	0.1628	0.000711	−1.661	−1.338	−1.013
β_{13}	3.037	0.5104	0.008999	2.031	3.038	4.037
β_{14}	−0.2705	0.3891	0.006878	1.035	−0.2704	0.4964
β_{15}	1.391	0.728	0.01622	−0.056361	1.387	2.852
β_{16}	0.8271	0.4269	0.009521	−0.03167	0.8293	1.675
β_{17}	2.441	0.4843	0.005597	1.471	2.44	3.4002
β_{18}	0.3023	0.1836	0.002122	−0.062160	0.3025	0.6692
β_{21}	4.483	0.1752	0.0007786	4.136	4.484	4.828
β_{22}	−0.6664	0.0781	0.000342	−0.8211	−0.6664	−0.5125
β_{23}	2.83	0.4099	0.004793	2.041	2.831	3.637
β_{24}	0.1816	0.1451	0.001702	−0.1039	0.1813	0.47

TABLE 5.20

Estimated Pearson Correlations of Plasma Inorganic
Phosphate between Six Hours for Control Group

Time Pairs	Correlation
(0,0.5)	0.612
(0,1)	0.677
(0,1.5)	0.670
(0,2)	0.326
(0,3)	0.445
(0.5,1)	0.542
(0.5,1.5)	0.324
(0.5,2)	0.133
(0.5,3)	0.234
(1,1.5)	0.304
(1,2)	−0.109
(1,3)	0.160
(1.5,2)	0.386
(1.5,3)	0.567
(2,3)	0.723

This example is concluded by estimating the variance–covariance matrix for both groups (Table 5.20).

For example, the estimated Pearson correlation of plasma inorganic phosphate between baseline and hour 2 is 0.326. The estimated correlations do appear to decrease with time, but it is difficult to detect a pattern in the correlation structure. Is the correlation structure the same for both groups? See Table 5.21 for the estimated correlations of plasma inorganic phosphate between hours 0, 0.5, 1, 1.5, 2, and 3.

As with the control group, the estimated correlations for the obese group appear to decrease with time. Various structures for the correlation matrix will be studied in Chapter 6. These estimates of the correlation matrices are not based on a Bayesian analysis; however, Table 5.22 does represent the Bayesian posterior distributions for the entries of the 5×5 variance–covariance matrix of plasma inorganic phosphate between the 5 time points for the first linear segment over the time points 0, 0.5, 1, 1.5, and 2 hours.

One notices the skewed distribution for the variances and covariances; thus, one should use the posterior medians as estimators for the entries of this matrix. Based on the medians, the estimated correlation between times 4 and 5 (corresponding to hours 1.5 and 2) is $0.4645/\sqrt{(0.5323)(0.6725)} = 0.77635$.

5.4.2 Hospital Study of a Dietary Regime

This example is quite different than the other presented in this chapter, in that different treatments are given to the same subject.

TABLE 5.21

Estimated Pearson Correlations of Plasma Inorganic Phosphate between Six Hours for Obese Group

Time Pairs	Correlation
(0,0.5)	0.919
(0,1)	0.769
(0,1.5)	0.742
(0,2)	0.595
(0,3)	0.657
(0.5,1)	0.704
(0.5,1.5)	0.671
(0.5,2)	0.601
(0.5,3)	0.625
(1,1.5)	0.896
(1,2)	0.860
(1,3)	0.609
(1.5,2)	0.848
(1.5,3)	0.769
(2,3)	0.731

TABLE 5.22

Posterior Distribution of the 5×5 Variance–Covariance Matrix Obese Group First Linear Segment

Parameter	Mean	SD	Error	2½	Median	97½
σ_{11}	0.7237	0.2581	0.001095	0.3868	0.6817	1.369
σ_{12}	0.6437	0.244	0.000986	0.3141	0.5972	1.247
σ_{13}	0.4633	0.1975	0.000795	0.1908	0.4272	0.9507
σ_{14}	0.4233	0.1852	0.000750	0.1645	0.3898	0.8814
σ_{15}	0.3862	0.196	0.000793	0.1005	0.3527	0.8646
σ_{22}	0.7746	0.2698	0.001053	0.4126	0.7232	1.44
σ_{23}	0.4432	0.1977	0.000796	0.166	0.4069	0.9295
σ_{24}	0.4003	0.1849	0.000742	0.1393	0.3677	0.8522
σ_{25}	0.4041	0.2013	0.000820	0.1128	0.3704	0.8878
σ_{33}	0.5844	0.2048	0.000776	0.3093	0.5466	1.088
σ_{34}	0.4537	0.1774	0.000681	0.2126	0.4197	0.8895
σ_{35}	0.4909	0.1962	0.000771	0.2234	0.4539	0.9726
σ_{44}	0.5323	0.1857	0.000718	0.2809	0.4965	0.9886
σ_{45}	0.4645	0.187	0.000762	0.209	0.4296	0.9261
σ_{55}	0.6725	0.2357	0.000981	0.357	0.6272	1.258

TABLE 5.23

Plasma Ascorbic Acid by Week

Subject	W1	W2	W6	W10	W14	W15	W16
1	0.22	0	1.03	0.67	0.75	0.65	0.59
2	0.18	0	0.96	0.96	0.98	1.03	0.70
3	0.73	0.37	1.18	0.76	1.07	0.80	1.10
4	0.30	0.25	0.74	1.10	1.48	0.39	0.36
5	0.54	0.42	1.33	1.32	1.30	0.74	0.56
6	0.16	0.30	1.27	1.06	1.39	0.63	0.40
7	0.30	1.09	1.17	0.90	1.17	0.75	0.88
8	0.70	1.30	1.80	1.80	1.60	1.23	0.41
9	0.31	0.54	1.24	0.56	0.77	0.28	0.40
10	1.40	1.40	1.64	1.28	1.12	0.66	0.77
11	0.60	0.80	1.02	1.28	1.16	1.01	0.67
12	0.73	0.50	1.08	1.26	1.17	0.91	0.87

Source: Crowder, M.J., and Hand, D.J., *Analysis of Repeated Measures*, Table 5.2, Chapman & Hall, London, p. 32, 1990.

In Table 5.23, 12 hospital patients were given a dietary regimen with observations of plasma ascorbic acid with two observations before administration of the dietary treatment, three during the treatment, and twice after. The objective of the study is to investigate the effect of diet over a 16-week period, with repeated measures at weeks 1, 2, 6, 10, 14, 15, and 16. Based on Table 5.23, the descriptive statistics for this study are portrayed in Table 5.24.

One should interpret the descriptive statistics using the scatter plot of Figure 5.8.

The scatter plot suggests a spline with three linear segments will suffice in modeling the mean profile of this study. This is a study with one treatment, namely a dietary supplement administered at weeks 6, 10, and 12, whereas pretreatment observations are made at weeks 1, 2, and 6 and post-treatment observations are made at weeks 15 and 16. The subjects are on the dietary supplement beginning at week 6 and ending at the beginning of week 14.

TABLE 5.24

Descriptive Statistics for Hospital Study

Week	Mean	SD	Median
1	0.5142	0.35387	0.4250
2	0.5808	0.47070	0.4600
6	1.2050	0.28965	1.1750
10	1.0792	0.34071	1.0800
14	1.1633	0.25776	1.165
15	0.7567	0.26749	0.7450
16	0.6425	0.23332	0.6399

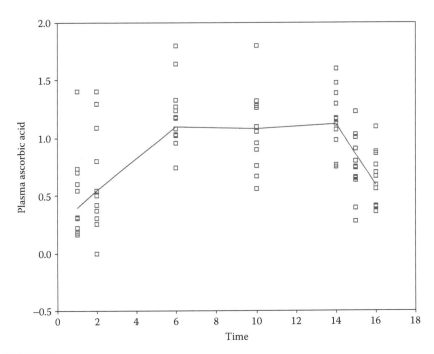

FIGURE 5.8
Plasma ascorbic acid for 12 hospital patients by week.

Tentatively, the mean profile will begin with a linear segment over time points 1, 2, and 6 weeks, the second over weeks 6, 10, and 14, and the last segment over weeks 14, 15, and 16. The Bayesian analysis is executed with **BUGS CODE 5.6** using 65,000 observations, with 5000 initial values and a refresh of 100.

A three-segment spline for the mean profile of plasma ascorbic acid is represented by the following:

$$E\left(Y_{ij}\right) = \beta_1 + \beta_2 t_{ij} \quad 1 \leq t_{ij} \leq 6$$

$$E\left(Y_{ij}\right) = \beta_3 + \beta_4 t_{ij} \quad 6 \leq t_{ij} \leq 14$$

and   (5.40)

$$E\left(Y_{ij}\right) = \beta_5 + \beta_6 t_{ij} \quad 14 \leq t_{ij} \leq 16$$

The Bayesian analysis is executed with **BUGS CODE 5.6** using 65,000 observations, with 5000 initial values and a refresh of 100. Note that the code closely follows the three equations of (5.40).

```
                          BUGS CODE 5.6
        model;
        {
     beta1 ~ dnorm(0.0, 0.001)
     beta2 ~ dnorm(0.0, 0.001)
     beta3 ~ dnorm(0.0,.0001)
     beta4 ~ dnorm(0.0,.0001)
     beta5 ~ dnorm(0.0,.0001)
     beta6 ~ dnorm(0.0,.0001)
# first segment
for(i in 1:N1){Y1[i,1:M1]~dmnorm(mu1[],Omega1[,])}
for (j in 1:M1){mu1[j]<-beta1+beta2*time1[j]}

Omega1[1:M1,1:M1]~dwish(R1[,],3)
Sigma1[1:M1,1:M1]<-inverse(Omega1[,])

# second segment
for(i in 1:N2){Y2[i,1:M2]~dmnorm(mu2[],Omega2[,])}
for (j in 1:M2){mu2[j]<-beta3+beta4*time2[j]}

Omega2[1:M2,1:M2]~dwish(R2[,],3)
Sigma2[1:M2,1:M2]<-inverse(Omega2[,])
# third segment
for(i in 1:N3){Y3[i,1:M3]~dmnorm(mu3[],Omega3[,])}
for (j in 1:M3){mu3[j]<-beta5+beta6*time3[j]}

Omega3[1:M3,1:M3]~dwish(R3[,],3)
Sigma3[1:M3,1:M3]<-inverse(Omega3[,])
}

list(N1 = 12,N2 = 12,N3 = 12,M1 = 3,M2 = 3,M3 = 3,time1
= c(1,2,6),
time2 = c(6,10,14),time3 = c(14,15,16),

# data for first segment
Y1 = structure(.Data =

c(.22,  .00,   1.03,.18,    .00,   .96,.73,    .37,
1.18,.30,   .25,  .74,.54,   .42,   1.33,.16,   .30,
1.27,.30,   1.09,  1.17,.70,   1.30,  1.80,.31,   .54,
1.24,1.40,   1.40,  1.64,.60,   .80,   1.02,.73,   .50,
1.08),.Dim = c(12,3)),
# data for second segment
Y2 = structure(.Data =

c(1.03,    .67,   .75,.96,   .96,   .98,1.18,   .76,
1.07,.74,   1.10,  1.48,1.33,   1.32,  1.30,1.27,
```

```
1.06,   1.39,1.17,     .90,   1.17,1.80,    1.80,   1.60,1.24,
.56,    .77,1.64,      1.28,  1.12,1.02,    1.28,   1.16,1.08,
1.26,   1.17),.Dim = c(12,3)),
# data for third segment
Y3 = structure(.Data =

c(75,   .65,    .59,.98,      1.03,  .70,1.07,     .80,
1.10,1.48,    .39,    .36,1.30,     .74,    .56,1.39,     .63,
.40,1.17,     .75,    .88,1.60,     1.23,   .41,.77,      .28,
.40,1.12,     .66,    .77,1.16,     1.01,   .67,1.17,     .91,
.87),.Dim = c(12,3)),
# un informative prior distribution for the precision
matrix
R1 = structure(.Data = c(1,0,0,
                         0,1,0,
                         0,0,1),.Dim = c(3,3)),

R2 = structure(.Data = c(1,0,0,
                         0,1,0,
                         0,0,1),.Dim = c(3,3)),

R3 = structure(.Data = c(1,0,0,
                         0,1,0,
                         0,0,1),.Dim = c(3,3)))

# initial values

list(beta1 = 0,beta2 = 0,beta3 = 0,beta4 = 0,beta5 =
0,beta6 = 0)
```

TABLE 5.25

Posterior Distribution for the Mean Profile of Hospital Study

Parameter	Mean	SD	Error	2½	Median	97½
β_1	0.3513	0.1491	0.001155	0.0562	0.3511	0.6486
β_2	0.1431	0.02871	0.000222	0.0861	0.1431	0.2002
β_3	1.194	0.2219	0.003421	0.7507	1.194	1.639
β_4	−0.002893	0.02007	0.000296	−0.04286	−0.002933	0.03714
β_5	2.966	2.507	0.1534	−2.897	3.081	7.626
β_6	−0.1456	0.1614	0.00987	−0.04453	−0.1531	0.2314

The posterior analysis is presented in Table 5.25 and portrays the intercept and slope for the three linear segments of the spline that represents the mean profile of plasma ascorbic acid of the hospital study.

For the first segment, the posterior mean for the intercept is 0.3513 and the posterior mean for the slope is 0.1431. For the second segment, the estimated

TABLE 5.26

Posterior Distribution of the Variance–Covariance Matrix

Parameter	Mean	SD	Error	2½	Median	97½
σ_{11}	0.2351	0.1158	0.000543	0.1011	0.2073	0.529
σ_{12}	0.1278	0.1043	0.000464	-0.014920	0.1083	0.3808
σ_{13}	0.06603	0.07432	0.000312	-0.050590	0.5558	0.2415
σ_{22}	0.3299	0.1601	0.000636	0.1444	0.2928	0.7344
σ_{23}	0.1086	0.09156	0.000355	-0.021290	0.09245	0.3321
σ_{33}	0.1908	0.09361	0.000413	0.08208	0.1685	0.4271

slope is almost 0 with −0.002893 for the posterior mean, while for the third segment the estimated slope is −0.1456. Do these estimates correspond to the lowess curve for the scatter plot in Figure 5.8? It appears to me that for the second segment, the curve is indeed flat corresponding to a zero slope.

Also for the third segment, the lowess curve depicts a linear association of plasma ascorbic acid with a negative slope; therefore, the posterior analysis of Table 5.25 aids in the interpretation of Figure 5.8. Thus, for every increase of one week, the average plasma ascorbic acid decreases by 0.1456.

Last, inferences for the variance–covariance matrix of the first segment are given in Table 5.26.

Notice that the variance of the plasma ascorbic acid at time 1 week is estimated as 0.2351, at time 2 as 0.3200, and at time 6 as 0.1908. Note that the posterior distribution of the entries are highly skewed; thus, I recommend using the posterior median as an estimate. Therefore, one's estimate of the variance at time 1 is 0.2072, at time 2 is 0.2928, and at time 6 is 0.1685. Note that times 1, 2, and 3 correspond to 1, 2, and 6 weeks, that is, time 3 corresponds to week 6. If one want to estimate the correlation

$$\rho_{12} = \frac{\sigma_{12}}{\sqrt{\sigma_{11}\sigma_{22}}}$$

between weeks 1 and 2 with the posterior medians, the estimate is $0.1083/\sqrt{(0.2073)(0.2928)} = 0.4396$. Remember an unstructured covariance matrix was assumed for the Bayesian analysis. If more structure is known for the variance–covariance matrix, the analysis needs to be modified.

5.5 Conclusions and Summary

Following the introduction, the first topic to be discussed in depth is the main idea of the chapter, namely using polynomials to fit the mean profile of the repeated measures for continuous and discrete observations. For continuous observations, the repeated measures for a given subject are assumed to follow approximately a multivariate normal distribution. In the case, the response is discrete, the various possible categories of the observations are assumed to have a multinomial distribution, which allows one to use a Dirichlet distribution for the Bayesian analysis.

The first case to be considered for a polynomial fit to the mean profile is the Van der Lende et al[2] study of chronic obstructive lung disease of two groups, namely smokers and nonsmokers, where the main response is FEV1, a measure of one's respiratory capability and measured with a spirometry. The subjects have been exposed to air pollution, and the investigator is interested in comparing the chronic obstructive lung disease (via FEV1) of the two groups. A scatter plot of FEV1 versus time reveals that two linear regressions are tentatively reasonable as a representation of the mean profile. A Bayesian analysis with **BUGS CODE 5.1** assumes that the variance–covariance matrix of FEV1 over the time points is unstructured, that each subject has the same covariance matrix, and that the joint distribution is approximately multivariate normal. Based on the scatter plot and descriptive statistics, formulas for the two linear regressions are specified.

Hematocrit measurements of hip-replacement patients are the foundation for the next example, where the response appears to be quadratic for each group. Using an unstructured covariance matrix and assuming a multivariate normal distribution for the repeated measures of each subject, a Bayesian analysis implies that the two groups (males and females) share the same quadratic response.

Last, the common variance–covariance matrix of the two groups is estimated with Bayesian methods, where a priori, an uninformative Wishart distribution is assumed for the distribution of the entries of the covariance matrix. See **BUGS CODE 5.3** for the program statements for determining the posterior distribution of the 4×4 matrix entries.

The Rhoads[4] study consists of two groups, a placebo and the other a treatment group receiving succimer and treatment for lowering the blood lead levels of children with elevated lead in the blood. A scatter plot reveals a linear response of the time points for the placebo, while a two-segment linear spline appears to be reasonable for the subjects receiving succimer.

Next to be considered is the investigation of Cornoni-Huntley,[5] where the response is binary, that is, "yes" by a subject is evidence that the person has attended church regularly the previous two years. Responses are observed at years 0, 3, and 6 where year 0 is baseline, and the main problem is comparing males and females with respect to church attendance. Since there are three years and the response is binary (yes or no), there are eight possible categories for each group, and one may assume that the eight categories follow a multinomial distribution. Assuming an uninformative prior for the eight probabilities corresponding to the eight categories allows one to perform a Bayesian analysis using the Dirichlet distribution of the joint distribution of the eight categories, which in turn allows one to determine the posterior distribution of the proportion of yes responses at years 0, 3, and 6. This is done for both groups, and the proportional responses can be compared between the two groups. As the distribution of the three yes proportions follow a Dirichlet distribution, one also knows the correlation between the proportion of yes responses for the three years. This follows because the joint distribution of the three yes proportions (corresponding to the years 0, 3, and 6) follow a Dirichlet distribution.

This chapter continues with several examples that serve to review the concepts of determining the mean profile for repeated measures.

The first example is from Zerbe[9] and analyzed by Davis[6] and involves 13 control subjects and 20 obese subjects where the main response is plasma inorganic phosphate. All subjects are given an oral glucose challenge, and the plasma inorganic phosphate is measured at hours 0, 0.5, 1, 1.5, 2, and 3. Of course our goal is to compare the mean profile of control versus obese patients. It appears from the scatter plot that the mean response for obese is a two-segment linear spline, while for the controls, the four-segment linear spline is appropriate. Thus, this is the most complex polynomial-type mean response considered so far. A Bayesian analysis presented in Table 5.19 estimates the intercept and slope for each linear segment of the two splines.

Our last example is a hospital study taken from Crowder and Hand[3] with 12 patients and plasma ascorbic acid is measured at 1, 2, 6, 14, 15, and 16 hours. The major goal is to see the effect of the dietary supplement administered at hour 6, and a scatter plot revealed a three-segment linear spline. Based on this scatter plot, the mean profile was defined by Equation 5.40, where the unknown coefficients are the intercept and slope for each of the three segments; thus, there are six regression coefficients. **BUGS CODE 5.6** was used to execute the Bayesian analysis of Table 5.25.

This chapter is very important, and the student should learn the concepts presented, especially the techniques of the scatter plot, descriptive statistics, and the model equations in order to determine the mean profile(s). It should be remembered that when the response is continuous, an unstructured covariance matrix was assumed, and that when the response is binary, a multinomial distribution was assumed for the possible categories.

Next to be considered in Chapter 6 is the problem of choosing an appropriate correlation pattern for the repeated measures.

Exercises

1. a. In Equation 5.1, if $g_i = 1$, to what group does subject i belong?

   b. In Equation 5.1, if $g_i = 0$, to what group does subject i belong?

2. Refer to Figure 5.1. Does the response of the smoker group appear to have the same slope as that for nonsmokers?

3. a. Refer to Table 5.2 for nonsmokers and locate the mean of FEV1 at time 12 on Figure 5.1.

   b. Refer to Table 5.2 for smokers and locate the mean of FEV1 at time 12 on Figure 5.2.

4. Using **BUGS CODE 5.1**, verify the posterior analysis of Table 5.3.

   a. What is the 95% credible interval for β_3?

   b. Based on the posterior distribution of d_{13}, is it reasonable to declare that $\beta_1 = \beta_3$?

5. With regard to the Bayesian analysis of Table 5.3, is it reasonable that the response versus time for smokers and nonsmokers have the same slope? Why?

6. Refer to Table 5.4.

   a. What is the posterior mean of the common slope?

   b. Do smokers and nonsmokers share the same intercept? Explain your answer by referring to the posterior distribution of d_{13}.

   c. Is the posterior distribution of β (the common slope) symmetric about its mean?

7. Refer to Figure 5.2. Does the scatter plot imply a quadratic association for hematocrit versus time for males and females?

8. Using **BUGS CODE 5.2**, verify the posterior analysis of Table 5.6. Use 65,000 observations for the MCMC simulation, with 5000 initial observations and a refresh of 100. Plot the posterior density of each parameter in Table 5.6.

9. Explain Equations 5.7 and 5.8 and interpret the six regression coefficients for the hip-replacement study.

10. If $\beta_4 = \beta_5 = \beta_6 = 0$, what is the implication for the two groups, males and females. Does the posterior analysis of Table 5.6 support the claim that $\beta_4 = \beta_5 = \beta_6 = 0$?

11. If $\beta_4 = \beta_5 = \beta_6 = 0$, then the two groups share the same quadratic response to hematocrit versus time.

    a. Refer to Table 5.7 and provide estimates for these three coefficients of the quadratic response.

    b. What is the equation for the expected hematocrit for the quadratic response at time t?

    c. Using 55,000 observations with 5000 initial values and a refresh of 100, verify the posterior analysis of Table 5.7.

    d. Determine the posterior means of β_1, β_2, and β_3.

12. Assuming a common quadratic response of hematocrit on time for both groups, what is the mean response at times 0 and 4?

13. Refer to Table 5.9.

    a. What is the posterior distribution of σ_i^2?

    b. Find the posterior median of σ_{24}.

    c. Using Table 5.9, estimate the correlation between times 2 and 4.

14. Refer to Figure 5.4 and the lowess curve for the median profile of the two groups.

    a. What type of response is suggested for the association between blood lead levels and time for placebo?

    b. Describe the spline type of representation for the median profile of the succimer group. Use two linear segments for your description of the first segment over time points 1 and 2 and the second segment for 5 and 7 weeks.

15. Explain why Equations 5.12 and 5.13 describe a two-segment linear spline.

16. Using **BUGS CODE 5.3** with 65,000 observations for the simulation and 5000 initial observations, verify the posterior analysis presented by Table 5.11.

17. Refer to Table 5.11.

    a. Show that the various posterior distributions are symmetric about their posterior means.

    b. Using Figure 5.4 and Table 5.11, explain the effect of succimer on the blood lead levels for the treatment group.

18. Refer to **BUGS CODE 5.3**.

    Determine the effect of using the identity matrices $R_1, R_2,$ and R_3 on the prior distribution of the precision matrices $\Omega_1, \Omega_2,$ and $\Omega_3,$ respectively.

    Do they induce an uninformative-type prior on these precision matrices?

19. Refer to Table 5.12 for the placebo group.

    a. Find the posterior mean of σ_{01}, the covariance between times 0 and 1.

    b. Find the posterior median of the variance of the response at time 0.

    c. At time 1.

    d. Use the entries of Table 5.12 to estimate the correlation between times 0 and 1.

20. Equation 5.16 is the equation for the probability mass function for the multinomial distribution of the frequencies $m_1, m_2,...,n$ corresponding to the eight categories, where $\theta_i, i = 1,2,...,8,$ is the probability a female belongs to category i.

a. Explicate the constraints on the θ_i, $i = 1,2,\ldots,8$.

b. Compute the estimate of θ_i.

c. Explain the constraint on the m_i.

21. Using **BUGS CODE 5.4**, verify the Bayesian analysis portrayed in Table 5.15. For the simulation, use 65,000 observations with 5,000 initial observations and a refresh of 100.

a. Show that the posterior distribution of δ_0, the proportion of males at baseline who answer yes to church attendance the previous three years, is symmetric about the posterior mean.

b. Describe the posterior distribution of η_6 (the proportion of females who respond yes to church attendance) at year 6.

22. Using **BUGS CODE 5.6** with 65,000 observations for the simulation, 5,000 initial values, and a refresh of 100, verify the posterior analysis of Table 5.25.

a. Find the statements in the code that correspond to the three equations of (5.40).

b. Interpret the six beta coefficients of Table 5.25.

c. Show that the posterior mean of the betas mimics the three-segment linear spline of Figure 5.8.

d. Verify that the posterior distribution of β_1 symmetric about its mean.

6

Correlation Patterns for Repeated Measures

6.1 Introduction

This chapter describes the choice for the correlation structure of repeated measures. One must know or have a good idea of the appropriate covariance structure of the repeated measures in order to have the correct standard errors of the estimated coefficients of the mean profile. In the Bayesian context, the correct covariance matrix will have appropriate standard deviations for the posterior distribution of the coefficients involved in the mean profile of the repeated response over time. It should be noted that the covariance matrix of the errors in the mean response depend on the mean response and conversely; thus, the two entities are related. We will take the approach of determining a good regression representation of the mean profile and use that to search for an appropriate covariance or correlation matrix.

It is to one's advantage to intelligently use the covariance structure of the repeated measures and to see that envision in two situations: (1) a repeated measure with the same individuals over two time points and (2) a cross-sectional study with two people, one at time point 1 and another at time point 2. Suppose the variance of the response at time point 1 is σ_1^2 and is σ_2^2 at time point 2 and that the main focus of the study is to compare the mean response Y_{i1} at time 1 with the mean response Y_{i2} at time 2. For the first scenario, consider

$$\mathrm{Var}\left(Y_{i1} - Y_{i2}\right) = \mathrm{Var}\left(Y_{i1}\right) + \mathrm{Var}\left(Y_{i2}\right) - 2\mathrm{Cov}\left(Y_{i1}, Y_{i2}\right)$$

$$= \sigma_1^2 + \sigma_2^2 - 2\sigma_1\sigma_2\rho_{12}$$

(6.1)

where:

ρ_{12} is the correlation between the response at time 1 with that of time 2

The correlation is almost always positive for a repeated measures study; thus, from (6.1), one sees the advantage of a repeated measures design compared to a cross-sectional, because for the latter the variance of the difference is

$$\mathrm{Var}\left(Y_{i1} - Y_{i2}\right) = \mathrm{Var}\left(Y_{i1}\right) + \mathrm{Var}\left(Y_{i2}\right)$$

$$= \sigma_1^2 + \sigma_2^2$$

(6.2)

One is assuming that the variance of the two responses is the same for the two scenarios, that the two individuals for second case give independent responses, and that the correlation is positive for the first, the repeated measures situation given by (6.1). Upon comparing (6.1) and (6.2), it is obvious that the repeated measures case provides a smaller variance for the difference in the two responses at times 1 and 2. In the special case, the variances have a common value σ^2, (6.1) reduces to

$$\text{Var}\left(Y_{i1} - Y_{i2}\right) = 2\sigma^2(1 - \rho) \tag{6.3}$$

while that of (6.2) reduces to

$$\text{Var}\left(Y_{i1} - Y_{i2}\right) = 2\sigma^2 \tag{6.4}$$

and the variance of the difference for repeated measures is reduced by the factor $(1 - \rho)$ compared to that for the cross-sectional, hence, the larger the correlation, the smaller the variance of the difference. This should convince one of the desirable properties of the repeated measures design.

Although the main focus is on the mean profile of the response over time, one cannot ignore the correlation structure, and the remaining part of the chapter is devoted to variance–covariance or correlation structures, including the unstructured, autoregressive, Toeplitz, and compound symmetry versions of the covariance. A rationale will be given for choosing the appropriate covariance structure and the main methods illustrated with an example from an exercise therapy trial.

The general model for the response at occasion j for subject i is

$$Y_{ij} = X'_{ij}\beta + e_{ij} \tag{6.5}$$

where:
X_{ij} is a p by 1 known vector of covariates
β is a p by 1 vector of unknown regression coefficients
the errors are normally distributed random variables with mean 0

Note for the ith subject, the errors of the response at times $t_1, t_2, ..., t_n$ has covariance matrix with the following entries: σ_j^2 is the variance of the response at time t_j ($j = 1,2,...,n$) and the covariance between the responses at times t_j and t_k is denoted by σ_{jk}. Last, the correlation between the two response Y_{ij} and Y_{ik} is

$$\rho_{jk} = \frac{\sigma_{jk}}{\sigma_j \sigma_k} \tag{6.6}$$

for all t_j and t_k, $t_j \neq t_k$.

Initially, our approach to determining an appropriate covariance matrix is based on the model (6.5) where the regression parameters β of the mean profile $X'_{ij}\beta$ for subject i is identified by descriptive statistics and scatter plots of the profile, an approach used in Chapter 5. Once the mean profile is chosen, the structure of the covariance matrix of the Y_{ij} ($j = 1,2,...,n$) determined.

Suppose Y_i is the n by 1 vector of responses for the ith subject, then Y_i is assumed to have a multivariate normal distribution with mean vector $X_i\beta$ and n by n precision matrix T; thus, the density of Y_i is

$$(2\pi)^{-n/2} |T|^{-1/2} \exp\left(-\frac{1}{2}\right)(y_i - X_i\beta)' T(y_i - X_i\beta) \qquad (6.7)$$

where:

$y_i \in R^n$
T is n by n positive definite and symmetric
β is p by 1 vector of unknown regression parameters

Note that the variance–covariance matrix is $\Sigma = T^{-1}$.

In addition, it is assumed that the responses from N subjects are independent and that the mean vector and precision matrix are the same for N subjects. It should also be emphasized that (6.7) is appropriate if the same times are common to N subjects and that if the times are not common to all subjects, the density (6.7) needs to be modified.

For the Bayesian approach, one assigns prior distributions to the regression parameters β and precision matrix T. When the variance–covariance matrix Σ is unstructured, a noninformative prior distribution will be assigned to T, but when structure is imposed on Σ, a different approach will be taken.

For additional information about the general topic of covariance structure for repeated measures, see Fitzmurice, Laird, and Ware.[1] Next to be discussed are the various forms that the covariance matrix one can adopt.

6.2 Patterns for Correlation Matrices

There is a wide variety of choices available for the structure of the covariance matrix Σ and the choice will be from among the following: (1) unstructured, (2) autoregressive, (3) Toeplitz, and (4) compound symmetry.

Of course, the first choice, unstructured, is the most general and involves $n(n + 1)/2$ parameters, where n is the number of time points, while for the autoregressive there are only two parameters. The Toeplitz has n variances and $n - 1$ covariances, while the compound symmetry covariance matrix has only two unknown parameters, the variance and the common covariance. Thus, among these four choices, the number of parameters to be estimated vary from a large number $n(n + 1)/2$ for unstructured to only two for compound symmetry and autoregressive parameters.

In order to show the reader the various Bayesian inference procedures in estimating the parameters of the mean response and the parameters of the covariance matrix, the following exercise therapy trial is explored, and it consists of two groups assigned to weightlifting programs. For the first group, the number of repetitions of exercise is increased as the subjects become stronger, and for the second, the number of repetitions is constant and the amount of weight increased as they become stronger. The main response is muscle strength measured at baseline (day 0) and at 2, 4, 6, 8, 10, and 12 days; thus, there are seven equally space time points. See Fitzmurice, Laird, and Ware (pp. 180–183)[1] for additional details about this study. Table 6.1 gives the information for the exercise therapy trial. There are 16 subjects in the first group designated as weight 1 and 19 in the second group designated as weight 2. The first column is the subject number, the second column is the time in days, the third column is the group index, while the four column is the muscle strength. This example also appears in Freund, Littel, and Spector[2] who analyzed it with the SAS package.

The descriptive statistics for this trial appears in Table 6.2, where the mean and standard deviation are shown.

TABLE 6.1

Exercise Therapy Trial

Subject	Time	Weight	Muscle Strength
1	0.00	1.00	79
1	2.00	1.00	.
1	4.00	1.00	79
1	6.00	1.00	80
1	8.00	1.00	80
1	10.00	1.00	78
1	12.00	1.00	80
2	0.00	1.00	83
2	2.00	1.00	83
2	4.00	1.00	85
2	6.00	1.00	85
2	8.00	1.00	86
2	10.00	1.00	87
2	12.00	1.00	87
3	0.00	1.00	81
3	2.00	1.00	83
3	4.00	1.00	82
3	6.00	1.00	82
3	8.00	1.00	83
3	10.00	1.00	83
3	12.00	1.00	82
4	0.00	1.00	81

(*Continued*)

TABLE 6.1 (*Continued*)

Exercise Therapy Trial

Subject	Time	Weight	Muscle Strength
4	2.00	1.00	81
4	4.00	1.00	81
4	6.00	1.00	82
4	8.00	1.00	82
4	10.00	1.00	83
4	12.00	1.00	81
5	0.00	1.00	80
5	2.00	1.00	81
5	4.00	1.00	82
5	6.00	1.00	82
5	8.00	1.00	82
5	10.00	1.00	.
5	12.00	1.00	86
6	0.00	1.00	76
6	2.00	1.00	76
6	4.00	1.00	76
6	6.00	1.00	76
6	8.00	1.00	76
6	10.00	1.00	76
6	12.00	1.00	75
7	0.00	1.00	81
7	2.00	1.00	84
7	4.00	1.00	83
7	6.00	1.00	80
7	8.00	1.00	85
7	10.00	1.00	85
7	12.00	1.00	85
8	0.00	1.00	77
8	2.00	1.00	78
8	4.00	1.00	79
8	6.00	1.00	79
8	8.00	1.00	81
8	10.00	1.00	82
8	12.00	1.00	81
9	0.00	1.00	84
9	2.00	1.00	85
9	4.00	1.00	87
9	6.00	1.00	89
9	8.00	1.00	.
9	10.00	1.00	.
9	12.00	1.00	86

(*Continued*)

TABLE 6.1 (*Continued*)

Exercise Therapy Trial

Subject	Time	Weight	Muscle Strength
10	0.00	1.00	74
10	2.00	1.00	75
10	4.00	1.00	78
10	6.00	1.00	78
10	8.00	1.00	79
10	10.00	1.00	78
10	12.00	1.00	78
11	0.00	1.00	76
11	2.00	1.00	77
11	4.00	1.00	77
11	6.00	1.00	77
11	8.00	1.00	77
11	10.00	1.00	76
11	12.00	1.00	76
12	0.00	1.00	84
12	2.00	1.00	84
12	4.00	1.00	86
12	6.00	1.00	85
12	8.00	1.00	86
12	10.00	1.00	86
12	12.00	1.00	86
13	0.00	1.00	79
13	2.00	1.00	80
13	4.00	1.00	79
13	6.00	1.00	80
13	8.00	1.00	80
13	10.00	1.00	82
13	12.00	1.00	82
14	0.00	1.00	78
14	2.00	1.00	78
14	4.00	1.00	77
14	6.00	1.00	.
14	8.00	1.00	75
14	10.00	1.00	75
14	12.00	1.00	76
15	0.00	1.00	78
15	2.00	1.00	80
15	4.00	1.00	77
15	6.00	1.00	77
15	8.00	1.00	75
15	10.00	1.00	.

(*Continued*)

TABLE 6.1 (*Continued*)

Exercise Therapy Trial

Subject	Time	Weight	Muscle Strength
15	12.00	1.00	75
16	0.00	1.00	84
16	2.00	1.00	85
16	4.00	1.00	85
16	6.00	1.00	85
16	8.00	1.00	85
16	10.00	1.00	83
16	12.00	1.00	.
17	0.00	2.00	84
17	2.00	2.00	85
17	4.00	2.00	84
17	6.00	2.00	83
17	8.00	2.00	83
17	10.00	2.00	83
17	12.00	2.00	84
18	0.00	2.00	74
18	2.00	2.00	75
18	4.00	2.00	75
18	6.00	2.00	78
18	8.00	2.00	75
18	10.00	2.00	76
18	12.00	2.00	76
19	0.00	2.00	83
19	2.00	2.00	84
19	4.00	2.00	82
19	6.00	2.00	81
19	8.00	2.00	.
19	10.00	2.00	83
19	12.00	2.00	.
20	0.00	2.00	86
20	2.00	2.00	87
20	4.00	2.00	87
20	6.00	2.00	87
20	8.00	2.00	87
20	10.00	2.00	87
20	12.00	2.00	86
21	0.00	2.00	82
21	2.00	2.00	83
21	4.00	2.00	84
21	6.00	2.00	85
21	8.00	2.00	84

(*Continued*)

TABLE 6.1 (*Continued*)

Exercise Therapy Trial

Subject	Time	Weight	Muscle Strength
21	10.00	2.00	85
21	12.00	2.00	86
22	0.00	2.00	79
22	2.00	2.00	80
22	4.00	2.00	79
22	6.00	2.00	79
22	8.00	2.00	80
22	10.00	2.00	79
22	12.00	2.00	80
23	0.00	2.00	79
23	2.00	2.00	79
23	4.00	2.00	79
23	6.00	2.00	81
23	8.00	2.00	81
23	10.00	2.00	83
23	12.00	2.00	.
24	0.00	2.00	87
24	2.00	2.00	89
24	4.00	2.00	91
24	6.00	2.00	90
24	8.00	2.00	91
24	10.00	2.00	87
24	12.00	2.00	89
25	0.00	2.00	81
25	2.00	2.00	81
25	4.00	2.00	81
25	6.00	2.00	82
25	8.00	2.00	82
25	10.00	2.00	83
25	12.00	2.00	83
26	0.00	2.00	82
26	2.00	2.00	82
26	4.00	2.00	.
26	6.00	2.00	84
26	8.00	2.00	86
26	10.00	2.00	85
26	12.00	2.00	.
27	0.00	2.00	79
27	2.00	2.00	79
27	4.00	2.00	80
27	6.00	2.00	81

(*Continued*)

TABLE 6.1 (*Continued*)

Exercise Therapy Trial

Subject	Time	Weight	Muscle Strength
27	8.00	2.00	81
27	10.00	2.00	81
27	12.00	2.00	.
28	0.00	2.00	79
28	2.00	2.00	80
28	4.00	2.00	81
28	6.00	2.00	82
28	8.00	2.00	83
28	10.00	2.00	.
28	12.00	2.00	.
29	0.00	2.00	83
29	2.00	2.00	84
29	4.00	2.00	84
29	6.00	2.00	84
29	8.00	2.00	84
29	10.00	2.00	.
29	12.00	2.00	83
30	0.00	2.00	81
30	2.00	2.00	81
30	4.00	2.00	82
30	6.00	2.00	84
30	8.00	2.00	83
30	10.00	2.00	82
30	12.00	2.00	85
31	0.00	2.00	78
31	2.00	2.00	78
31	4.00	2.00	79
31	6.00	2.00	79
31	8.00	2.00	78
31	10.00	2.00	79
31	12.00	2.00	79
32	0.00	2.00	83
32	2.00	2.00	82
32	4.00	2.00	82
32	6.00	2.00	84
32	8.00	2.00	84
32	10.00	2.00	83
32	12.00	2.00	84
33	0.00	2.00	80
33	2.00	2.00	79
33	4.00	2.00	79

(*Continued*)

TABLE 6.1 (*Continued*)

Exercise Therapy Trial

Subject	Time	Weight	Muscle Strength
33	6.00	2.00	81
33	8.00	2.00	.
33	10.00	2.00	.
33	12.00	2.00	80
34	0.00	2.00	80
34	2.00	2.00	82
34	4.00	2.00	82
34	6.00	2.00	82
34	8.00	2.00	81
34	10.00	2.00	81
34	12.00	2.00	81
35	0.00	2.00	85
35	2.00	2.00	86
35	4.00	2.00	87
35	6.00	2.00	86
35	8.00	2.00	86
35	10.00	2.00	86
35	12.00	2.00	86
36	0.00	2.00	77
36	2.00	2.00	78
36	4.00	2.00	80
36	6.00	2.00	81
36	8.00	2.00	82
36	10.00	2.00	82
36	12.00	2.00	82
37	0.00	2.00	80
37	2.00	2.00	81
37	4.00	2.00	80
37	6.00	2.00	81
37	8.00	2.00	81
37	10.00	2.00	82
37	12.00	2.00	83

Figure 6.1 shows a scatter plot of the exercise therapy trial, and it appears as if the mean muscle strength varies little over time for each group and that the muscle strength for the second group seems to exceed (very slightly) that for group 1.

In what is to follow the various patterns for the covariance matrix are defined, and the covariance matrix is estimated via Bayesian techniques.

TABLE 6.2

Descriptive Statistics Exercise Study

Time	Mean	Standard Deviation
Group 1: Muscle Strength		
0	79.69	3.114
2	80.67	3.309
4	80.81	3.582
6	81.13	3.662
8	80.80	3.840
10	81.08	4.030
12	81.07	4.317
Group 2: Muscle Strength		
0	81.05	3.106
2	81.67	3.352
4	81.90	3.567
6	82.62	2.854
8	82.74	3.462
10	82.61	2.852
12	82.94	3.235

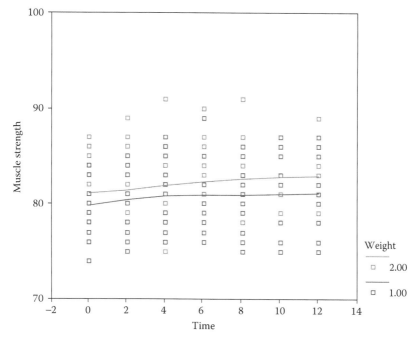

FIGURE 6.1

Scatter plot for exercise therapy trial muscle strength versus day.

In order to estimate the covariance matrix, I will assume the model for the mean profile of group 1 is

$$y_{ij} = \beta_1 + \beta_2 t_{ij} + e_{ij} \tag{6.8}$$

where:
$i = 1,2,\ldots,16$
$t_{ij} = 0,2,4,6,8,10,12$

That is, it is assumed that the mean profile for group 1 is a linear regression over the seven time points with intercept β_1 and slope β_2. After examining the scatter plot and descriptive statistics, this assertion appears to be reasonable. In order to estimate the covariance matrix, it is necessary to adopt a mean profile for the repeated measures. Note that the covariance matrix of the errors e_{ij} for $j = 1,2,3,4,5,6,7$ is yet to be specified for the exercise therapy trial. The first possibility is the unstructured covariance matrix.

6.2.1 Unstructured

When there is no structure for the covariance matrix, one assumes that the n by n covariance matrix is given by

$$\Sigma = (\sigma_{jk}) \tag{6.9}$$

where:
σ_{jj} is the variance for the response at occasion j (= $1,2,\ldots,n$)
σ_{jk} is the covariance of the main response between occasions j and k, where $j \neq k$ and $j,k = 1,2,\ldots,n$

Thus, the variances and covariances can only assume values such that Σ is symmetric and positive definite.

How is the correlation and covariance matrix to be estimated with a Bayesian approach? One must place prior distributions on the regression parameters β and β_2 and the precision matrix $T = \Sigma^{-1}$. For group 1 of the exercise therapy trial, the code for estimating Σ is given by **BUGS CODE 6.1**.

Note that the regression coefficients have noninformative normal prior distributions with mean 0 and precision 0.0001 and that the precision matrix omega is assumed to have a constant density. Also be aware that the sigma matrix is the unstructured covariance matrix and the rho values are the corresponding correlations.

The analysis is executed with 75,000 observations using 5,000 for the burn in and a refresh of 100, and the Bayesian analysis is reported in Table 6.3.

One should look for patterns among the various correlations and variances of Table 6.6. One sees that the variances are increasing over time and that the correlations are decreasing with time as the time increases that the correlations tend to become smaller. These patterns will be referred to when investigating covariance matrices with a pattern.

BUGS CODE 6.1

```
model;
{

# prior distribution for the regression coefficients.
            beta1 ~ dnorm(0.0, 0.001)
            beta2 ~ dnorm(0.0, 0.001)

for(i in 1:N1){Y[i,1:M1]~dmnorm(mu1[],Omega[,])}
for (j in 1:M1){mu1[j]<-beta1+beta2*age[j]}
# non-informative precision matrix
Omega[1:M1,1:M1]~dwish(R[,],7)
Sigma[1:M1,1:M1]<-inverse(Omega[,])
rho[1,2]<-Sigma[1,2]/sqrt(Sigma[1,1]*Sigma[2,2])
rho[1,3]<-Sigma[1,3]/sqrt(Sigma[1,1]*Sigma[3,3])

rho[1,4]<-Sigma[1,4]/sqrt(Sigma[1,1]*Sigma[4,4])
rho[1,5]<-Sigma[1,5]/sqrt(Sigma[1,1]*Sigma[5,5])
rho[1,6]<-Sigma[1,6]/sqrt(Sigma[1,1]*Sigma[6,6])
rho[1,7]<-Sigma[1,7]/sqrt(Sigma[1,1]*Sigma[7,7])
rho[2,3]<-Sigma[2,3]/sqrt(Sigma[2,2]*Sigma[3,3])
rho[2,4]<-Sigma[2,4]/sqrt(Sigma[2,2]*Sigma[4,4])
rho[2,5]<-Sigma[2,5]/sqrt(Sigma[2,2]*Sigma[5,5])
rho[2,6]<-Sigma[2,6]/sqrt(Sigma[2,2]*Sigma[6,6])
rho[2,7]<-Sigma[2,7]/sqrt(Sigma[2,2]*Sigma[7,7])
rho[3,4]<-Sigma[3,4]/sqrt(Sigma[3,3]*Sigma[4,4])
rho[3,5]<-Sigma[3,5]/sqrt(Sigma[3,3]*Sigma[5,5])
rho[3,6]<-Sigma[3,6]/sqrt(Sigma[3,3]*Sigma[6,6])
rho[3,7]<-Sigma[3,7]/sqrt(Sigma[3,3]*Sigma[7,7])
rho[4,5]<-Sigma[4,5]/sqrt(Sigma[4,4]*Sigma[5,5])
rho[4,6]<-Sigma[4,6]/sqrt(Sigma[4,4]*Sigma[6,6])
rho[4,7]<-Sigma[4,7]/sqrt(Sigma[4,4]*Sigma[7,7])
rho[5,6]<-Sigma[5,6]/sqrt(Sigma[5,5]*Sigma[6,6])
rho[5,7]<-Sigma[5,7]/sqrt(Sigma[5,5]*Sigma[7,7])
rho[6,7]<-Sigma[6,7]/sqrt(Sigma[6,6]*Sigma[7,7])
}
# weight = 1

list(N1 = 16,M1 = 7,Y = structure(.Data = c(79,NA,79,
80,     80,     78,     80,
83,     83,     85,     85,     86,     87,     87,
81,     83,     82,     82,     83,     83,     82,
81,     81,     81,     82,     82,     83,     81,
80,     81,     82,     82,     82,     NA, 86,
76,76, 76,     76,     76,     76,     75,
81,84, 83,     83,     85,     85,     85,
77,     78,     79,     79,     81,     82,     81,
```

```
84,      85,      87,      89,NA,NA,86,
74,      75,      78,      78,      79,      78,      78,
76,      77,      77,      77,      77,      76,      76,
84,      84,      86,      85,      86,      86,      86,
79,      80,      79,      80,      80,      82,      82,
78,      78,      77, NA, 75,      75,      76,
78,      80,      77,      77,      75,NA, 75,
84,      85,      85,      85,      85,      83,      NA),.Dim =
c(16,7)),

age = c(0,2,4,6,8,10,12),

R = structure(.Data = c(1,0,0,0,0,0,0,
                        0,1,0,0,0,0,0,
                        0,0,1,0,0,0,0,
                        0,0,0,1,0,0,0,
                        0,0,0,0,1,0,0,
                        0,0,0,0,0,1,0,
                        0,0,0,0,0,0,1),.Dim = c(7,7)))

# initial values

list(beta1 = 0,beta2 = 0)
```

It should be noted that there are 21 variances and covariances but only 16 subjects and 16 times 7 = 112 total observations; thus, one perhaps is hesitant to accept these estimates (posterior means and medians) as stable. This is one disadvantage of using an unstructured covariance matrix in that the number of parameters increases dramatically with the number of time points. With seven time points, there are 21 entries in the covariance matrix.

Finally, concluding the analysis, the posterior density of the correlation between times 2 and 3 days is portrayed in Figure 6.2.

This clearly shows that the distribution of ρ_{23} is skewed to the right with a posterior mean of 0.895 but a posterior median of 0.906.

6.2.2 Autoregressive

What is an autoregressive structure for the covariance matrix? Consider the model (6.5), where

$$e_{ij} = \rho e_{ij-1} + u_{ij}$$

with $i = 1,2,...,N$ and $j = 1,2,...,n$, u_{ij} are independent and normally distributed with mean 0 and variance $\sigma^2 = 1/\tau$, τ is the precision, and correlation ρ

TABLE 6.3

Posterior Analysis for the Exercise Therapy Trial for Group 1

Parameter	Mean	SD	Error	2½	Median	97½
β_1	79.86	1.064	0.007906	77.95	79.87	81.76
β_2	0.1415	0.05601	0.0004104	0.03044	0.0418	0.2519
ρ_{12}	0.9349	0.03557	0.0001619	0.8448	0.9425	0.9799
ρ_{13}	0.9026	0.05034	0.0002155	0.7767	0.913	0.9684
ρ_{14}	0.8913	0.0566	0.0002513	0.7487	0.9031	0.9648
ρ_{15}	0.8127	0.09628	0.001686	0.5889	0.8308	0.9364
ρ_{16}	0.776	0.1115	0.002013	0.5197	0.797	0.9231
ρ_{17}	0.7941	0.09994	0.0004676	0.5456	0.8132	0.9313
ρ_{23}	0.895	0.05348	0.0002256	0.7608	0.906	0.9655
ρ_{24}	0.8829	0.05852	0.0002364	0.7375	0.8945	0.9606
ρ_{25}	0.8204	0.09252	0.00178	0.6097	0.8377	0.9383
ρ_{26}	0.7566	0.1218	0.002105	0.4707	0.7795	0.9179
ρ_{27}	0.7705	0.1115	0.0005392	0.4933	0.7919	0.9233
ρ_{34}	0.977	0.01225	0.000498	0.946	0.9796	0.9927
ρ_{35}	0.9546	0.0515	0.002143	0.8964	0.9616	0.9855
ρ_{36}	0.8929	0.0767	0.002491	0.7451	0.9076	0.967
ρ_{37}	0.8907	0.06119	0.0003285	0.7351	0.9046	0.9671
ρ_{45}	0.9662	0.05512	0.002459	0.9209	0.9727	0.9903
ρ_{46}	0.8966	0.08521	0.002783	0.7384	0.9132	0.971
ρ_{47}	0.8822	0.06193	0.0003043	0.7256	0.8955	0.9625
ρ_{56}	0.9398	0.03872	0.0002823	0.839	0.9493	0.9838
ρ_{57}	0.8981	0.06497	0.001559	0.7566	0.9117	0.97
ρ_{67}	0.9417	0.05819	0.002088	0.841	0.9533	0.9853
σ_{11}	11.12	45.29	0.3479	5.146	9.771	22.25
σ_{22}	12.0	43.92	0.3433	5.511	10.55	24.35
σ_{33}	14.49	46.32	0.3557	6.77	12.84	29.18
σ_{44}	15.76	45.24	0.3466	7.397	14.0	31.71
σ_{55}	21.23	45.11	0.457	9.898	18.92	43.36
σ_{66}	22.92	45.41	0.9101	10.22	20.01	46.53
σ_{77}	20.13	48.3	0.368	9.469	17.99	40.52

satisfied $-1 < \rho < 1$. Thus, for each subject, the e_{ij} follows a first-order autoregressive process with correlation ρ.

The model (6.5) is rearranged to give

$$Y_{ij} = \rho Y_{ij-1} + \left(X'_{ij} - \rho X'_{ij-1} \right) \beta + u_{ij} \qquad (6.10)$$

for $i = 1,2,\dots,N$ and $j = 2,\dots,n$. Also assumed is that the time periods are equally spaced and each subject has n occasions at which the response is measured. This will later be generalized to the case where the time points

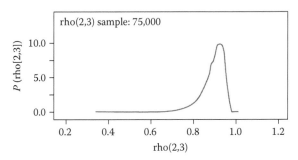

FIGURE 6.2
Posterior density of ρ_{23}.

TABLE 6.4

Correlation Pattern for Autoregressive Structure

1	ρ	ρ^2	ρ^3	$\cdots$	ρ^{n-2}	ρ^{n-1}
ρ	1	ρ	ρ^2		ρ^{n-3}	ρ^{n-2}
ρ^2		1				
ρ^3			1			
$\vdots$						
ρ^{n-2}					1	ρ
ρ^{n-1}	ρ^{n-2}					1

are not equally spaced and the number of occasions the response is measured is not necessarily the same number n.

It can be shown that the correlation is ρ for responses one unit apart, is ρ^2 for responses two units apart, and so on. This is represented by the correlation matrix (Table 6.4).

Note there are only two parameters σ and ρ for the covariance matrix with autoregressive structure.

Consider the exercise therapy trial assuming an autoregressive structure for the correlation, then corresponding to (6.10), the model is

$$Y_{ij} = \rho Y_{ij-1} + \beta_1(1-\rho) + \beta_2\left(t_{ij} - \rho t_{ij-1}\right) + u_{ij} \tag{6.11}$$

where for each i, $t_{ij} = 0,2,4,6,8,10,12$ and $j = 1,2,3,4,5,6,7$.

BUGS CODE 6.2 closely follows the formula (6.11), and the Bayesian analysis is executed with 75,000 observations for the simulation, with a burn in of 5,000 and a refresh of 100. Normal distributions with mean 0 and variances 1000 are the prior distributions for the two regression coefficients, while the correlation parameter rho is given a uniform distribution over (0,1).

BUGS CODE 6.2

```
model;
{
        beta1 ~ dnorm(0.0, 0.001)
        beta2 ~ dnorm(0.0, 0.001)

for(i in 1:N1){for (j in 2:M1){Y[i,j]~dnorm(mu[i,j],tau)}}
for(i in 1:N1){for (j in 2:M1)
{mu[i,j]<-beta1*(1-rho)+beta2*(age[j]-rho*age[j-
1])+rho*Y[i,j-1]}}
for(i in 1:N1){Y[i,1]~dnorm(mu[i,1],tau)}
for(i in 1:N1){mu[i,1]<-beta1+beta2*age[1]}
tau~dgamma(.001,.001)
rho~dbeta(1,1)
sigma<-1/tau
}


# weight = 1

list(N1 = 16,M1 = 7,Y = structure(.Data = c(79,NA,79,
80,     80,     78,     80,
83,     83,     85,     85,     86,     87,     87,
81,     83,     82,     82,     83,     83,     82,
81,     81,     81,     82,     82,     83,     81,
80,     81,     82,     82,     82,     NA, 86,
76,76,  76,     76,     76,     76,     75,
81,84,  83,     83,     85,     85,     85,
77,     78,     79,     79,     81,     82,     81,
84,     85,     87,     89,NA,NA,86,
74,     75,     78,     78,     79,     78,     78,
76,     77,     77,     77,     77,     76,     76,
84,     84,     86,     85,     86,     86,     86,
79,     80,     79,     80,     80,     82,     82,
78,     78,     77, NA, 75,     75,     76,
78,     80,     77,     77,     75,NA, 75,
84,     85,     85,     85,     85,     83,     NA),.Dim =
c(16,7)),

age = c(0,2,4,6,8,10,12))

# initial values

list(beta1 = 0,beta2 = 0, rho =.5,tau = 1)
```

TABLE 6.5

Bayesian Analysis for Exercise Therapy Trial for Group 1, Autoregressive Structure

Parameter	Mean	SD	Error	2½	Median	97½
β_1	79.68	0.4079	0.005595	78.88	79.67	80.46
β_2	0.1278	0.075	0.001565	−0.020540	0.129	0.2755
ρ	0.9626	0.02839	0.00125	0.8933	0.9693	0.9985
σ^2	2.654	0.02839	0.002488	2.013	2.618	3.499

The posterior analysis for the exercise therapy trial is reported in Table 6.5.

What are the implications when using the autoregressive structure for the error of the linear regression model (6.10)? One sees that the intercept is estimated as 79.68 with the posterior mean and as 0.1278 for the slope, and the correlation between observations (muscle strength) measured two days apart is estimated as 0.9626. It is interesting to observe that the slope has a 95% credible interval (−0.02054, 0.2755), which includes 0. This is also implied by the scatter plot in Figure 6.1. This implies the correlation between responses four days apart is estimated as ρ^2, namely as 0.9265, and as ρ^3 or 0.8919.

Also, the Markov chain Monte Carlo (MCMC) errors appear to be "small" enough to conclude that 70,000 observations for the simulation is sufficient in order to estimate the parameters by their posterior means.

The autoregressive process can be generalized to the case when there is unequal spacing between the responses at the various times; thus, consider the model

$$Y_{ij} = \rho^{t_{ij}-t_{ij-1}} Y_{ij-1} + \beta_1\left(1 - \rho^{t_{ij}-t_{ij-1}}\right) + \beta_2\left(t_{ij} - \rho^{t_{ij}-t_{ij-1}} t_{ij-1}\right) + u_{ij} \qquad (6.12)$$

where for each i, $t_{ij} = 0,4,6,8,10,12$ and $j = 1,2,3,4,5,6,7$. That is to say the observation at time day 2 is omitted from consideration for the analysis, and **BUGS CODE 6.3** will execute the Bayesian analysis with 75,000 observations for the simulation, with a burn in of 5,000 and a refresh of 100. One would expect the analysis to be very similar to that given in Table 6.5. The code closely follows the notation of formula (6.12).

Upon on comparing Table 6.5 to Table 6.6, one sees the similarity. For example, the posterior mean for ρ is 0.9626 for the case when day 1 response is not omitted, but when day 2 is omitted the posterior mean is 0.9845, a difference of 0.02, very small indeed.

It is interesting to observe that these exponential regression models can be generalized to quite complicated patterns in the response over time, and this will be considered next using the exercise therapy trial. The number of responses for each subject will vary as will the spacing between the time points. Thus, in an arbitrary way, I take the original data set for the exercise therapy trial and vary the spacing and the number of responses from patient to patient.

BUGS CODE 6.3

```
model;
     {
                beta1 ~ dnorm(0.0, 0.001)
                beta2 ~ dnorm(0.0, 0.001)

for(i in 1:N1){for (j in 2:M1){Y[i,j]~dnorm(mu[i,j],tau)}}
for(i in 1:N1){for (j in 2:M1){mu[i,j]<-beta1*(1-
pow(rho,(age[j]-age[j-1])))+beta2*(age[j]-
pow(rho,(age[j]-age[j-1]))*age[j-1])+
pow(rho,(age[j]-age[j-1]))*Y[i,j-1]}}

for(i in 1:N1){Y[i,1]~dnorm(mu[i,1],tau)}
for(i in 1:N1){mu[i,1]<-beta1+beta2*age[1]}

rho~dbeta(1,1)
tau~dgamma(.001,.001)
sigmasq<-1/tau
}


# weight = 1

list(N1 = 16,M1 = 6,Y = structure(.Data =
c(79,79,80,80,78,80,
83,    85,    85,    86,    87,    87,
81,    82,    82,    83,    83,    82,
81,    81,    82,    82,    83,    81,
80,    82,    82,    82,    NA, 86,
76,76,76,    76,    76,    75,
81, 83,83,    85,    85,    85,
77,    79,    79,    81,    82,81,
84,    87,    89, NA,NA,86,
74,    78,    78,    79,    78,78,
76,    77,    77,    77,    76,76,
84,    86,    85,    86,    86,86,
79,    79,    80,    80,    82,82,
78,    77, NA, 75,    75,    76,
78,    77,    77,    75,NA, 75,
84,    85,    85,    85,    83,    NA),.Dim = c(16,6)),

age = c(0,4,6,8,10,12))


# initial values
list(beta1 = 0,beta2 = 0,tau =.5,rho =.5)
list(N1 = 16,M1 = 7,Y = structure(.Data =
c(79,NA,79,80,80,78,80
```

TABLE 6.6

Posterior Analysis for Exercise Therapy Trial for Group 1 with Muscle Strength at Day 2 Omitted

Parameter	Mean	SD	Error	2½	Median	97½
β_1	79.66	0.4056	0.00798	78.85	79.65	80.46
β_2	0.1615	0.07288	0.002063	0.02422	0.1595	0.3134
ρ	0.9845	0.01266	0.000373	0.9519	0.9871	0.9995
σ^2	2.917	0.453	0.003143	2.159	2.873	3.935

TABLE 6.7

Revised Information for the Exercise Therapy Trial for Group 1

```
83,  NA,  85,  85,  86,  NA,  87,
81,  83,  NA,  NA,  83,  83,  82,
81,  NA,  81,  82,  NA,  NA,  81,
80,  81,  82,  82,  82,  NA,  86,
NA,  76,  76,  NA,  76,  76,  75,
NA,  84,  83,  83,  85,  85,  NA,
77,  NA,  NA,  79,  81,  NA,  81,
84,  85,  87,  NA,  NA,  NA,  86,
74,  NA,  78,  78,  79,  NA,  78,
76,  NA,  77,  77,  NA,  76,  NA,
84,  84,  86,  85,  86,  86,  86,
79,  NA,  79,  80,  80,  NA,  82,
78,  78,  77,  NA,  75,  75,  76,
78,  80,  77,  77,  NA,  NA,  75,
84,  85,  85,  NA,  85,  83,  NA),.Dim = c(16,7)),
```

Refer to list statement of **BUGS CODE 6.3**, and compare to the following list statement, where I have changed the original information of the exercise therapy trial to that appearing in Table 6.7. Comparing the two tables, one sees that the third patient did not have muscle strength responses measured on days 4 and 6, whereas information for the third patient was initially complete for all seven days. Thus, the number of measurements was reduced from seven to five, and there is a space of six days between days 2 and 8. Similar changes were made to the other patients resulting in a data set where the number of responses varies from a complete set to one involving only four.

Using **BUGS CODE 6.3**, a Bayesian analysis is performed with 75,000 observations for the simulation, with a burn in of 5,000 and a refresh of 100, and the results are reported in Table 6.8.

Comparing Table 6.6 with Table 6.8, one sees very little difference in the posterior means of the four parameters, although there is a slight increase in

TABLE 6.8

Posterior Analysis of Exercise Therapy Trial for Modified Data, Group 1

Parameter	Mean	SD	Error	2½	Median	97½
β_1	79.83	0.469	0.007077	78.98	79.83	80.74
β_2	0.1149	0.08104	0.00167	−0.043440	0.1151	0.273
ρ	0.976	0.01805	0.000679	0.9302	0.9794	0.9992
σ^2	3.04	0.5054	0.004425	2.211	2.996	4.186

the estimated variance from 2.917 to 3.04. In the exercises, the data set is further modified so that significant changes in the posterior analysis do in fact occur. There are even more messy spacing of the time points and the number of observations for the patients, which will be described in exercises.

An alternative to the unstructured and autoregressive patterns is compound symmetry, which is introduced in the next section.

6.2.3 Compound Symmetry

The compound symmetry pattern places a constant correlation ρ for all possible correlations among n time points of the main response. The model takes into account the individual random effect of each of the N subjects.

Let Y_{ij} denote the response for subject i on occasion j, and let

$$Y_{ij} = \beta_1 + \beta_2 t_{ij} + a_i + b_j + e_{ij} \qquad (6.13)$$

where:
β_1 is the average response at baseline
β_2 is the slope
a_i is the random effect of subject i
b_j is the random effect of occasion j
e_{ij} are independent normally distributed variables with mean 0 and variance
σ^2, a_i are individual random effects that are normally distributed with mean 0 and variance σ_a^2, and last b_j are normally distributed with mean 0 and variance σ_b^2. In addition, a_i, b_j, and e_{ij} are mutually independent

Consider the covariance between the jth and kth occasions, that is

$$\mathrm{Cov}\left(Y_{ij}, Y_{ik}\right) = \sigma_a^2 \qquad (6.14)$$

and

$$\mathrm{Var}\left(Y_{ij}\right) = \sigma_a^2 + \sigma_b^2 + \sigma^2 \qquad (6.15)$$

therefore,

$$\rho\left(Y_{ij}, Y_{ik}\right) = \frac{\sigma_a^2}{\sigma_a^2 + \sigma_b^2 + \sigma^2} \tag{6.16}$$

for all $i = 1,2,\ldots,N$, $j,k = 1,2,\ldots,n$ for $j \neq k$.

Using this formulation, a Bayesian analysis is executed for the exercise therapy trial and based on **BUGS CODE 6.4**.

BUGS CODE 6.4

```
model;

{
for(i in 1:P){for(j in 1:M){y[i,j]~dnorm(mu[i,j],tau)}}
for(i in 1:P){for(j in 1:M){Z[i,j]~dnorm(mu[i,j],tau)}}
for(i in 1:P){for(j in 1:M)
{mu[i,j]<-beta1+beta2*age[j]+a[i]+b[j]}}
for(i in 1:P){a[i]~dnorm(0,taua)}
for(j in 1:M){b[j]~dnorm(0,taub)}
beta1~dnorm(0.0,.0001)
beta2~dnorm(0.0,.0001)
theta~dnorm(0,.001)
taua~dgamma(.0001,.0001)
taub~dgamma(.0001,.0001)
tau~dgamma(.0001,.0001)
sigmaa<-1/taua
sigmab<-1/taub
sigma<-1/tau
rho<-(sigmaa)/(sigmaa+sigmab+sigma)

}

list(P = 16,M = 7,
age = c(0,2,4,6,8,10,12),

y = structure(.Data = c(79,NA,79,80,    80,    78,    80,
83,     83,    85,    85,    86,    87,    87,
81,     83,    82,    82,    83,    83,    82,
81,     81,    81,    82,    82,    83,    81,
80,     81,    82,    82,    82,    NA, 86,
76,76, 76,    76,    76,    76,    75,
81,84, 83,    83,    85,    85,    85,

77,     78,    79,    79,    81,    82,    81,
84,     85,    87,    89,NA,NA,86,
74,     75,    78,    78,    79,    78,    78,
76,     77,    77,    77,    77,    76,    76,
84,     84,    86,    85,    86,    86,    86,
```

```
79,     80,     79,     80,     80,     82,     82,
78,     78,     77, NA, 75,     75,     76,
78,     80,     77,     77,     75, NA, 75,
84,     85,     85,     85,     85,     83,     NA), .Dim =
c(16,7)))

list(taua = 1, taub = 1, tau = 1, beta1 = 0, beta2 = 0)
```

TABLE 6.9

Posterior Analysis for Exercise Therapy Trial for Compound Symmetry

Parameter	Mean	SD	Error	2½	Median	97½
ρ	0.8877	0.04644	0.001131	0.7897	0.8903	0.9977
σ^2	1.633	0.2521	0.001149	1.21	1.609	2.195
σ_a^2	14.32	6.248	0.05811	6.602	12.95	30.12
σ_b^2	0.02836	0.1011	0.001251	0.0000842	0.003773	0.2076
β_1	78.93	6.349	0.3224	54.16	80.05	82.08
β_2	0.1212	0.0349	0.0002572	0.05247	0.1213	0.1893

The posterior analysis is executed with 75,000 observations for the simulation, with a burn in of 5,000 and a refresh of 100.

The prior distributions for the slope and intercept are normal with mean 0 and variance 1000, and the three precision parameters are given gamma distributions with parameters 0.0001, 0.0001. This in effect assigns noninformative priors to the parameters of the model (6.13) (Table 6.9).

The correlation between the muscle strength measurements is estimated as 0.8877 (with the posterior mean) for all possible time pairs among the seven days 0,2,4,6,8,10,12. This is the deficiency of compound symmetry, namely the correlations do not die out or decrease with time but stay the same at ρ.

The three variance components are estimated as 1.609, 12.95, and 0.003773 for σ^2, σ_a^2, and σ_b^2, respectively, demonstrating that between-subject variability σ_a^2 is the largest median among the three, while the smallest posterior median is the between-times component σ_b^2. The estimates of the intercept and slope are comparable to those given under the unstructured pattern (see Table 6.3) and the autoregressive patterns (see Table 6.5).

We have considered three structures for the covariance matrix: (1) unstructured, (2) autoregressive, and (3) compound symmetry. The last pattern to be introduced is the Toeplitz, a form that lists somewhat between the autoregressive and compound symmetry.

6.2.4 Toeplitz

Another pattern for the covariance matrix is Toeplitz, which is defined for the correlation matrix shown in Table 6.10.

TABLE 6.10

Correlation Matrix for Toeplitz Pattern

1	ρ_1	ρ_2	$\cdots$	ρ_{n-1}
ρ_1	1	ρ_1		ρ_{n-2}
ρ_2	ρ_1	1		ρ_{n-3}
$\vdots$				
ρ_{n-1}	ρ_{n-2}			1

The variance is assumed constant over the n time points for each subject, and the correlation is ρ_1 when the space between two time points is one unit, is ρ_2 when the space between two time points is two units, and so on.

First, let us consider the estimation of ρ_2, the correlation between a response and the response separated by two time units, then it can be shown that the model is

$$Y_{i,j} = \beta_1\left(1-\rho_2\right) + \beta_2\left(t_j - \rho_2 t_{j-2}\right) + \rho_2 Y_{i,i-2} + u_{ij} \qquad (6.17)$$

where:

t_j is the time point at occasion j

β_1 are independent and distributed $n(0,\sigma^2)$

Note that $j = 3,4,5,6,7$. Consider the exercise therapy trial, then the code appearing in **BUGS CODE 6.5** is used for the Bayesian analysis and the code is quite similar to the notation of formula (6.17). The analysis is executed with 75,000 observations for the simulation, with a burn in of 5,000 and a refresh of 100.

The main focus is the estimation of β_1, the intercept; β_2, the slope; σ^2, the variance of the residual; and ρ_2, the correlation.

The posterior analysis appears in Table 6.11.

Thus, the correlation ρ_2 is estimated as 0.9952 with the posterior mean with a corresponding 95% credible interval (0.8642, 0.9985), that is, for responses four days apart, the estimated correlation is 0.9643. Also, the estimates for the intercept and slope are very similar to previous estimates using other patterns (unstructured, autoregressive, and compound symmetry) for the correlation matrix.

What is the posterior mean of ρ_1? ρ_1 is the correlation between response one unit apart in time. Consider **BUGS CODE 6.6**, where the prior distributions

TABLE 6.11

Bayesian Analysis of Exercise Therapy Trial, for Estimation of ρ_2

Parameter	Mean	SD	Error	2½	Median	97½
β_1	80.0	0.3929	0.005134	79.3	80.0	80.79
β_2	0.1335	0.0599	0.001047	0.01754	0.1341	0.2513
ρ_2	0.9552	0.03721	0.001803	0.8642	0.9643	0.9985
σ^2	4.674	0.6682	0.004944	3.543	4.612	6.154

BUGS CODE 6.5

```
model;
      {
            beta1 ~ dnorm(0.0, 0.001)
            beta2 ~ dnorm(0.0, 0.001)

for(i in 1:N1){for (j in 3:M1){Y[i,j]~dnorm(mu[i,j],tau)}}
for(i in 1:N1){for (j in 3:M1){mu[i,j]<-beta1*
(1-rho2)+beta2*(age[j]-rho2*age[j-2])+rho2*Y[i,j-2]}}
for(i in 1:N1){Y[i,1]~dnorm(mu[i,1],tau)}
for(i in 1:N1){mu[i,1]<-beta1+beta2*age[1]}
for(i in 1:N1){Y[i,2]~dnorm(mu[i,2],tau)}
for(i in 1:N1){mu[i,2]<-beta1+beta2*age[2]}

rho2~dbeta(1,1)
tau~dgamma(.001,.001)
sigmasq<-1/tau

}
# weight = 1
list(N1 = 16,M1 = 7, Y = structure(.Data = c(79,NA,79,
80,    80,    78,    80,
83,    83,    85,    85,    86,    87,    87,
81,    83,    82,    82,    83,    83,    82,
81,    81,    81,    82,    82,    83,    81,
80,    81,    82,    82,    82,    NA, 86,
76,76, 76,    76,    76,    76,    75,
81,84, 83,    83,    85,    85,    85,

77,    78,    79,    79,    81,    82,    81,
84,    85,    87,    89, NA,NA,86,
74,    75,    78,    78,    79,    78,    78,
76,    77,    77,    77,    77,    76,    76,
84,    84,    86,    85,    86,    86,    86,
79,    80,    79,    80,    80,    82,    82,
78,    78,    77, NA,    75,    75,    76,
78,    80,    77,    77,    75,NA, 75,
84,    85,    85,    85,    85,    83,    NA),.Dim =
c(16,7)),

age = c(0,2,4,6,8,10,12))

# initial values

list(beta1 = 0,beta2 = 0,tau =.5,rho2 =.5)
```

```
                        BUGS CODE 6.6
model;
        {
                beta1 ~ dnorm(0.0, 0.001)
                beta2 ~ dnorm(0.0, 0.001)

for(i in 1:N1){for (j in 2:M1){Y[i,j]~dnorm(mu[i,j],tau)}}
for(i in 1:N1){for (j in 2:M1)
{mu[i,j]<-beta1*(1-rho1)+beta2*(age[j]-rho1*age[j-
1])+rho1*Y[i,j-1]}}
for(i in 1:N1){Y[i,1]~dnorm(mu[i,1],tau)}
for(i in 1:N1){mu[i,1]<-beta1+beta2*age[1]}

rho1~dbeta(1,1)
tau~dgamma(.001,.001)
sigmasq<-1/tau
}
# weight = 1
list(N1 = 16,M1 = 7, Y = structure(.Data = c(79,NA,79,
80,      80,      78,      80,
83,      83,      85,      85,      86,      87,      87,
81,      83,      82,      82,      83,      83,      82,
81,      81,      81,      82,      82,      83,      81,
80,      81,      82,      82,      82,      NA, 86,
76,76, 76,      76,      76,      76,      75,
81,84, 83,      83,      85,      85,      85,
77,      78,      79,      79,      81,      82,      81,
84,      85,      87,      89,NA,NA,86,
74,      75,      78,      78,      79,      78,      78,
76,      77,      77,      77,      77,      76,      76,
84,      84,      86,      85,      86,      86,      86,
79,      80,      79,      80,      80,      82,      82,
78,      78,      77, NA, 75,      75,      76,
78,      80,      77,      77,      75,NA, 75,
84,      85,      85,      85,      85,      83,      NA),.Dim =
c(16,7)),


age = c(0,2,4,6,8,10,12))
# initial values
list(beta1 = 0,beta2 = 0,tau =.5,rho1 =.5)
```

for the intercept and slope are $n(0,0.001)$, that is normal with mean 0 and precision 0.001 (or variance 1000), and the prior distribution for ρ_1 is uniform over (0,1).

Based on **BUGS CODE 6.6**, the posterior analysis will estimate the four parameters β_1, β_2, ρ_1, and σ^2.

TABLE 6.12

Bayesian Analysis of Exercise Therapy Trial for Estimation of ρ_1

Parameter	Mean	SD	Error	2½	Median	97½
β_1	79.69	0.4145	0.00479	78.87	79.69	80.51
β_2	0.1273	0.07704	0.001528	−0.022849	0.1274	0.2788
ρ_1	0.9611	0.02854	0.001242	0.894	0.9659	0.9985
σ^2	2.66	0.3813	0.002425	2.017	2.626	3.51

Using 75,000 observations for the simulation, with a burn in of 5,000 and a refresh of 100, the posterior analysis is reported in Table 6.12.

Thus, the correlation between the muscle strengths two days apart (1 time unit) is estimated as 0.9611 via the posterior mean, the intercept as 79.69, the slope as 0.1273, and the variance of the errors as 2.66. Note that the 95% credible interval for the slope is (−0.02284, 0.2788), which contains 0.

In a similar way, the other entries $\rho_3, \rho_4, \rho_5, \ldots$ in the Toeplitz matrix can be estimated by Bayesian techniques.

6.3 Choosing a Pattern for the Covariance Matrix

In the previous section, various patterns for the variance–covariance matrix were introduced, assuming a multivariate normal distribution for the repeated measures (see Equation 6.7).

Note that the variance–covariance matrix is $\Sigma = T^{-1}$ and, for example, that the covariance matrix Σ can assume various patterns such as unstructured, autoregressive, compound symmetry, and Toeplitz.

Two approaches will be used to choose the appropriate variance–covariance matrix: (1) graphical procedures that plot the observed values of the repeated response versus the predicted values and (2) an analytical procedure called the Deviance Information Criterion (DIC).

With regard to the predictive approach, the predicted values are generated from the Bayesian predictive distribution with density of Z, namely

$$f(z \mid Y = y) = E_{\beta,T}\big[g(y \mid \beta, T)\big] \tag{6.18}$$

where:

$E_{\beta,T}$ is the expectation with respect to the posterior distribution of β and T of the conditional density of Y given β and T, which is defined by (6.7)

The future n values of Z are generated from (6.18), where the n components of Z match those of y, the observed values of Y, which have the multivariate normal distribution defined by (6.7). WinBUGS allows one to generate observations from the predictive distribution (6.18) for various patterns of the covariance matrix $\Sigma = T^{-1}$. Additional information about the Bayesian predictive density (6.18) and its use in model checking can be found in Gelman et al. (Chapter 6)[3] and in Chapter 1.

The second approach to assess the goodness of fit and to choose a pattern for the covariance (correlation) matrix is based on DIC. Consider model m with unknown parameters θ_m, then the DIC is defined by

$$DIC(m) = 2E_{\theta_m}\left[D(\theta_m, m)\right] - D\left[E_\theta(\theta), m\right]$$

$$= D\left[E_\theta(\theta), m\right] + 2p_m \qquad (6.19)$$

where:

E_{θ_m} is the expectation with respect to the posterior distribution of θ_m

$$D(\theta_m, m) = -2\log L(\theta_m, m \mid Y = y) \qquad (6.20)$$

is the usual deviance measure and

$$L(\theta_m, m \mid Y = y) \qquad (6.21)$$

is the likelihood function (as a function of θ_m).

In addition, the number of effective parameters for model m is given by

$$p_m = E_{\theta_m}\left[D(\theta_m, m)\right] - D\left[E_{\theta_m}(\theta_m, m)\right] \qquad (6.22)$$

Note that the likelihood function is based on the multivariate normal density (6.7) and m is the index of the model under consideration; thus, the deviance measure is minus twice the log of the likelihood function.

Smaller DIC values indicate better fits to the data with the model under consideration, but one should make sure the distribution of DIC is symmetric about its posterior mean.

For additional information about the DIC (6.19), see Ntzoufras (p. 140)[4]

Based on the information from the exercise therapy trial, the graphical approach and the DIC approach will be employed to choose a pattern, among three patterns, the unstructured, the autoregressive, and compound symmetry for the covariance matrix of the repeated measure (muscle strength).

6.3.1 Unstructured Covariance Pattern

See section 6.2.1 and **BUGS CODE 6.1** for a review of the Bayesian analysis of the exercise therapy trial, assuming an unstructured covariance matrix Σ.

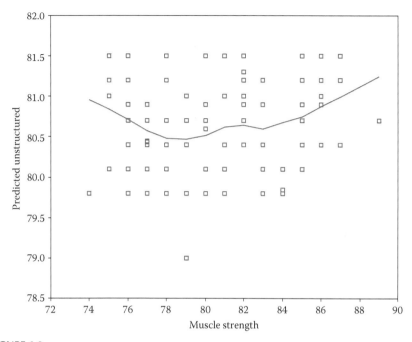

FIGURE 6.3
Predicted versus observed muscle strength values for exercise therapy trial with unstructured covariance matrix.

In order to generate future observations from the predictive distribution, the following code should be inserted into the program:

$$\text{for(i in 1:N1)\{Z[i,1:M1]\sim dmnorm(mu1[],Omega[,])\}} \qquad (6.23)$$

The dimension (112 by 1) of the predicted values Z is the same as that of the actual observations y of muscle strength, and a scatter plot of the observed versus predicted values is shown in Figure 6.3.

The lowess curve shows a weak association between the observed and predicted values; thus, one might question the choice of an unstructured matrix for the correlation between the seven responses (over the seven time periods).

Refer to formulas (6.19) through (6.22) for the definition of DIC, then using **BUGS CODE 6.1**, the DIC computations for the exercise therapy trial are presented in Table 6.13. Of course model m refers to the unstructured covariance matrix. When using WinBUGS, use the inference menu, then pull down the DIC label and place variable Y and press set during the execution of the Bayesian analysis. Table 6.13 will appear, where

TABLE 6.13

DIC Estimates for the Exercise Therapy Trial for Unstructured
Covariance Matrix

$E_{\theta_m}[D(\theta_m, m)]$	$D[E_{\theta_m}(\theta_m), m]$	DIC	p_m
379.7	373.4	386.1	6.352

$$E_{\theta_m}\left[D(\theta_m, m)\right] = \text{Dbar} \tag{6.24}$$

$$D\left[E_{\theta_m}(\theta_m, m)\right] = \text{Dhat} \tag{6.25}$$

and of course DIC is given by (6.19) with p_m given by (6.22); thus, the Bayesian analysis produces a value of 386.1 for DIC.

6.3.2 Autoregressive Covariance Pattern

See section 6.2.2 and **BUGS CODE 6.2** for a review of the Bayesian analysis of the exercise therapy trial assuming an autoregressive matrix Σ. In order to generate future observations from the predictive distribution, the following code should be inserted into the program:

```
for(i in 1:N1){for (j in 1:M1){Z[i,j]~dnorm(mu[i,j],tau)}}  (6.26)
```

The dimension 112 by 1 of the predicted values Z is the same as that of the actual observations y of muscle strength, and a scatter plot of the observed versus predicted values is shown in Figure 6.4.

A lowess plot reveals a fairly strong relationship between the observed and predicted values for muscle strength. Compared to the association between the observed and predicted values assuming an unstructured pattern depicted in Figure 6.2, the association is much stronger when assuming an autoregressive pattern.

Refer to formulas (6.19) through (6.22) for the definition of DIC, then using **BUGS CODE 6.2**, the DIC computations for the exercise therapy trial are presented in Table 6.14. Of course model m refers to the autoregressive pattern for the covariance matrix. When using WinBUGS, use the inference menu, then pull down the DIC label and place variable Y and press set during the execution of the Bayesian analysis. Table 6.14 will appear.

6.3.3 Compound Symmetry

See section 6.2.3 and **BUGS CODE 6.4** for a review of the Bayesian analysis of the exercise therapy trial assuming compound symmetry for the matrix Σ.

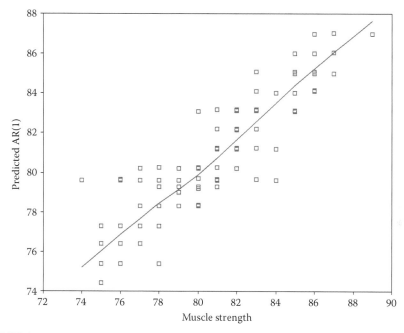

FIGURE 6.4
Predicted muscle strength versus observed values for exercise therapy trail with autoregressive pattern for covariance matrix.

TABLE 6.14

DIC Estimates for the Exercise Therapy Trial for Autoregressive Covariance Matrix

$E_{\theta_m}[D(\theta_m, m)]$	$D[E_{\theta_m}(\theta_m), m]$	DIC	p_m
399.1	393.3	404.9	5.822

In order to generate future observations from the predictive distribution, the following code should be inserted into the program:

```
for(i in 1:P){for(j in 1:M){Z[i,j]~dnorm(mu[i,j],tau)}} (6.27)
```

The dimension (112 by 1) of the predicted values Z is the same as that of the actual observations y of muscle strength, and a scatter plot of the observed versus predicted values is shown in Figure 6.5.

The lowess plot reveals a somewhat strong association between the predicted and observed muscle strength of the exercise therapy trial similar to that reported in Figure 6.3 for the autoregressive pattern.

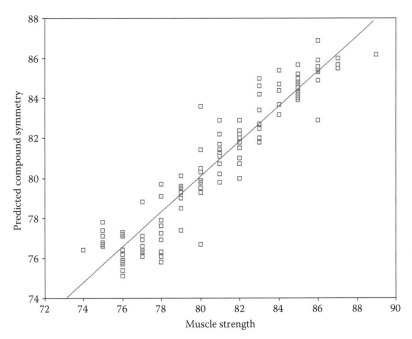

FIGURE 6.5
Predicted versus observed values of muscle strength with compound symmetry pattern.

TABLE 6.15

DIC Estimates for the Exercise Therapy Trial for Compound
Symmetry Covariance Matrix

$E_{\theta_m}\left[D(\theta_m,m)\right]$	$D\left[E_{\theta_m}(\theta_m),m\right]$	DIC	p_m
348.2	329.6	366.9	18.67

Refer to formulas (6.19) through (6.22) for the definition of DIC, then using
BUGS CODE 6.4, the DIC computations for the exercise therapy trial are
presented in Table 6.15. Of course model m refers to the compound symmetry pattern for the covariance matrix. When using WinBUGS, use the inference menu, then pull down the DIC label and place variable Y and press set
during the execution of the Bayesian analysis. Table 6.15 will appear.

Thus, the estimated DIC for the exercise therapy trial assuming compound
symmetry for the covariance pattern is 374.4, the smallest among the three
models. This is also confirmed by Figure 6.4, which shows a strong association between the observed and predicted for the compound symmetry
pattern, compared to the unstructured and autoregressive, although the
association is quite similar to that for the autoregressive pattern depicted by
Figure 6.3.

TABLE 6.16

Spearman's Correlation of Observed Muscle Strength Predictions
Assuming Three Covariance Patterns

Unstructured	Autoregressive	Compound Symmetry
0.095	0.890	0.923

Finally, let us consider the correlation between observed muscle strength values and those predicted assuming unstructured, autoregressive, and compound symmetry for the covariance pattern (Table 6.16).

It is seen that the largest Spearman correlation corresponds to the compound symmetry pattern.

To summarize, it appears that to choose a suitable pattern for the correlation matrix, one must employ a variety of techniques including graphical, including scatter plots revealing the association between observed and predicted values, and analytical methods, including use of the DIC.

6.4 More Examples

This section will focus on the topic of this chapter, namely, choosing the pattern of the covariance matrix of the repeated measures over time; thus, the choice will be made among three alternative covariance patterns: (1) the unstructured, (2) the first-order autoregressive, and (3) the compound symmetry pattern.

In order to do this, an example introduced in Chapter 5 will be utilized. For this example, a quadratic is proposed to model the mean profile of two groups, which was analyzed by Crowder and Hand (p. 79),[5] where the hematocrit of hip-replacement patients is measured on four occasions. Also measured is the age of each subject at the beginning of the study, and there are two groups, male and female patients. Hematocrit is a measure of how much volume the red blood cells take up in the blood and is measured as a percentage; thus, an observed value of 33 means red blood cells comprise 33% by volume of the blood. For our study, one follows the hematocrit of hip-replacement patients in order to see if they are becoming anemic. The data for Crowder and Hand is found in Table 6.5, where the four times are denoted by t_1, t_2, t_3, and t_4, and missing values are denoted by periods.

Recall from Figure 6.2 that a quadratic mean response is estimated for the mean profile of the combined group. The expectation for the mean profile of this study is described by

$$E(Y_{ij}) = \beta_1 + \beta_2 t_{ij} + \beta_3 t_{ij}^2 + \beta_4 g_i + \beta_5 t_{ij} g_i + \beta_6 t_{ij}^2 g_i \qquad (6.28)$$

where:

Y_{ij} is the hematocrit of subject i at time t_{ij}

$g_i = 0$ if subject i is a male, otherwise $g_i = 1$

Therefore, the expectation (6.28) reduces to

$$E(Y_{ij}) = \beta_1 + \beta_2 t_{ij} + \beta_3 t_{ij}^2 \tag{6.29}$$

for males and reduces to

$$E(Y_{ij}) = \beta_1 + \beta_4 + (\beta_2 + \beta_5)t_{ij} + (\beta_3 + \beta_6)t_{ij}^2 \tag{6.30}$$

for females.

Note that there are 13 male and 17 female hip-replacement patients in the Crowder and Hand (p. 79)[5] study. Also be aware that not all patients have a complete set of four measurements. Indeed, some have four, some three, and some two, but all have at least two repeated values.

6.4.1 Unstructured Covariance Matrix

The Bayesian analysis for this example was based on **BUGS CODE 6.2**, but for the present I have added the following statement in order to generate the predictive values:

```
for(i in 1:N1){Z1[i,1:M1]~dmnorm(mu1[],Omega[,])}    (6.31)
```

The analysis assumes that the repeated measures (the hematocrit values) follow a multivariate normal distribution with a quadratic mean response for males and females and the same unstructured covariance matrix for each group. The reader should refer to **BUGS CODE 6.2**, where the analysis assumes that the three regression coefficients have a noninformative normal prior distribution and the unknown precision matrix Omega has a noninformative uniform distribution.

In order to generate the predicted values for the two groups, the analysis is based on **BUGS CODE 6.7**, with 75,000 observations are generated for the simulation, with a burn in of 5,000 for and a refresh of 100. Note that the Bayesian analysis assumes an unstructured covariance matrix for the observations at the four time points and inferences are focused only on the combined group of 30 males and females.

The predicted versus observed values are reported in Figure 6.6 and reveal a poor association between the two variables.

The graph demonstrates a weak association between observed and predicted hematocrit values, implying the model is not a good fit to the observed hematocrit values.

The posterior analysis shown in Table 6.17 provides estimates of the coefficients of the quadratic fit for the mean profile of hematocrit for the 30 subjects.

BUGS CODE 6.7

```
model;
{
        beta1 ~ dnorm(0.0, 0.001)
        beta2 ~ dnorm(0.0, 0.001)
beta3 ~ dnorm(0.0,.0001)

for(i in 1:N1){Y[i,1:M1]~dmnorm(mu[],Omega[,])}
for (j in 1:M1){mu[j]<-beta1+beta2*age[j]+beta3*age[j]*ag
e[j]}
for(i in 1:N1){Z[i,1:M1]~dmnorm(mu[],Omega[,])}


Omega[1:M1,1:M1]~dwish(R[,],4)
Sigma[1:M1,1:M1]<-inverse(Omega[,])


}

list(N1 = 30,M1 = 4,

Y = structure(.Data =

c(47.1,31.1,NA,32.8,
44.1,31.5,NA,37.0,
39.7,33.7,NA,24.5,
43.3,18.4,NA,36.6,
37.4,32.3,NA,29.1,
45.7,35.5,NA,39.8,
44.9,34.1,NA,32.1,
42.9,32.1,NA,NA,
46.1,28.8,NA,37.8,
42.1,34.4,34.0,36.1,
38.3,29.4,32.9,30.5,
43.0,33.7,34.1,36.7,

37.8,26.6,26.7,30.6,
37.3,26.5,NA,38.5,
NA,28.0,NA,33.9,
27.0,32.5,NA,32.0,
38.4,32.3,NA,37.9,
38.8,32.6,NA,26.9,
44.7,32.2,NA,34.2,
38.0,27.1,NA,37.9,
34.0,23.2,NA,26.0,
44.8,37.2,NA,29.7,
46.0,29.1,NA,26.7,
```

```
41.9,32.0,37.1,37.6,
38.0,31.7,38.4,35.7,
42.2,34.0,32.9,33.3,
39.7,33.5,26.6,32.7,
37.5,28.2,28.8,30.3,
34.6,31.0,30.1,28.7,

35.5,24.7,28.1,29.8),.Dim = c(30,4)),
age = c(1,2,3,4),
R = structure(.Data = c(1,0,0,0,
                        0,1,0,0,
                        0,0,1,0,
                        0,0,0,1),.Dim = c(4,4)))
# initial values
list(beta1 = 0,beta2 = 0,beta3 = 0)
```

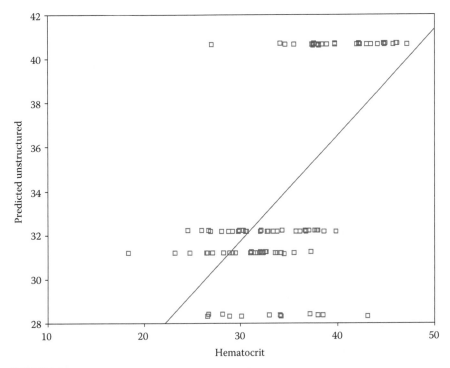

FIGURE 6.6
Predicted versus observed hematocrit values with unstructured covariance matrix.

TABLE 6.17

Quadratic Mean Profile for Hip-Replacement Study for Unstructured Covariance Matrix

Parameter	Mean	SD	Error	2½	Median	97½
β_1	56.77	2.535	0.1385	51.63	56.86	61.61
β_2	−19.43	2.362	0.1333	−24.06	−19.49	−14.66
β_3	3.32	0.4551	0.02567	2.419	3.331	4.223

The estimated DIC for the hip-replacement model is not computed because the posterior distribution of the DIC is extremely skewed about the posterior mean of DIC; thus, the only evidence about the goodness of fit of the model with unstructured pattern is graphical, namely Figure 6.6.

6.4.2 Autoregressive Pattern

When the pattern of the covariance matrix is first-order autoregressive, it can be shown that the expectation of the hematocrit of subject i at time t_j is

$$E\left[Y_{ij}\right] = \rho Y_{i,j-1} + \beta_1\left(1-\rho\right) + \beta_2\left(t_j - \rho t_{j-1}\right) + \beta_3\left(t_j^2 - \rho t_{j-1}^2\right) \qquad (6.32)$$

where the first-order autoregressive coefficient is ρ, which satisfies the constraint $-1 < \rho < 1$. It can also be shown that the variance σ^2 of the Y_{ij} is constant for all occasions. Based on (6.32), the Bayesian analysis is executed with **BUGS CODE 6.8** using 75,000 observations for the simulation, with a burn in of 5,000 and a refresh of 100.

BUGS CODE 6.8

```
model;
        {
            beta1 ~ dnorm(0.0, 0.001)
            beta2 ~ dnorm(0.0, 0.001)
            beta3 ~ dnorm(0.0, 0.001)

    for(i in 1:N1){for (j in 2:M1){Y[i,j]~dnorm(mu[i,j],tau)}}
    # future values for hematocrit denoted by Z
    for(i in 1:N1){for (j in 1:M1){Z[i,j]~dnorm(mu[i,j],tau)}}
    for(i in 1:N1){for (j in 2:M1)
    {mu[i,j]<-beta1*(1-rho)+beta2*(age[j]-rho*age[j-
    1])+beta3*(
    age[j]*age[j]-rho*age[j-1]*age[j-1])+rho*Y[i,j-1]}}
    for(i in 1:N1){Y[i,1]~dnorm(mu[i,1],tau)}
    for(i in 1:N1){mu[i,1]<-beta1+beta2*age[1]}
    rho~dbeta(1,1)
```

```
tau~dgamma(.001,.001)
sigmasq<-1/tau

}
# hematocrit values
list(N1 = 30,M1 = 4, Y = structure(.Data =
c(47.1,31.1,NA,32.8,
44.1,31.5,NA,37.0,
39.7,33.7,NA,24.5,
43.3,18.4,NA,36.6,
37.4,32.3,NA,29.1,
45.7,35.5,NA,39.8,
44.9,34.1,NA,32.1,
42.9,32.1,NA,NA,
46.1,28.8,NA,37.8,
42.1,34.4,34.0,36.1,
38.3,29.4,32.9,30.5,
43.0,33.7,34.1,36.7,
37.8,26.6,26.7,30.6,
37.3,26.5,NA,38.5,
NA,28.0,NA,33.9,
27.0,32.5,NA,32.0,
38.4,32.3,NA,37.9,
38.8,32.6,NA,26.9,
44.7,32.2,NA,34.2,
38.0,27.1,NA,37.9,
34.0,23.2,NA,26.0,
44.8,37.2,NA,29.7,
46.0,29.1,NA,26.7,
41.9,32.0,37.1,37.6,
38.0,31.7,38.4,35.7,
42.2,34.0,32.9,33.3,
39.7,33.5,26.6,32.7,
37.5,28.2,28.8,30.3,
34.6,31.0,30.1,28.7,
35.5,24.7,28.1,29.8),.Dim = c(30,4)),
age = c(1,2,3,4))
# initial values
list(beta1 = 0,beta2 = 0,tau =.5,rho =.5,beta3 = 0)
```

The Bayesian analysis generates predicted values for hematocrit, and these are plotted versus observed values in Figure 6.6, which demonstrates a moderate-to-weak association between the two variables. Table 6.18 is the posterior analysis.

The quadratic fit has estimated coefficients 60.97, −21.1, and 3.535 using the posterior mean, and each posterior distribution appears to be symmetric about

TABLE 6.18

Posterior Analysis for Hip-Replacement Study for Autoregressive Correlation Pattern

Parameter	Mean	SD	Error	2½	Median	97½
β_1	60.97	8.229	0.47	54.29	61.89	68.13
β_2	−21.1	6.99	0.3985	−27.87	−21.87	−15.39
β_3	3.535	1.24	0.07178	2.489	3.674	4.693
ρ	0.3095	0.1207	0.0053	0.04656	0.3147	0.5383
σ^2	19.57	7.448	0.4141	14.01	18.46	27.16

the posterior mean. The autoregressive coefficient ρ is estimated as 0.3095 with a 95% credible interval of (0.0465, 0.5383), and the common variance at the four time points is estimated as 18.46 (using the median because of the skewness). Such a "low" value for the first-order correlation implies a moderate association between the observed and predicted hematocrit values as portrayed in Figure 6.7.

Figure 6.7 implies a moderate to fairly strong association between the observed and predicted hematocrit values, which suggest a moderately good fit to the data using the model with an autoregressive error structure. Compared to the model with unstructured pattern for the covariance

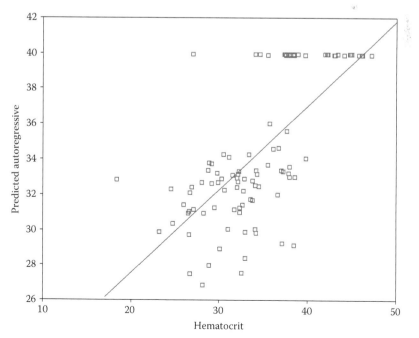

FIGURE 6.7
Observed versus predicted hematocrit values with autoregressive correlation pattern.

TABLE 6.19

DIC Estimates for the Hip-Replacement Study for Autoregressive Covariance Matrix

$E_{\theta_m}[D(\theta_m, m)]$	$D[E_{\theta_m}(\theta_m), m]$	DIC	P_m	Minimum Deviance
569.3	562.8	575.7	6.436	552.5

matrix, the model with autoregressive errors provides a "better" fit to the hip-replacement data.

Consider Table 6.19, a report of the estimated DIC for the hip-replacement study, which shows an estimate of 246.8 for the DIC.

6.4.3 Compound Symmetry

Refer to 6.2.3 for a general introduction to the Bayesian analysis assuming compound symmetry for the repeated measures over the time points of a subject.

Let Y_{ij} denote the hematocrit of subject i on occasion j, and let

$$Y_{ij} = \beta_1 + \beta_2 t_{ij} + \beta_3 t^2_{ij} + a_i + b_j + e_{ij} \qquad (6.33)$$

where:

a_i is the random effect of subject i

b_j is the random effect of occasion j

e_{ij} are independent normally distributed variables with mean 0 and variance σ^2, a_i are individual random effects that are normally distributed variables with mean 0 and variance σ_a^2, and last b_j are normally distributed variables with mean 0 and variance σ_b^2. In addition, a_i, b_j, and e_{ij} are mutually independent. The mean profile is quadratic, where β_1 is the average hematocrit at time 0, β_2 is the linear effect, and β_3 is the quadratic effect on the average hematocrit.

It can be shown that the correlation between two distinct values of hematocrit measured at occasions j and k is referred to as Equation 6.16.

Estimation of ρ will be based on its posterior distribution assuming normality for the observations, using model (6.13), and utilizing **BUGS CODE 6.4**. See section 6.2.3 and the ensuing explanation about the compound symmetry pattern for the covariance matrix. Note that P is a function of the three variance components σ_a^2, σ_b^2, and σ^2, and these components will also be estimated via Bayesian techniques.

The posterior analysis is based on **BUGS CODE 6.9** and uses 450,000 observations for the simulation, with a burn in of 5,000 and a refresh of 100.

A Bayesian analysis for generating future hematocrit values produces the graph of the predicted versus observed values in Figure 6.8.

It is seen that the graph reveals somewhat of a strong association between the observed and predicted hematocrit values. It appears that the association

BUGS CODE 6.9

```
model;

{

for(i in 1:P){for(j in 1:M){y[i,j]~dnorm(mu[i,j],tau)}}
for(i in 1:P){for(j in 1:M){z[i,j]~dnorm(mu[i,j],tau)}}
for(i in 1:P){for(j in 1:M){mu[i,j]<-beta1+beta2*age[j]+b
eta3*age[j]*age[j]+a[i]+b[j]}}


for(i in 1:P){a[i]~dnorm(0,taua)}

for(j in 1:M){b[j]~dnorm(0,taub)}

beta1~dnorm(0.0,.001)
beta2~dnorm(0.0,.001)
beta3~dnorm(0.0,.001)


theta~dnorm(0,.001)
taua~dgamma(.001,.001)
taub~dgamma(.001,.001)
tau ~dgamma(.001,.001)

sigmaa<-1/taua
sigmab<-1/taub

sigma<-1/tau


rho<-(sigmaa)/(sigmaa+sigmab+sigma)

}

list(P = 30,M = 4, age = c(1,2,3,4),

y = structure(.Data = c(47.1,31.1,NA,32.8,
44.1,31.5,NA,37.0,
39.7,33.7,NA,24.5,
43.3,18.4,NA,36.6,
37.4,32.3,NA,29.1,
45.7,35.5,NA,39.8,
44.9,34.1,NA,32.1,
42.9,32.1,NA,NA,
46.1,28.8,NA,37.8,
42.1,34.4,34.0,36.1,
```

```
38.3,29.4,32.9,30.5,
43.0,33.7,34.1,36.7,
37.8,26.6,26.7,30.6,
37.3,26.5,NA,38.5,
NA,28.0,NA,33.9,
27.0,32.5,NA,32.0,
38.4,32.3,NA,37.9,
38.8,32.6,NA,26.9,
44.7,32.2,NA,34.2,
38.0,27.1,NA,37.9,
34.0,23.2,NA,26.0,
44.8,37.2,NA,29.7,
46.0,29.1,NA,26.7,
41.9,32.0,37.1,37.6,
38.0,31.7,38.4,35.7,
42.2,34.0,32.9,33.3,
39.7,33.5,26.6,32.7,
37.5,28.2,28.8,30.3,
34.6,31.0,30.1,28.7,
35.5,24.7,28.1,29.8),.Dim = c(30,4)))
list(taua = 1,taub = 1,tau = 1, beta1 = 0,beta2 = 0,beta3
= 0)
```

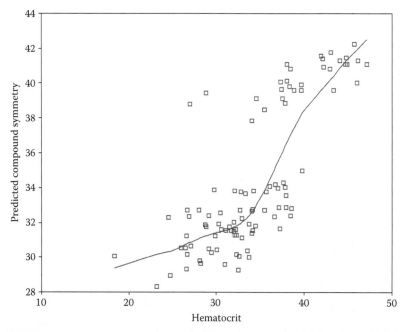

FIGURE 6.8
Predicted versus observed hematocrit compound symmetry pattern for correlation matrix.

TABLE 6.20

Posterior Distributions for Compound Symmetry Pattern

Parameter	Mean	SD	Error	2½	Median	97½
ρ	0.09357	0.1008	0.002248	0.0000659	0.05622	0.3442
σ^2	15.18	3.023	0.03597	10.3	14.82	21.96
σ_a^2	3.366	2.739	0.03776	0.003505	3.027	9.741
σ_b^2	134.7	666.2	10.22	0.005988	12.11	984.3
β_1	37.45	22.58	0.8605	−17.45	47.27	61.74
β_2	−2.31	20.26	0.7802	−25.19	−11.08	51.18

assuming a compound symmetry covariance matrix is somewhat stronger than that assuming an autoregressive pattern.

Of course, the main interest is with the estimated mean quadratic response and the common correlation ρ, and the posterior analysis is reported in Table 6.20.

It is important to note that the posterior distributions for the correlation and variance components are highly skewed; thus, it is imperative to use the posterior medians instead of the posterior means as estimates of central location for the distribution. For example, the posterior median for the correlation is 0.056622 and is the sensible estimate for the main parameter of interest ρ. The posterior analysis shows that the largest source of variation is that between the four time points, while the smallest is between subjects. The posterior median of β_1 is 47.27 compared to a posterior mean of 37.45, showing the skewness of its posterior distribution. Similar assertions can be made of the other two coefficients for the mean profile.

Also estimated is the DIC with a value of 241.8 and with a minimum value of 208.2 (Table 6.21).

Last, the Spearman correlation coefficient between the observed hematocrit values and those forecast with an unstructured covariance matrix is estimated at 0.550, 0.653 using an autoregressive pattern and is 0.785 with a compound symmetry alternative. This implies that compound symmetry is giving the best fit, which is also corroborated to some extent by the graphical evidence and the DIC analysis.

With the hip-replacement study, it is difficult to choose an optimum pattern, among the three alternatives, unstructured, autoregressive, and compound symmetry. This is because of the relatively large variance of the error terms and that there are only four time points and a large number of

TABLE 6.21

DIC Estimates for the Hip-Replacement Study for Compound Symmetry Pattern

$E_{\theta_m}[D(\theta_m, m)]$	$D[E_{\theta_m}(\theta_m), m]$	DIC	p_m	Minimum Deviance
545.3	527.7	563.0	17.67	513.5

observations that are missing. This especially makes it difficult to estimate the autoregressive coefficient because there are only three time points for the estimation via the Bayesian analysis. The graphical evidence portrayed in Figures 6.5 through 6.7 implies that the compound symmetry pattern gives the best fit, but that the autoregressive option is also plausible. In summary, the dearth of sample information poses problems to choosing an appropriate covariance matrix for the repeated hematocrit observations.

6.5 Comments and Conclusions

The chapter begins with a general explanation of the various possible patterns for the covariance matrix of repeated measures over time. The various patterns described in this chapter are unstructured, autoregressive, compound symmetry, and Toeplitz. The choice of an appropriate pattern depends on various factors, and these choices are illustrated with the exercise therapy trial, where Table 6.1 lists the data, and Figure 6.5 presents the scatter plot with lowess curve. The approach to choosing the pattern is to first determine the mean profile of the various groups with descriptive statistics and graphical techniques.

A scatter plot of the muscle strength versus time (with seven time points) implies a linear regression of average muscle strength over time. Assuming an unstructured pattern for the covariance, a Bayesian analysis is executed, which provides estimates of the posterior distribution of the various parameters including the coefficients of the linear regression and the entries of the 7 by 7 covariance matrix. See Table 6.1 for the analysis.

In a similar way, the analysis is repeated for the autoregressive structure for the covariance pattern. The model for the autoregressive model is defined by (6.5), and Table 6.4 portrays the general pattern for a repeated measures autoregressive pattern. It is seen that there are only two parameters for such a pattern, the common variance of the response at the seven time points and the first-order correlation ρ. Thus, the correlation is the same for responses one time unit apart and is ρ^2 for responses two time units apart. The Bayesian analysis is based on **BUGS CODE 6.2** and reported in Table 6.5 with a posterior mean of 0.9626 for the first-order correlation. This example is repeated but for the case for unequal spacing between the time points.

A third alternative for the pattern of the covariance matrix is compound symmetry, which depends on only two parameters, a common variance and a common correlation ρ, that is the correlation is always ρ regardless of the time separation between the time points.

Last, the Toeplitz pattern for the covariance is discussed, where the correlation between units one time unit apart is the same and where the correlation for responses two time units apart is the same.

How is a particular pattern chosen? This subject is explored in section 6.3. Two components are essential in choosing an appropriate covariance matrix, namely: (1) a plot of the predicted versus the observed repeated response and (2) calculating the DIC. For example, for the unstructured pattern, **BUGS CODE 6.1** generates the predicted values of muscle strength and the predicted versus observed values plotted in Figure 6.1. The estimated DIC value is reported in Table 6.13. In the same way, but assuming an autoregressive pattern, the predicted values are generated and plotted against the corresponding observed muscle strengths shown in Figure 6.3. The plot shows a strong association between the predicted and observed values of muscle strength. Last, assuming compound symmetry for the covariance pattern and using **BUG CODE 6.4** to generate the predicted muscle strength, the goodness of fit of this model can be assessed from Figure 6.4.

For additional information about the DIC and related topics, see Leonard and Hsu[6] and Congdon.[7] Jones[8] contains interesting material about modeling repeated measures with errors that follow an autoregressive process. For a topic related to choosing a pattern for the covariance matrix, Carroll and Ruppert[9] present ideas concerning the choice of the variances of a repeated measure response over time.

As mentioned earlier, the repeated measures design has certain advantages over the cross-sectional study; see Jones and Kenward[10] for additional information. Last, the example using the hematocrit of the hip-replacement study had many missing values a topic to be explored in depth in Chapter 10 and Daniels and Hogan.[11]

Exercises

1. Compare Equation 6.1 to 6.2 and describe the advantages of a repeated measures design versus a cross-sectional study.

2. Consider formula (6.7), the multivariate normal density, and define the unknown parameters β and T of the distribution.

3. Verify Figure 6.1, the scatter plot of the muscle strength versus time. Use the muscle strength values of Table 6.1 and verify the descriptive statistics of Tables 6.2a and 6.2b.

4. Based on Figure 6.5, what polynomial would you use to explain the mean profile of the two groups?

5. Using **BUGS CODE 6.1**, verify Table 6.3, the posterior analysis for the exercise therapy trial. Note the analysis assumes an unstructured covariance matrix for the muscle strength value over time (days).

a. What is the posterior mean of the intercept β_1?

b. What is the 95% credible interval for the slope β_2?

c. With **BUGS CODE 6.1**, generate the posterior density of ρ_{13}. Is the posterior distribution of ρ_{13} skewed? If so, describe the skewness.

d. Do the posterior medians of the various correlations ρ_{ij} $(i,j = 1,2,...,7)$ suggest a pattern for the covariance matrix?

6. Based on formula (6.9), the linear model with autoregressive errors, derive formula (6.10). The latter is an equivalent form of the model.

7. Using **BUGS CODE 6.2**, verify Table 6.5. Based on the posterior medians, what is the estimated correlation of muscle strength?

a. Between days 4 and 6?

b. Between days 0 and 6?

c. Between days 2 and 12?

d. Do the correlations decrease as the length between time points increase?

8. Based on **BUGS CODE 6.3**, verify Table 6.6.

9. From (6.13), derive (6.16).

10. Based on **BUGS CODE 6.4**, verify Table 6.9, the posterior analysis for the exercise therapy trial assuming a compound symmetry covariance matrix.

a. What is the posterior mean and median of ρ?

b. What is the posterior mean of the intercept β_1?

c. What is the 95% credible interval for the slope β_2?

d. Based on the posterior median of ρ, estimate the compound symmetry covariance matrix.

11. Describe the correlation pattern of the Toeplitz matrix. Derive the model (6.17) and explain what ρ_2 is.

12. Based on **BUGS CODE 6.5**, verify Table 6.11, the posterior analysis of the exercise therapy trial, assuming a Toeplitz for the covariance matrix.

a. Is the posterior distribution of ρ_2? Is it skewed?

b. What are your estimates of β_1 and β_2?

c. What is the 95% credible interval for σ^2?

13. Based on **BUGS CODE 6.6**, verify Table 6.12.

a. Describe the skewness of the posterior distribution of ρ_1.

b. What is the MCMC simulation error for estimating the posterior mean of ρ_1?

c. The 95% credible interval for β_2 contains 0. What does this imply about the slope of the model?

14. Choosing the appropriate covariance matrix is based on two components. What are they?

15. Describe formula (6.18) for predicting future observations.

16. Using **BUGS CODE 6.1**, generate predicted muscle strength values and verify Figure 6.2. This assumes an unstructured pattern for the covariance matrix. Does the graph of predicted versus observed muscle strength values suggest a weak or strong association? What does this imply about how well the model fits the data?

17. Using **BUGS CODE 6.1**, verify the DIC value of 386.1 of Table 6.13.

18. Using **BUGS CODE 6.2**, generate predicted muscle strength values and verify Figure 6.3. Does Figure 6.3 imply a strong or weak association between observed and predicted values?

19. Based on Table 6.14, what is your estimate of the DIC for the model, assuming an autoregressive pattern for the covariance matrix?

20. Based on **BUGS CODE 6.4**, generate predicted muscle strength values assuming a compound symmetry pattern for the covariance structure.

    Does Figure 6.4 imply a strong association between the observed and predicted muscle strength? Also, does this imply a good goodness of fit for the model?

21. Verify Table 6.15. What is the estimated DIC value assuming compound symmetry?

22. Among the three options (unstructured, autoregressive, and compound symmetry), and based on the graphical scatter plot of predicted versus observed values) and analytical (estimated DIC values) evidence, what is your choice for the pattern of the covariance matrix?

7

General Mixed Linear Model

7.1 Introduction and Definition of the Model

Recall from Chapter 6 that the assumed mean for the response of the ith subject at occasion j was

$$E(Y_{ij}) = X'_{ij}\beta \tag{7.1}$$

where:
  X'_{ij} is a 1 by p vector of covariate values
  β is a p by 1 vector of unknown regression parameters

In addition, it was assumed that the responses for subject i is a multivariate normal with covariance matrix Σ_i for $i = 1, 2, \ldots, N$. Various patterns for the covariance matrix were described, and Bayesian techniques were employed to estimate the unknown parameters. Several patterns for the covariance matrix were included: unstructured, autoregressive, compound symmetry, and Toeplitz.

In this chapter an additional component will be introduced in the model, which will allow one to describe the variation from one subject to the other comprising the study. These additional components are called random effects, and the model (7.1) is extended to the ith subject as

$$Y_i = X_i\beta + Z_ib_i + \varepsilon_i \tag{7.2}$$

where:
  Y_i is a n_i by 1 vector of responses over the n_i time points
  X_i is an n_i by p design matrix
  β is the p by 1 vector of unknown regression parameters
  Z_i is an n_i by q known design matrix
  b_i is a q by 1 random vector with mean vector 0 and covariance matrix G
  ε_i is a random n_i by 1 vector with mean vector 0 and covariance matrix $\sigma^2 I_{n_i}$

In addition, it is assumed that the N random vectors b_i ($i = 1,2,\ldots,N$) are independent, that the N random vectors ε_i ($i = 1,2,\ldots,N$) are independent, and that the b_i and ε_i are mutually independent. The parameters β are called fixed effects and the b_i random effects.

Note that given the random effects, the mean response of subject i is

$$E\left(Y_i \mid b_i\right) = X_i\beta + Z_ib_i \tag{7.3}$$

and that averaging over the random effects, the population mean is

$$E\left(Y_i\right) = X_i\beta$$

Thus, for subject i, the deviation of the mean response from the population mean is the random effect Z_ib_i.

In addition, the marginal dispersion matrix (averaging over the random effects) of the n_i responses for the ith subject is

$$D(Y_i) = Z_iD\left(b_i\right)Z_i' + \sigma^2 I_{n_i}$$
$$= Z_iGZ_i' + \sigma^2 I_{n_i} \tag{7.4}$$

The addition of random effects to the dispersion matrix of the model allows the variances and covariances of the repeated response to vary with time, via the design matrix Z_i. As the name implies, the generalized linear model is generalized from a fixed model (with mean $X_i\beta$) to one that includes both random effects (via the b_i) and fixed effects.

It is important to realize that the generalized linear model allows for an unbalanced allotment of time points and that the time points are denoted by t_{ij} for $j = 1,2,\ldots,n_i$ for subject i.

When describing the compound symmetry pattern for the covariance matrix in Chapter 6, a special case of the generalized linear model was introduced. Recall that this model is defined as

$$Y_{ij} = \beta_1 + \beta_2 t_{ij} + b_i + e_{ij} \tag{7.5}$$

that is, the population mean is a linear regression with intercept β_1 and slope β_2, but in addition there is a random intercept term b_i with mean 0 and variance σ_b^2, and ε_{ij} ($j = 1,2,\ldots,n_i$) are independent with mean 0 and variance σ^2. The deviation of the responses from the population mean $\beta_1 + \beta_2 t_{ij}$ is the random effect b_i, while the ε_{ij} represent the measurement errors. Of course, additional terms can be added to the random effects such as a random slope for the subjects.

The following sections will introduce the reader to the appropriate interpretation of the generalized linear model, a description of the covariance matrix induced by the model and some examples that illustrate the Bayesian analysis for estimating the mean profile and covariance parameters.

See Fitzmaurice, Laird, and Ware[1] for a general introduction to the non-Bayesian analysis of the generalized linear model.

7.2 Interpretation of the Model

What is the appropriate interpretation of the elements of the model defined by (7.5). To do this, consider two subjects, then for the first subject with $i = 1$, the intercept is $\beta_1 + b_1$, and for the second subject with $i = 2$, the intercept is $\beta_1 + b_2$, but the slope is the same, namely, β_2 for both subjects. On the other hand, the population mean profile, obtained by averaging the model over the random effects, is

$$E\left(Y_{ij}\right) = \beta_1 + \beta_2 t_{ij} \qquad (7.6)$$

for $i = 1,2$

$$t_{1j} = \left(t_{11}, t_{12}, \ldots, t_{1m_1}\right)$$

the time points for subject 1 and

$$t_{2j} = \left(t_{21}, t_{22}, \ldots, t_{2n_2}\right)$$

the time points for subject 2.

Consider the following example with common time points 1,2,3,...,9,10 for both individuals, and let

$$\beta_1 = 5$$

$$\beta_2 = 2$$

$$b_1 = 4$$

and

$$b_2 = -4$$

then a graphical representation for the two subject is given by Figure 7.1.

The intercept is 9 for the first subject, while it is 1 for the second, but the slope is 2 for both. On the other hand, the mean profile (averaged over two random effects) has an intercept of 5 and a slope of 2.

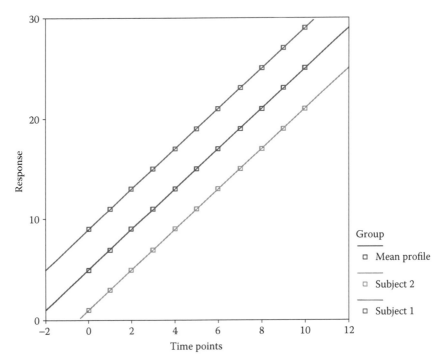

FIGURE 7.1
Two subjects and the mean profile.

It should be noted that the response of (7.5) has been averaged over the measurement errors e_{ij} for $i = 1,2$ and $j = 0,1,...,10$. When the measurement errors are accounted four, Figure 7.1 perhaps looks like Figure 7.2. Note that the measurement errors account for the deviations from the linear regressions of the response for the two subjects.

7.3 General Linear Mixed Model Notation

This section introduces the reader to the notation used for the linear mixed model. Recall the general model defined in Equation 7.2.

Consider a special case of the general model (7.2), where for the ith subject

$$Y_{ij} = \beta_1 + \beta_2 t_{ij} + b_{1i} + b_{2i} t_{ij} + \varepsilon_{ij} \tag{7.7}$$

for $j = 1,2,...,n_i$ and $i = 1,2,...,N$.

The b_{1i} and b_{2i} have bivariate normal distribution with 2 by 1 mean vector 0 and 2 by 2 covariance matrix G, and ε_{ij} are independent normal random

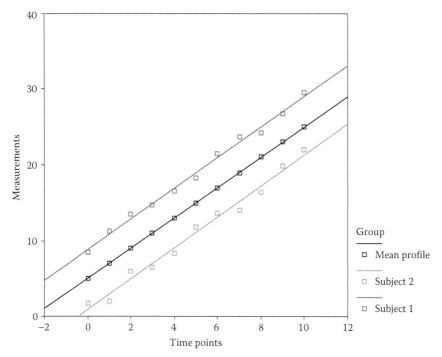

FIGURE 7.2
Measurement error for the two subjects.

variables with mean 0 and variance σ^2. Note that for this model the intercept and slope for each subject are represented by random effects, and the mean profile is given by

$$E\left(Y_{ij}\right) = \beta_1 + \beta_2 t_{ij}$$

In terms of the general model (7.2), the special case has $n_i \times 2$ design matrices $X_i = Z_i$, which are given by Table 7.1.

That is, the two are the same first column consisting of ones and the second column consisting of the time points for the ith subject.

TABLE 7.1

Design Matrix $X_i = Z_i$

1	t_{i1}
1	t_{i2}
.	
.	
.	
1	t_{in_i}

TABLE 7.2

Design Matrix X_i for the Fixed Effects

1	t_{i1}	0	0
1	t_{i2}	0	0
$\vdots$	$\vdots$	$\vdots$	$\vdots$
1	t_{in_i}	0	0

TABLE 7.3

The Design Matrix for the Random Effects

1	t_{i1}	1	t_{i1}
1	t_{i2}	1	t_{i2}
$\vdots$	$\vdots$	$\vdots$	$\vdots$
1	t_{in_i}	1	t_{in_i}

For the more general case of two treatment groups

$$Y_{ij} = \beta_1 + \beta_2 t_{ij} + \beta_3 grp_i + \beta_4 t_{ij} grp_i + b_{i1} + b_{2i} t_{ij} + \varepsilon_{ij} \tag{7.8}$$

where grp = 1 if subject i is in the treatment group, and when grp = 0, individual i belongs to the control group. Now, what are the design matrices for the fixed and random effects?

It can be shown that the n_i by 4 design matrix X_i for the fixed effects is given by Table 7.2, whereas the n_i by 4 design matrix Z_i for the random effects is given by Table 7.3.

7.4 Pattern of the Covariance Matrix

The pattern of the covariance matrix is much more general than that previously considered in Chapter 6, but recall that the compound symmetry pattern was based on a random model.

Thus, consider a more general case of (7.5), namely,

$$Y_{ij} = X'_{ij}\beta + b_i + \varepsilon_{ij} \tag{7.9}$$

where:
X'_{ij} is a 1 by p known design matrix
β is a p by 1 unknown parameter vector
b_i is a random subject effect
ε_{ij} are random variables representing the measurement error for subject i at occasion j

It is observed that

$$\text{Var}\left(Y_{ij}\right) = \text{Var}\left(b_i\right) + \text{Var}\left(\varepsilon_{ij}\right)$$

$$= \sigma_b^2 + \sigma^2$$

(7.10)

where the $b_i \sim nid(0,\sigma_b^2)$ and the $\varepsilon_{ij} \sim nid(0,\sigma^2)$.

Our objective is to determine the correlation between Y_{ij} and Y_{ik}; one must know the covariance

$$\text{Cov}\left(Y_{ij}, Y_{ik}\right) = \text{Cov}\left(X'_{ij}\beta + b_i + \varepsilon_{ij}, X'_{ij}\beta + b_i + \varepsilon_{ik}\right)$$

$$= \text{Var}\left(b_i\right)$$

$$= \sigma_b^2$$

(7.11)

Therefore, the correlation is

$$\rho\left(Y_{ij}, Y_{ik}\right) = \frac{\sigma_b^2}{\sigma_b^2 + \sigma^2}$$

(7.12)

a compound symmetry pattern for the covariance matrix of the responses over time.

Now consider the variance of the observations (7.7), given by

$$\text{Var}\left(Y_{ij}\right) = \text{Var}\left(b_{i1} + b_{i2}t_{ij} + \varepsilon_{ij}\right)$$

$$= \text{Var}\left(b_{i1}\right) + 2t_{ij}\text{Cov}\left(b_{i1}, b_{i2}\right) + t_{ij}^2\text{Var}\left(b_{2i}\right) + \text{Var}\left(\varepsilon_{ij}\right)$$

$$= g_{11} + 2t_{ij}g_{12} + t_{ij}^2 g_{22} + \sigma^2$$

(7.13)

where the variances g_{11} and g_{22} and the covariance g_{12} are unknown. In addition, the covariance is given by

$$\text{Cov}(Y_{ij}, Y_{ik}) = \text{Cov}(b_{i1} + b_{2i}t_{ij} + \varepsilon_{ij}, b_{i1} + b_{2i}t_{ik})$$

(7.14)

Thus, it is interesting to observe that the variance of the observation is increasing (if g_{12} is nonnegative) and that the variances and covariances are functions of time. Additional information about the pattern of the covariance matrix for the generalized linear model can be found in Fitzmaurice, Laird, and Ware.[1]

7.5 Bayesian Approach

Consider the model defined in Equation 7.2, where

$$Y_i = X_i\beta + Z_i b_i + \varepsilon_i$$

for subject i, the Bayesian approach is to first assign prior distributions to the unknown fixed parameters β, the components of the variance covariance matrix of the random effects b_i, and last to the variance σ^2 of the measurements errors.

The regression coefficients will be assigned noninformative normal distributions, and the precision parameter $\tau = 1/\sigma$ is a noninformative gamma distribution. The q_i components of the random vector will be expressed as linear combinations of independent $n(0,1)$ random variables with unknown coefficients, and the latter will be assigned uninformative normal distributions. The linear combinations allow one to investigate the covariance between the components of the random vector b_i and to see if the covariance can be considered 0. Of course, as before, the posterior analysis will be executed with WinBUGS, which produces the posterior distribution of the unknown parameters of the model (7.2), plots of the posterior densities, and Markov chain Monte Carlo (MCMC) errors, which allow us to determine the number of observations for the simulation that will provide reliable posterior means.

7.6 Examples

The first example is based on a study by Van der Lende et al.[2] and is an epidemiological study in two areas of Holland, a rural area and an urban area. The subjects were followed over time to ascertain data on the prevalence of various factors that might have an impact on chronic obstructive lung disease. The sample is based on males and females as well as smokers and nonsmokers, but for our purpose, only 100 smokers are used in the example. Forced expiratory volume (FEV1) was obtained by spirometry and determined every 3 years for the 15 years of the study. A Bayesian analysis is performed where FEV1 is modeled as a generalized linear model with the population mean profile represented by a linear regression and two random effects, one for the intercept and another for the slope. This example is fairly well balanced in that the number of time points for each subject is approximately the same as well as the spacing between time points.

In the previous section, a linear relation between the main response and time was assumed for the mean profile of the subjects. From experience one would expect the mean profile to not be too complex and that simple polynomials or linear shift point models should be able to indeed model the mean profile. For our next case, a quadratic will be proposed to model the mean profile of two groups. A good example of this is the study selected from Table 5.2 of Crowder and Hand (p. 79),[3] where the hematocrit of hip-replacement patients is measured on four occasions.

Also measured is the age of each subject at the beginning of the study, and there are two groups, male and female patients. Hematocrit is a measure of how much volume the red blood cells take up in the blood and is measured as a percentage, thus an observed value of 33 means red blood cells comprise 33% by volume of the blood. For our study, one follows the hematocrit of hip-replacement patients in order to see if they are becoming anemic. The data for Crowder and Hand is found in Table 5.5, where the four times are denoted by t_1, t_2, t_3, and t_4, and missing values are denoted by periods.

An interesting study using a spline with one join point is analyzed by Fitzmarurice, Laird, and Ware[1] and is based on blood levels from lead-exposed children randomized study with placebo and a treatment arm using succimer, a chelating agent. See Rhoads[4] for additional information. It is hoped that succimer will decrease the blood levels of lead in children, where the lead level in µg/dL is measured at baseline, and at weeks 0, 1, 4, and 6, thus, of interest is comparing the two groups. The comparison will be based on fitting a linear association for the placebo group over all time points and fitting a spline with one known join point at week 1. That is, a linear association between blood level and time will be assumed for weeks 0 and 1 and another linear association over weeks 1, 4, and 6.

It is seen that the examples increase in complexity from a linear regression model, to a quadratic mean profile, followed by a three-segment linear spline, and finally with two groups and a two-segment linear spline.

7.6.1 Study of Air Quality in the Netherlands

The first example is based on the Van der Lende et al.[2] investigation of pulmonary function in Holland, and the main response is FEVI, as determined by spirometry, and is modeled as Equation 7.7, where the "fixed" effects represent the mean population profile as a linear regression with intercept β_1 and slope β_2, while the random effects b_{1i} and b_{2i} for subject i represent that individuals intercept and slope. Thus, for subject i, the intercept is $\beta_1 + b_{1i}$ and the slope is $\beta_2 + b_{2i}$. The random vector $b = (b_{1i}, b_{2i})'$ is assumed to have a bivariate normal distribution with mean vector 0 and 2 by 2 covariance matrix G.

The components of the variance–covariance matrix $G = (g_{ij})$ are determined by the covariance matrix of the joint normal distribution of the two random effects, b_{1i} and b_{2i}, and the Bayesian analysis will consist of estimating the elements of the covariance matrix. Of interest is the covariance g_{12}, which if 0 implies zero correlation between the intercept and slope random effects.

The Bayesian analysis assumes that the precision matrix of the two random effects b_{1i} and b_{2i} has a noninformative Wishart distribution and the following prior information for the parameters: (1) β_1 and β_2 are assigned independent noninformative $N(0.00, 0001)$, (2) the precision matrix of the two random effects b_{1i} and b_{2i} has a noninformative Wishart distribution, and (3) the precision τ of the measurement errors ε_{ij} of (7.7) is assigned a noninformative gamma $(0.0001, 0.0001)$ distribution. The Bayesian analysis is based on **BUGS CODE 7.1**.

BUGS CODE 7.1

```
    model;
    {
            for(i in 1:2){beta[i]~dnorm(0.0, 0.001)}

 for(i in 1:N){for(j in 1:M){Y[i,j]~dnorm(mu[i,j],tau)}}
 for(i in 1:N){for(j in 1:M){mu[i,j]<-
beta[1]+beta[2]*age[j]+b[i,1]+b[i,2]*age[j]}}
 for(i in 1:N){b[i,1:2]~dmnorm(mu1[],omega[,])}
 for(j in 1:2){mu1[j]<-0}
 omega[1:2,1:2]~dwish(R[,],2)
 Sigma[1:2,1:2]<-inverse(omega[,])

sigma<-1/tau

tau~dgamma(.001,.001)

}

list(N = 100,M = 7,
# Van der Lende et al. data
Y = structure(.Data = c(3.1,3.2,3.5,3.0,2.9,NA,NA,
3.6,3.5,3.5,3.1,NA,2.8,2.7,
NA,2.7,2.9,2.7,2.7,2.5,NA,
3.4,3.3,2.9,2.3,2.5,2.4,NA,
NA,2.5,2.5,2.1,2.4,NA,2.3,
3.9,4.0,4.1,3.8,4.0,NA,NA,
2.7,NA,3.3,2.4,2.2,2.2,2.3,
3.0,2.9,3.0,2.8,2.8,2.7,2.8,
2.8,2.7,2.1,1.9,1.8,NA,1.7,
3.9,3.8,3.5,3.3,3.5,3.2,3.3,
3.2,NA,3.0,2.8,2.9,2.5,2.2,
3.3,3.4,3.4,3.1,3.5,3.0,2.8,
NA,3.1,3.4,3.1,3.2,NA,2.6,
3.2,3.2,3.3,3.1,3.2,2.8,NA,
```

```
4.0,3.6,3.7,3.7,3.2,3.0,NA,
3.3,3.6,3.3,3.3,3.0,2.9,NA,
NA,3.0,3.2,3.0,2.9,2.4,NA,
2.8,2.5,2.6,2.5,2.4,2.3,2.2,
3.6,3.9,3.7,NA,3.3,3.1,NA,
3.6,3.4,3.4,NA,3.0,NA,2.7,
NA,2.0,1.8,1.7,1.8,1.6,1.5,
3.6,3.4,3.2,3.3,3.3,NA,2.6,
2.6,3.6,NA,2.9,3.2,3.1,2.2,
3.3,3.0,2.9,3.0,2.9,2.8,2.4,
3.6,3.6,3.4,NA,NA,2.8,2.8,
1.8,1.5,1.7,NA,1.8,1.5,1.2,
3.6,3.8,3.2,3.1,3.2,3.2,2.8,
4.2,3.8,3.5,3.8,3.5,3.4,3.3,
NA,3.9,4.0,NA,3.7,3.3,3.4,
3.7,NA,4.1,2.7,2.9,2.9,2.5,
2.6,2.6,2.7,2.6,2.6,2.4,2.5,
3.5,3.6,3.1,3.0,2.9,3.1,2.7,
2.2,2.3,2.3,2.1,NA,NA,1.5,
NA,3.5,3.3,3.8,3.6,3.9,3.4,
NA,3.1,3.3,3.1,NA,2.8,2.7,
3.7,3.7,3.7,3.3,3.1,NA,NA,
NA,2.4,2.3,2.1,1.9,2.2,2.0,
4.4,4.2,4.1,4.1,NA,4.0,NA,
2.9,2.7,2.8,NA,2.5,NA,2.1,
3.1,2.6,2.7,2.5,2.4,2.1,2.3,
2.5,2.4,2.0,NA,NA,2.1,2.3,
3.3,3.1,3.0,3.2,NA,3.3,2.9,
2.8,2.7,2.7,2.4,NA,2.1,1.8,
2.7,2.4,3.0,1.9,2.6,2.5,2.4,
3.4,3.4,3.4,3.1,2.9,NA,2.5,
3.7,3.4,3.3,3.2,2.7,NA,2.8,
NA,2.5,2.4,2.6,2.0,NA,2.0,
3.6,3.4,3.5,3.5,3.6,3.3,3.2,
2.6,2.7,2.7,2.7,NA,NA,2.4,
3.4,3.5,3.7,3.4,3.6,NA,3.0,
3.4,3.1,3.0,2.9,NA,2.6,2.4,
3.6,3.3,NA,3.2,3.0,2.7,NA,
4.1,3.6,3.8,3.9,3.6,3.5,NA,
3.7,3.6,3.3,3.1,3.3,3.1,NA,
4.8,3.8,3.8,3.5,3.6,NA,3.3,
3.7,4.2,3.9,3.3,3.2,3.5,3.6,
2.8,2.9,NA,NA,2.7,2.4,2.5,
3.4,3.4,3.3,2.7,3.0,3.1,3.1,
3.0,3.1,2.8,1.7,2.9,2.7,2.2,
3.0,2.5,NA,2.1,2.3,NA,1.9,
3.5,3.4,2.9,3.1,NA,2.8,NA,
```

```
2.9,2.5,NA,2.4,2.7,2.3,2.1,
NA,3.5,NA,3.0,3.0,3.2,3.2,
2.3,2.6,2.6,NA,2.4,2.3,2.2,
3.5,3.6,3.4,3.3,NA,2.8,2.8,
3.4,3.3,NA,NA,2.8,2.7,2.6,
3.5,3.1,2.9,2.6,2.6,2.3,NA,
3.8,3.7,3.6,3.4,3.2,3.0,2.8,
3.9,3.1,3.6,3.4,3.3,NA,3.0,
3.5,3.8,4.0,3.8,NA,3.2,3.0,
3.4,3.0,3.1,3.1,NA,2.8,NA,
2.4,2.8,2.7,2.1,NA,2.0,1.9,
1.9,1.9,1.8,1.6,1.6,NA,NA,
3.0,2.8,2.5,NA,NA,2.2,2.2,
2.7,3.0,NA,NA,2.3,1.9,1.6,
2.8,3.4,NA,3.0,2.5,2.5,NA,
NA,3.1,NA,2.8,2.5,2.0,2.7,
NA,4.4,3.5,3.6,NA,3.4,3.2,
4.1,3.7,3.9,2.8,3.4,3.0,3.2,
2.1,2.2,2.4,2.3,2.2,NA,NA,
3.2,2.9,2.8,2.7,2.6,2.4,2.3,
3.5,3.5,3.9,3.3,NA,2.6,2.5,
3.2,3.7,3.4,3.4,3.5,NA,NA,
3.5,3.5,3.2,3.0,3.1,2.8,2.6,
4.3,NA,4.1,4.1,3.8,3.7,3.5,
2.9,3.1,2.7,2.8,NA,2.2,2.2,
2.3,2.2,2.0,2.0,1.9,NA,NA,
3.8,3.4,3.3,3.1,3.1,NA,NA,
4.0,3.3,3.5,NA,3.3,2.8,2.7,
2.8,2.9,2.8,2.4,2.5,2.4,2.3,
3.0,NA,3.2,2.5,2.5,2.7,NA,
NA,2.6,3.0,2.4,2.5,2.1,NA,
3.0,3.1,2.9,2.8,2.5,2.4,2.1,
3.0,2.7,NA,2.4,2.2,2.2,2.3,
2.4,2.3,2.3,2.0,1.9,NA,NA,
NA,2.6,2.5,NA,2.1,2.2,1.7,
NA,3.1,2.9,NA,2.4,2.7,2.2,
3.7,3.5,3.3,2.9,3.1,NA,2.6,
2.9,3.1,2.9,2.8,2.6,NA,NA,
3.0,3.0,2.7,2.5,2.1,2.0,2.2),.Dim = c(100,7)),
R = structure(.Data = c(1,0,0,1),.Dim = c(2,2)),
age = c(0,3,6,9,12,15,19))
list(beta = c(0,0),tau = 1)
```

TABLE 7.4

Posterior Analysis for Van der Lende et al. Air Pollution Study

Parameter	Mean	SD	Error	2½	Median	97½
σ_1^2	0.2889	0.04528	0.000221	0.2135	0.2859	0.3897
σ_{12}	−0.002887	0.00575	0.0000291	−0.01461	−0.00278	0.008153
σ_2^2	0.01069	0.001541	0.00000826	0.008103	0.01056	0.01409
β_1	3.251	0.05596	0.001762	3.144	3.25	3.361
β_2	−0.03878	0.01021	0.000504	−0.05829	−0.03905	−0.01844
σ^2	0.04225	0.00306	0.0000169	0.03663	0.0421	0.04869

Source: Van der Lende, R. et al., *Bull. Eur. Physiopath. Respir.*, 17, 775–792, 1981.

The analysis is executed with 75,000 observations for the simulation, with the burn in of 5,000 and a refresh of 100, and the Bayesian analysis is reported in Table 7.4.

The 95% credible interval for $g_{21} = \sigma_{12}$ is (−0.01461, 0.008153), which contains 0 and implies that the covariance between the intercept and slope random effects is 0, thus the Bayesian analysis is revised assuming the random effects are not correlated.

Thus, consider **BUGS CODE 7.2**, a revision of **BUGS CODE 7.1**, where the former uses the same data as the list statement in **BUGS CODE 7.1**.

A Bayesian analysis is executed with 75,000 observations for the simulation, with the burn in of 5,000 and a refresh of 100, and the posterior analysis is reported in Table 7.5.

BUGS CODE 7.2

```
model;
{
        beta1 ~ dnorm(0.0, 0.001)
        beta2 ~ dnorm(0.0, 0.001)

for(i in 1:N){for(j in 1:M){Y[i,j]~dnorm(mu[i,j],tau)}}
for(i in 1:N){for(j in 1:M){mu[i,j]<- beta1+beta2*age[j]+
b1[i]+b2[i]*age[j]}}
for (i in 1:N){b1[i]~dnorm(.000,tau1)}
for (i in 1:N){b2[i]~dnorm(.000,tau2)}
tau~dgamma(.0001,.0001)
tau1~dgamma(.0001,.0001)
tau2~dgamma(.0001,.0001)
sigma<-1/tau
sigma1<-1/tau1
sigma2<-1/tau2
}
```

TABLE 7.5

Posterior Analysis for Van der Lende et al. Air Pollution Study Assuming Uncorrelated Random Effects

Parameter	Mean	SD	Error	2½	Median	97½
β_1	3.256	0.05319	0.001615	3.152	3.255	3.36
β_2	−0.03821	0.001733	0.00001768	−0.04162	−0.03822	−0.0348
σ^2	0.04266	0.00298	0.0000206	0.03717	0.04254	0.04889
σ_1^2	0.2664	0.04018	0.0001917	0.1968	0.2627	0.3554
σ_2^2	0.000088	0.0000358	0.000000634	0.0000339	0.0000842	0.0001717

Source: Van der Lende, R. et al., *Bull. Eur. Physiopath. Respir.*, 17, 775–792, 1981.

Note that $g_{11} = \sigma_1^2$ and $g_{22} = \sigma_2^2$, and that based on the model (7.7), it can be shown that the variance function is given by

$$\mathrm{Var}(Y_{ij}) = \mathrm{Var}(b_{i1} + b_{i2}t_{ij} + \varepsilon_{ij})$$

$$= \mathrm{Var}(b_{i1}) + 2t_{ij}\mathrm{Cov}(b_{i1}, b_{i2}) + t_{ij}^2\mathrm{Var}(b_{2i}) + \mathrm{Var}(\varepsilon_{ij}) \qquad (7.15)$$

$$= g_{11} + 2t_{ij}g_{12} + t_{ij}^2 g_{22} + \sigma^2$$

which is estimated (using the posterior means) as

$$\mathrm{Var}(Y_{ij}) = 0.2664 + t_{ij}^2 \times 0.000088 + 0.04266 \qquad (7.16)$$

where $t_{ij} = c(0,3,6,9,12,15,19)$. Thus, the estimated variance increases from 0.2664 at baseline to 0.340828 at time $t = 19$. In a similar way, the estimated variance at time 15 is $0.2664 + 15 \times 15 \times 0.000088 + 0.04266 = 0.32886$.

In addition, it can be shown that the covariance is given by

$$\mathrm{Cov}(Y_{ij}, Y_{ik}) = \mathrm{Cov}(b_{i1} + b_{2i}t_{ij} + \varepsilon_{ij}, b_{i1} + b_{2i}t_{ik})$$

$$= g_{11} + (t_{ij} + t_{ik})g_{12} + t_{ij}t_{ik}g_{22} \qquad (7.17)$$

and, thus is estimated as

$$\mathrm{Cov}(Y_{ij}, Y_{ik}) = 0.2664 + t_{ij}t_{ik} \times 0.0000888 \qquad (7.18)$$

For example, the covariance between times 15 and 19 is estimated as 0.2664 + 0.02505 = 0.2914. Last, the correlation between times 15 and 19 is estimated as $(0.2914)/\mathrm{sqrt}(0.3408 \times 0.32886) = 0.8704$.

Note that the mean profile for this study is given by

$$E(Y_{ij}) = 3.256 - 0.03821 \times t_{ij} \qquad (7.19)$$

where the intercept and slope are estimated by their posterior means appearing in Table 7.5.

7.6.2 Hip-Replacement Study

Recall from the previous section the quadratic fit for hematocrit for both groups over time points 1 2, 3, and 4 depicted by Figure 7.3.

Thus, the model for males and females consists of a fixed part that portrays the quadratic fit plus random effects for the intercept, slope, and quadratic effect of individual subjects.

For males, the model for hematocrit is

$$Y_{1ij} = \beta_1 + \beta_2 t_{ij} + \beta_3 t_{ij}^2 + b_{1i} + b_{2i}t_{ij} + b_{3i}t_{ij}^2 + \varepsilon_{ij}, \tag{7.20}$$

where β_1, β_2, and β_3 are the coefficients for the intercept, linear, and quadratic effects for the mean profile (averaged over the random effects), and the random effects b_{1i}, b_{2i}, b_{3i} form a trivariate normal distribution with 3 by 1 mean vector 0 and 3 by 3 covariance matrix $G = (g_{ij})$. The components of G are the variances and covariances of the three random effects. Last, the measurement errors are independent normal random variables with mean 0 and variance denoted by σ^2.

In a similar fashion, the model for females is given by

$$Y_{2ij} = \beta_4 + \beta_5 t_{ij} + \beta_6 t_{ij}^2 + b_{4i} + b_{5i}t_{ij} + b_{6i}t_{ij}^2 + \varepsilon_{ij} \tag{7.21}$$

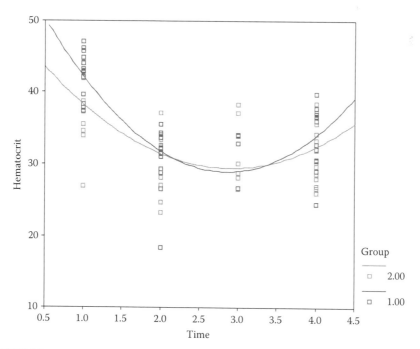

FIGURE 7.3
Hematocrit versus time for males and females.

with the definition for the terms of the model similar to that given to the model (7.20) for males, where β_4, β_5, and β_6 are the coefficients for the intercept, linear, and quadratic effects for the mean profile (averaged over the random effects).

The random effects c_{1i}, c_{2i}, c_{3i} form a trivariate normal distribution with 3 by 1 mean vector 0 and 3 by 3 covariance matrix $H = (h_{ij})$. The components of H are the variances and covariances of the three random effects. Last, the measurement errors are independent normal random variables with mean 0 and variance denoted by σ^2.

BUGS CODE 7.3 closely follows the notation given by formulas (7.21) and (7.22).

BUGS CODE 7.3

```
model;
      {

# Y1 is hematocrit for males
for(i in 1:N1){for(j in 1:M1)
{Y1[i,j]~dnorm(mu1[i,j],tau)}}
for(i in 1:N1){for(j in 1:M1){mu1[i,j]<-beta[1]+beta[2]
*age[j]+beta[3]*age[j]*age[j]
+b[i,1]+b[i,2]*age[j]+b[i,3]*age[j]*age[j]}}
# prior distribution for the regression coefficients
   for(i in 1:6){beta[i]~dnorm(0,.0001)}
# prior distribution for the random effects
   for (i in 1:N1){b[i,1:3]~dmnorm(vu1[], tau1[,])}
   for (j in 1:3){vu1[j]<-0}
# prior Wishart distribution for the precision matrix of
random effects
   tau1[1:3,1:3]~dwish(R[,],3)
# Sigma1 is variance covariance matrix of random effects
   Sigma1[1:3,1:3]<-inverse(tau1[,])

# Y2 is hematocrit for females

for(i in 1:N2){for(j in 1:M2)
{Y2[i,j]~dnorm(mu2[i,j],tau)}}
for(i in 1:N2){for(j in 1:M2){mu2[i,j]<-beta[4]+beta[5]
*age[j]+beta[6]*age[j]*age[j]+
c[i,1]+c[i,2]*age[j]+c[i,3]*age[j]*age[j]}}
# prior distribution for random effects for females
   for (i in 1:N2){c[i,1:3]~dmnorm(vu2[], tau2[,])}
   for (j in 1:3){vu2[j]<-0}

# prior Wishart distribution for precision matrix of
random effects
```

```
    tau2[1:3,1:3]~dwish(R[,],3)
# Sigma2 is the variance covariance matrix for random
effects
    Sigma2[1:3,1:3]<-inverse(tau2[,])
# tau is precision of measurement errors,
tau~dgamma(.0001,.0001)


sigma<-1/tau


}


# Males
list(N1 = 13,N2 = 17,M1 = 4,M2 = 4,
Y1 = structure(.Data =
c(47.1,31.1,NA,32.8,
44.1,31.5,NA,37.0,
39.7,33.7,NA,24.5,
43.3,18.4,NA,36.6,
37.4,32.3,NA,29.1,
45.7,35.5,NA,39.8,
44.9,34.1,NA,32.1,
42.9,32.1,NA,NA,
46.1,28.8,NA,37.8,
42.1,34.4,34.0,36.1,
38.3,29.4,32.9,30.5,
43.0,33.7,34.1,36.7,
37.8,26.6,26.7,30.6),.Dim = c(13,4)),
# females
Y2 = structure(.Data =
c(37.3,26.5,NA,38.5,
NA,28.0,NA,33.9,
27.0,32.5,NA,32.0,
38.4,32.3,NA,37.9,
38.8,32.6,NA,26.9,
44.7,32.2,NA,34.2,
38.0,27.1,NA,37.9,
34.0,23.2,NA,26.0,
44.8,37.2,NA,29.7,
46.0,29.1,NA,26.7,
41.9,32.0,37.1,37.6,
38.0,31.7,38.4,35.7,
42.2,34.0,32.9,33.3,
39.7,33.5,26.6,32.7,
37.5,28.2,28.8,30.3,
34.6,31.0,30.1,28.7,
35.5,24.7,28.1,29.8),.Dim = c(17,4)),
```

```
R = structure(.Data = c(1,0,0,0,1,0,0,0,1),.Dim =
c(3,3)),
age = c(1,2,3,4))

# initial values
list(beta = c(0,0,0,0,0,0),
tau = 1)
```

The Bayesian analysis is based on **BUGS CODE 7.3** and is executed with 75,000 observations, with a burn in of 5,000 and a refresh of 100 (Table 7.6).

The reader should note the notation for the variance covariance matrix H for the random effects c_{1i}, c_{2i}, c_{3i} of females as follows: δ_1^2 is the variance of $c_{1i} = h_{11}$; $h_{12} = \delta_{12}$, the covariance between c_{1i} and c_{2i}; and so on.

This analysis is considered preliminary in the sense that our primary interest is in estimating the covariances between the various random effects. Note the 95% credible interval for the six covariances contain 0, implying they are in fact 0, thus it is plausible to consider a Bayesian analysis with uncorrelated random effects.

TABLE 7.6

Posterior Distribution for the Hip-Replacement Study with Correlated Random Effects

Parameter	Mean	SD	Error	2½	Median	97½
σ_1^2	1.558	5.28	0.09934	0.1436	0.6951	7.464
σ_{12}	−0.611	5.567	0.115	−5.008	−0.03463	1.465
σ_{13}	0.08753	1.168	0.02393	−0.6107	−0.00605	1.102
σ_2^2	1.796	6.576	0.1384	0.1406	0.6991	9.051
σ_{23}	−0.3759	1.41	0.02901	−2.147	−0.1269	0.1167
σ_3^2	0.2691	0.3318	0.006155	0.0797	0.2009	0.8312
δ_1^2	2.532	5.065	0.09685	0.158	1.032	14.23
δ_{12}	−0.1552	2.706	0.0465	−5.201	0.06477	2.863
δ_{13}	−0.1154	0.5694	0.008143	−1.215	0.06884	0.8064
δ_2^2	1.68	2.817	0.04632	0.1505	0.8271	8.255
δ_{23}	−0.3818	0.6687	0.01061	−2.05	−0.1782	0.0704
δ_3^2	0.2472	0.1953	0.00266	0.07593	0.193	0.7292
β_1	60.58	3.291	0.1518	53.94	60.06	66.87
β_2	−22.24	3.143	0.1517	−28.24	−22.24	−15.87
β_3	3.898	0.6281	0.02897	2.64	3.896	5.088
β_4	50.53	2.717	0.1226	−19.69	−14.94	−9.684
β_5	−14.86	2.521	0.1211	−19.69	−14.94	−9.684
β_6	2.604	0.5006	0.02368	1.595	2.618	3.554
σ^2	13.9	2.655	0.02724	9.439	13.65	19.82

For now, it is assumed that the covariances between the random effects for males, namely, b_{1i}, b_{2i}, b_{3i} are uncorrelated, and the same for the three random effects for females, namely, b_{4i}, b_{5i}, b_{6i}, thus the Bayesian analysis is executed with **BUGS CODE 7.4** using 75,000 observations, a burn in of 5,000 and a refresh of 100. **BUGS CODE 7.4** is a slight revision of **BUGS CODE 7.3**, and the data for the present analysis is found in the list statement of **BUGS CODE 7.3**. The main parameters of interest are the beta parameters that define the mean profile of hematocrit for males and females, as well as the variances of the six random effects.

BUGS CODE 7.4

```
        model;
        {
        beta1 ~ dnorm(0.0, 0.001)
        beta2 ~ dnorm(0.0, 0.001)
  beta3 ~ dnorm(0.0,.0001)
  beta4 ~ dnorm(0.0,.0001)
  beta5 ~ dnorm(0.0,.0001)
  beta6 ~ dnorm(0.0,.0001)
# Y1 is hematocrit for males
for(i in 1:N1){for(j in 1:M1)
{Y1[i,j]~dnorm(mu1[i,j],tau)}}
for(i in 1:N1){for(j in 1:M1){mu1[i,j]<-beta1+beta2
*age[j]+beta3*age[j]*age[j]
+b1[i]+b2[i]*age[j]+b3[i]*age[j]*age[j]}}

# Y2 is hematocrit for females
for(i in 1:N2){for(j in 1:M2)
{Y2[i,j]~dnorm(mu2[i,j],tau)}}
for(i in 1:N2){for(j in 1:M2){mu2[i,j]<-beta4+beta5
*age[j]+beta6*age[j]*age[j]
+b4[i]+b5[i]*age[j]+b6[i]*age[j]*age[j]}}

for(i in 1:N1){b1[i]~dnorm(0,tau1)}
for(i in 1:N1){b2[i]~dnorm(0,tau2)}
for(i in 1:N1){b3[i]~dnorm(0,tau3)}
for(i in 1:N2){b4[i]~dnorm(0,tau4)}
for(i in 1:N2){b5[i]~dnorm(0,tau5)}
for(i in 1:N2){b6[i]~dnorm(0,tau6)}

tau~dgamma(.0001,.0001)
tau1~dgamma(.0001,.0001)
tau2~dgamma(.0001,.0001)
tau3~dgamma(.0001,.0001)
tau4~dgamma(.0001,.0001)
tau5~dgamma(.0001,.0001)
```

```
tau6~dgamma(.0001,.0001)
sigma<-1/tau
sigma1<-1/tau1
sigma2<-1/tau2
sigma3<-1/tau3
sigma4<-1/tau4
sigma5<-1/tau5
sigma6<-1/tau6
}
```

The mean profile for males is given by

$$E\left(Y_{1ij}\right) = \beta_1 + \beta_2 t_{ij} + \beta_3 t_{ij}^2 \tag{7.22}$$

and is estimated as

$$59.36 - 21.06t_{ij} + 3.673t_{ij}^2 \tag{7.23}$$

using the posterior means of Table 7.7.

Note the notation of Table 7.7. σ^2 is the variance of the measurement error, σ_1^2 is the variance of the constant random effect b_{1i} of males, σ_2^2 is the variance of the linear effect for males, and σ_3^2 is the variance of the random quadratic effect for males. On the other hand, σ_4^2 is the variance for the random constant effect for females, σ_5^2 is the variance of the random linear effect for females, and last σ_6^2 is the variance of the random quadratic effect for females. It is obvious that the posterior distribution of the variances of the random effects

TABLE 7.7

Posterior Analysis of Hip-Replacement Study with Uncorrelated Random Effects

Parameter	Mean	SD	Error	2½	Median	97½
β_1	59.36	3.365	0.1535	52.75	59.31	65.96
β_2	−21.06	3.183	0.1519	−27.33	−21.0	−14.88
β_3	3.673	0.6214	0.02939	2.464	3.663	4.892
β_4	50.53	3.01	0.1359	44.66	50.53	56.51
β_5	−14.83	2.8	0.1322	−20.39	−14.84	−9.358
β_6	2.596	0.5437	0.02542	1.531	2.598	3.671
σ^2	15.99	2.841	0.04503	11.14	15.73	22.24
σ_1^2	0.9864	2.284	0.05745	0.0001163	0.05098	7.581
σ_2^2	0.1586	0.3329	0.00709	0.0001094	0.01992	1.117
σ_3^2	0.01547	0.02081	0.0004799	0.00009301	0.004313	0.09299
σ_4^2	1.565	2.8	0.08291	0.0001328	0.1686	9.472
σ_5^2	0.1396	0.2777	0.00761	0.00009242	0.018	0.96
σ_6^2	0.01257	0.02216	0.0004268	0.00007897	0.003464	0.0745

are highly skewed, thus I recommend using the posterior median to estimate the variances.

Now consider estimating the variance and covariance functions for males.

Assuming the covariances are 0, it can be demonstrated that the variance function is

$$\text{Var}\left(Y_{1ij}\right) = g_{11} + t_{ij}^2 g_{22} + t_{ij}^4 g_{33} \tag{7.24}$$

and the covariance function is

$$\text{Cov}\left(Y_{1ij}, Y_{1ik}\right) = g_{11} + t_{ij} t_{ik} g_{22} + t_{ij}^2 t_{ik}^2 g_{33} \tag{7.25}$$

Note how the variance and covariance functions increase with increasing values of the time points!

Based on the estimated of Table 7.7, (7.24) and (7.25) allow one to estimate the variance and covariance functions as

$$0.9864 + t_{ij}^2(0.1586) + t_{ij}^4(0.01547) \tag{7.26}$$

for the variance function and

$$0.9864 + t_{ij} t_{ik}(0.1586) + t_{ij}^2 t_{ik}^2(0.01547) \tag{7.27}$$

for the covariance function.

Consider the correlation of hematocrit between time points 1 and 4, then the variance at time point 1 is estimated as $0.9864 + 0.1586 + 0.01547 = 1.16047$ and at time point 4 is estimated as $0.9864 + 16(0.1586) + 256(0.01547) = 7.482$. The covariance between time point 1 and time point 4 is estimated as

$$0.9864 + 4(0.1586) + 16(0.01547) = 1.866 \tag{7.28}$$

Therefore, the correlation of the hematocrit value at time point 1 and its value at time point 4 is estimated as

$$\frac{1.866}{\text{sqrt}[1.16047 \times 7.482]} = 0.6332 \tag{7.29}$$

How well do the models (7.20) and (7.21) fit the hematocrit data?

To investigate the goodness of fit, it is possible to predict future hematocrit values using **BUGS CODE 7.4**, which results in Figure 7.4, a plot of the predicted hematocrit values versus the observed values by males and females separately. I calculated the Pearson correlation between observed and predicted hematocrit values and estimated the correlation as 0.876 for males and 0.826 for females. It appears that the generalized linear model defined by (7.20) does a "good" job in modeling the hematocrit values.

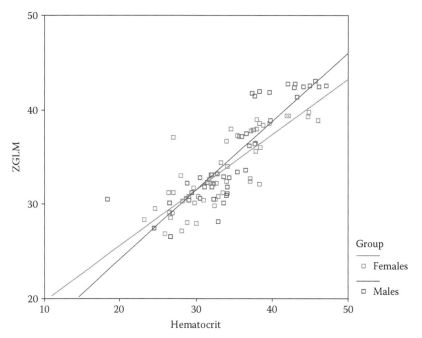

FIGURE 7.4
Predicted versus observed values of hematocrit for males and females.

7.6.3 Hospital Study of a Dietary Regime

Recall the example from Section 5.4.2 pertaining to a dietary study of Crowder and Hand.[4] This example is quite different than the other presented in this chapter, in that different treatments are given to the same subject (see Table 7.8).

Twelve hospital patients were given a dietary regimen with observations of plasma ascorbic acid with two observations before administration of the dietary treatment, three during the treatment, and twice after. The objective of the study is to investigate the effect of diet over a 16-week period, with repeated measures at weeks 1, 2, 6, 10, 14, 15, and 16. Based on Table 7.8, the descriptive statistics for this study are portrayed in Table 7.9.

One should interpret the descriptive statistics using the scatter plot of Figure 7.5.

The scatter plot suggests a spline with three linear segments will suffice in modeling the mean profile of this study. This is a study with one treatment, namely a dietary supplement administered at weeks 6, 10, and 12, where pretreatment observations are made at weeks 1, 2, and 6 and post-treatment observations are made at weeks 15 and 16. The subjects are on the dietary supplement beginning at week 6 and ending at the beginning of week 14. Tentatively, the mean profile will begin with a linear segment over time

TABLE 7.8

Posterior Analysis for Hospital Visit Dietary Regime

Parameter	Mean	SD	Error	2½	Median	97½
σ_1^2	0.7568	1.067	0.01188	0.122	0.4623	3.234
σ_{12}	−0.09508	0.3465	.002881	−0.8864	−0.04034	0.3696
σ_{13}	0.01212	0.8482	0.008886	−1.256	0.000232	1.302
σ_{14}	−0.000868	0.1694	0.000915	−0.3303	−0.000089	0.3225
σ_{15}	−0.667	38.64	0.5854	−2.038	0.0000886	1.75
σ_{16}	−0.4154	2.473	0.03726	−0.5965	0.000272	0.6472
σ_2^2	0.3569	0.2683	0.00101	0.09891	0.2845	1.051
σ_{23}	−0.00399	0.3509	0.002256	−0.6617	−0.000738	0.6338
σ_{24}	−0.00399	0.3509	0.002256	−0.6617	−0.000738	0.6338
σ_{25}	0.06269	10.78	0.1224	−0.895	−0.000949	0.9579
σ_{26}	−0.006347	0.7008	0.007816	−0.3871	−0.001933	0.3634
σ_3^2	0.9142	2.155	0.04385	0.1242	0.4833	4.228
σ_{34}	−0.06558	0.2579	0.004028	−0.5431	−0.02972	0.2121
σ_{35}	−0.6109	62.02	1.828	−2.139	0.000614	2.071
σ_{36}	0.03767	3.96	0.1164	−0.6496	−0.000884	0.6563
σ_4^2	0.1809	0.1023	0.000422	0.06828	0.1561	0.441
σ_{45}	0.0108	6.698	0.1658	−0.4902	−0.000368	0.4936
σ_{46}	−0.003252	0.4358	0.01056	−0.2345	−0.002135	0.2288
σ_5^2	754.5	4871	193.8	0.126	0.5016	14720
σ_{56}	−48.07	310.6	12.35	−934.9	−0.01044	0.4218
σ_6^2	3.337	19.76	0.7858	0.1023	0.246	59.56
σ^2	43.85	8.557	0.2331	16.58	43.82	59.99
β_1	0.3324	1.904	0.008651	−3.425	0.3347	4.08
β_2	0.1456	0.5392	0.00225	−.9184	0.1466	1.205
β_3	1.209	3.555	0.03248	−5.763	1.19	8.251
β_4	−0.00634	0.3585	0.003323	−0.7141	−0.00486	0.6952
β_5	53.13	21.66	0.8096	8.848	53.36	95.9
β_6	−3.35	1.46	0.05396	−6.209	−3.361	−0.3977

points 1, 2, and 6 weeks, the second over weeks 6, 10, and 14, and the last over weeks 14, 15, and 16.

A three-segment spline for the mean profile of plasma ascorbic acid is represented by the following mean profiles:

$$E\left(Y_{ij}\right) = \beta_1 + \beta_2 t_{ij} \quad 1 \leq t_{ij} \leq 6 \tag{7.30}$$

$$E\left(Y_{ij}\right) = \beta_3 + \beta_4 t_{ij} \quad 6 \leq t_{ij} \leq 14 \tag{7.31}$$

and

$$E\left(Y_{ij}\right) = \beta_5 + \beta_6 t_{ij} \quad 14 \leq t_{ij} \leq 16 \tag{7.32}$$

TABLE 7.9

Posterior Analysis for Dietary Regimen with Uncorrelated Random Effects

Parameter	Mean	SD	Error	2½	Median	97½
β_1	0.3262	1.81	0.01079	−3.213	0.3261	3.899
β_2	0.1461	0.4891	0.002893	−0.8159	0.147	1.109
β_3	1.211	3.435	0.03992	−5.496	1.189	7.999
β_4	−0.005897	0.3266	0.003801	−0.6498	−0.004871	0.6319
β_5	51.64	19.68	0.8469	11.9	51.76	89.82
β_6	−3.249	1.307	0.05625	−5.786	−3.255	−0.5991
σ^2	39.99	6.32	0.06409	29.57	39.33	54.27
σ_1^2	0.257	0.9566	0.01431	0.0000115	0.004798	2.453
σ_2^2	0.02404	0.07576	0.001046	0.0000099	0.001374	0.2085
σ_3^2	0.26	0.9523	0.01448	0.0000114	0.005253	2.44
σ_4^2	0.00386	0.0108	0.0001309	0.0000083	0.000443	0.03125
σ_5^2	34.1	32.96	0.7388	0.0000721	28.39	114.5
σ_6^2	0.04111	0.08417	0.00243	0.0000102	0.00265	0.2772

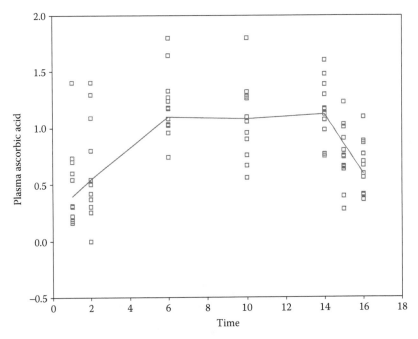

FIGURE 7.5

Plasma ascorbic acid for 12 hospital patients by week.

However, the model defined by Equations 7.30 through 7.32 is now generalized to include random effects.

$$Y_{1ij} = \beta_1 + \beta_2 t_{ij} + b_{1i} + b_{2i}t_{ij} + \varepsilon_{ij} \quad 1 \le t_{ij} \le 6 \tag{7.33}$$

$$Y_{2ij} = \beta_3 + \beta_4 t_{ij} + b_{3i} + b_{4i}t_{ij} + \varepsilon_{ij} \quad 6 \le t_{ij} \le 14 \tag{7.34}$$

and

$$Y_{3ij} = \beta_5 + \beta_6 t_{ij} + b_{5i} + b_{6i}t_{ij} + \varepsilon_{ij} \quad 14 \le t_{ij} \le 16 \tag{7.35}$$

where the six random effects $b_{1i}, b_{2i}, b_{3i}, b_{4i}, b_{5i}, b_{6i}$ follow a multivariate normal distribution with 6 by 1 mean vector 0 and 6 by 6 variance covariance matrix $G = (g_{ij})$, $i,j = 1,2,\ldots,6$. It should be emphasized that the subject i is the same for the three regimes defined by (7.33) through (7.35). Note that for the first segment (7.33), b_{1i} is the random intercept for subject i, but b_{5i} is the random intercept for the same subject i for the third segment (7.35).

The Bayesian analysis is executed with **BUGS CODE 7.5** using 250,000 observations, with 5,000 initial values and a refresh of 100.

Notice the asymmetry in the posterior distributions of all the parameters except for the regression coefficients, thus the posterior median should be

BUGS CODE 7.5

```
        model;
{

# prior distribution for beta

for(i in 1:6){beta[i]~dnorm(0,.0001)}

# first segment

for(i in 1:N1){for(j in 1:M1)
{Y1[i,j]~dnorm(mu1[i,j],tau)}}
for(i in 1:N1){for(j in 1:M1)
{mu1[i,j]<- beta[1]+beta[2]*time1[j]+
b[i,1]+b[i,2]*time1[j]}}

# second segment

for(i in 1:N2){for (j in 1:M2)
{Y2[i,j]~dnorm(mu2[i,j],tau)}}
for(i in 1:N2){for (j in 1:M2)
{mu2[i,j]<-beta[3]+beta[4]*time2[j]+
b[i,3]+b[i,4]*time2[j]}}
```

```
# third segment

for(i in 1:N3){for (j in 1:M3)
{Y3[i,j]~dnorm(mu3[i,j],tau)}}
for (i in 1:N3){for (j in 1:M3){mu3[i,j]<- beta[5]
+beta[6]*time3[j]+b[i,5]+b[i,6]*time3[j]}}

# prior distribution of random effects

for(i in 1:N1){b[i,1:6]~dmnorm(vu[],Omega[,])}
for (
i in 1:6){vu[i]<-0}
Omega[1:6,1:6]~dwish(R[,],6)
Sigma[1:6,1:6]<-inverse(Omega[,])

tau~dgamma(.00001,.00001)

Var<-1/tau

}

list(N1 = 12,N2 = 12,N3 = 12,M1 = 3,M2 = 3,M3 = 3,time1
= c(1,2,6),
time2 = c(6,10,14),time3 = c(14,15,16),

# data for first segment
Y1 = structure(.Data =

c(.22, .00, 1.03,.18, .00, .96,.73, .37, 1.18,.30, .25,
  .74,.54, .42, 1.33,.16, .30, 1.27,.30, 1.09, 1.17,.70,
  1.30, 1.80,.31, .54, 1.24,1.40, 1.40, 1.64,.60, .80,
  1.02,.73, .50, 1.08),.Dim = c(12,3)),
# data for second segment
Y2 = structure(.Data =

c(1.03, .67, .75,.96, .96, .98,1.18, .76, 1.07,.74, 1.10,
  1.48,1.33, 1.32, 1.30,1.27, 1.06, 1.39,1.17, .90,
  1.17,1.80, 1.80, 1.60,1.24, .56, .77,1.64, 1.28, 1.12,1.02,
  1.28, 1.16,1.08, 1.26, 1.17),.Dim = c(12,3)),
# data for third segment
Y3 = structure(.Data =

c(75, .65, .59,.98, 1.03, .70,1.07, .80, 1.10,1.48, .39,
  .36,1.30, .74, .56,1.39, .63, .40,1.17, .75, .88,1.60,
  1.23, .41,.77, .28, .40,1.12, .66, .77,1.16, 1.01,
  .67,1.17, .91, .87),.Dim = c(12,3)),
```

```
# un informative prior distribution for the precision
matrix
R = structure(.Data = c(1,0,0,0,0,0,
                        0,1,0,0,0,0,
                        0,0,1,0,0,0,
                        0,0,0,1,0,0,
                        0,0,0,0,1,0,
                        0,0,0,0,0,1
),.Dim = c(6,6)))
# initial values
list(beta = c(0,0,0,0,0,0),tau = 1)
```

used as estimates instead of the posterior mean. It is observed that the estimated slopes via the posterior means appear to conform to Figure 7.5, and
it is interesting to see that all the 95% credible intervals for the covariances
between the random effects contain 0. Thus, the relevant posterior analysis
is to assume the random effects are uncorrelated; therefore, the appropriate analysis is based on **BUGS CODE 7.6** using 150,000 observations, with a
burn in of 5,000 and a refresh of 100.

BUGS CODE 7.6

```
        model;
{

# prior distribution for beta

for(i in 1:6){beta[i]~dnorm(0,.0001)}

# first segment

for(i in 1:N1){for(j in 1:M1)
{Y1[i,j]~dnorm(mu1[i,j],tau)}}
for(i in 1:N1){for(j in 1:M1)
{mu1[i,j]<- beta[1]+beta[2]*time1[j]+
b1[i]+b2[i]*time1[j]}}

# second segment

for(i in 1:N2){for (j in 1:M2)
{Y2[i,j]~dnorm(mu2[i,j],tau)}}
for(i in 1:N2){for (j in 1:M2)
{mu2[i,j]<-beta[3]+beta[4]*time2[j]+
b3[i]+b4[i]*time2[j]}}
```

```
# third segment

for(i in 1:N3){for (j in 1:M3)
{Y3[i,j]~dnorm(mu3[i,j],tau)}}
for (i in 1:N3){for (j in 1:M3){mu3[i,j]<- beta[5]+
beta[6]*time3[j]+b5[i]+b6[i]*time3[j]}}

# prior distribution of random effects

for(i in 1:N1){b1[i]~dnorm(0,tau1}
for(i in 1:N1){b2[i]~dnorm(0,tau1}
for(i in 1:N1){b3[i]~dnorm(0,tau1}
for(i in 1:N1){b4[i]~dnorm(0,tau1}
for(i in 1:N1){b5[i]~dnorm(0,tau1}
for(i in 1:N1){b6[i]~dnorm(0,tau1}

tau~dgamma(.00001,.00001)

sigma<-1/tau

}

list(N1 = 12,N2 = 12,N3 = 12,M1 = 3,M2 = 3,M3 = 3,time1
= c(1,2,6),
time2 = c(6,10,14),time3 = c(14,15,16),

# data for first segment
Y1 = structure(.Data =

c(.22, .00,   1.03,.18,     .00,   .96,.73,     .37,
  1.18,.30,   .25,   .74,.54,     .42,   1.33,.16,    .30,
  1.27,.30,   1.09, 1.17,.70,   1.30,  1.80,.31,    .54,
  1.24,1.40, 1.40,  1.64,.60,    .80,   1.02,.73,    .50,
  1.08),.Dim = c(12,3)),
# data for second segment
Y2 = structure(.Data =

c(1.03, .67, .75,.96, .96, .98,1.18, .76, 1.07,.74, 1.10,
  1.48,1.33, 1.32, 1.30,1.27, 1.06, 1.39,1.17, .90,
  1.17,1.80, 1.80, 1.60,1.24, .56, .77,1.64, 1.28,
  1.12,1.02, 1.28, 1.16,1.08, 1.26, 1.17),.Dim = c(12,3)),
# data for third segment
Y3 = structure(.Data =

c(75, .65, .59,.98, 1.03, .70,1.07, .80, 1.10,1.48, .39,
  .36,1.30, .74, .56,1.39, .63, .40,1.17, .75, .88,1.60,
  1.23, .41,.77, .28, .40,1.12, .66, .77,1.16, 1.01,
  .67,1.17, .91, .87),.Dim = c(12,3)),

# initial values

list(beta = c(0,0,0,0,0,0),tau = 1, tau1 = 1,tau2 =
1,tau3 = 1,tau4 = 1,tau5 = 1, tau6 = 1)
```

The Bayesian analysis for the hospital study is shown in Table 7.9.

Recall that σ_1^2 is the variance of the intercept for the first regime of the three-segment linear spline over time points 1, 2, and 6 and is estimated as 0.004789 with the posterior median, while σ_2^2 is the variance over the individual slopes over the same three time points, estimated as 0.001374 via the posterior median. Also, for the third linear segment with time points 14, 15, and 16, the slope β_6 is estimated as -3.255, via the posterior median, but the variability σ_6^2 of the individual slopes over the same segment is estimated as 0.00265 via the posterior median.

How do the estimates of the regression coefficients and variances of the random effects of Table 7.9 (which assumes uncorrelated random effects) compare to the corresponding estimates of Table 7.8 (which assumes correlated random effects)?

Variance and covariance functions are quite involved because they must be defined over a three-segment linear spline. For example, there is the covariance function over each segment separately as well as three covariance functions between the three segments. In addition, there are three variance functions, one for each segment.

First, consider the covariance function for the first segment. One may show for subject i that

$$\text{Cov}\left(Y_{1ij}, Y_{1ik}\right) = g_{11} + g_{12}\left(t_{ij} + t_{ik}\right) + g_{22}t_{ij}t_{ik} \tag{7.36}$$

where:
$t_{ij} \neq t_{ik}$
$t_{ij}, t_{ik} = 1, 2, 6$
$g_{11} = \text{Var}(b_{1i})$
$g_{22} = \text{Var}(b_{2i})$
$g_{12} = \text{Cov}(b_{1i}, b_{2i})$

Of course similar expressions apply for the other two segments of the spline. For example, consider the covariance between a value of ascorbic acid measured at a time point of the first segment and a value of ascorbic acid measured at a time point of the second segment, and then it can be shown as

$$\text{Var}\left(Y_{1ij}\right) = g_{11} + 2t_{ij}g_{12} + t_{ij}^2 g_{22} + \sigma^2 \tag{7.37}$$

where:

$$g_{23} = \text{Cov}(b_{2i}, b_{3i}) \tag{7.38}$$

and

$$g_{24} = \text{Cov}(b_{2i}, b_{4i}) \tag{7.39}$$

Also, note that for the second segment with time points 6, 10, and 14

$$\text{Var}\left(Y_{2ij}\right) = g_{33} + 2t_{ij}g_{34} + t_{ij}^2 g_{44} + \sigma^2 \tag{7.40}$$

where:

$$g_{33} = \text{Var}\left(b_{3i}\right) \tag{7.41}$$

$$g_{34} = \text{Cov}\left(b_{3i}, b_{4i}\right) \tag{7.42}$$

and

$$g_{44} = \text{Var}\left(b_{4i}\right) \tag{7.43}$$

Estimates of the various variance and covariance functions will be estimated based on formulas (7.36) through (7.43) and the posterior analysis of Table 7.9. First, consider the covariance function for the first segment given by (7.36) and estimated as

$$0.257 + 0.02402 t_{ij} t_{ik} \tag{7.44}$$

where $j \neq k$ and $t_{ij} = 1, 2, 6$.

Now consider the correlation of ascorbic acid between time points 1 and 6,

$$\rho\left(Y_{1i1}, Y_{1i6}\right) = \frac{\text{Cov}\left(Y_{1i1}, Y_{1i6}\right)}{\sqrt{\text{Var}\left(Y_{1i1}\right) \text{Var}\left(Y_{1i6}\right)}}$$

which is estimated by $0.40112/\text{sqrt}(39.611 \times 39.73) = 0.01095$. Note that the estimate is based on the assumption of uncorrelated random effects, that is, $g_{ij} = 0$ for all i and j.

7.6.4 Succimer Study

According to the succimer study, such a mean profile is implied by the scatter plot of Figure 7.6, where the mean profile of the two groups is delineated by a lowess plot.

Group 2 is the one using the chelating agent succimer and 1 denotes placebo. Note that there are 50 children in each group.

Consider the following mean profile:

1. The placebo group

$$E\left(Y_{1ij}\right) = \beta_1 + \beta_2 t_{ij} \tag{7.45}$$

where:

Y_{1ij} is the lead level for subject i at time $j = 0, 1, 4, 6$

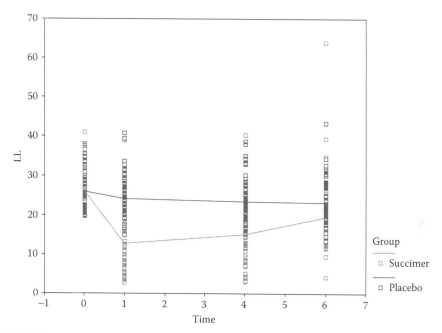

FIGURE 7.6
Blood lead levels versus time for placebo and treatment groups.

2. The succimer group
    a. For the first segment with time points 0 and 1 weeks.

$$E\left(Y_{2ij}\right) = \beta_3 + \beta_4 t_{ij} \qquad (7.46)$$

    where:
        Y_{ij} is the lead level for subject i at time $j = 0,1$

    b. For the second segment with time points 1, 4, and 6 weeks.

$$E\left(Y_{3ij}\right) = \beta_5 + \beta_6 t_{ij} \qquad (7.47)$$

    where:
        Y_{ij} is the lead level for subject i at time $j = 1,4,6$

The mean profiles expressed by (7.45) through (7.47) are averaged over the random effects, thus consider the generalized linear model representation of the three mean profiles. For example, corresponding to (7.45), let

$$Y_{1ij} = \beta_1 + \beta_2 t_{ij} + b_{1i} + b_{2i} t_{ij} + \varepsilon_{ij} \qquad (7.48)$$

where, then let

$$Y_{2ij} = \beta_3 + \beta_4 t_{ij} + b_{3i} + b_{4i} t_{ij} + \varepsilon_{ij} \tag{7.49}$$

where $t_{ij} = 0,1$ be the generalization of (7.46) to include random effects for the intercept and slope of the ith subject for the first segment of the linear spine for the succimer group.

Finally, for the second segment of the linear spline for the succimer group, let

$$Y_{3ij} = \beta_5 + \beta_6 t_{ij} + b_{5i} + b_{6i} t_{ij} + \varepsilon_{ij} \tag{7.50}$$

where $t_{ij} = 1,4,6$. See the graph in Figure 7.6 for the association between lead level and time for the two groups, placebo versus succimer. It is assumed that the joint distribution of $(b_{1i}b_{2i},b_{3i},b_{4i},b_{5i},b_{6i})$ for each subject is a multivariate normal with 6 by 1 mean vector 0 and unknown 6 by 6 covariance matrix Σ. It is also assumed that the N random vectors $(b_{1i},b_{2i},b_{3i},b_{4i},b_{5i},b_{6i})$ are independent. In addition, the regression coefficients $\beta_1,\beta_2,\beta_3,\beta_4,\beta_5,\beta_6$ are assumed to be unknown random variables along with the variance σ^2 of the measurement errors ε_{ij}.

For the Bayesian analysis, it will be assumed that the prior distribution of the regression coefficients are independent normal random variables with mean 0 and variance 1000, while the prior distribution of the measurement precision $\tau = 1/\sigma^2$ is assumed to be an uninformative gamma. In addition, it is assumed that the prior distribution of the random effects $(b_{1i},b_{2i},b_{3i},b_{4i},b_{5i},b_{6i})$ is a multivariate normal with mean vector 0 and covariance matrix Σ and that the prior distribution of Σ is an uninformative Wishart.

The code in **BUGS CODE 7.7** closely follows the notation used in formulas (7.45) through (7.50) and the analysis is executed with 150,000 observations, with a burn in of 5,000 and a refresh of 100.

BUGS CODE 7.7

```
        model;
{
# prior distribution for beta

for(i in 1:6){beta[i]~dnorm(0,.0001)}

# first segment

for(i in 1:N1){for(j in 1:M1)
{Y1[i,j]~dnorm(mu1[i,j],tau)}}
for(i in 1:N1){for(j in 1:M1)
{mu1[i,j]<- beta[1]+beta[2]*time1[j]+
b[i,1]+b[i,2]*time1[j]}}
```

```
# second segment
for(i in 1:N2){for (j in 1:M2)
{Y2[i,j]~dnorm(mu2[i,j],tau)}}
for(i in 1:N2){for (j in 1:M2)
{mu2[i,j]<-beta[3]+beta[4]*time2[j]+
b[i,3]+b[i,4]*time2[j]}}

# third segment
for(i in 1:N3){for (j in 1:M3)
{Y3[i,j]~dnorm(mu3[i,j],tau)}}
for (i in 1:N3){for (j in 1:M3){mu3[i,j]<- beta[5]
+beta[6]*time3[j]+b[i,5]+b[i,6]*time3[j]}}

# prior distribution of random effects
for(i in 1:N1){b[i,1:6]~dmnorm(vu[],Omega[,])}
for (
i in 1:6){vu[i]<-0}
Omega[1:6,1:6]~dwish(R[,],6)
Sigma[1:6,1:6]<-inverse(Omega[,])

tau~dgamma(.00001,.00001)

Var<-1/tau

}

list(N1 = 50,N2 = 50,N3 = 50,M1 = 4,M2 = 2,M3 = 3,time1
= c(0,1,4,6),
time2 = c(0,1),time3 = c(1,4,6),

# data for first segment
Y1 = structure(.Data =

c(30.8,26.9,25.8,23.8,
24.7,24.5,22.0,22.5,
28.6,20.8,19.2,18.4,
33.7,31.6,28.5,25.1,19.7,14.9,15.3,14.7,31.1,31.2,29.2,30.1,
19.8,17.5,20.5,27.5,21.4,26.3,19.5,19.0,21.1,20.3,18.4,
20.8,20.6,23.9,19.0,17.0,24.0,16.7,21.7,20.3,37.6,33.7,34.4,
31.4,31.9,27.9,27.3,34.2,26.2,26.8,25.3,24.8,20.5,21.1,
17.4,21.1,33.3,26.2,34.0,28.2,27.9,21.6,23.6,27.7,24.7,21.2,
22.9,21.9,28.8,26.4,23.8,22.0,32.0,30.2,30.2,27.5,21.8,19.3,
16.4,17.6,24.9,20.9,22.2,19.8,19.8,18.9,18.9,15.5,35.4,
30.4,26.5,28.1,25.3,23.9,22.2,27.2,20.3,21.0,16.7,13.5,
20.4,17.2,15.9,17.7,24.1,20.1,17.9,18.7,28.5,32.6,27.5,
22.8,26.6,22.4,21.8,21.0,20.5,17.5,19.6,18.4,25.2,25.1,
23.4,22.2,34.7,39.5,38.6,43.3,30.3,29.4,33.1,28.4,26.6,
25.3,25.1,27.9,20.7,19.3,21.9,21.8,28.9,28.9,32.8,31.8,
27.2,28.5,35.0,30.5,22.4,22.0,19.1,18.7,32.5,25.1,27.8,
27.3,24.9,23.6,21.2,21.1,24.6,25.0,21.7,23.9,23.1,20.9,
```

```
21.7,19.9,25.8,21.9,23.6,24.8,30.0,27.6,24.0,23.7,20.0,
22.7,21.2,20.5,38.1,40.8,38.0,32.7,25.1,28.1,27.5,24.8,
22.1,21.1,21.5,20.6,25.4,24.3,22.7,20.1),.Dim = c(50,4)),
# data for second segment

Y2 = structure(.Data =

c(26.5,14.8,25.8,23.0,20.4,2.8,20.4,5.4,24.8,23.1,27.9,6.3,
35.3,25.5,28.6,15.8,29.6,15.8,
21.5,6.5,21.8,12.0,23.0,4.2,22.2,11.5,25.0,3.9,26.0,21.4,
19.7,13.2,29.6,17.5,24.4,16.4,33.7,14.9,26.7,6.4,26.8,20.4,
20.2,10.6,20.2,17.5,24.5,10.0,27.1,14.9,34.7,39.0,24.5,5.1,
27.7,4.0,24.3,24.3,36.6,23.3,34.0,10.7,32.6,19.0,29.2,9.2,
26.4,15.3,21.8,10.6,21.1,5.6,
22.1,21.0,28.9,12.5,19.8,11.6,23.5,7.9,29.1,16.8,30.3,3.5,
30.6,28.2,22.4,7.1,31.2,10.8,
31.4,3.9,41.1,15.1,29.4,22.1,21.9,7.6,20.7,8.1),.Dim =
c(50,2)),
# data for third segment

Y3 = structure(.Data =

c(14.8,19.5,21.0,23.0,19.1,23.2,2.8,3.2,9.4,5.4,4.5,11.9,
23.1,24.6,30.9,6.3,18.5,16.3,25.5,26.3,30.3,15.8,22.9,25.9,
15.8,23.7,23.4,6.5,7.1,16.0,12.0,16.8,19.2,4.2,4.0,16.2,
11.5,9.5,14.5,3.9,12.8,12.7,21.4,21.0,22.4,13.2,14.6,11.6,
17.5,21.0,24.2,16.4,11.6,16.6,14.9,14.5,63.9,6.4,5.1,15.1,
20.4,19.3,23.8,10.6,9.0,16.0,17.5,17.4,18.6,10.0,15.6,15.2,
14.9,18.1,21.3,39.0,28.8,34.7,5.1,8.2,23.6,4.0,4.2,11.7,
24.3,18.4,27.8,23.3,40.4,39.3,10.7,12.6,21.2,19.0,16.3,18.6,
9.2,8.3,18.4,15.3,24.6,32.4,10.6,14.4,18.7,5.6,7.3,12.3,
21.0,8.6,24.6,12.5,16.7,22.2,11.6,13.0,23.1,7.9,12.4,18.9,
16.8,15.1,18.8,3.5,3.0,11.5,28.2,27.0,25.5,7.1,17.2,18.7,
10.8,19.8,22.2,3.9,7.0,17.8,15.1,10.9,27.1,22.1,25.3,4.1,
7.6,10.8,13.0,8.1,25.7,12.3),.Dim = c(50,3)),

# un informative prior distribution for the precision
matrix
R = structure(.Data = c(1,0,0,0,0,0,
                        0,1,0,0,0,0,
                        0,0,1,0,0,0,
                        0,0,0,1,0,0,
                        0,0,0,0,1,0,
                        0,0,0,0,0,1

),.Dim = c(6,6)))

# initial values

list(beta = c(0,0,0,0,0,0),tau = 1)
```

TABLE 7.10

Posterior Analysis for the Succimer Study with Correlated Random Effects

Parameter	Mean	SD	Error	2½	Median	97½
σ_1^2	19.06	5.408	0.04072	10.57	18.36	31.67
σ_{12}	0.3973	0.5387	0.005073	−0.7207	0.412	1.421
σ_{13}	1.268	3.133	0.06295	−4.757	1.187	7.743
σ_{14}	−10.31	5.291	0.09533	−22.03	−9.842	−1.159
σ_{15}	−11.39	6.75	0.05683	−26.05	−10.92	0.747
σ_{16}	1.119	1.177	0.01136	−1.108	1.073	3.593
σ_2^2	0.1828	0.07327	0.000592	0.07954	0.1697	0.3601
σ_{23}	0.08211	0.357	0.004105	−0.6179	0.07548	0.8162
σ_{24}	0.04555	0.6764	0.006836	−1.286	0.03495	1.423
σ_{25}	0.1928	1.001	0.01014	−1.774	0.1836	2.217
σ_{26}	−0.09248	0.1627	0.001641	−0.4315	−0.08691	0.2176
σ_3^2	7.719	3.284	0.07526	2.769	7.234	15.45
σ_{34}	8.66	3.652	0.0641	1.591	8.559	16.23
σ_{35}	15.51	6.096	0.1383	5.083	14.97	29.17
σ_{36}	0.08056	0.9712	0.02021	−1.755	0.03379	2.14
σ_4^2	28.18	9.9	0.2535	12.65	26.88	51.09
σ_{45}	40.13	11.09	0.174	22.47	38.73	65.66
σ_{46}	−3.279	1.649	0.02305	−7.045	−3.09	−0.598
σ_5^2	62.16	16.32	0.1728	36.42	60.04	100
σ_{56}	−5.04	2.406	0.02791	−10.55	−4.766	−1.123
σ_6^2	1.564	0.5325	0.005946	0.7187	1.498	2.791
σ^2	11.92	1.102	0.009855	9.955	11.85	14.27
β_1	25.69	0.7216	0.007704	24.27	25.69	27.12
β_2	−0.3719	0.1184	0.000748	−0.6048	−0.3714	−0.1398
β_3	26.54	0.63	0.005864	25.3	26.54	27.77
β_4	−13.03	1.011	0.01198	−15.02	−13.03	−11.04
β_5	11.5	1.25	0.02001	9.031	11.5	13.96
β_6	1.385	0.2233	0.00272	0.9471	1.385	1.826

The Bayesian analysis reported in Table 7.10 consists of the posterior characteristics (mean, standard deviation, MCMC error, the 2½ percentile, the median, and the 97½ percentile). Their important parameters are the regression coefficients, the variance of the measurement error, and the entries of the variance–covariance matrix of the random effects.

A plot of the posterior density of a model parameter adds additional information, thus consider σ_1^2, the variance of the random intercept of placebo subjects (Figure 7.7).

As before, the covariance and variance function will be estimated for the succimer study, thus consider the second segment for the treatment group

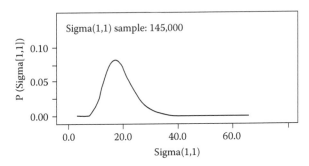

FIGURE 7.7
Posterior density of σ^2, the variance of measurement error.

portrayed by formula (7.50) over the time points 1, 4, and 6. Based on formula (7.50), it can be demonstrated that

$$\mathrm{Cov}\left(Y_{3ij}, Y_{3ik}\right) = g_{55} + g_{56}\left(t_{ij} + t_{ik}\right) + g_{66}t_{ij}t_{ik} \qquad (7.51)$$

where

$t_{ij} \neq t_{ik}, t_{ij} = 1,4,6$
$g_{55} = \mathrm{Var}\left(b_{5i}\right)$
$g_{56} = \mathrm{Cov}\left(b_{5i}, b_{6i}\right)$
$g_{66} = \mathrm{Var}\left(b_{6i}\right)$

Based on the estimates shown in Table 7.10 and formula (7.51), the covariance between the lead level at time 1 and the lead level at time 6 is

$$60 - 4.766(1+6) + (1.498 \times 1 \times 6) = 60 - 33.362 + 8.988 = 65.62$$

Now the variance function for the second segment of the treatment group is

$$\mathrm{Var}\left(Y_{3ij}\right) = g_{55} + t_{ij}^2 g_{66} + 2t_{ij}g_{56} + \sigma^2 \qquad (7.52)$$

Thus, the estimated variance at time 1 is

$$60 + 1.498 - 9.532 + 11.85 = 63.816$$

and the estimated variance at time 6 is

$$60 + 36(1.498) - 12(4.766) + 11.85 = 68.616$$

Consequently, the correlation is

$$\rho\left(Y_{3i1}, Y_{3i6}\right) = \frac{65.63}{\mathrm{sqrt}[63.816 \times 68.616]} = 0.9916$$

7.7 Diagnostic Procedures for Repeated Measures

How does one investigate the goodness of fit for repeated measures models? This last section of this chapter introduces methods to measure how well the model fits the data. Informal graphical techniques have been used to assess the goodness of fit by plotting the predicted versus the observed repeated measures. This approach is somewhat subjective, thus another approach utilized by Fitzmaurice et al.[1] will be introduced. It consists of transformed residuals that are uncorrelated and have constant variance and then plotted against various covariates in the model. The transformed residuals are based on the Cholesky decomposition, which in turn is founded on the estimated variance covariance matrix of the measurement error of the model. This section consists of two parts for a repeated measures model: (1) the model with only fixed effects and (2) the generalized linear model using both fixed and random effects.

7.7.1 Transformed Residuals for Fixed Effects Model

Consider the response Y_i for subject i

$$Y_i = X_i\beta + e_i$$

where:
  Y_i is an n by 1 vector of responses
  X_i is an n by p matrix of p independent variables
  β is a p by 1 unknown parameter vector
  e_i is an n by 1 vector of measurement errors

Note there are n time points for each subject.
    The variance–covariance matrix of the measurement vector e_i is the same as that for the vector of n responses Y_i, that is, the dispersion matrix is

$$D(e_i) = D(Y_i - X_i\beta) = \Sigma_i \tag{7.53}$$

As before, it is assumed that the responses are normally distributed and that the N responses Y_i ($i = 1,2,\ldots,N$) are independent and that the n occasions are the same for all subjects.
    The transformed residuals are defined by the Cholesky decomposition explained in Golub et al.[5], where

$$\Sigma_i = L_i L_i' \tag{7.54}$$

and $L_i = (L[i,j])$ is a lower triangular matrix. Now the n transformed residuals are defined by

$$TR_i = L_i R_i \tag{7.55}$$

where R is the n by 1 vector of residuals and

$$R_i = Y_i - \tilde{Y}_i \tag{7.56}$$

and $\tilde{Y}_i$ is the n by 1 vector of predicted responses for the ith subject. For the time being it will be assumed that L_i is the same for all subjects, thus the subject subscript i can be suppressed. In order to illustrate the construction of the transformed residuals, consider the hip-replacement study where the main response is hematocrit with a quadratic response.

$$Y_{ij} = \beta_1 + \beta_2 t_{ij} + \beta_3 t_{ij}^2 + \varepsilon_{ij} \tag{7.57}$$

for subject i at time t_{ij}. Recall that for this study, there were two groups, males and females, but it was determined that the two groups have mean profile given by (7.57). There are four time points 1, 2, 3, and 4, thus the Cholesky decomposition is defined by the matrix L_i given by Table 7.11.

Note that $L_i L_i'$ is the 4 by 4 symmetric matrix (Table 7.12).

Suppose the entries of Σ_i are denoted by (σ_{ij}), then based on the relationship (7.54), it can be shown that the entries of the matrix L_i are given by

$$L_{11} = \sqrt{\sigma_{11}} \tag{7.58}$$

$$L_{21} = \frac{\sigma_{21}}{L_{11}} \tag{7.59}$$

$$L_{31} = \frac{\sigma_{13}}{L_{11}} \tag{7.60}$$

TABLE 7.11

The Matrix L_i

L_{11}	0	0	0
L_{21}	L_{22}	0	0
L_{31}	L_{32}	L_{33}	0
L_{41}	L_{42}	L_{43}	L_{44}

TABLE 7.12

The Matrix $L_i L_i'$

L_{11}^2	$L_{11}L_{21}$	$L_{11}L_{31}$	$L_{11}L_{41}$
$L_{21}L_{11}$	$L_{21}^2 + L_{22}^2$	$L_{21}L_{31} + L_{22}L_{32}$	$L_{21}L_{41} + L_{22}L_{42}$
$L_{31}L_{11}$	$L_{31}L_{21} + L_{32}L_{22}$	$L_{31}^2 + L_{32}^2 + L_{33}^2$	$L_{31}L_{41} + L_{32}L_{42} + L_{33}L_{43}$
$L_{11}L_{41}$	$L_{21}L_{41} + L_{22}L_{42}$	$L_{31}L_{41} + L_{32}L_{42} + L_{33}L_{43}$	$L_{41}^2 + L_{42}^2 + L_{43}^2 + L_{44}^2$

$$L_{41} = \frac{\sigma_{14}}{L_{11}} \tag{7.61}$$

$$L_{22} = \sqrt{\sigma_{22} - L_{21}^2} \tag{7.62}$$

$$L_{32} = \frac{\sigma_{23} - L_{21}L_{31}}{L_{22}} \tag{7.63}$$

$$L_{42} = \frac{\sigma_{24} - L_{21}L_{41}}{L_{22}} \tag{7.64}$$

$$L_{33} = \sqrt{\sigma_{33} - L_{31}^2 - L_{32}^2} \tag{7.65}$$

$$L_{43} = \frac{\sigma_{34} - L_{31}L_{41} - L_{32}L_{42}}{L_{33}} \tag{7.66}$$

and

$$L_{44} = \sqrt{\sigma_{44} - L_{41}^2 - L_{42}^2 - L_{43}^2} \tag{7.67}$$

The other entries above the main diagonal are 0. Based on formulas (7.54) through (7.67), the transformed residuals are estimated using **BUGS CODE 7.8**. Also performed is the Bayesian analysis for the hip-replacement study, which provides estimates of the coefficients $\beta_1, \beta_2, \beta_3$ for the mean profile, the estimated variance covariance matrix Σ_i, the entries of the matrix L_i, and the vector of transformed residuals TR.

The Bayesian analysis is executed with 150,000 observations, with a burn in of 5,000 and a refresh of 100.

BUGS CODE 7.8

```
        model;
        {
              beta1 ~ dnorm(0.0, 0.001)
              beta2 ~ dnorm(0.0, 0.001)
beta3 ~ dnorm(0.0,.0001)

for(i in 1:N1){Y[i,1:M1]~dmnorm(mu[],Omega[,])}
for(i in 1:N1){Z[i,1:M1]~dmnorm(mu[],Omega[,])}

for(i in 1:N1){for(j in 1:M1){Res[i,j]<-Y[i,j]-Z[i,j]}}

for (j in 1:M1){mu[j]<-beta1+beta2*time[j]+beta3*time[j]*
time[j]}
```

```
Omega[1:M1,1:M1]~dwish(R[,],4)
Sigma[1:M1,1:M1]<-inverse(Omega[,])

L[1,1]<-sqrt(Sigma[1,1])
L[2,1]<-Sigma[1,2]/L[1,1]
L[3,1]<-Sigma[1,3]/L[1,1]
L[4,1]<-Sigma[1,4]/L[1,1]
L[2,2]<-sqrt(Sigma[2,2]-L[2,1]*L[2,1])
L[3,2]<-(Sigma[2,3]-L[2,1]*L[3,1])/L[2,2]
L[4,2]<-(Sigma[2,4]-L[2,1]*L[4,1])/L[2,2]
L[3,3]<-sqrt(Sigma[3,3]-L[3,1]*L[3,1]-L[3,2]*L[3,2])
L[4,3]<-(Sigma[3,4]-L[3,1]*L[4,1]-L[3,2]*L[4,2])/L[3,3]
L[4,4]<-sqrt(Sigma[4,4]-L[4,1]*L[4,1]-L[4,2]*L[4,2]-
L[4,3]*L[4,3])
L[1,2]<-0
L[1,3]<-0
L[1,4]<-0
L[2,3]<-0
L[2,4]<-0
L[3,4]<-0
T[1:M1,1:M1]<-inverse(L[1:M1,1:M1])

for(i in 1:N1){for(j in 1:M1){TR[i,j]<-Res[i,j]*T[j,j]}}

}

list(N1 = 30,M1 = 4,
Y = structure(.Data =
c(47.1,31.1,NA,32.8,
44.1,31.5,NA,37.0,
39.7,33.7,NA,24.5,
43.3,18.4,NA,36.6,
37.4,32.3,NA,29.1,
45.7,35.5,NA,39.8,
44.9,34.1,NA,32.1,
42.9,32.1,NA,NA,
46.1,28.8,NA,37.8,
42.1,34.4,34.0,36.1,
38.3,29.4,32.9,30.5,
43.0,33.7,34.1,36.7,
37.8,26.6,26.7,30.6,
37.3,26.5,NA,38.5,
NA,28.0,NA,33.9,
27.0,32.5,NA,32.0,
38.4,32.3,NA,37.9,
38.8,32.6,NA,26.9,
44.7,32.2,NA,34.2,
38.0,27.1,NA,37.9,
```

```
34.0,23.2,NA,26.0,
44.8,37.2,NA,29.7,
46.0,29.1,NA,26.7,
41.9,32.0,37.1,37.6,
38.0,31.7,38.4,35.7,
42.2,34.0,32.9,33.3,
39.7,33.5,26.6,32.7,
37.5,28.2,28.8,30.3,
34.6,31.0,30.1,28.7,

35.5,24.7,28.1,29.8),.Dim = c(30,4)),
time = c(1,2,3,4),
R = structure(.Data = c(1,0,0,0,
                        0,1,0,0,
                        0,0,1,0,
                        0,0,0,1),.Dim = c(4,4)))
# initial values
list(beta1 = 0,beta2 = 0,beta3 = 0)
```

The posterior distribution of the nonzero entries of the L_i matrix is shown in Table 7.13. Recall that the entries of this matrix are determined by the relationship $\Sigma_i = L_i L_i'$, and for a 4 by 4 matrix, Equations 7.58 through 7.67 determine the entries $L[i,j]$ of Table 7.13. Also shown are the estimated coefficient of the quadratic mean profile for hematocrit, which is portrayed in Figure 7.8.

Based on the L_i entries of Table 7.13, the posterior distribution of the transformed residuals is computed by $TR_i = L_i R_i$. What are the statistical properties of the transformed residuals? To determine the correlation of the transformed residuals, consider **BUGS CODE 7.9**, which when executed will provide the posterior distribution of the first-order correlation of the transformed values.

There are 99 transformed residuals assumed to follow a first-order autocorrelation model

$$TR[j] = \rho TR[j-1] + \varepsilon[j] \tag{7.68}$$

where:
$j = 2,3,\ldots,99$
ρ is the first-order autocorrelation
the independent $\varepsilon[j]$ follow a normal distribution with mean 0 and variance σ^2

Some 75,000 observations are generated for the simulation, with a burn in of 5,000 and a refresh of 100.

Note the purpose of transforming the residuals is to use them for diagnostic purposes for the goodness of fit of the model, and they should have zero correlation and a constant variance. The posterior analysis of Table 7.14

TABLE 7.13

Posterior Analysis for L_i, Σ_i, and Mean Profile

Parameter	Mean	SD	Error	2½	Median	97½
$L[1,1]$	4.554	0.6148	0.003519	3.536	4.489	5.935
$L[2,1]$	1.014	0.7589	0.005318	−0.4305	0.9955	2.571
$L[2,2]$	3.868	0.5177	0.005217	3.016	3.812	5.035
$L[3,1]$	−2.908	2.095	0.04752	−7.301	−2.825	1.019
$L[4,1]$	1.223	0.8097	0.00578	−0.3232	1.203	2.889
$L[4,2]$	−0.3089	0.7609	0.007353	−1.836	−0.3041	1.182
$L[4,3]$	3.75	0.5868	0.003689	2.741	3.708	5.027
$L[4,4]$	1.336	0.3538	0.004788	0.8541	1.271	2.201
σ_{11}	21.11	5.891	0.03393	12.5	20.15	35.22
σ_{12}	4.071	3.789	0.02449	−1.971	4.405	13.15
σ_{13}	−13.49	10.51	0.2245	−37.02	−12.51	4.609
σ_{14}	5.66	4.076	0.02789	−1.458	5.308	14.81
σ_{22}	16.83	4.753	0.0394	9.984	16.04	28.33
σ_{23}	−0.9225	8.806	0.1133	−17.35	−0.8515	15.06
σ_{24}	−0.0194	0.503	0.03181	−7.129	−0.005787	6.985
σ_{33}	67.87	32.31	0.6246	26.46	61.13	148.4
σ_{34}	24.95	10.18	0.1038	9.331	23.52	49.09
σ_{44}	19.14	5.478	0.03224	11.24	18.22	32.42
β_1	56.66	2.51	0.1316	51.69	56.7	61.54
β_2	−19.31	2.339	0.1269	−24.0	−19.34	−14.75
β_3	3.298	0.4508	0.02446	2.438	3.301	4.207

reveals a small correlation estimated as 0.159 with the posterior mean, and the variance of the transformed residuals is estimated as 0.439 using the posterior median. Now consider a plot of the transformed residuals versus the ages of the 30 subjects of the hip-replacement study. If the model provides an adequate goodness of fit to the hematocrit data, there should be no discernable pattern, as portrayed in Figure 7.9.

A lowess plot of the transformed residuals reveals no apparent pattern with respect to the mean, and the range of the residuals is apparently constant over age.

7.7.2 Transformed Residuals for the General Mixed Linear Model

Consider the hip-replacement study where the hematocrit Y_{ij} for subject i at time t_{ij} is

$$Y_{ij} = \beta_1 + \beta_2 t_{ij} + \beta_3 t_{ij}^2 + b_{1i} + b_{2i} t_{ij} + b_{3i} t_{ij}^2 + \varepsilon_{ij} \qquad (7.69)$$

where the three random effects correspond to the individual effects for subject i with (b_{1i}, b_{2i}, b_{3i}) having a multivariate normal distribution with a 3 by

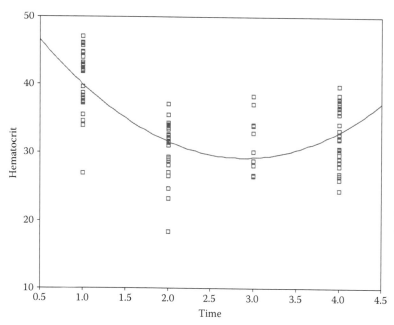

FIGURE 7.8
Mean profile of hematocrit of hip-replacement study.

BUGS CODE 7.9

```
# first-order autocorrelation model
for(i in 2:N1){Y[i]~dnorm(mu[i],tau)}
for(i in 2:N1){mu[i]<-rho*Y[i-1]}

tau~dgamma(.001,.001)
sigmasq<-1/tau
rho~dbeta(1,1)
}
# transformed residuals
list(N1 = 30,Y = c(

1.452,-.039,NA,.505,
.784,.072,NA,3.832,
-.203,.645,NA,-6.094,
.606,-3.367,NA,3.503,
-.708,.275,NA,-2.432,
1.136,1.119,NA,1.165,
.958,.752,NA,-.052,
.514,.232,NA,NA,
1.226,-.642,NA,4.468,
.335,.837,.830,3.123,
-.514,-.482,.665,-1.312,
```

```
.537,.645,.844,3.596,
-.622,-1.214,.270,-1.250,
-.736,-1.244,NA,5.023,
NA,NA,NA,1.377,
-3.038,.326,NA,-.124,
-.488,.281,NA,4.538,
-.400,.357,NA,-4.167,
.910,.253,NA,1.609,
-.581,-1.087,NA,4.554,
-1.473,-2.111,NA,-4.889,
.941,1.564,NA,-1.942,
1.205,-.563,NA,-4.343,
.290,.197,1.291,4.229,
-.576,.122,1.488,2.807,
.355,.724,.667,.907,
-.202,.591,-.278,.437,
-.696,-.797,.054,-1.478,
-1.336,-.059,.244,-2.734,
-1.136,-1.712,-.058,-1.876))
# initial values
list(tau = 1, rho =.5)
```

TABLE 7.14

Posterior Distribution of the First-Order Correlation of the Transformed Residuals

Parameter	Mean	SD	Error	2½	Median	97½
ρ	0.159	0.1286	0.000748	0.005371	0.1284	0.4759
σ^2	4.68	1.59	0.007847	2.518	4.379	8.587

1 mean vector 0 and unknown variance–covariance matrix Σ_i. Note that the time points are at 1, 2, 3, and 4 weeks and that the individuals have a common covariance matrix Σ_i. In the previous section, the transformed residuals were based on the variance–covariance matrix of the residuals for model (7.58). The same approach is taken for the generalized linear model (7.69), where the residuals are now given by

$$R_{ij} = Y_{ij} - Y_{ij}^{\sim} \tag{7.70}$$

where:
$Y_{ij}^{\sim}$ is the predicted value of Y_{ij}

The transformed residuals are based on the variance covariance matrix of the residuals (7.70). Let

$$Tigma = L_i L_i' \tag{7.71}$$

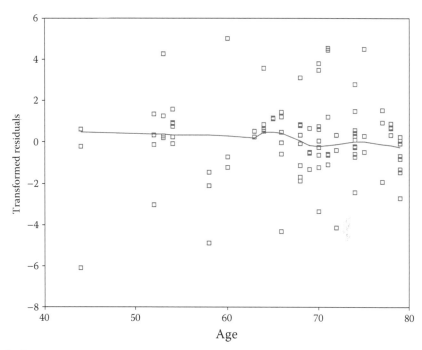

FIGURE 7.9
Transformed residuals versus age for hip-replacement study with fixed effects model.

where:

Tigma is the variance–covariance matrix of the residuals (7.70)

This is the Cholesky decomposition of the variance–covariance matrix Tigma, where L_i is a lower triangular 4 by 4 matrix. The entries of L_i are determined by

$$L_{11} = \sqrt{Tig_{11}} \tag{7.72}$$

$$L_{21} = \frac{Tig_{21}}{L_{11}} \tag{7.73}$$

$$L_{31} = \frac{Tig_{13}}{L_{11}} \tag{7.74}$$

$$L_{41} = \frac{Tig_{14}}{L_{11}} \tag{7.75}$$

$$L_{22} = \sqrt{Tig_{22} - L_{21}^2} \tag{7.76}$$

$$L_{32} = \frac{Tig_{23} - L_{21}L_{31}}{L_{22}} \tag{7.77}$$

$$L_{42} = \frac{\text{Tig}_{24} - L_{21}L_{41}}{L_{22}} \tag{7.78}$$

$$L_{33} = \sqrt{\text{Tig}_{33} - L_{31}^2 - L_{32}^2} \tag{7.79}$$

$$L_{43} = \frac{\text{Tig}_{34} - L_{31}L_{41} - L_{32}L_{42}}{L_{33}} \tag{7.80}$$

and

$$L_{44} = \sqrt{\text{Tig}_{44} - L_{41}^2 - L_{42}^2 - L_{43}^2} \tag{7.81}$$

where:

Tig_{ij} is the ijth entry of the variance–covariance matrix Tigma (7.71)

The entries of the Tigma matrix are computed from the residuals that are estimated from the Bayesian analysis, executed by **BUGS CODE 7.10**.

BUGS CODE 7.10

```
model;
    {

for(i in 1:N1){for(j in 1:M1){Y[i,j]~dnorm(mu[i,j],tau)}}
for(i in 1:N1){for(j in 1:M1){Z[i,j]~dnorm(mu[i,j],tau)}}
# these are the residuals for the model
for(i in 1:N1){for(j in 1:M1){RES[i,j]<-Y[i,j]-Z[i,j]}}
for(i in 1:N1){for(j in 1:M1){mu[i,j]<-beta[1]+beta[2]
*age[j]+beta[3]*age[j]*age[j]
+b[i,1]+b[i,2]*age[j]+b[i,3]*age[j]*age[j]}}
# prior distribution for the regression coefficients
   for(i in 1:3){beta[i]~dnorm(0,.0001)}
# prior distribution for the random effects
   for (i in 1:N1){b[i,1:3]~dmnorm(vu[], tau1[,])}
   for (j in 1:3){vu[j]<-0}
# prior Wishart distribution for the precision matrix of
random effects
   tau1[1:3,1:3]~dwish(R[,],3)
# Sigma is variance covariance matrix of random effects
   Sigma[1:3,1:3]<-inverse(tau1[,])

# tau is precision of measurement errors,
tau~dgamma(.0001,.0001)

sigma<-1/tau
# tigma values computed from the residuals RES
```

```
# tigma[1,1] is the variance of the residuals at time
point 1
tigma[1,1]<-(15.2802)
# tigma[1,2] is the covariance between the residuals at
time versus # those at time 2
tigma[1,2]<-(-.160119)
tigma[1,3]<-(-1.0286)
tigma[1,4]<-(1.3889)
tigma[2,2]<-(11.65177)
tigma[2,3]<-(.10265)
tigma[2,4]<-(-3.12843)
tigma[3,3]<-(3.533)
timga[3,4]<- (-.815536)
tigma[4,4]<-(3.188808)


L[1,1]<-sqrt(15.2802)
L[2,1]<-.160119/L[1,1]
L[3,1]<-1.0286/L[1,1]
L[4,1]<-1.3889/L[1,1]
L[2,2]<-sqrt(11.65177-L[2,1]*L[2,1])
L[3,2]<-(.10265-L[2,1]*L[3,1])/L[2,2]
L[4,2]<-(-3.12843-L[2,1]*L[4,1])/L[2,2]
L[3,3]<-sqrt(3.533-L[3,1]*L[3,1]-L[3,2]*L[3,2])
L[4,3]<-(-.815536-L[3,1]*L[4,1]-L[3,2]*L[4,2])/L[3,3]
L[4,4]<-sqrt(3.1888-L[4,1]*L[4,1]-L[4,2]*L[4,2]-
L[4,3]*L[4,3])
L[1,2]<-0
L[1,3]<-0
L[1,4]<-0
L[2,3]<-0
L[2,4]<-0
L[3,4]<-0
T[1:M1,1:M1]<-inverse(L[1:M1,1:M1])

for(i in 1:N1){for(j in 1:M1)
{TRGLM[i,j]<-RES[i,j]*T[j,j]}}


}
list(N1 = 30,M1 = 4,
Y = structure(.Data =
c(47.1,31.1,NA,32.8,
44.1,31.5,NA,37.0,
39.7,33.7,NA,24.5,
43.3,18.4,NA,36.6,
37.4,32.3,NA,29.1,
45.7,35.5,NA,39.8,
44.9,34.1,NA,32.1,
```

```
42.9,32.1,NA,NA,
46.1,28.8,NA,37.8,
42.1,34.4,34.0,36.1,
38.3,29.4,32.9,30.5,
43.0,33.7,34.1,36.7,
37.8,26.6,26.7,30.6,
37.3,26.5,NA,38.5,
NA,28.0,NA,33.9,
27.0,32.5,NA,32.0,
38.4,32.3,NA,37.9,
38.8,32.6,NA,26.9,
44.7,32.2,NA,34.2,
38.0,27.1,NA,37.9,
34.0,23.2,NA,26.0,
44.8,37.2,NA,29.7,
46.0,29.1,NA,26.7,
41.9,32.0,37.1,37.6,
38.0,31.7,38.4,35.7,
42.2,34.0,32.9,33.3,
39.7,33.5,26.6,32.7,
37.5,28.2,28.8,30.3,
34.6,31.0,30.1,28.7,
35.5,24.7,28.1,29.8),.Dim = c(30,4)),
R = structure(.Data = c(1,0,0,0,1,0,0,0,1),.Dim = c(3,3)),

age = c(1,2,3,4))

# initial values
list(beta = c(0,0,0),
tau = 1)
```

The Bayesian analysis for the general linear mixed model for the hip-replacement study is executed with **BUGS CODE 7.10** using 150,000 observations for the simulation, with a burn in of 5,000 and a refresh of 100. Our objective is to estimate the coefficient of the mean profile, the variance–covariance matrix of the three random effects, the variance about regression curve, and the transformed residuals (Table 7.15).

Of course the transformed residuals are used for diagnostic purposes to assess the goodness of fit of the mixed model. Now how well do the predicted values forecast the actual hematocrit values? A scatter plot of predicted and observed hematocrit values is reported in Figure 7.10.

The plot implies that the mixed linear model is a good fit to the hematocrit data, and furthermore it can be verified that the Pearson correlation between the two is 0.889, which gives us additional information about the model's goodness of fit. How does this compare to the plot of the predicted versus observed values of hematocrit assuming a fixed model with the quadratic mean profile model (7.57)?

TABLE 7.15

Posterior Analysis for Hip-Replacement Study with Generalized Mixed Model

Parameter	Mean	SD	Error	2½	Median	97½
σ_{11}	2.063	4.594	0.07346	0.1504	0.8517	11.7
σ_{12}	−0.4115	3.138	0.04905	−5.79	0.02418	1.803
σ_{13}	0.001526	0.5986	0.009015	−0.683	−0.03449	1.025
σ_{22}	1.575	2.983	0.04396	0.1469	0.7654	7.931
σ_{23}	−0.3431	0.6296	0.0090	−1.786	−0.1643	0.02657
σ_{33}	0.1852	0.1498	0.001967	0.0632	0.1446	0.5529
β_1	55.14	2.26	0.08015	50.57	55.16	59.47
β_2	−18.25	2.131	0.07925	−22.37	−18.27	−13.93
β_3	3.197	0.4196	0.01541	2.346	3.2	4.006
σ^2	14.82	2.789	0.01973	10.08	14.57	21.02

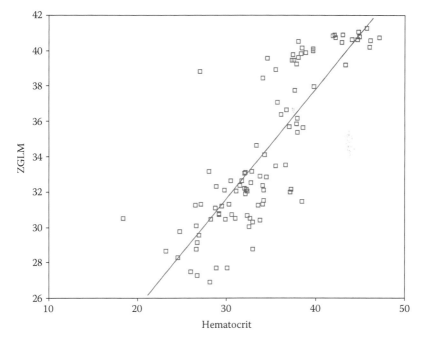

FIGURE 7.10
Predicted versus observed hematocrit values for hip-replacement study with mixed linear model.

When Figure 7.11 is compared to Figure 7.10, it appears that the mixed model (7.69) with fixed and random effects is a better fit to the data.

What do the transformed values tell us about the goodness of fit for the mixed linear model? First, consider a plot expressed in Figure 7.12 of the transformed residuals versus age. It is apparent from the lowess plot that there is no apparent association between the transformed residuals and age.

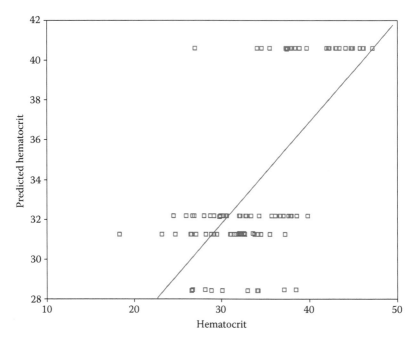

FIGURE 7.11
Predicted versus observed hematocrit values assuming a fixed quadratic mean profile.

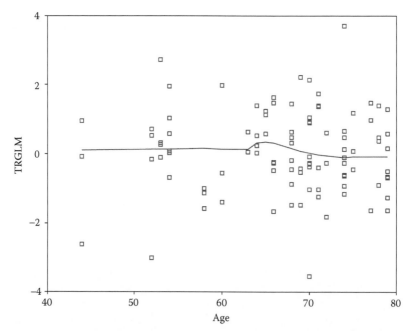

FIGURE 7.12
Transformed residuals versus age for hip-replacement study with mixed model.

7.8 Comments and Conclusions

This chapter begins with the definition of the mean profile of the linear mixed model, followed by the full model with random effects. Then follows the derivation of the conditional (given the random effects) and marginal mean of repeated measures and the conditional (given the random effects) and marginal dispersion matrix of the vector of repeated responses. It is important to know the interpretation of the model terms including the fixed effects, the random effects, the design matrices of the fixed and random effects, and the measurement error.

The interpretation is represented graphically with two random subjects for a model with a linear regression for the fixed effects and a random subject intercept effect. Using graphs, the interpretation of the random intercepts, the mean profile, and the measurement errors are easily displayed and understood.

The mixed model was introduced in Chapter 6 where compound symmetry was the pattern of the variance–covariance matrix of the observations. Recall that compound symmetry implies a constant correlation between responses over all time points and a constant variance. The introduction of random effects provides for a more complex covariance pattern.

Section 7.3 introduces the notation used for the linear mixed model and defines the variance and covariance functions (of the repeated measures), which are derived from the specification of the model. Section 7.4 discusses the pattern of the covariance matrix of the mixed linear model, while Section 7.5 reviews the Bayesian approach to making inferences about the fixed effects, the random effects, and the prediction of future observations.

Section 7.6 is about examples that illustrate the techniques for analyzing general mixed models with fixed and random effects. The first example is a study by Van der Lende et al.[2] based on a study of air pollution where the repeated measure is a measure of chronic obstructive disease observed a FEV1 (forced expiratory volume) via spirometry. The fixed effects are a linear regression and the random effects are a random slope and intercept for each subject. The two random effects are given a prior a bivariate normal distribution with zero mean vector but an unknown dispersion matrix, which is given a prior an uninformative Wishart distribution. The second example is concerned with hematocrit values in a hip-replacement study reported by Crowder and Hand[3] with two groups, males and females. A mean profile with a quadratic regression for the two groups is used along with three random effects for each group, the random subject effects consisting of a constant, a linear effect, and a quadratic subject effect. The three effects for each group are given a multivariate normal with a zero mean vector and an unknown dispersion matrix, given an uninformative Wishart distribution.

The third study was analyzed by Rhoads[4] is a clinical trial with two groups with blood lead levels as the repeated measure. Succimer is a chelating agent that hopefully will lower the blood lead levels, and a novel part of the analysis is the use of a spline to fit the mean profile of the lead levels. The last example is a hospital dietary regime study where the main response is plasma ascorbic acid. The mean profile of the two groups is represented as a spline, and the time points are at 1, 2, 6, 10, 14, 15, and 16 weeks.

In all examples, the covariance between the random effects is 0, thus the analyses are repeated assuming that the random effects are independent.

This chapter is concluded with a section that emphasizes diagnostic procedure to assess the goodness of fit of the models. There are two sections: (1) a model assuming no random effects and (2) a model including both random and fixed effects. Diagnostic procedures are graphical and consist of plotting the observed versus the predicted values of the repeated measures. A second graphical procedure is to calculate the transformed residuals based on the Cholesky[5] decomposition and then to plot the transformed residuals versus various covariates in the model. The hip-replacement study was used to illustrate the methodology for both scenarios, and it was demonstrated that the mixed model containing both random and fixed effects provided a better fit to the hematocrit data. Chapter 8 of Fitzmaurice, Laird, and Ware should be consulted for additional information on diagnostic procedures.

Exercises

1. The general mixed model for subject i is expressed as

$$Y_i = X_i\beta + Z_i b_i + \varepsilon_i$$

where:

Y_i is the n_i by 1 vector of repeated measures over n_i time points

X_i is an n_i by p known matrix for the fixed effects

β is a p by 1 vector of unknown fixed effects

Z_i is an n_i by q design matrix for the random effects

b_i is a q by 1 vector of random effects

ε_i is an n_i by 1 vector of measurement errors

In addition, the vector b_i is assumed to have a multivariate normal distribution with mean vector 0 and unknown dispersion matrix Σ_i, and b_i are independent and independent variables from the ε_i, and last ε_i have a normal distribution with mean vector 0 and common unknown variance σ^2.

Show that the dispersion matrix of Y_i is given by

$$D(Y_i) = Z_i D(b_i) Z_i' + \sigma^2 I_{n_i}$$

2. Consider

$$Y_{ij} = X_{ij}\beta + b_i + \varepsilon_{ij}$$

where:

Y_{ij} is the response of subject i at time t_{ij}

X_{ij} is a 1 by p vector of covariate values

β is a p by 1 vector of fixed effects

b_i is a normal random variable with mean vector 0 and variance σ

ε_{ij} is independent normal random variables with mean vector 0 and variance σ^2

a. Show that $\mathrm{Cov}(Y_{ij}, Y_{ik}) = \sigma_b^2$.

b. Show that $\mathrm{Var}(Y_{ij}) = \sigma^2 + \sigma_b^2$.

c. What is the correlation of the response between occasion j and occasion k?

3. Based on the model

$$Y_{ij} = \beta_1 + \beta_2 t_{ij} + b_{1i} + b_{2i}t_{ij} + \varepsilon_{ij}$$

where:

the vector (b_{1i}, b_{2i}) is distributed as a bivariate normal distribution with mean vector 0 and unknown dispersion matrix G

a. Using **BUGS CODE 7.1**, verify the posterior analysis of Table 7.4.

b. Estimate with the posterior median $\mathrm{Cov}(b_{1i}, b_{2i})$.

c. What is the posterior mean of the variance $\mathrm{Var}(b_{2i})$?

d. Estimate the intercept and slope of the mean profile with the posterior mean.

4. Using **BUGS CODE 7.2**, verify the posterior analysis of Table 7.5.

a. What are the posterior means of β_1 and β_2?

b. What is the posterior median of the variance of the random intercept term?

c. What is the posterior mean of the variance of the random slope?

5. Derive the variance function (7.15).

6. Derive the covariance function (7.17).

7. Based on Figure 7.1, is it reasonable to assume that the mean profile for hematocrit is quadratic for both groups?

8. Based on the mixed model

$$Y_i = X_i\beta + Z_ib_i + \varepsilon_i$$

a. What is the conditional mean of the $n_i \times 1$ vector Y_i given the random effects b_i?

b. What is the marginal mean of the $n_i \times 1$ vector Y_i, averaged over the random effects b_i?

c. What is the marginal dispersion matrix of the $n_i \times 1$ vector Y_i, averaged over the random effects b_i and ε_i?

9. Based on **BUGS CODE 7.3**, verify Table 7.6.

a. What is the estimated mean profile for males?

b. What is the estimated mean profile for females?

c. For females, what is the posterior mean of $\text{Cov}(c_{i1}, c_{i2})$, the covariance between the two random effects c_{i1} and c_{i2}?

d. Based on the 95% credible interval for $\text{Cov}(c_{i1}, c_{i2})$, is this covariance 0?

10. Based on **BUGS CODE 7.4**, verify Table 7.7.

a. Does the mean profile for males change from that reported in Table 7.6?

b. Does the mean profile for females change from that reported in Table 7.6?

c. Based on the posterior median, estimate the random effect b_{3i}, the quadratic effect for males.

d. Based on the posterior median, estimate the random effect c_{3i}, the quadratic effect for females.

e. What is the main difference between the posterior analysis reported in Table 7.7 and that reported in Table 7.6?

11. a. Derive the variance function (7.24) for males.

b. Derive the covariance function (7.25) for males.

c. How does the variance function (7.24) depend on the time points 1, 2, 3, and 4?

d. Based on Figure 7.3, does the mixed model for males and females provide a good fit to the hematocrit data?

12. For the dietary hospital study, the mean profile is a three-segment spline, where each spline is linear. Interpret the coefficient of the mean profile for each linear segment. See (7.33) through (7.35).

13. Based on **BUGS CODE 7.5**, verify the posterior analysis reported in Table 7.8.

a. What is the posterior median of the covariance between the first and fourth random effects, $\text{Cov}(b_{1i}, b_{4i})$?

b. What is the posterior mean of the covariance between the first and fourth random effects, $\text{Cov}(b_{1i}, b_{4i})$?

c. Is the posterior distribution of $\text{Cov}(b_{1i}, b_{4i})$ skewed? Explain the skewness.

d. Using the posterior means, estimate the mean profile for the three-segment linear spline.

14. Based on **BUGS CODE 7.6**, verify the posterior analysis of Table 7.9.

a. Compare the mean profile of the three-segment linear spline of Table 7.5 to that reported in Table 7.8.

b. What is the posterior mean of the two random effects of the first of the three-segment linear spline?

c. Consider the posterior distribution of the six random effects. Which ones are skewed?

d. Based on Table 7.9, are the MCMC errors based on 150,000 observations sufficient to estimate the posterior mean of the six beta coefficients?

15. a. Derive the covariance function $\text{Cov}(Y_{1ij}, Y_{1ik})$, (7.36), for the first segment of the three-segment spline.

b. Derive the variance function $\text{Var}(Y_{1ij})$, (7.37), for the first segment of the three-segment spline.

c. Derive the variance function $\text{Var}(Y_{2ij})$, (7.14), with time points 6, 12, and 14 for the second segment of the three-segment spline.

16. Refer to Figure 7.6 that reports the association between blood lead levels versus time by group. Describe the association between the two variables using a spline for the treatment group. Give the time points for each segment of the spline and the known join points. Equations 7.46 and 7.47 define two linear segments for the treatment group (succimer group). Interpret the four beta coefficients $\beta_1, \beta_2, \beta_3, \beta_4$. What coefficient is the slope for the first segment?

17. a. Refer to Equation 7.49. What do the two random effects b_{3i} and b_{4i} represent?

b. Refer to Equation 7.50. What do b_{5i} and b_{6i} represent?

18. Using **BUGS CODE 7.7**, verify Table 7.10.

a. What is the posterior median of $\sigma_{12} = \text{Cov}(b_{1i}, b_{2i})$?

b. What is the posterior mean of $\sigma_{12} = \text{Cov}(b_{1i}, b_{2i})$?

c. Is the posterior distribution of $\sigma_{12} = \text{Cov}(b_{1i}, b_{2i})$ skewed? Why?

d. What is the 95% credible interval for $\sigma_{12} = \text{Cov}(b_{1i}, b_{2i})$?

e. What is the prior distribution of $(b_{1i}, b_{2i}, b_{3i}, b_{4i}, b_{5i}, b_{6i})$?

f. Why is the posterior distribution of β_4 negative?

19. a. Derive the covariance function $\text{Cov}(Y_{3ij}, Y_{3ik})$, (7.51), over the time points 1, 4, and 6 of the treatment group.

b. Derive the variance function $\text{Var}(Y_{3ij})$, (7.52), for the second segment of the succimer group.

c. What do the 95% credible intervals imply about the covariances $g_{ij} = \sigma_{ij}$ among the six random effects?

20. Consider the mean profile

$$Y_i = X_i\beta + \varepsilon_i$$

for subject i of the general mixed model. Let the dispersion matrix of the residuals be

$$D(\varepsilon_i) = D(Y_i - X_i\beta)$$

$$= \Sigma_i$$

a. Describe the Cholesky[5] decomposition

$$L_iL_i' = \Sigma_i,$$

where:

L_i is a lower triangular matrix with ijth entry $L[i,j]$

b. Express the entries $L[i,j]$ in terms of the entries σ_{ij} of Σ_i. See Tables 7.11 and 7.12.

c. For the hip-replacement study, verify Equations 7.58 through 7.67.

Note that the hip-replacement study has 30 subjects and four time points.

21. Execute **BUGS CODE 7.8** with 150,000 observations for the simulation, with a burn in of 5,000 and a refresh of 100, and verify the posterior analysis of Table 7.13.

a. Describe the statement $L[1,1]$<-sqrt(Sigma[1,1]) of the code.

b. What role does $L[1,1]$ play in the Cholesky decomposition of the variance covariance matrix Σ_i of the residuals?

c. Refer to Table 7.13. There are four time points in the hip-replacement study. What are the 95% credible intervals for the covariances Sigma[i,j]?

d. Interpret the posterior mean of the three betas $\beta_1, \beta_2, \beta_3$.

e. In view of your answers in part c, should the Bayesian analysis be modified?

22. Refer to Equation 7.69, the general mixed linear model for the hip-replacement study. Tigma is the 4 by 4 dispersion matrix of the residuals (7.70).

a. Verify (7.72) through (7.81).

b. Using **BUGS CODE 7.10**, verify Table 7.15.

    c. Refer to Table 7.15. What are the 95% credible intervals for the covariances σ_{ij}?

    d. What is the posterior mean of the quadratic effect β_3?

    e. What does Figure 7.10 imply about the goodness of fit of the mixed linear model?

23. Based on Figures 7.10 and 7.11, does the fixed model (7.57) fit the data better than the mixed linear model (with random effects) (7.68)?

24. Using **BUGS CODE 7.10**, verify Table 7.15, the posterior analysis of the hip-replacement study with the mixed model Z(7.64).

    a. Are the posterior distribution of $\beta_1, \beta_2, \beta_3$ of Table 7.15 different from that reported in Table 7.14?

    b. What is the 95% credible interval of σ_{13}?

8

Repeated Measures for Categorical Data

This chapter introduces students to the basic concepts of the Bayesian approach to the analysis of repeated measures for categorical information.

The chapter will be presented in three parts: (1) the analysis of categorical data using a multinomial distribution for the various patterns displayed by the repeated variable, (2) a Bayesian analysis for categorical data with covariates using a Bayesian version of the generalized estimating equations (GEE), and (3) the generalized linear model with fixed and random effects.

Approach (1) does not involve covariates, while approaches (2) and (3) will accommodate discrete and continuous covariates. The reader is advised to refer to Davis[1] for the conventional treatment of using the multinomial model for categorical data, similar to part (1) and to Fitzmaurice, Laird, and Ware[2] for the conventional approach using GEE, similar to part (2), and using generalized linear model for categorical data, similar to part (3).

Of course, the approach is Bayesian, and for part (1) the posterior distribution of the various patterns of the categorical variables is Dirichlet, which is the conjugate distribution to the multinomial distribution of the pattern counts. For part (2), a Bayesian version of the GEE is proposed with uninformative priors for the model parameters, and last for part (3) the Bayesian approach of using a multivariate normal distribution for the prior distribution of the random effects is proposed.

Additional information about the conventional approach to the analysis of categorical repeated measures is contained in the following references. Hand and Crowder[3] and Crowder and Hand[4] employ for information stressing the conventional analysis of variance, Jones and Kenward[5] emphasize the analysis of binary and categorical data for repeated and cross-over studies, and finally Aerts et al.[6] employ GEE for clustered data with categorical responses.

8.1 Introduction to the Bayesian Analysis with a Dirichlet Posterior Distribution

The following three sections present the Bayesian analysis for categorical data using a multinomial model for the cell frequencies and a posterior distribution for the cell probabilities that is Dirichlet. Noninformative priors will be specified for the cell probabilities, and the main focus of the analysis will be to determine the profile of the probabilities at various time points.

The methodology is illustrated with three examples: (1) the first is a single population where each subject is observed under three treatment conditions, (2) the second example is one population with a binary response over four time points, and (3) the third example features two populations with a binary response, where the goal is to compare the mean profile over the time points. It should be observed that the approach is restricted in the sense that continuous covariates are not accommodated. Continuous covariates will be included using a Bayesian version of GEE and with the generalized linear model for categorical data.

8.1.1 Bayesian Analysis Using the Multinomial Distribution for Categorical Data

The Bayesian posterior analysis involving 46 subjects given three drugs A, B, and C presented in Chapter 3 is based on the Dirichlet distribution for the various possible patterns of the repeated measures, whose counts are assumed to follow a multinomial distribution, but the analysis will not allow continuous covariates. Discrete covariates can be included by stratifying on the levels of the covariate values.

Let X_i denote the number of subjects in the ith category or profile, then $(X_1, X_2.X_3, X_4, X_5, X_6, X_7, X_8)$ has a multinomial distribution with parameters n and $\theta = (\theta_1, \theta_2, \theta_3, \theta_4, \theta_5, \theta_6, \theta_7, \theta_8)$, where θ_i is the probability that a subject will belong to category $i = 1,2,...,8$, the sum of the θs is 1, and n is the total number of observations.

The properties of the multinomial mass function

$$f(x \mid \theta) \propto \prod_{i=1}^{i=8} \theta_i^{x_i} \tag{8.1}$$

where $0 \le x_i \le n, 0 < \theta_i < 1$, are easily found.

Also $\sum_{i=1}^{i=8} x_i = n$ and $\sum_{i=1}^{i=8} \theta_i = 1$. One should be aware that the normalizing constant for the mass function (8.1) is the multinomial coefficient $(n!/x_1!x_2!...x_8!)$.

The properties of the multinomial distribution are well known, for example, the mean of the number of observations in category i is

$$E(X_i) = n\theta_i \tag{8.2}$$

the variance is given by

$$\text{Var}(X_i) = n\theta_i(1 - \theta_i) \tag{8.3}$$

and the covariance is

$$\text{Cov}(X_i, X_j) = -n\theta_i\theta_j \tag{8.4}$$

Note that (8.1) is the conditional distribution of the eight counts corresponding to the eight patterns given θ.

The Bayesian analysis will use the prior distribution that is conjugate to the multinomial distribution (8.1); this implies that the prior density should also belong to the conjugate family, namely the Dirichlet density. Thus, consider the uniform density

$$g(\theta) = 1 \tag{8.5}$$

where $0 < \theta_i < 1$, $\sum_{i=1}^{i=8} \theta_i = 1$.

Then it can be shown that the joint posterior distribution of the θ_i, $i = 1, 2, \ldots, 8$ is Dirichlet with density

$$g(\theta \mid x) \propto \prod_{i=1}^{i=8} \theta_i^{x_i} \tag{8.6}$$

where $0 < \theta_i < 1$, $\sum_{i=1}^{i=8} \theta_i = 1$.

That is, the posterior distribution of θ is Dirichlet with parameter vector $x = (x_1 + 1, x_2 + 1, \ldots, x_8 + 1)$.

On the other hand, assume the prior density is given by

$$g(\theta \mid x) = \prod_{i=1}^{i=8} \theta_i^{x_i - 1} \tag{8.7}$$

where $0 < \theta_i < 1$, $\sum_{i=1}^{i=8} \theta_i = 1$.

Thus, the posterior density of θ is

$$g(\theta \mid x) \propto \prod_{i=1}^{i=8} \theta_i^{x_i - 1} \tag{8.8}$$

where $0 < \theta_i < 1$, $\sum_{i=1}^{i=8} \theta_i = 1$.

Therefore, the posterior distribution of θ is Dirichlet with parameter vector $x = (x_1, x_2, \ldots, x_8)$, and the mean of θ_i is

$$E(\theta_i \mid x) = \frac{x_i}{x_0} \tag{8.9}$$

where

$$x_0 = \sum_{i=1}^{i=8} x_i$$

In addition, the posterior variance is given by

$$\text{Var}(\theta_i \mid x) = \frac{x_i(x_0 - x_i)}{x_0^2(x_0 + 1)} \tag{8.10}$$

and the posterior covariance by

$$\text{Cov}(\theta_i, \theta_j | x) = -\frac{x_i x_j}{x_0^2(x_0+1)} \qquad (8.11)$$

Because of the constraint $\sum_{i=1}^{i=8} \theta_i = 1$, the covariance (8.11) is negative.

Now refer to Table 3.15, which reports the eight patterns for the response to the three drugs. Of interest is estimating the probability that a subject responds to drug A. Referring to the table, one sees that a subject responds favorably to drug A if and only if pattern 1 or 2 or 3 or 4 occurs. In a similar way, the probability a subject responds favorably to drug B is the probability of pattern 1, plus the probability of pattern 2, plus the probability of pattern 5, plus the probability of pattern 6. Last, the probability of pattern 1, plus the probability of pattern 3, plus the probability of pattern 5, plus the probability of pattern 6 is the probability a subject responds favorably to drug C.

Consider the following probabilities:

$$\pi_a = \theta_1 + \theta_2 + \theta_3 + \theta_4 \qquad (8.12)$$

$$\pi_b = \theta_1 + \theta_2 + \theta_5 + \theta_6 \qquad (8.13)$$

$$\pi_c = \theta_1 + \theta_3 + \theta_5 + \theta_7 \qquad (8.14)$$

Then π_a is the probability of a favorable response to drug A.

Of primary interest is the mean profile of the three drugs. In order to investigate the profile, consider

$$d_{ab} = \pi_a - \pi_b \qquad (8.15)$$

$$d_{ab} = \pi_a - \pi_c \qquad (8.16)$$

and

$$d_{bc} = \pi_b - \pi_c \qquad (8.17)$$

In order to perform the Bayesian analysis, one must assign a prior distribution to the eight parameters θ_i ($i = 1,2,...,8$), which have the likelihood function given by (8.6). Consider the vague uninformative prior (8.7), then we know that the posterior distribution of θ is Dirichlet with parameter vector $x = (x_1, x_2,..., x_8)$. The Bayesian analysis will be based on **BUGS CODE 8.1**.

The Bayesian analysis is executed with 65,000 observations for the simulation, with a burn in of 5,000 and a refresh of 100, and the results appear in Table 8.1.

Using the posterior mean, one sees that the estimated probability of a favorable response to drug A is 0.6087 with a 95% credible interval of (0.4657, 0.7423). Note that the sum of the eight posterior means of the θ is indeed 1, as it should be, thus serving as a check on **BUGS CODE 8.1**.

BUGS CODE 8.1

```
model;
{
for(i in 1:8){g[i]~dgamma(a[i],2)}
sg<-sum(g[])
for (i in 1:8){theta[i]<-g[i]/sg}
# the thetas have a Dirichlet distribution
pa<-theta[1]+theta[2]+theta[3]+theta[4]
# pa is the probability of a favorable response to A
pb<-theta[1]+theta[2]+theta[5]+theta[6]
# pb is the probability of a favorable response to B
pc<-theta[1]+theta[3]+theta[5]+theta[7]

# pc is the probability of a favorable response to C
}
# a is the parameter vector of the Dirichlet
list(a = c(6,16,2,4,2,4,6,6))
```

TABLE 8.1

Bayesian Analysis for Drug Study

Parameter	Mean	SD	Error	2½	Median	97½
π_a	0.6087	0.07078	0.0002727	0.4657	0.6102	0.7423
π_b	0.6089	0.7131	0.0003035	0.4645	0.611	0.7429
π_c	0.3478	0.06952	0.0003002	0.2178	0.3458	0.4896
θ_1	0.1304	0.0493	0.0002159	0.0509	0.125	0.2411
θ_2	0.348	0.06592	0.000296	0.2192	0.3456	0.4892
θ_3	0.04325	0.02965	0.0001184	0.005392	0.03668	0.1174
θ_4	0.087	0.04109	0.0001693	0.02499	0.08106	0.183
θ_5	0.04369	0.02979	0.0001228	0.00539	0.0372	0.1182
θ_6	0.08672	0.04106	0.0001638	0.02479	0.0805	0.1829
θ_7	0.1304	0.04925	0.0002005	0.05026	0.125	0.2404
θ_8	0.1304	0.04908	0.0002075	0.0507	0.1251	0.2402
d_{ab}	−0.000154	0.07462	0.00026	−0.1479	0.000073	0.1492
d_{ac}	0.2609	0.1068	0.000437	0.0459	0.2628	0.4637
d_{bc}	0.2611	0.1073	0.000454	0.04399	0.2638	0.4644

Based on the posterior distribution of d_{ab}, d_{ac}, and d_{bc}, it appears that the mean profile of the three drugs is as follows: (1) drugs A and B give the same posterior mean, but both A and B differ from drug C. Note that the 95% credible interval for d_{ac} and d_{bc} is the same, namely (0.0459, 0.4637), and does not contain 0, implying that the probability of a favorable response to drug C is different from the corresponding probability of a favorable response to drugs A and B.

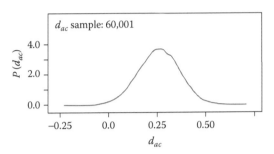

FIGURE 8.1
Posterior density of d_{ac}.

Figure 8.1 displays the posterior density of d_{ac}, the difference in the probability of a favorable response to drug A minus the probability of a favorable response to drug C. The distribution appears to be symmetric as confirmed by Table 8.1, which reports the same mean and median for the posterior distribution of d_{ac}.

Notice from (8.10) and (8.11), the correlation is given by

$$\rho(\theta_i, \theta_j) = -\frac{x_i x_j}{\sqrt{x_i(x_0 - x)x_j(x_0 - x_j)}} \tag{8.18}$$

Thus, in particular the estimated correlation is

$$\tilde{\rho}(\theta_1, \theta_2) = -\frac{(6)(16)}{\sqrt{6(46-6)16(46-16)}} = -0.28$$

where $x_0 = 46$, $x_1 = 6$, and $x_2 = 16$.

Of course, using formula (8.18) and Table 3.15, one may compute estimates of the other correlations.

8.1.2 Categorical Data for Repeated Measures with Multinomial Distribution

Davis (p. 187)[1] presents a conventional weighted least squares approach for a study of wheezing children, which involves a binary response and a repeated measurement factor that is ordered. See Ware[7] and Agresti[8] for previous non-Bayesian analyses of this data set; however, the present approach will be Bayesian and similar to that used in the previous section for the drug study. Table 7.4 of the David account is portrayed as Table 8.2. This is a study of the effect on respiratory health on children due to air pollution, where the response is either a yes if the child experiences wheezing or a no if the subject does not experience wheezing over a given period of time.

Note that there are two responses (yes or no) at each of four time points giving a total of 16 patterns for the study.

TABLE 8.2

Wheezing Events of 1019 Children

Age 9	Age 10	Age 11	Age 12	Number of Children	Pattern
Yes	Yes	Yes	Yes	94	1
Yes	Yes	Yes	No	30	2
Yes	Yes	No	Yes	15	3
Yes	Yes	No	No	28	4
Yes	No	Yes	Yes	14	5
Yes	No	Yes	No	9	6
Yes	No	No	Yes	12	7
Yes	No	No	No	63	8
No	Yes	Yes	Yes	19	9
No	Yes	Yes	No	15	10
No	Yes	No	Yes	10	11
No	Yes	No	No	44	12
No	No	Yes	Yes	17	13
No	No	Yes	No	42	14
No	No	No	Yes	35	15
No	No	No	No	572	16

Source: Davis, C.S., *Statistical Methods for the Analysis of Repeated Measures*, Springer-Verlag, New York, p. 188, 2002.

In this scenario, it is assumed that the frequencies $X = (X_1, X_2, ..., X_{16})$ of the 16 patterns of Table 8.2 follow a multinomial distribution, where θ_i $(i = 1, 2, ..., 16)$ is the probability of pattern i. Also, the observed frequencies of the multinomial are given by the vector $x = $ (94, 30, 15, 28, 14, 9, 12, 63, 19, 15, 10, 44, 17, 42, 35, 572). Assuming the improper prior density

$$g(\theta|x) = \prod_{i=1}^{i=16} \theta_i^{-1}, \ 0 < \theta_i < 1, \sum_{i=1}^{i=16} \theta_i = 1 \tag{8.19}$$

for the cell probabilities, one may show the posterior distribution of the θ_i is Dirichlet with parameter vector $x = $ (94, 30, 15, 28, 14, 9, 12, 63, 19, 15, 10, 44, 17, 42, 35, 572).

What is of primary interest in this study of air pollution in children? For example, what is the probability of no change in wheezing over the four ages? What is the probability of exactly one wheezing event over four years? The former question is the probability

$$p_a = \theta_1 + \theta_{16} \tag{8.20}$$

while the latter enquiry is given by

$$p_b = \theta_9 + \theta_{12} + \theta_{14} + \theta_{14} \tag{8.21}$$

In addition, questions concerning the mean profile of the probability of wheezing over the four ages would be of major concern.

For example, the probability of wheezing at age 9 is

$$p_9 = \theta_1 + \theta_2 + \theta_3 + \theta_4 + \theta_5 + \theta_6 + \theta_7 + \theta_8 \tag{8.22}$$

while

$$p_{10} = \theta_1 + \theta_2 + \theta_3 + \theta_4 + \theta_9 + \theta_{10} + \theta_{11} + \theta_{12} \tag{8.23}$$

is the probability of wheezing at age 10.

In addition, the probability of wheezing at age 11 is

$$p_{11} = \theta_1 + \theta_2 + \theta_5 + \theta_6 + \theta_9 + \theta_{10} + \theta_{13} + \theta_{14} \tag{8.24}$$

and that at age 12 is

$$p_{12} = \theta_1 + \theta_3 + \theta_5 + \theta_7 + \theta_9 + \theta_{11} + \theta_{13} + \theta_{15} \tag{8.25}$$

Equations 8.22 through 8.25 allow one to investigate the mean profile of the probability of wheezing over the four ages.

For example, one would be interested in

$$H : p_9 = p_{10} = p_{11} = p_{12} \tag{8.26}$$

where H represents the scenario that the probability of wheezing does not change over the four ages.

A Bayesian approach will be based on computing the 95% credible intervals for the six differences

$$d_{9,10} = p_9 - p_{10} \tag{8.27}$$

$$d_{9,11} = p_9 - p_{11} \tag{8.28}$$

$$d_{9,12} = p_9 - p_{12} \tag{8.29}$$

$$d_{10,11} = p_{10} - p_{11} \tag{8.30}$$

$$d_{10,12} = p_{10} - p_{12} \tag{8.31}$$

and

$$d_{11,12} = p_{11} - p_{12} \tag{8.32}$$

BUGS CODE 8.2 will be used for the Bayesian analysis of the Davis wheezing study, and the code closely follows formulas (8.20) through (8.32).

BUGS CODE 8.2

```
model;
{

for(i in 1:16){g[i]~dgamma(a[i],2)}

sg<-sum(g[])

for (i in 1:16){theta[i]<-g[i]/sg}

# the thetas have a Dirichlet distribution

pa<-theta[1]+theta[16]

# pa is the probability of the response does not change

pb<-theta[8]+theta[12]+theta[14]+theta[15]

# pb is the probability of exactly one wheezing event.

pc<-theta[16]

# pc is the probability of no wheezing event for four
years.

P9<-theta[1]+theta[2]+theta[3]+theta[4]+theta[5]+theta[6]
+theta[7]+theta[8]

# P9 is the probability of wheezing at age 9
P10<-theta[1]+theta[2]+theta[3]+theta[4]+theta[9]+theta[10]
+theta[11]+theta[12]
P11<-theta[1]+theta[2]+theta[5]+theta[6]+theta[9]+theta[10]
+theta[13]+theta[14]
P12<-theta[1]+theta[3]+theta[5]+theta]7]+theta[9]+theta[11]
+theta[13]+theta[15]

d910<-p9-p10
d911<-p9-p11
d912<-p9-p12
d1011<-p10-p11
d1012<-p10-p12
d1112<-p11-p12
}
list(a = c(94,30,15,28,14,9,12,63,19,15,10,44,17,42,35,
572))
```

TABLE 8.3

Bayesian Analysis for Wheezing Study

Parameter	Mean	SD	Error	2½	Median	97½
$d_{9,10}$	0.009794	0.01333	0.0000366	−0.001144	0.009737	0.03609
$d_{9,11}$	0.02459	0.01422	0.0000405	−0.003341	0.02455	0.05257
$d_{9,12}$	0.04812	0.01414	0.0000368	0.020558	0.04804	0.07589
$d_{10,11}$	0.009794	0.01333	0.0000366	−0.01634	0.009737	0.03609
$d_{10,12}$	0.03832	0.01365	0.0000377	0.01158	0.03831	0.06512
$d_{11,12}$	0.02353	0.01266	0.0000330	−0.00114	0.02352	0.04847
p_9	0.2601	0.01365	0.0000526	0.2337	0.2599	0.2873
p_{10}	0.2504	0.01358	0.0000517	0.2244	0.2501	0.2774
p_{11}	0.2355	0.01325	0.0000490	0.2102	0.2353	0.262
p_{12}	0.212	0.01273	0.0000475	0.1875	0.2119	0.2373
p_a	0.6536	0.01493	0.0000434	0.6241	0.6537	0.6824
p_b	0.1806	0.01208	0.0000335	0.1575	0.1804	0.2048
p_c	0.5613	0.01551	0.0000444	0.5307	0.5614	0.5917

A Bayesian analysis is executed with 65,000 observations, with a burn in of 5,000 and a refresh of 100, and the results are reported in Table 8.3.

Across the four ages, it appears that the only difference in the wheezing probabilities is from ages 9 to 12 and 10 to 12. These assertions are based on the 95% credible intervals for the wheezing probabilities. Of course this is apparent from the four posterior means for the four wheezing probabilities $p_9, p_{10}, p_{11}, p_{12}$. However, the trend of the four probabilities is decreasing from 0.2601 at age 9 to 0.212 at age 12. In addition, Table 8.3 reveals that the probability p_a of no change in the wheezing event (either no wheezing over the four ages or one wheezing event at each age) is 0.6536.

8.1.3 Categorical Response and Two Groups

The previous sections presented the Bayesian analysis of categorical data using a multinomial model for the cell frequencies and a Dirichlet distribution for the posterior distribution of the cell probabilities. The main goal for this scenario was to model the cell frequencies over time and determine if or if not there was a trend. A similar approach is taken here and applied to the comparison of two groups, where the repeated measure is categorical. The example is about church attendance in Iowa in the 1970s (Table 8.4) and is based on the analysis of Cornoni-Huntley et al.[9]

There are 1311 females and 662 males, and subjects respond yes if they attended church on a regular basis the previous year. Thus, the response is binary and there are three time points yielding eight patterns for each gender. Let θ_i be the probability of the ith pattern for females and ϕ_i the corresponding probability for males for $i = 1, 2, \dots, 8$. Following in the fashion of the two

TABLE 8.4

Iowa Church Attendance

Sex	Year 0	Year 3	Year 6	Count
Female				
	Yes	Yes	Yes	904
	Yes	Yes	No	88
	Yes	No	Yes	25
	Yes	No	No	51
	No	Yes	Yes	33
	No	Yes	No	22
	No	No	Yes	30
	No	No	No	158
Male				
	Yes	Yes	Yes	391
	Yes	Yes	No	36
	Yes	No	Yes	12
	Yes	No	No	26
	No	Yes	Yes	15
	No	Yes	No	21
	No	No	Yes	18
	No	No	No	143

previous sections, suppose the frequencies for females for the eight patterns follow a multinomial distribution, then assuming an improper prior density for θ, namely

$$g(\theta) = \prod_{i=1}^{i=8} \theta_i^{-1} \tag{8.33}$$

with $0 < \theta_i < 1$, $\sum_{i-1}^{i=8} \theta_i = 1$

then the posterior distribution of θ is Dirichlet with parameter vector $x = (904, 88, 25, 51, 33, 22, 30, 158)$.

In a similar fashion, the posterior distribution of $\phi = (\phi_1, \phi_2, ..., \phi_8)$ is Dirichlet with parameter vector $y = (391, 36, 12, 26, 15, 21, 18, 143)$, where ϕ_i is the probability of pattern i (= 1,2,...,8) for males. Note that one is assuming an improper prior for ϕ as

$$g(\phi) = \prod_{i=1}^{i=8} \phi_i^{-1} \tag{8.34}$$

with $0 < \phi_i < 1$, $\sum_{i-1}^{i=8} \phi_i = 1$.

Our main interests are (1) determining the posterior distribution of the probability of a yes response at each time for the three time points for females, (2) determining the posterior distribution of the probability of a yes response at each time of the three time points for males, and (3) comparing the mean profiles between males and females and comparing them across the three years.

Let

$$p_0 = \theta_1 + \theta_2 + \theta_3 + \theta_4 \tag{8.35}$$

then p_0 is the probability of a yes response for females at year 0. Now let

$$p_3 = \theta_1 + \theta_2 + \theta_5 + \theta_6 \tag{8.36}$$

then p_3 is the probability of a yes response for females at year 3. In a similar fashion, let

$$p_5 = \theta_1 + \theta_3 + \theta_5 + \theta_7 \tag{8.37}$$

then p_5 is the probability of a yes response (of regular church attendance) for females at year 5.

Let

$$q_0 = \phi_1 + \phi_2 + \phi_3 + \phi_4 \tag{8.38}$$

then q_0 is the probability of a yes response for males at year 0. Now let

$$q_3 = \phi_1 + \phi_2 + \phi_5 + \phi_6 \tag{8.39}$$

then q_3 is the probability of a yes response for males at year 3.

In a similar fashion, let

$$q_5 = \phi_1 + \phi_3 + \phi_5 + \phi_7 \tag{8.40}$$

then q_5 is the probability of a yes response (of regular church attendance) for males at year 5.

The mean response for the probability of a yes response to regular church attendance over years 0, 3, and 5 for males is given by the vector $p = (p_0, p_3, p_5)$ and for females given by the vector $q = (q_0, q_3, q_5)$.

In order to compare the mean response between females and males, consider the three differences

$$d_0 = p_0 - q_0 \tag{8.41}$$

$$d_3 = p_3 - q_3 \tag{8.42}$$

and

$$d_5 = p_5 - q_5 \tag{8.43}$$

then d_0 is the difference in the probability of a yes response at year 0 between females and males. Also, d_3 is the difference in the probability of a yes response at year 3 between females and males, and d_5 is the difference in the probability of a yes response at year 5 between females and males. The program statements of **BUGS CODE 8.3** closely follow the notation of formulas (8.33) through (8.43).

Based on **BUGS CODE 8.3**, a Bayesian analysis is executed with 65,000 observations for the simulation, with a burn in of 5,000 and a refresh of 100, and the results are portrayed in Table 8.5.

BUGS CODE 8.3

```
model;

{

for(i in 1:8){g[i]~dgamma(a[i],2)}

sg<-sum(g[])

for (i in 1:8){theta[i]<-g[i]/sg}

for(i in 1:8){h[i]~dgamma(b[i],2)}

sh<-sum(h[])

for (i in 1:8){phi[i]<-h[i]/sh}

# the thetas have a Dirichlet distribution for females
# the phis have a Dirichlet distribution for males.

p0<-theta[1]+theta[2]+theta[3]+theta[4]
q0<-phi[1]+phi[2]+phi[3]+phi[4]

p3<-theta[1]+theta[2]+theta[5]+theta[6]
q3<-phi[1]+phi[2]+phi[5]+phi[6]

p5<-theta[1]+theta[3]+theta[5]+theta[7]
q5<-phi[1]+phi[3]+phi[5]+phi[7]

d0<-p0-q0
d3<-p3-q3
d5<-p5-q5

}

list(a = c(904,88,25,51,33,22,30,158),
       b = c(391,36,12,26,15,21,18,143))
```

TABLE 8.5

Posterior Analysis for Church Attendance

Parameter	Mean	SD	Error	2½	Median	97½
d_0	0.1123	0.0206	0.0000896	0.07206	0.1122	0.1529
d_3	0.09925	0.02091	0.0000909	0.05884	0.0993	0.1405
d_5	0.09797	0.02189	0.000989	0.05537	0.09782	0.141
p_0	0.8146	0.01065	0.0000451	0.7934	0.8148	0.8349
p_3	0.7986	0.01108	0.0000439	0.7766	0.7988	0.8198
p_5	0.7566	0.01188	0.0000484	0.733	0.7568	0.7795
q_0	0.7024	0.01767	0.0000761	0.6673	0.7026	0.7365
q_3	0.6994	0.01777	0.0000799	0.6641	0.6995	0.7336
q_5	0.6587	0.01841	0.0000829	0.6221	0.6589	0.6941

Upon inspection of Table 8.5, it appears that the frequency of regular church attendance is decreasing from year 0 to 3 for both males and females. For example, the probability of attendance at year 0 for females is estimated as 0.8146 but is estimated as 0.7566 at year 5. A similar pattern is followed by males with attendance at year 0 is estimated as 0.7024 but as 0.6587 at year 5. Also, it appears that females have regular attendance at a higher rate than males at each of the three years. Based on the 95% credible interval, there is a meaningful difference between the female and male frequencies at each year. For example for year 0, the 95% credible interval for the difference d_0 is (0.07206, 0.1529), implying a real difference in the two probabilities of regular church attendance.

8.2 Bayesian GEE

A Bayesian version of GEE is introduced in order to analyze repeated measures with categorical data and accommodating covariates, such as treatment groups, and other subject information.

The approach using GEE is a form of weighted least squares, where the deviations of the residuals is weighted by the inverse of the estimated variance–covariance matrix of the response over the various time points.

The solution $\tilde{\beta}$ to the GEE identity

$$\sum_{i=1}^{i=N} D_i V_i^{-1}(Y_i - \mu_i) = 0 \tag{8.44}$$

is the GEE estimate of the unknown regression parameters. Note that the left-hand side of (8.44) is a function of β, which follows from the following: N is the number of subjects, Y_i is an n by 1 vector of observations over n time points, μ_i is the n by 1 corresponding mean vector of Y_i, and is a function of unknown regression parameters β and known covariates. It is usually assumed that for some link function g that

$$g(\mu_i) = X_i \beta \tag{8.45}$$

where:
  X_i is an n by p known design matrix of various covariates
  β is the p by 1 vector of unknown regression parameters

Note that D_i is a p by n matrix of partial derivatives

$\partial\mu_{i1}/\partial\beta_1$	$\partial\mu_{i2}/\partial\beta_1$	$\ldots$	$\ldots$	$\partial\mu_{in}/\partial\beta_1$
$\vdots$				
$\partial\mu_{i1}/\partial\beta_p$	$\partial\mu_{i2}/\partial\beta_p$			$\partial\mu_{in}/\partial\beta_p$

Also, V_i is the n by n matrix whose elements consist of the estimated variances and covariances, with jk-th entry

$$\sigma(j,k) = \text{Cov}\left(Y_{ij}, Y_{ik}\right) \tag{8.46}$$

where:
  Y_{ij} is the response at time point j for subject i

Of course,

$$\sigma(j,j) = \text{Var}(Y_{ij}) \tag{8.47}$$

Note that each component of the GEE is a function of a prior distribution for β, then the posterior estimate of β is a solution to the GEE (8.44). The examples that illustrate the Bayesian GEE approach will be based on the binomial and Poisson distribution for the repeated measures response.
  First, consider the Bernoulli model where for subject i at time j

$$P\left[Y_{ij} = 1\right] = \mu_{ij} \tag{8.48}$$

with link function

$$g(\mu_{ij}) = \log \frac{\mu_{ij}}{1 - \mu_{ij}} = \text{logit}(\mu_{ij}) \tag{8.49}$$

which is referred as the logistic model.
  It is usually assumed that

$$\text{logit}(\mu_{ij}) = X_{ij}\beta \tag{8.50}$$

where:

   X_{ij} is a 1 by p vector of covariates
   β is a p by 1 vector of unknown regression parameters

Thus, it is seen that the link function g expresses a function of the mean parameter μ_{ij} as a linear function of the components of the regression parameter β. Note that the logistic link can also be expressed as

$$\mu_{ij} = \frac{\exp(X_{ij}\beta)}{1 - \exp(X_{ij}\beta)} \tag{8.51}$$

Also note that for the binomial model

$$\mathrm{Var}\left(Y_{ij}\right) = \mu_{ij}\left(1 - \mu_{ij}\right) \tag{8.52}$$

Thus, the variance of an observation is a function of the mean of the observation. It should also be stressed that the distribution of the response at each time point of a subject is known, but that the joint distribution of n observations is not usually known. Of course, the attraction of GEE (8.44) is that the joint distribution is not required for this weighted least squares approach.

The Poisson case is also studied, where for subject i at time j

$$P\left[Y_{ij} = k\right] = \mu_{ij} \exp\frac{-\mu_{ij}}{k!} \tag{8.53}$$

for $k = 0,1,2,\ldots$ with canonical link function

$$\log\left(\mu_{ij}\right) = X_{ij}\beta \tag{8.54}$$

where:

   X_{ij} is a 1 by p vector of known covariate values
   β is a p by 1 vector of unknown regression parameters

Thus, the mean and variance of Y_{ij} is μ_{ij} and in terms of the mean

$$\mu_{ij} = \exp\left(X_{ij}\beta\right) \tag{8.55}$$

The essence of a generalized linear model is that for categorical data, the link function is represented as a liner function of the regression parameters.

8.2.1 Bayesian GEE for the Poisson Distribution

The example involving the Poisson distribution is taken from Hand and Taylor[10] and analyzed by Hand and Crowder[3] and is a study of Alzheimer's disease. The main response is the subject's number of correctly recalled words, and there are two groups, one is placebo and the other where a treatment is administered to the subjects. The repeated measures (number of correctly recalled words) are observed at times 0, 1, 2, 4, and 6. In what is to be presented the only difference is that the placebo group with 26 subjects is included in the analysis, and observations are listed in Table 8.6. The first column is the time, the second the number of correctly recalled words, and the third the subject identification.

TABLE 8.6

Alzheimer's Study with Placebo Group

Time	W	Subject
0.00	20.00	1.00
1.00	19.00	1.00
2.00	20.00	1.00
4.00	20.00	1.00
6.00	18.00	1.00
0.00	14.00	2.00
1.00	15.00	2.00
2.00	16.00	2.00
4.00	9.00	2.00
6.00	6.00	2.00
0.00	7.00	3.00
1.00	5.00	3.00
2.00	8.00	3.00
4.00	8.00	3.00
6.00	5.00	3.00
0.00	6.00	4.00
1.00	10.00	4.00
2.00	9.00	4.00
4.00	10.00	4.00
6.00	10.00	4.00
0.00	9.00	5.00
1.00	7.00	5.00
2.00	9.00	5.00
4.00	4.00	5.00
6.00	6.00	5.00
0.00	9.00	6.00
1.00	10.00	6.00
2.00	9.00	6.00
4.00	11.00	6.00
6.00	11.00	6.00
0.00	7.00	7.00
1.00	3.00	7.00
2.00	7.00	7.00
4.00	6.00	7.00

(*Continued*)

TABLE 8.6 (*Continued*)

Alzheimer's Study with Placebo Group

Time	W	Subject
6.00	3.00	7.00
0.00	18.00	8.00
1.00	20.00	8.00
2.00	20.00	8.00
4.00	23.00	8.00
6.00	21.00	8.00
0.00	6.00	9.00
1.00	10.00	9.00
2.00	10.00	9.00
4.00	13.00	9.00
6.00	14.00	9.00
0.00	10.00	10.00
1.00	15.00	10.00
2.00	15.00	10.00
4.00	15.00	10.00
6.00	14.00	10.00
0.00	5.00	11.00
1.00	9.00	11.00
2.00	7.00	11.00
4.00	3.00	11.00
6.00	12.00	11.00
0.00	11.00	12.00
1.00	11.00	12.00
2.00	8.00	12.00
4.00	10.00	12.00
6.00	9.00	12.00
0.00	10.00	13.00
1.00	2.00	13.00
2.00	9.00	13.00
4.00	3.00	13.00
6.00	2.00	13.00
0.00	17.00	14.00
1.00	12.00	14.00
2.00	14.00	14.00
4.00	15.00	14.00
6.00	13.00	14.00
0.00	16.00	15.00
1.00	15.00	15.00
2.00	13.00	15.00
4.00	7.00	15.00

(*Continued*)

TABLE 8.6 (*Continued*)

Alzheimer's Study with Placebo Group

Time	W	Subject
6.00	9.00	15.00
0.00	7.00	16.00
1.00	10.00	16.00
2.00	4.00	16.00
4.00	10.00	16.00
6.00	5.00	16.00
0.00	5.00	17.00
1.00	0.00	17.00
2.00	5.00	17.00
4.00	0.00	17.00
6.00	0.00	17.00
0.00	16.00	18.00
1.00	7.00	18.00
2.00	7.00	18.00
4.00	6.00	18.00
6.00	10.00	18.00
0.00	5.00	19.00
1.00	6.00	19.00
2.00	9.00	19.00
4.00	5.00	19.00
6.00	6.00	19.00
0.00	2.00	20.00
1.00	1.00	20.00
2.00	1.00	20.00
4.00	2.00	20.00
6.00	2.00	20.00
0.00	7.00	21.00
1.00	11.00	21.00
2.00	7.00	21.00
4.00	5.00	21.00
6.00	11.00	21.00
0.00	9.00	22.00
1.00	16.00	22.00
2.00	17.00	22.00
4.00	10.00	22.00
6.00	6.00	22.00
0.00	2.00	23.00
1.00	5.00	23.00
2.00	6.00	23.00
4.00	7.00	23.00

(*Continued*)

TABLE 8.6 (*Continued*)

Alzheimer's Study with Placebo Group

Time	W	Subject
6.00	6.00	23.00
0.00	7.00	24.00
1.00	3.00	24.00
2.00	5.00	24.00
4.00	5.00	24.00
6.00	5.00	24.00
0.00	19.00	25.00
1.00	13.00	25.00
2.00	19.00	25.00
4.00	17.00	25.00
6.00	17.00	25.00
0.00	7.00	26.00
1.00	5.00	26.00
2.00	8.00	26.00
4.00	8.00	26.00
6.00	6.00	26.00

Source: Crowder, M.J., and Hand, D.J., *Analysis of Repeated Measures*, Table 8.10, Chapman & Hall, New York, 1990.

Let Y_{ij} denote the number of correctly recalled words for subject i at time j, where $i = 1,2,\ldots,26$ and $j = 0,1,2,4,6$; thus, for example, $Y_{26,4} = 8$, the number of correctly recalled words by subject 26 at time 4 is 8. It is assumed that the marginal distribution of Y_{ij} is Poisson with mean μ_j and probability mass function

$$P\left[Y_{ij} = k\right] = \mu_j \exp\frac{-\mu_j}{k!} \qquad (8.56)$$

where:
$i = 1,2,\ldots,26$
$j = 0,1,2,4,6$
$\mu_j > 0$

Note the distribution of Y_{ij} is the same for all subjects with mean and variance μ_j, $j = 0,1,2,4,6$.

It is also assumed that the log link function

$$\log\left(\mu_j\right) = \beta_1 + \beta_2 t_{ij} \qquad (8.57)$$

is a linear function of time, or equivalently

$$\mu_j = \exp(\beta_1 + \beta_2 t_{ij}) \tag{8.58}$$

where for each i, $t_{ij} = 0,1,2,4,6$. Our goal is to estimate the regression coefficients β_1 and β_2 using the GEE

$$\sum_{i=1}^{i=N} D_i V_i^{-1}(Y_i - \mu_i) = 0 \tag{8.59}$$

It can be shown that the estimated covariance matrix V_i is given by the 5 by 5 matrix (Table 8.7).
And the 2 by 5 matrix $D_i \lambda$ is given by

$$D[1,1] = \sum_{j=1}^{j=5} \lambda[j,1] \exp(\beta_1 + \beta_2 t_j)$$

$$D[1,2] = \sum_{j=1}^{j=5} \lambda[j,2] \exp(\beta_1 + \beta_2 t_j)$$

$$D[1,3] = \sum_{j=1}^{j=5} \lambda[j,3] \exp(\beta_1 + \beta_2 t_j) \tag{8.60}$$

$$D[1,4] = \sum_{j=1}^{j=5} \lambda[j,4] \exp(\beta_1 + \beta_2 t_j)$$

$$D[1,5] = \sum_{j=1}^{j=5} \lambda[j,5] \exp(\beta_1 + \beta_2 t_j)$$

TABLE 8.7

Covariance Matrix V_i

26.8777541	19.264	18.608	17.226	13.880
19.264	29.7846	21.872	23.406	21.380
18.608	21.872	26.07389	20.047	17.397
17.226	23.406	20.047	31.11939	20.990
13.880	21.380	17.397	20.990	27.56460

$$D[2,1] = \sum_{j=1}^{j=5} t_j \lambda[j,1] \exp\left(\beta_1 + \beta_2 t_j\right)$$

$$D[2,2] = \sum_{j=1}^{j=5} t_j \lambda[j,2] \exp\left(\beta_1 + \beta_2 t_j\right)$$

$$D[2,3] = \sum_{j=1}^{j=5} t_j \lambda[j,3] \exp\left(\beta_1 + \beta_2 t_j\right) \qquad (8.61)$$

$$D[2,4] = \sum_{j=1}^{j=5} t_j \lambda[j,4] \exp\left(\beta_1 + \beta_2 t_j\right)$$

$$D[2,5] = \sum_{j=1}^{j=5} t_j \lambda[j,5] \exp\left(\beta_1 + \beta_2 t_j\right)$$

where:

$t_1 = 0, t_2 = 1, t_3 = 2, t_4 = 4, t_5 = 6$

ijth element of the matrix V_i^{-1} is denoted as λ_{ij}

It can be shown that the 2 by 1 matrix $D_i V_i^{-1}(Y_i - \mu_i)$ is given by the following with first entry

$$\sum_{j=1}^{j=5} D[1,j]\left[Y[i,j] - \exp\left(\beta_1 + \beta_2 t_j\right)\right] \qquad (8.62)$$

while the second entry is given by

$$\sum_{j=1}^{j=5} D[2,j]\left[Y[i,j] - \exp\left(\beta_1 + \beta_2 t_j\right)\right] \qquad (8.63)$$

Note that these two entries apply to subject i, where y_{i1} is the time t_1 observation for subject i and y_{i5} is the time t_5 observation for subject i. Also, it should be noted that the only thing that changes from one subject to the next are the observations for those subjects.

The program statements of **BUGS CODE 8.4** follow closely the notation of formulas (8.56) through (8.63). Refer to the remarks indicated by # for an explanation of the various code statements.

The Bayesian analysis is based on **BUGS CODE 8.4** and executed with 150,000 observations for the simulation, with a burn in of 5,000 and a refresh of 100 (Table 8.8).

BUGS CODE 8.4

```
model;

{
# Poisson distribution for the observations
for(i in 1:N){for(j in 1:M){Y[i,j]~dpois(mu[j])}}

# predicted values
for(i in 1:N){for(j in 1:M){Z[i,j]~dpois(mu[j])}}

# on the log scale, regression is linear
for(j in 1:M){mu[j]<-exp(beta1+beta2*t[j])}
# intercept at t = 0
int<-exp(beta1)

# delta is the inverse of the estimated variance
covariance matrix V of observations

delta[1:M,1:M]<-inverse(V[,])

D[1,1]<-delta[1,1]*exp(beta1+beta2*t[1])+
delta[2,1]*exp(beta1+beta2*t[2])+delta[3,1]*exp(beta1+bet
a2*t[3])+
delta[4,1]*exp(beta1+beta2*t[4])+delta[5,1]*exp(beta1+bet
a2*t[5])

D[1,2]<-delta[1,2]*exp(beta1+beta2*t[1])+
delta[2,2]*exp(beta1+beta2*t[2])+delta[3,2]*exp(beta1+
beta2*t[3])+
delta[4,2]*exp(beta1+beta2*t[4])+delta[5,2]*exp(beta1+
beta2*t[5])

D[1,3]<-delta[1,3]*exp(beta1+beta2*t[1])+
delta[2,3]*exp(beta1+beta2*t[2])+delta[3,3]*exp(beta1+
beta2*t[3])+
delta[4,3]*exp(beta1+beta2*t[4])+delta[5,3]*exp(beta1+
beta2*t[5])

D[1,4]<-delta[1,4]*exp(beta1+beta2*t[1])+
delta[2,4]*exp(beta1+beta2*t[2])+delta[3,4]*exp(beta1+
beta2*t[3])+
delta[4,4]*exp(beta1+beta2*t[4])+delta[5,4]*exp(beta1+
beta2*t[5])

D[1,5]<-delta[1,5]*exp(beta1+beta2*t[1])+
delta[2,5]*exp(beta1+beta2*t[2])+delta[3,5]*exp(beta1+
beta2*t[3])+
```

```
delta[4,5]*exp(beta1+beta2*t[4])+delta[5,5]*exp(beta1+
beta2*t[5])

D[2,1]<-t[1]*delta[1,1]*exp(beta1+beta2*t[1])+
t[2]*delta[2,1]*exp(beta1+beta2*t[2])+t[3]*delta[3,1]*exp
(beta1+beta2*t[3])+t[4]*delta[4,1]*exp(beta1+beta2*t[4])+
t[5]*delta[5,1]*exp(beta1+beta2*t[5])

D[2,2]<-t[1]*delta[1,2]*exp(beta1+beta2*t[1])+
t[2]*delta[2,2]*exp(beta1+beta2*t[2])+t[3]*delta[3,2]*exp
(beta1+beta2*t[3])+t[4]*delta[4,2]*exp(beta1+beta2*t[4])+
t[5]*delta[5,2]*exp(beta1+beta2*t[5])

D[2,3]<-t[1]*delta[1,3]*exp(beta1+beta2*t[1])+
t[2]*delta[2,3]*exp(beta1+beta2*t[2])+t[3]*delta[3,3]*exp
(beta1+beta2*t[3])+t[4]*delta[4,3]*exp(beta1+beta2*t[4])+
t[5]*delta[5,3]*exp(beta1+beta2*t[5])

D[2,4]<-t[1]*delta[1,4]*exp(beta1+beta2*t[1])+
t[2]*delta[2,4]*exp(beta1+beta2*t[2])+t[3]*delta[3,4]*exp
(beta1+beta2*t[3])+t[4]*delta[4,4]*exp(beta1+beta2*t[4])+
t[5]*delta[5,4]*exp(beta1+beta2*t[5])

D[2,5]<-t[1]*delta[1,5]*exp(beta1+beta2*t[1])+
t[2]*delta[2,5]*exp(beta1+beta2*t[2])+t[3]*delta[3,5]*exp
(beta1+beta2*t[3])+t[4]*delta[4,5]*exp(beta1+beta2*t[4])+
t[5]*delta[5,5]*exp(beta1+beta2*t[5])

# gee equation (8.60)

for (i in 1:N){A1[i]<-
D[1,1]*(Y[i,1]-exp(beta1+beta2*t[1]))+D[1,2]*(Y[i,2]-exp(
beta1+beta2*t[2]))+D[1,3]*(Y[i,3]-exp(beta1+beta2*t[3]))+
D[1,4]*(Y[i,4]-exp(beta1+beta2*t[4]))+
D[1,5]*(Y[i,5]-exp(beta1+beta2*t[5]))}

# gee equation (8.61)

for (i in 1:N){A2[i]<-D[2,1]*(Y[i,1]-exp(beta1+beta2*t[1]))
+D[2,2]*(Y[i,2]-exp(beta1+beta2*t[2]))+D[2,3]*(Y[i,3]-exp(b
eta1+beta2*t[3]))+D[2,4]*(Y[i,4]-exp(beta1+beta2*t[4]))+
D[2,5]*(Y[i,5]-exp(beta1+beta2*t[5]))}

sa1<-sum(A1[])

sa2<-sum(A2[])
```

```
beta1~dnorm(0,.001)
beta2~dnorm(0,.001)

}

list(N = 26,M = 5, t = c(0,1,2,4,6),
# estimated VC matrix of the observations
V = structure(.Data = c(

26.875,19.264,18.608,17.226,13.889,
19.264,29.784,21.872,23.406,21.830,
18.608,21.872,26.073,20.047,17.397,
17.226,23.406,20.047,31.119,20.990,
13.880,21.830,17.397,20.990,27.564),.Dim = c(5,5)),

Y = structure(.Data = c(20,19,20,20,18,

14,15,16,9,6,

7,5,8,8,5,

6,10,9,10,10,

9,7,9,4,6,

9,10,9,11,11,

7,3,7,6,3,

18,20,20,23,21,

6,10,10,13,14,

10,15,15,15,14,

5,9,7,3,12,

11,11,8,10,9,

10,2,9,3,2,

17,12,14,15,13,

16,15,13,7,9,

7,10,4,10,5,
```

```
5,0,5,0,0,

16,7,7,6,10,

5,6,9,5,6,

2,1,1,2,2,

7,11,7,5,11,

9,16,17,10,6,

2,5,6,7,6,

7,3,5,5,5,

19,13,19,17,17,

7,5,8,8,6),.Dim = c(26,5))))
list(beta1 = 0,beta2 = 0)
```

TABLE 8.8

Posterior Analysis for Alzheimer Study

Parameter	Mean	SD	Error	2½	Median	97½
β_1	2.276	0.0448	0.000573	2.188	2.276	2.263
β_2	−0.01736	0.01362	0.000168	−0.04409	−0.01722	0.00924
$\exp(\beta_1)$	9.747	0.4372	0.005638	8.918	9.738	10.62

It appears that the regression coefficients have symmetric posterior distributions with a 95% credible interval for β_2 as $(-0.04409, 0.00924)$, implying that β_2 is 0. The y-intercept $\exp(\beta_1)$ on the original scale is estimated as 9.747, which aggress quite well with the lowess plot of the number of correctly recalled words versus time for the placebo group and reported in Figure 8.2.

It appears that the association between the word count and time is flat with an intercept estimated with the posterior mean estimated as 9.747. It also appears that the slope is 0. Note that the derivative of the Poisson mean at time j is $\beta_2 \exp(\beta_1 + \beta_2 t_j)$, but if β_2 is 0 (implied by its 95% credible interval), the derivative is 0, which agree with the slope portrayed by the lowess plot of Figure 8.2.

The Bayesian version of the approach using GEE differ from the conventional way. It depends on a "good" estimate of the variance covariance

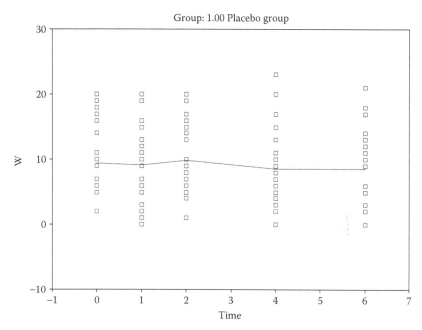

FIGURE 8.2
Number of correctly recalled words versus time.

matrix V of the observations over the time points, and in addition, the Bayesian inference is not iterative, as is the conventional approach. Thus, it is imperative that the Bayesian approach employs a "good" estimate of the variance–covariance matrix. The estimate V for the Alzheimer study is very "good" because of the "large" number of observations and the relatively large range of this categorical data set; however, this would not be the case for many studies, including those with a binary response (a small number of possible values) and those with a "small" number of observation over time. In those cases, there are alternative ways to estimate the working covariance matrix V of (8.44) of the GEE, and the reader is referred to Davis (p. 297)[1] and to Fitzmaurice, Laird, and Ware (p. 356).[2]

8.2.2 Bayesian GEE for One Population and a Continuous Covariate

The discussion of a Bayesian version of the GEE approach is to generalize the previous example to one population with a continuous covariate.

Consider a placebo-controlled clinical trial of 59 epileptics. Patients with partial seizures were enrolled in a randomized clinical trial of the anti-epileptic drug, progabide. Participants in the study were randomized to either progabide or a placebo, as an adjuvant to the standard anti-epileptic chemotherapy. Progabide is an anti-epileptic drug whose primary mechanism of action is to enhance gamma-aminobutyric acid (GABA) content; GABA is the primary inhibitory neurotransmitter in the brain.

Prior to receiving treatment, baseline data on the number of epileptic seizures during the preceding eight-week interval were recorded. Counts of epileptic seizures during two-week intervals before each of four successive post-randomization clinic visits were recorded.

Let Y_{ij} denote the number of epileptic seizures for subject i at time j, where $i = 1,2,...,28$, and $j = 0,2,4,6,8$; thus, for example, $Y_{26,4} = 1$, the number of seizures of subject 26 at time 4 weeks. It is assumed that the marginal distribution of Y_{ij} is Poisson with mean μ_j and probability mass function

$$P\left[Y_{ij} = k\right] = \mu_j \exp\frac{-\mu_j}{k!} \tag{8.64}$$

where:
$i = 1,2,...,26$
$j = 0,2,4,6,8$
$\mu_j > 0$

Note the distribution of Y_{ij} is the same for all subjects with mean and variance μ_j, $j = 0,2,4,6,8$.

It is also assumed that the log link function

$$\log\left(\mu_j\right) = \beta_1 + \beta_2 t_{ij} + \beta_3 \text{age}_i \tag{8.65}$$

is a linear function of time, where t_{ij} is the time for subject i at time j, and age_i is the age of subject i.

Equivalently

$$\mu_j = \exp\left(\beta_1 + \beta_2 t_{ij} + \beta_3 \text{age}_i\right) \tag{8.66}$$

where for each i, $t_{ij} = 0,2,4,6,8$. Our goal is to estimate the regression coefficients β_1, β_2, and β_3 using the GEE

$$\sum_{i=1}^{i=N} D_i V_i^{-1}\left(Y_i - \mu_i\right) = 0 \tag{8.67}$$

It can be shown that the estimated covariance matrix V_i is given by the 5 by 5 matrix (Table 8.9).

Suppose $\lambda = V_i^{-1}$ and suppose the elements of the 3 by 5 matrix $D_i\lambda$ are given by

TABLE 8.9

Covariance Matrix V_i of Epileptic Study

42.5895	46.3073	31.8372	66.7496	39.9297
46.3073	102.7580	62.8981	92.5129	43.7643
31.8372	62.8981	66.6508	76.1993	36.8045
66.7496	92.5129	76.1993	215.2675	65.8087
39.9297	43.7643	36.8045	65.8087	58.1833

$$D_{11}[i] = \sum_{j=1}^{j=5} \lambda[j,1]\exp(\beta_1 + \beta_2 t_j + \beta_3 \text{age}_i)$$

$$D_{12}[i] = \sum_{j=1}^{j=5} \lambda[j,2]\exp(\beta_1 + \beta_2 t_j + \beta_3 \text{age}_i)$$

$$D_{13}[i] = \sum_{j=1}^{j=5} \lambda[j,3]\exp(\beta_1 + \beta_2 t_j + \beta_3 \text{age}_i) \qquad (8.68)$$

$$D_{14}[i] = \sum_{j=1}^{j=5} \lambda[j,4]\exp(\beta_1 + \beta_2 t_j + \beta_3 \text{age}_i)$$

$$D_{15}[i] = \sum_{j=1}^{j=5} \lambda[j,5]\exp(\beta_1 + \beta_2 t_j + \beta_3 \text{age}_i)$$

$$D_{21}[i] = \sum_{j=1}^{j=5} t_j \lambda[j,1]\exp(\beta_1 + \beta_2 t_j + \beta_3 \text{age}_i)$$

$$D_{22}[i] = \sum_{j=1}^{j=5} t_j \lambda[j,2]\exp(\beta_1 + \beta_2 t_j + \beta_3 \text{age}_i)$$

$$D_{23}[i] = \sum_{j=1}^{j=5} t_j \lambda[j,3]\exp(\beta_1 + \beta_2 t_j + \beta_3 \text{age}_i) \qquad (8.69)$$

$$D_{24}[i] = \sum_{j=1}^{j=5} t_j \lambda[j,4]\exp(\beta_1 + \beta_2 t_j + \beta_3 \text{age}_i)$$

$$D_{25}[i] = \sum_{j=1}^{j=5} t_j \lambda[j,5]\exp(\beta_1 + \beta_2 t_j + \beta_3 \text{age}_i)$$

$$D_{31}[i] = age_i \sum_{j=1}^{j=5} \lambda[j,1]\exp(\beta_1 + \beta_2 t_j + \beta_3 age_i)$$

$$D_{32}[i] = age_i \sum_{j=1}^{j=5} \lambda[j,2]\exp(\beta_1 + \beta_2 t_j + \beta_3 age_i)$$

$$D_{33}[i] = age_i \sum_{j=1}^{j=5} \lambda[j,3]\exp(\beta_1 + \beta_2 t_j + \beta_3 age_i) \qquad (8.70)$$

$$D_{34}[i] = age_i \sum_{j=1}^{j=5} \lambda[j,4]\exp(\beta_1 + \beta_2 t_j + \beta_3 age_i)$$

$$D_{35}[i] = age_i \sum_{j=1}^{j=5} \lambda[j,5]\exp(\beta_1 + \beta_2 t_j + \beta_3 age_i)$$

where $t_1 = 0, t_2 = 2, t_3 = 4, t_4 = 6, t_5 = 8$.

Note that the ijth element of the matrix V_i^{-1} is denoted as λ_{ij}.

It can be shown that the 3 by 1 matrix $D_i V_i^{-1}(Y_i - \mu_i)$ is given by the following: The first component is

$$\sum_{j=1}^{j=5} D_{1j}[i]\left[Y[i,j] - \exp(\beta_1 + \beta_2 t_j + \beta_3 age_i) \right] \qquad (8.71)$$

the second entry is

$$\sum_{j=1}^{j=5} D_{2j}[i]\left[Y[i,j] - \exp(\beta_1 + \beta_2 t_j + \beta_3 age_i) \right] \qquad (8.72)$$

and the third is

$$\sum_{j=1}^{j=5} D_{3j}[i]\left[Y[i,j] - \exp(\beta_1 + \beta_2 t_j + \beta_3 age_i) \right] \qquad (8.73)$$

Note that these three entries apply to subject i, where $Y[i,j]$ is the number of seizures at week t_j for subject i. Also, it should be noted that the only thing that changes from one subject to the next are the observations $Y[i,j]$ for those subjects.

The Bayesian analysis is based on formulas (8.64) through (8.72), **BUGS CODE 8.5** closely follows the notation of formulas (8.64) through (8.72), and the data for the epileptic study is included in the first list statement. Note that the observations (the number of seizures) are assumed to follow a Poisson distribution and the log of the mean is the link function.

BUGS CODE 8.5

```
model;

{
# Poisson distribution for the observations
for(i in 1:N){for(j in 1:M){Y[i,j]~dpois(mu[i,j])}}

# on the log scale, regression is linear
for(i in 1:N){for(j in 1:M){mu[i,j]<-exp(beta1+beta2*t[j]
+beta3*age[i])}}

# delta is the inverse of the estimated variance
covariance matrix V of observations

delta[1:M,1:M]<-inverse(V[,])

for(i in 1:N){
D11[i]<-delta[1,1]*exp(beta1+beta2*t[1]+beta3*age[i])+
delta[2,1]*exp(beta1+beta2*t[2]+beta3*age[i])+delta[3,1]*
exp(beta1+beta2*t[3]+beta3*age[i])+
delta[4,1]*exp(beta1+beta2*t[4]+beta3*age[i])+delta[5,1]*
exp(beta1+beta2*t[5]+beta3*age[i])

D12[i]<-delta[1,2]*exp(beta1+beta2*t[1]+beta3*age[i])+
delta[2,2]*exp(beta1+beta2*t[2]+beta3*age[i])+delta[3,2]*
exp(beta1+beta2*t[3]+beta3*age[i])+
delta[4,2]*exp(beta1+beta2*t[4]+beta3*age[i])+delta[5,2]*
exp(beta1+beta2*t[5]+beta3*age[i])

D13[i]<-delta[1,3]*exp(beta1+beta2*t[1]+beta3*age[i])+
delta[2,3]*exp(beta1+beta2*t[2]+beta3*age[i])+delta[3,3]*
exp(beta1+beta2*t[3]+beta3*age[i])+
delta[4,3]*exp(beta1+beta2*t[4]+beta3*age[i])+delta[5,3]*
exp(beta1+beta2*t[5]+beta3*age[i])

D14[i]<-delta[1,4]*exp(beta1+beta2*t[1]+beta3*age[i])+
delta[2,4]*exp(beta1+beta2*t[2]+beta3*age[i])+delta[3,4]*
exp(beta1+beta2*t[3]+beta3*age[i])+
delta[4,4]*exp(beta1+beta2*t[4]+beta3*age[i])+delta[5,4]*
exp(beta1+beta2*t[5]+beta3*age[i])

D15[i]<-delta[1,5]*exp(beta1+beta2*t[1]+beta3*age[i])+
delta[2,5]*exp(beta1+beta2*t[2]+beta3*age[i])+delta[3,5]*
exp(beta1+beta2*t[3]+beta3*age[i])+
delta[4,5]*exp(beta1+beta2*t[4]+beta3*age[i])+delta[5,5]*
exp(beta1+beta2*t[5]+beta3*age[i])

D21[i]<-t[1]*delta[1,1]*exp(beta1+beta2*t[1]+beta3*age[i])+
t[2]*delta[2,1]*exp(beta1+beta2*t[2]+beta3*age[i])+t[3]*
```

```
delta[3,1]*exp(beta1+beta2*t[3]+beta3*age[i])+t[4]*delta[4,1]*
exp(beta1+beta2*t[4]+beta3*age[i])+t[5]*delta[5,1]*exp(beta1+
beta2*t[5]+beta3*age[i])

D22[i]<-t[1]*delta[1,2]*exp(beta1+beta2*t[1]+beta3*age[i])+
t[2]*delta[2,2]*exp(beta1+beta2*t[2]+beta3*age[i])+t[3]*
delta[3,2]*exp(beta1+beta2*t[3]+beta3*age[i])+t[4]*delta[4,2]*
exp(beta1+beta2*t[4]+beta3*age[i])+t[5]*delta[5,2]*exp(beta1+
beta2*t[5]+beta3*age[i])

D23[i]<-t[1]*delta[1,3]*exp(beta1+beta2*t[1]+beta3*age[i])+
t[2]*delta[2,3]*exp(beta1+beta2*t[2]+beta3*age[i])+t[3]*
delta[3,3]*exp(beta1+beta2*t[3]+beta3*age[i])+t[4]*delta[4,3]*
exp(beta1+beta2*t[4]+beta3*age[i])+t[5]*delta[5,3]*exp
(beta1+ beta2*t[5]+beta3*age[i])

D24[i]<-t[1]*delta[1,4]*exp(beta1+beta2*t[1]+beta3*age[i])+
t[2]*delta[2,4]*exp(beta1+beta2*t[2]+beta3*age[i])+t[3]*
delta[3,4]*exp(beta1+beta2*t[3]+beta3*age[i])+t[4]*
delta[4,4]*exp(beta1+beta2*t[4]+beta3*age[i])+t[5]*
delta[5,4]*exp(beta1+beta2*t[5]+beta3*age[i])

D25[i]<-t[1]*delta[1,5]*exp(beta1+beta2*t[1]+beta3*age[i])+
t[2]*delta[2,5]*exp(beta1+beta2*t[2]+beta3*age[i])+t[3]*
delta[3,5]*exp(beta1+beta2*t[3]+beta3*age[i])+t[4]*
delta [4,5]*exp(beta1+beta2*t[4]+beta3*age[i])+t[5]*delta[5,5]*
exp(beta1+beta2*t[5]+beta3*age[i])

D31[i]<-age[i]*(delta[1,1]*exp(beta1+beta2*t[1]+beta3*
age[i])+
delta[2,1]*exp(beta1+beta2*t[2]+beta3*age[i])+delta[3,1]*
exp(beta1+beta2*t[3]+beta3*age[i])+delta[4,1]*exp(beta1+
beta2*t[4]+beta3*age[i])+delta[5,1]*exp(beta1+beta2*t[5]+
beta3*age[i]))

D32[i]<-age[i]*(delta[1,2]*exp(beta1+beta2*t[1]+beta3*ag
e[i])+
delta[2,2]*exp(beta1+beta2*t[2]+beta3*age[i])+delta[3,2]*
exp(beta1+beta2*t[3]+beta3*age[i])+delta[4,2]*exp(beta1+
beta2*t[4]+beta3*age[i])+delta[5,2]*exp(beta1+beta2*t[5]+
beta3*age[i]))

D33[i]<-age[i]*(delta[1,3]*exp(beta1+beta2*t[1]+beta3*ag
e[i])+
delta[2,3]*exp(beta1+beta2*t[2]+beta3*age[i])+delta[3,3]*
exp(beta1+beta2*t[3]+beta3*age[i])+delta[4,3]*exp(beta1+
beta2*t[4]+beta3*age[i])+delta[5,3]*exp(beta1+beta2*t[5]+
beta3*age[i]))
```

```
D34[i]<-age[i]*(delta[1,4]*exp(beta1+beta2*t[1]+beta3*
age[i])+
delta[2,4]*exp(beta1+beta2*t[2]+beta3*age[i])+delta[3,4]*
exp(beta1+beta2*t[3]+beta3*age[i])+delta[4,4]*exp(beta1+
beta2*t[4]+beta3*age[i])+delta[5,4]*exp(beta1+beta2*t[5]+
beta3*age[i]))

D35[i]<-age[i]*(delta[1,5]*exp(beta1+beta2*t[1]+beta3*
age[i])+
delta[2,5]*exp(beta1+beta2*t[2]+beta3*age[i])+delta[3,5]*
exp(beta1+beta2*t[3]+beta3*age[i])+delta[4,5]*exp(beta1+
beta2*t[4]+beta3*age[i])+delta[5,5]*exp(beta1+beta2*t[5]+
beta3*age[i]))

}

# gee equation
for (i in 1:N){A1[i]<-
D11[i]*(Y[i,1]-exp(beta1+beta2*t[1]+beta3*age[i]))+D12[i]
*(Y[i,2]-exp(beta1+beta2*t[2]+beta3*age[i]))+D13[i]*(
Y[i,3]-exp(beta1+beta2*t[3]+beta3*age[i]))+D14[i]*(
Y[i,4]-exp(beta1+beta2*t[4]+beta3*age[i]))+
D15[i]*(Y[i,5]-exp(beta1+beta2*t[5]+beta3*age[i]))}

# gee equation
for (i in 1:N){A2[i]<-D21[i]*(Y[i,1]-exp(beta1+beta2*t[1]
+beta3*age[i]))+D22[i]*(Y[i,2]-exp(beta1+beta2*t[2]+beta3*
age[i]))+D23[i]*(Y[i,3]-exp(beta1+beta2*t[3]+beta3*age[i]))+
D24[i]*(Y[i,4]-exp(beta1+beta2*t[4]+beta3*age[i]))+
D25[i]*(Y[i,5]-exp(beta1+beta2*t[5]+beta3*age[i]))}

# gee
for (i in 1:N){A3[i]<-D31[i]*(Y[i,1]-exp(beta1+beta2*t[1]
+beta3*age[i]))+D32[i]*(Y[i,2]-exp(beta1+beta2*t[2]+beta3
*age[i]))+D33[i]*(Y[i,3]-exp(beta1+beta2*t[3]+beta3*age[i]))+
D34[i]*(Y[i,4]-exp(beta1+beta2*t[4]+beta3*age[i]))+
D35[i]*(Y[i,5]-exp(beta1+beta2*t[5]+beta3*age[i]))}

sa1<-sum(A1[])

sa2<-sum(A2[])

sa3<-sum(A3[])

# the prior distributions
beta1~dnorm(0.0,.001)
beta2~dnorm(0.0,.001)
beta3~dnorm(0.0,.001)

}
```

```
list(N = 27,M = 5, t = c(0,2,4,6,8),
age = c(31,30,25,36,22,29,31,42,37,28,36,24,23,
36,26,26,28,31,32,21,29,21,32,25,30,40,19,22),

# V is the estimated variance covariance matrix of the
observations

V = structure(.Data = c(42.5895,46.3073,31.8372,66.7496,
39.9297,
46.3073,102.7580,62.8981,92.5129,43.7643,
31.8372,62.8981,66.6508,76.1993,36.8045,
66.7496,92.5129,76.1993,215.2675,65.8087,
39.9297,43.7643,36.8045,65.8087,58.1833
),.Dim = c(5,5)),

# observations for the number of seizures for placebo
Y = structure(.Data = c(
11,5,3,3,3,
11,3,5,3,3,
6,2,4,0,5,
8,4,4,1,4,
66,7,18,9,21,
27,5,2,8,7,
12,6,4,0,2,
52,40,20,23,12,
10,14,13,6,0,
52,26,12,6,22,
33,12,6,8,4,
18,4,4,6,2,
42,7,9,12,14,
87,16,24,10,9,
50,11,0,0,5,
18,0,0,3,3,
111,37,29,28,29,
18,3,5,2,5,
20,3,0,6,7,
12,3,4,3,4,
9,3,4,3,4,
17,2,3,3,5,
28,8,12,2,8,
55,18,24,76,25,
9,2,1,2,1,
10,3,1,4,2,
47,13,15,13,12),.Dim = c(27,5)))

list(beta1 = 0, beta2 = 0, beta3 = 0)
```

TABLE 8.10

Posterior Analysis for the Placebo Group of the Epileptic Study with
Generalized Estimating Equations

Parameter	Mean	SD	Error	2½	Median	97½
β_1	3.359	0.1226	0.005441	3.134	3.359	3.582
β_2	−0.1872	0.009073	0.000125	−0.205	−0.1872	−0.1695
β_3	−0.005767	0.003751	0.000181	−0.0129	−0.005758	0.001648

The posterior analysis uses **BUGS CODE 8.5** and is executed with 65,000 observations for the simulation, with a burn in of 5,000 and a refresh of 100 (Table 8.10).

It is seen that the log mean appears to have a slope that is decreasing, and that the effect of age on the log mean is minimal, with a 95% credible interval (−0.0129, 0.001648). The reader is referred to the exercises where the main objective of the analysis is to compare the placebo group with a treatment group. Also, the simulation errors for the three regression coefficients appear to be sufficiently small and the posterior distributions symmetric about the posterior mean.

The model with a binary response was not considered, but a problem in the exercises will inform the student of the Bayesian GEE approach for estimating the parameters of the logistic link function of the mean proportion.

8.3 Generalized Mixed Linear Models for Categorical Data

Two approaches for the analysis of repeated categorical data have been presented, namely, one based on the multinomial distribution and the other based on GEE. This section presents a third alternative, one referred to a generalized linear models, because the model includes both fixed and random effects. The random effects model the heterogeneity between the various subjects of the study and, at the same time, induce a correlation between the responses at several time points. Several cases will be considered: (1) The distribution of count data will assumed to be Poisson with a log link function, (2) the distribution of the response is binary and assumed to have a binomial distribution with a logistic link function for the probability of an event, and (3) a proportional odds model applicable to ordinal categorical data. It should be stated that the random subject effects induce a complexity to the interpretation of the effects of the model. The analysis of the generalized linear model for a given set of data begins with three specifications: (1) the distribution of the observations, (2) a systematic component, and (3) the link function.

In general for the generalized linear mixed effect mode, consider:

1. The conditional distribution of the response Y_{ij} given a q by 1 vector of random effects b_i as belonging to a member of the exponential family, where

$$\text{Var}\left(Y_{ij}|b_i\right) = h\left[E\left(Y_{ij}|b_i\right)\phi\right]$$

   where:

   h is a known function of the conditional mean of Y_{ij} given b_i

   ϕ is a scalar

   Also it is assumed that given the random effects b_i, the n observations Y_{ij} ($j = 1,2,...,n$) are independent.

2. The conditional mean $E\left(Y_{ij}|b_i\right)$ depends on fixed and random effects

$$v_{ij} = X_{ij}\beta + Z_{ij}b_i \tag{8.74}$$

   where:

   X_{ij} is a 1 by p vector of known covariates
   β is a p by 1 unknown parameter vector
   Z_{ij} is a 1 by q known vector of covariate values
   b_i is a q by 1 vector of random effects

   and

$$h\left[E\left(Y_{ij}|b_i\right)\right] = v_{ij} \tag{8.75}$$

   This definition of the generalized linear model closely follows Fitzmaurice, Laird, and Ware (pp. 397–401).[2]

3. The random effects $b_1, b_2, ..., b_N$ are assumed to follow a multivariate normal distribution with q by 1 mean vector 0 and unknown q by q unknown variance–covariance matrix Σ. Also it is assumed that the random effects are independent of the covariates.

Thus, it is assumed that the covariance matrix Σ is the same for all N subjects and that the time points are the same and spaced the same way for all subjects. Of course, this can be generalized to the situation where the spacing of the time points varies from subject to subject and that the number of time points can also vary from individual to individual.

Three special cases of the exponential family of distributions will be presented, namely (1) the Poisson for count data, (2) the Bernoulli for binary data, and (3) a multinomial distribution for ordered categorical data.

8.3.1 Generalized Linear Models for Count Data

For this particular case, the model is defined as follows:

1. Conditional on the random effects b_i, the distribution of Y_{ij} is Poisson and the observations over n time points are independent. The conditional variance is

$$\mathrm{Var}\left(Y_{ij}\,|\,b_i\right) = E\left(Y_{ij}\,|\,b_i\right) \tag{8.76}$$

2. The canonical link is the log function

$$\log\left[E\left(Y_{ij}\,|\,b_i\right)\right] = v_{ij} = X_{ij}\beta + Z_{ij}b_i \tag{8.77}$$

3. The random effects are independent multivariate normal with mean vector 0 and variance–covariance matrix Σ. In addition, the random effects are independent of the covariates.

As an example, consider the experimental use of dopamine. In this study, 25 patients with stage II–IV Parkinson's disease were randomized to five groups: 8.4, 16.8, 33.5 mg of dopamine (Table 8.11). The response is a clinical global rating, which was measured at days 2, 7, and 14. This is an ordered categorical variable interpreted as follows:

1 = very much worse,

2 = much worse,

3 = slightly worse,

4 = no change,

5 = slightly improved,

6 = much improved,

and

7 = very much improved.

Note that $X_1 = 1$ if dose = 8.4 mg, otherwise 0; $X_2 = 1$ if dose = 16.8 mg, otherwise 0; $X_3 = 1$ if dose = 33.5 mg, otherwise 0; and $X_4 = 1$ if dose = 67 mg, otherwise 0. Also, when $X_1 = X_2 = X_3 = X_4 = 0$, the subject is assigned to placebo. Based on this information, the proposed model is

$$Y_{ij} \sim \mathrm{Poisson}\left(\mu_{ij}\right) \tag{8.78}$$

$$\log\left(\mu_{ij}\,|\,b_{1i}, b_{2i}\right) = \beta_1 + \beta_2 t[j] + \beta_3 X_1 + \beta_4 X_2 + X_3\beta_5 + \beta_6 X_4 + b_{1i} + b_{2i}t[j]$$

where (b_{1i}, b_{2i}), $i = 1, 2, \ldots, N$ are independent and have a two-dimensional normal distribution with a 2 by 1 mean vector 0 and unknown 2 by 2 covariance matrix Σ. Therefore, the proposed model is a generalized linear model

TABLE 8.11

Parkinson's Disease Study with Clinical Ratings

T_2	T_7	T_{14}	X_1	X_2	X_3	X_4	Patient	Dose
2	4	4	0	0	1	0	1	33.50
2	3	4	0	0	0	1	2	67.00
3	3	4	0	0	0	0	3	0.00
3	3	2	1	0	0	0	4	8.40
3	3	3	0	1	0	0	5	16.80
5	5	5	0	0	0	1	6	67.00
4	4	4	0	1	0	0	7	16.80
4	3	2	1	0	0	0	8	8.40
3	3	3	0	0	1	0	9	33.50
4	4	3	0	0	0	0	10	0.00
4	4	4	1	0	0	0	11	8.40
5	4	3	0	0	1	0	12	33.50
5	3	4	0	0	0	0	13	0.00
3	7	6	0	0	0	1	14	67.00
5	5	5	0	0	1	0	15	33.50
3	6	4	1	0	0	0	16	8.40
3	3	3	0	1	0	0	17	16.80
5	6	5	0	0	0	0	18	0.00
5	3	5	0	0	0	1	19	67.00
3	3	4	0	0	0	1	20	67.00
3	5	6	0	0	1	0	21	33.50
6	5	5	0	1	0	0	22	16.80
3	3	3	1	0	0	0	23	8.40
4	4	5	0	0	0	0	24	0.00
3	3	3	0	0	0	0	25	25.00

Source: Davis, C.S., *Statistical Methods for the Analysis of Repeated Measures*, Springer-Verlag, New York, p. 344, 2002.

where the fixed effects designate a linear response with respect to time and dose effects, and with two random effects, one corresponding to the constant term and the other b_{2i} representing the subject-to-subject variation of the time effect. The effect of the first dose 8.5 mg of the log response is β_3, while the effect of the highest dose 67 mg is β_6; thus, it is of interest to compare the various doses using the contrasts

$$d_{34} = \beta_3 - \beta_4$$

$$d_{35} = \beta_3 - \beta_5$$

$$\vdots$$

$$d_{56} = \beta_5 - \beta_6$$

(8.79)

For example, d_{34} compares the effect of dose 8.4 mg on the log response with the effect of dose 33.5 mg on the clinical response.

The Bayesian analysis will be based on the study information of Table 8.9 and formulas (8.78) and (8.79). Note that the study information is contained in the first list statement of **BUGS CODE 8.6**. It should also be noted that the random effects b_{1i} and b_{2i} are assumed to be independent. Of course, this need not be the case, they can be correlated, but it has been my experience when assuming they are correlated, the data always implies they are in fact independent! In one of the following exercises, the student will be asked to perform the analysis when the random effects are correlated.

BUGS CODE 8.6

```
model;

{

for (i in 1:N){for (j in 1:M){Y[i,j]~dpois(mu[i,j])

log(mu[i,j])<-beta[1]+beta[2]*tim[j]+beta[3]*X1[i]+beta[4]*
X2[i]+beta[5]*X3[i]+beta[6]*X4[i]+b1[i]+b2[i]*tim[j]}}

for(i in 1:6){beta[i]~dnorm(0,.0001)}

for (i in 1:N){b1[i]~dnorm(0,tau1)}

for (i in 1:N){b2[i]~dnorm(0,tau2)}

tau1~dgamma(0.001,.001)
tau2~dgamma(0.001,.001)

sigma1<-1/tau1
sigma2<-1/tau2

}

list(N = 25,M = 3,

Y = structure(.Data =
c(2,4,4,
  2,3,4,
  3,3,4,
  3,3,2,
  3,3,3,
  5,5,5,
  4,4,4,
  4,3,2,
```

```
        3,3,3,
        4,4,3,
        4,4,4,
        5,4,3,
        5,3,4,
        3,7,6,
        5,5,5,
        3,6,4,
        3,3,3,
        6,5,6,
        5,3,3,
        3,3,4,
        3,5,5,
        6,5,5,
        3,3,3,
        4,4,5,
        3,3,3),.Dim = c(25,3)),

X1 = c(0,0,0,1,0,0,0,1,0,0,1,0,0,0,0,1,0,0,0,0,0,0,1,0,0),

X2 = c(0,0,0,0,1,0,1,0,0,0,0,0,0,0,0,0,1,0,0,0,0,1,0,0,0),

X3 = c(1,0,0,0,0,0,0,0,1,0,0,1,0,0,1,0,0,0,0,0,1,0,0,0,0),

X4 = c(0,1,0,0,0,1,0,0,0,0,0,0,0,1,0,0,0,0,1,1,0,0,0,0,0),

tim = c(2,7,14))

list(beta = c(0,0,0,0,0,0), tau1 = 1, tau2 = 1)
```

Execution of the Bayesian analysis will be with 150,000 observations for the simulation, with a burn in of 5,000 and a refresh of 100. Table 8.12 reports the Bayesian analysis.

This clinical response is measured on days 2, 7, and 14. Table 8.10 depicts the descriptive statistics for the Parkinson's disease study. One should compare the descriptive statistics of Table 8.13 with the Bayesian posterior analysis of Table 8.12.

Based on Table 8.12, note the effect of $t[j]$ on $E\left[\log\left(\mu_{ij}\right)\right]$ is estimated as -0.00041 via the posterior mean, that is, for every unit increase in time $t[j]$, the average log mean decreases by the amount -0.000410, a miniscule amount.

When comparing the effect of the various doses on the clinical response, it appears obvious that there are no differences. The 95% credible intervals for the six differences d_{ij}, $i = 3,4,5$ and $j = 4,5,6$ with $i < j$, implying the effects of the various doses on the clinical response are the same. Beta regression coefficients appear to have symmetric posterior distributions; however, the variance components do have asymmetric posterior distribution.

TABLE 8.12

Bayesian Analysis of Parkinson's Disease Study with Dopamine

Parameter	Mean	SD	Error	2½	Median	97½
β_1	1.364	0.1667	0.000857	1.033	1.366	1.686
β_2	−0.000410	0.01293	0.0000632	−0.02585	−0.000360	0.02464
β_3	−0.1549	0.2188	0.001209	−0.5851	−0.1544	0.2731
β_4	−0.03853	0.227	0.01183	−0.4855	−0.03794	0.4068
β_5	−0.02289	0.2123	0.001113	−0.4417	−0.022360	0.3926
β_6	−0.000766	0.2111	0.001117	−0.4184	0.0007905	0.4094
d_{34}	−0.1163	0.2421	0.001245	−0.592	−0.1164	0.3609
d_{35}	−0.132	0.271	0.001143	−0.5793	−0.1318	0.3147
d_{36}	−0.1541	0.227	0.001099	−0.6015	−0.1547	0.2945
d_{45}	−0.01564	0.2354	0.001135	−0.4757	−0.01612	0.447
d_{46}	−0.03777	0.2346	0.001185	−0.4973	−0.03793	0.4265
d_{56}	−0.022120	0.2192	0.00114	−0.448	−0.022840	0.4103
σ_1^2	0.00972	0.01361	0.000167	0.0005258	0.004851	0.04795
σ_2^2	0.0005477	0.000167	0.000525	0.004851	0.04795	0.001303

TABLE 8.13

Descriptive Statistics for Parkinson's Disease:
Log Clinical Response versus Time and Log
Clinical Response versus Dose

Time	Mean	SD	Median
2 days	1.2813	0.30094	1.0986
7 days	1.3314	0.26176	1.3863
14 days	1.3363	0.29617	1.3862
Dose			
0 mg	1.4083	0.23934	1.3863
8.40 mg	1.1867	0.28117	1.0985
16.80 mg	1.2705	0.28675	1.3863
33.50 mg	1.3334	0.28765	1.3863
67.00 mg	1.3829	0.34023	1.3863

In light of the posterior analysis, the various doses and age do not have an effect of the log response; thus, the model needs to be modified to:

$$Y_{ij} \sim \text{Poisson}\left(\mu_{ij}\right) \tag{8.80}$$

$$\log\left(\mu_{ij}\,|\,b_{1i},b_{2i}\right) = \beta_1 + b_{1i} \tag{8.81}$$

where b_{1i} are independent normal random variables with mean 0 and variance $\sigma_1^2 = 1/\tau_1$. The prior distribution for β_1 is normal with mean 0 and

TABLE 8.14

Posterior Analysis of Parkinson's Disease Study with Dopamine for Reduced Model

Parameter	Mean	SD	Error	2½	Median	97½
β_1	1.344	0.06178	0.000195	1.221	1.345	1.463
σ_1^2	0.007736	0.01007	0.000112	0.000512	0.004193	0.03538

precision 0.0001, while the prior distribution of τ_1 is gamma with 0.0001 for the first parameter and 0.0001 for the second. Table 8.14 reports the posterior analysis for reduced model.

Comparing Table 8.10 to 8.11, one sees very little difference in the posterior means of β_1 and σ_1^2, implying that the effect of the doses and age on the log clinical response is extremely small. The student is asked to verify Tables 8.10 and 8.11 as an exercise.

The effect of dose on clinical response is best understood by referring to Figure 8.3, which is a scatter plot of clinical response versus time by dose group. From the graph, does it appear that dose has little effect on the clinical response over time. That is, does Figure 8.3 corroborate the Bayesian analysis of Table 8.10?

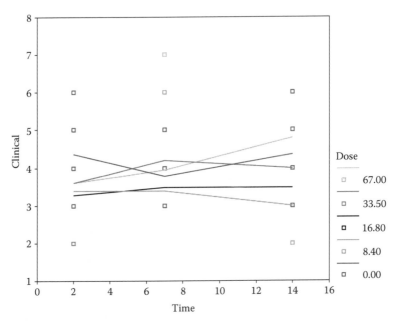

FIGURE 8.3
Clinical response versus time by dose.

8.3.2 Generalized Linear Mixed Model for Binary Data

Let $Y_{ij} = 0,1$ be a binary response for subject i at time j, then the logistic generalized mixed model is defined as:

1. Conditional on the random effect b_i, Y_{ij} are independent and are Bernoulli distributed with mean $E(Y_{ij}|b_i)$ and variance

$$\text{Var}(Y_{ij}|b_i) = E(Y_{ij}|b_i)\left[1 - E(Y_{ij}|b_i)\right] \qquad (8.82)$$

2. The logistic link function is given by

$$v_{ij} = X_{ij}\beta + b_i \qquad (8.83)$$

where

$$v_{ij} = \frac{\log\left[\Pr(Y_{ij} = 1|b_i)\right]}{\Pr(Y_{ij} = 0|b_i)} \qquad (8.84)$$

3. The random effects $b_i(i = 1,2,\ldots,n)$ are independent and have a normal distribution with mean 0 and variance σ_1^2.

This example is based on the study of Stokes, Davis, and Koch[11] of a clinical trial for subjects with pulmonary disease and consists of 111 patients recruited from two clinics and are randomized to receive placebo or an active treatment. Responses were recorded as 1 (good) or 0 (poor) and measured at baseline and visits 1, 2, 3, and 4. Also known are the gender of subjects, their age, and if they have received placebo or not.

Suppose that the response at time j (= 0,1,2,3,4) for subject i is given by

$$Y_{ij} \sim \text{Bernoulla}(\mu_{ij}) \qquad (8.85)$$

where

$$\log \text{it}(\mu_{ij}) = \beta_1 + \beta_2 t[j] + \beta_3 t^2[j] + \beta_4 \text{age}[i] + \beta_5 \text{gen}[i] + \beta_6 \text{grp}[i] + b_{1i} \qquad (8.86)$$

and $i = 1,2,\ldots,N$, $j = 1,2,\ldots,n$. N is the number of subjects and n is the number of time points per subject. A quadratic response is proposed for the proportion at each time point $t[j]$, and there is one random effect, which accounts for the subject-to-subject variation of the intercept on the logistic scale. The Bayesian analysis is based on **BUGS CODE 8.7** and the data from Stokes, Davis, and Koch[11] is included in the first list statement of the code. For this example, only 56 patients from center 1 are included.

The Bayesian analysis is executed with 65,000 observations for the simulation, with a burn in of 5,000 and a refresh of 100, and it is reported in Table 8.15.

BUGS CODE 8.7

```
model;

{

for(i in 1:N){for (j in 1:M){Y[i,j]~dbern(mu[i,j])}}

for(i in 1:N){for(j in 1:M){logit(mu[i,j])<-beta[1]+beta[
2]*t[j]+beta[3]*t[j]*t[j]+beta[4]*age[i]+beta[5]*gen[i]+b
eta[6]*grp[i]+b1[i]}}
# probability of a good response
for(i in 1:N){for (j in 1:M){p[i,j]<-(exp(beta[1]+beta[2]*
t[j]+beta[3]*t[j]*t[j]+beta[4]*age[i]+beta[5]*gen[i]+beta[6]*
grp[i]+b1[i]))/(1+exp(beta[1]+beta[2]*t[j]+beta[3]*t[j]*t[j]+
beta[4]*age[i]+beta[5]*gen[i]+beta[6]*grp[i]+b1[i]))}}

for(i in 1:6){beta[i]~dnorm(0,.0001)}

for(i in 1:N){b1[i]~dnorm(0,tau1)}

tau1~dgamma(.0001,.0001)

sig1<-1/tau1

}

list(N = 56,M = 5,

Y = structure(.Data = c(
0,      0,      0,      0,      0,
0,      0,      0,      0,      0,
1,      1,      1,      1,      1,
1,      1,      1,      1,      0,
1,      1,      1,      1,      1,
0,      0,      0,      0,      0,
0,      1,      0,      1,      1,
0,      0,      0,      0,      0,
1,      1,      1,      1,      1,
1,      0,      1,      1,      0,
1,      1,      1,      1,      1,
0,      1,      1,      1,      0,
1,      1,      0,      0,      0,
0,      0,      0,      0,      0,
1,      1,      1,      1,      1,
0,      0,      0,      0,      1,
0,      0,      0,      0,      0,
0,      0,      1,      1,      1,
```

```
0,      0,      0,      0,      0,
1,      1,      0,      1,      0,
0,      1,      0,      1,      0,
1,      1,      1,      1,      1,
1,      1,      1,      1,      1,
0,      1,      0,      0,      0,
0,      0,      0,      0,      0,
0,      0,      0,      0,      0,
1,      1,      0,      1,      1,
0,      1,      1,      0,      0,
0,      1,      1,      0,      0,
0,      0,      0,      0,      0,
0,      0,      0,      0,      0,
0,      1,      1,      1,      1,
1,      1,      1,      1,      0,
0,      0,      1,      1,      0,
0,      0,      0,      1,      0,
0,      0,      0,      0,      0,
0,      0,      1,      0,      0,
0,      0,      0,      0,      0,
0,      0,      0,      0,      0,
0,      0,      0,      0,      0,
0,      0,      0,      0,      1,
0,      0,      1,      0,      0,
1,      1,      1,      1,      1,
0,      0,      0,      0,      0,
0,      1,      0,      1,      1,
0,      0,      0,      0,      0,
0,      0,      0,      0,      0,
1,      1,      1,      1,      1,
1,      1,      1,      1,      1,
0,      0,      1,      1,      1,
0,      0,      1,      1,      1,
0,      0,      0,      1,      0,
0,      0,      0,      1,      0,
1,      1,      1,      1,      0,
1,      1,      1,      1,      1,
0,      1,      1,      0,      1),.Dim = c(56,5)),

age = c(4,4,6,4,2,8,2,3,4,4,1,3,3,4,4,3,2,8,3,1,3,7,3,0,1,4,
2,3,3,0,2,0,2,2,2,5,4,7,3,1,2,0,2,6,4,6,3,2,4,8,3,5,2,6,2,3,
3,6,1,9,2,8,3,7,2,3,3,0,1,5,2,6,4,5,3,1,5,0,2,8,2,6,1,4,3,1
,1,3,2,7,2,6,4,9,6,3,5,7,2,7,2,2,1,5,4,3,3,2,1,1,2,4,2,5),

grp = c(0,0,1,0,0,1,0,1,1,0,1,1,0,0,0,1,0,1,0,1,1,1,1,1,0
,1,0,0,0,1,0,1,1,0,1,0,1,1,0,0,0,1,0,0,0,0,0,1,0,1,1,0,1,
1,0,1),
```

```
gen = c(0,0,0,0,1,0,0,0,0,0,0,0,0,0,0,0,0,0,0,1,1,0,0,0,0,0,0,1
,0,0,1,0,0,0,0,0,0,0,1,0,0,0,0,0,0,0,0,0,0,0,0,0,0,0,0,0,1,
0,0,0),

t = c(0,1,2,3,4))

list(beta = c(0,0,0,0,0,0), tau1 = 1)
```

TABLE 8.15

Posterior Analysis of Respiratory Study

Parameter	Mean	SD	Error	2½	Median	97½
β_1	−2.135	1.273	0.02478	−4.754	−2.09	0.2833
β_2	1.586	0.4956	0.00344	0.6456	1.577	2.584
β_3	−0.357	0.1172	0.000804	−0.5927	−0.3548	−0.1346
β_4 (age)	−0.07730	0.2787	0.005475	−0.6392	−0.07514	0.4639
β_5 (sex)	−1.041	1.722	0.02923	−4.513	−1.023	2.285
β_6 (treat)	1.533	1.145	0.02241	−0.655	1.489	3.899
σ_1^2	13.31	5.871	0.08935	5.53	12.06	28.12

It can be shown for subject 1 that the posterior mean is −3.164 with a standard deviation of 2.417, and 95% credible interval (−8.746, 0.6977); thus, ones estimate of the constant term of subject 1 is −2.135, −3.164.

Consider a change in the visit number from $t[j]$ to $t[j + 1]$, then what is the effect on the difference $\log(\mu_{i,j+1}|b_i) - \log(\mu_{ij}|b_i)$, in the log conditional mean? It can be confirmed that it is

$$\log(\mu_{i,j+1}|b_i) - \log(\mu_{ij}|b_i) = \beta_2\{t[j+1] - t[j]\} + \beta_5\{t^2[j+1] - t^2[j]\} \qquad (8.87)$$

Note that if $t[j+1] = 2$ and $t[j] = 1$, then the difference in the conditional mean is $\beta_2 + 3\beta_3$.

The Bayesian analysis of the respiratory study confirms that the mean is quadratic clinical response (0 = poor and 1 = good) over visits 0, 1, 2, 3, and 4. It is also confirmed by Table 8.16 and Figure 8.4, which clearly shows a quadratic response of the probability of a good response versus visit number.

Note that the probability of a "good" response for subject i at time j is given by

$$p[i,j] = \frac{\xi(i,j)}{1 + \xi(i,j)} \qquad (8.88)$$

TABLE 8.16

Descriptive Statistics of Respiratory Study

Visit	Mean	SD
0	0.32	0.471
1	0.46	0.503
2	0.63	1.244
3	0.52	0.504
4	0.38	0.489

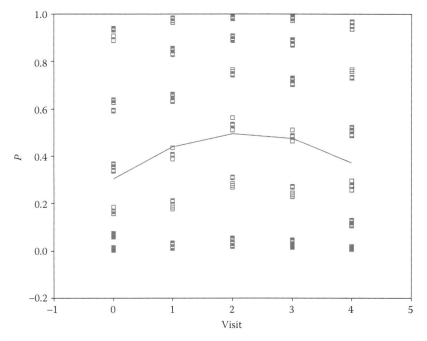

FIGURE 8.4
Probability of a good response versus visit.

where $\xi(i,j) = \exp(\beta_1 + \beta_2 t[j] + \beta_3 t^2[j] + \beta_4 \text{age}[i] + \beta_5 \text{gen}[i] + \beta_6 \text{grp}[i] + b_{1i})$, which shows the dependence of the probability on all the factors taken into account by the logistic model (8.83).

The Bayesian analysis implies that there is a treatment effect, that is, the active treatment has a higher probability of a good response at visits 1, 2, 3, and 4, and this is implied by Figure 8.5, which is a scatter plot of the probability of a good response versus visit by group (placebo, active treatment, and overall). The three graphs are depicted as a lowess plot.

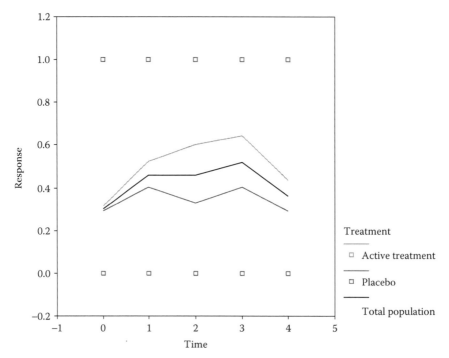

FIGURE 8.5
The probability of a good response versus visit by treatment.

8.3.3 Generalized Linear Mixed Model for Ordinal Data

Consider the response for subject i at time j. $Y_{ij} = 1, 2, ..., K$, the generalized linear mixed model for ordinal categorical data, is specified by:

1. The conditional distribution of Y_{ij} given a q by 1 vector of random effects b_i is multinomial.
2. The link function is the cumulative response probabilities given as

$$\log \frac{\Pr\left(Y_{ij} \leq k \mid b_i\right)}{\Pr\left(Y_{ij} > k \mid b_i\right)} = v_{ij} \tag{8.89}$$

where

$$v_{ij} = \alpha_k + X_{ij}\beta + Z_{ij}b_i \tag{8.90}$$

where:

X_{ij} is a 1 by p known vector of covariates
β is a p by 1 unknown vector of regression parameters

Z_{ij} is a 1 by q vector of known covariate values

b_i is a q by 1 vector of random effects

α_k is an unknown scalar for the ordinal value k

3. The random effects b_i are independent and have a multivariate normal distribution with mean vector 0 and unknown covariance matrix Σ. For example, refer to the experimental use of dopamine.

Table 8.11 contains the information about this study, where the first three columns give the clinical rating at each time point ($t = 2, 7$, and 14 days) for each of 25 patients. When employing the proportional odds model of (8.89) and (8.90), it is useful to express the information in Table 8.17.

Thus, at day 2, 2 subjects scored 2, 11 scored 3, 4 scored 4, 5 scored 5, 2 scored 6, and last 0 scored 7. There were no 1 scores, thus it will not be included in the analysis.

Consider the model

$$Y_{ij} \sim \text{multinomial}\left(p_{i1}, p_{i2}, \ldots, p_{i6}; n_i\right) \tag{8.91}$$

where $p_{ij} = \Pr(Y_{ij} = j)$ with $i = 2, 7, 14$ and $j = 1, 2, 3, 4, 5, 6$. That is for time i and score j, the probability a subject scores j is p_{ij}. Thus, for each time point, the various scores are assumed to follow a multinomial distribution. Note n_i is the total number of scores at time I, summed over the six scores.

The proportional odds assumptions are expressed by

$$\text{logit}\left(\gamma_{ij}\right) = \text{alpha}_j - \left(\beta_1 + \beta_2 i + b_i\right) \tag{8.92}$$

where:

i denotes time i (= 2, 7, and 14 days)

$j = 1, 2, \ldots, K - 1$, and the ordinal scores are $1, 2, \ldots, K$

$$\gamma_{ij} = \Pr\left(Y_{ij} \leq j | b_i\right) \tag{8.93}$$

and

$$b_i \sim nid(0, \tau) \tag{8.94}$$

TABLE 8.17

Multinomial Outcomes for Parkinson's Disease Study

Time (days)	2 Rating	3 Rating	4 Rating	5 Rating	6 Rating	7 Rating
2	2	11	4	5	2	0
7	0	12	6	5	1	1
14	2	8	8	5	2	0
Total	4	31	18	15	5	1

where $\tau > 0$.

Note that the proportional odds assumption (8.92) assumes that $\beta_1 + \beta_2 i + b_i$ does not depend on the ordinal score j.

The random effects measure the heterogeneity over the three time points with variance $\sigma^2 = 1/\tau$. For the Bayesian analysis, β_1 and β_2 are given noninformative normal distributions $n(0, 0.0001)$ and τ is specified a noninformative gamma distribution $(0.001, 0.001)$. Note $\beta_1 + b_i$ denotes the constant term at time point i. Consider two time points i and $i + 5$, then the difference of the logits at ordinal score j is

$$\operatorname{logit}\left(\gamma_{i+5,j}\right) - \operatorname{logit}\left(\gamma_{i,j}\right) = 5\beta_2 \tag{8.95}$$

which measures the change in the logit over a period of five days.

BUGS CODE 8.8 will execute the analysis for the Parkinson's disease study using the proportional odds model and is based on the code found in Congdon (p. 102).[12] For additional information about the analysis of categorical data including the proportional odds model, see Agresti (p. 164),[13] and for information from a Bayesian view using the proportional odds model for ordered categorical data, see Broemeling.[14]

Table 8.18 presents the Bayesian analysis for the Parkinson's disease study and is based on **BUGS CODE 8.8**, which is executed with 650,000 observations for the simulation, with a burn in of 5,000 and a refresh of 100.

The Bayesian analysis consists of estimates of the parameters of the proportional odds model. For example, the estimate of the constant term is −0.8539 using the posterior median, while the slope over time is estimated

---

BUGS CODE 8.8

```
model;

{

for (i in 1:M){Y[i,1:K]~dmulti(p[i,1:K],N[i])

N[i]<-sum(Y[i,])}

for(i in 1:M){for(j in 1:K-1){logit(gamma[i,j])<-alpha[j]
-mu[i]}}

for(i in 1:M){p[i,1]<-gamma[i,1]}

for(i in 1:M){for(j in 2:K-1)
{p[i,j]<-gamma[i,j]-gamma[i,j-1]}}
```

```
for(i in 1:M){p[i,K]<-1-gamma[i,K-1]}

for(i in 1:M){mu[i]<-beta[1]+beta[2]*tim[i]+b[i]}

for(i in 1:2){beta[i]~dnorm(0,.0001)}

alpha[1]~dnorm(0,1)
alpha[2]~dnorm(0,1)
alpha[3]~dnorm(0,1)
alpha[4]~dnorm(0,1)
alpha[5]~dnorm(0,1)I(theta[4],)
for (i in 1:M){b[i]~dnorm(0,tau)}
tau~dgamma(.001,.001)
sigma<-1/tau

}

list(K = 6,M = 3,

Y = structure(.Data =
c(2,11,4,5,2,0,
0,12,6,5,1,1,
2,8,8,5,2,0),.Dim = c(3,6)),

tim = c(2,7,14))

list(beta = c(0,0), theta = c(0,0
```

TABLE 8.18

Bayesian Analysis for Parkinson's Disease Study

Parameter	Mean	SD	Error	2½	Median	97½
β_1	−0.7004	1.346	0.04109	−2.647	−0.8539	3.279
β_2	−0.003592	0.1505	0.004798	−0.4722	0.01686	0.1902
σ^2	2.826	52.53	0.3712	0.00082	0.03812	19.98
α_1	−3.11	0.5504	0.007554	−4.198	−3.107	−2.037
α_2	−.8242	0.486	0.008285	−1.77	−0.8233	0.1296
α_3	0.126	0.4802	0.008107	−0.8047	0.1254	1.064
α_4	1.329	0.4929	0.007419	0.3689	1.33	2.298
α_5	2.456	0.5809	0.006396	1.349	2.433	3.631
b_1	−0.0474	0.8002	0.03588	−1.464	−0.004308	1.307
b_2	0.001245	0.4865	0.01721	−1.108	0.007953	0.9199
b_3	−0.04022	0.6875	0.02623	−1.712	−0.003659	1.299

as 0.01686, via the posterior median. The variance of the constant random effect is estimated as 0.03812, using the posterior median. Note the skewness of the posterior distributions, and especially for the variance σ^2. The constant term for the first time point (day 2) is estimated as -0.8539, -0.004308, using the posterior median for β_1 and b_1. For the second cutoff point α_2, the proportional odds at time $t[i] = 2$ is measured as $\alpha_2 - (\beta_1 + \beta_2 t[j] + b_1)$ and estimated as $-0.8233 - (-0.8539 + 0.01686 \times 2 - 0.0043) = 0.001$. In Figure 8.6, the estimated multinomial probabilities at days 2, 7, and 14 are displayed, while Table 8.17 exhibit the actual multinomial probabilities.

Figure 8.5 is the graphical analog of Table 8.19.

This analysis used data that was aggregated over 25 patients. Consider the list statement for **BUGS CODE 8.8**, where the first row of six counts corresponds to day 2, while the second to day 7, and the third to day 14.

```
Y = structure(.Data =
c(2,11,4,5,2,0,
0,12,6,5,1,1,
2,8,8,5,2,0),.Dim = c(3,6)),
```

An alternative analysis will be based on the individual patient data, namely, where there are 25 patients, averaged over three time points, and for each patient a score from 2 to 7 is assigned with the following interpretation:

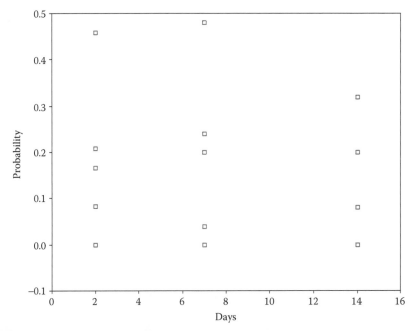

FIGURE 8.6
Computed multinomial probabilities versus days 2, 7, and 14.

TABLE 8.19

Multinomial Probabilities of Parkinson's Disease by Day

Day	Score 1	Score 2	Score 3	Score 4	Score 5	Score 6
2	0.0833	0.4583	0.1666	0.2083	0.0833	0.0000
7	0.0000	0.4800	0.2400	0.2000	0.0400	0.0400
14	0.0800	0.3200	0.3200	0.2000	0.0800	0.0000

2 = much worse, 3 = slightly worse, 4 = no change, 5 = slightly improved, 6 = much improved, and 7 = very much improved. No patient was assigned a score of 1; thus, the model for a patient's observation is assumed to follow a multinomial with six categories, corresponding to the six scores (Table 8.20).

The cumulative proportional odds model for this alternative is defined by the following equations;

TABLE 8.20

Multinomial Response for the Parkinson's Disease Study

Subject	Score 2	Score 3	Score 4	Score 5	Score 6	Score 7
1	1	0	2	0	0	0
2	1	1	1	0	0	0
3	0	2	1	0	0	0
4	1	2	0	0	0	0
5	0	3	0	0	0	0
6	0	0	0	3	0	0
7	0	0	3	0	0	0
8	1	1	1	0	0	0
9	0	3	0	0	0	0
10	0	1	2	0	0	0
11	0	0	3	0	0	0
12	0	1	1	1	0	0
13	0	1	1	1	0	0
14	0	1	0	0	1	1
15	0	0	0	3	0	0
16	0	1	1	0	1	0
17	0	3	0	0	0	0
18	0	0	0	1	2	0
19	0	2	0	1	0	0
20	0	0	2	1	0	0
21	0	1	0	2	0	0
22	0	0	0	2	1	0
23	0	3	0	0	0	0
24	0	0	2	1	0	0
25	0	3	0	0	0	0

$$Y_{ij} \sim \text{Multinomial}\left(p_1, p_2, \ldots, p_6; n(i)\right) \tag{8.95}$$

where

$$p_j = \Pr\left(Y_{ij} = 1 \big| b_i\right) \tag{8.96}$$

and

$$\text{logit}\left(Y_{ij} \le j \big| b_i\right) = \alpha_j - \left\{\beta_1 + \beta_2 j + \sum_{l=3}^{l=6} \beta_l X_l(i) + b_i\right\} \tag{8.97}$$

where:
$i = 1, 2, \ldots, 25$
$j = 1, 2, \ldots, K (= 6)$

In addition, the covariates are defined as $X_1(i) = 1$ if for subject i, dose = 8.4 mg, otherwise 0; $X_2(i) = 1$ if dose = 16.8 mg, otherwise 0; $X_3(i) = 1$ if dose = 33.5 mg, otherwise 0; and $X_4(i) = 1$ if dose = 67 mg, otherwise 0. Also, when $X_1 = X_2 = X_3 = X_4 = 0$, the subject is assigned to placebo. The random subject effects are independent and distributed normal with mean 0 and variance σ^2.

For the Bayesian analysis, the precision $\tau > 0$ is assigned an uninformative gamma distribution, β_l are assigned uninformative normal, as well as α_j, $j = 1, 2, \ldots, 6$. **BUGS CODE 8.9** closely follows formulas (8.95) through (8.97).

BUGS CODE 8.9

```
model;
{

for (i in 1:N){Y[i,1:K]~dmulti(p[i,1:K],n[i])}

for(i in 1:N){n[i]<-sum(Y[i,])}


for(i in 1:N){for(j in 1:K-1){logit(gamma[i,j])<-alpha[j]-
mu[i]}}

for(i in 1:N){p[i,1]<-gamma[i,1]}

for(i in 1:N){for(j in 2:K-1){p[i,j]<-gamma[i,j]-gamma[i,j-1]}}

for(i in 1:N){p[i,K]<-1-gamma[i,K-1]}

for(i in 1:N){mu[i]<-beta[1]+beta[2]*i+beta[3]*X1[i]+beta[4]*
X2[i]+beta[5]*X3[i]+beta[6]*X4[i]+b[i]}
```

```
for(i in 1:6){beta[i]~dnorm(0,.0001)}
alpha[1]~dnorm(0,1)
alpha[2]~dnorm(0,1)
alpha[3]~dnorm(0,1)
alpha[4]~dnorm(0,1)
alpha[5]~dnorm(0,1)
for (i in 1:N){b[i]~dnorm(0,tau)}
tau~dgamma(.001,.001)
sigma<-1/tau


}

list(K = 6,N = 25,
Y = structure(.Data =
c(

1,0,2,0,0,0,
1,1,1,0,0,0,
0,2,1,0,0,0,
1,2,0,0,0,0,
0,3,0,0,0,0,
0,0,0,3,0,0,
0,0,3,0,0,0,
1,1,1,0,0,0,
0,3,0,0,0,0,
0,1,2,0,0,0,
0,0,3,0,0,0,
0,1,1,1,0,0,
0,1,1,1,0,0,
0,1,0,0,1,1,
0,0,0,3,0,0,
0,1,1,0,1,0,
0,3,0,0,0,0,
0,0,0,1,2,0,
0,2,0,1,0,0,
0,2,1,0,0,0,
0,1,0,2,0,0,
0,0,0,2,1,0,
0,3,0,0,0,0,
0,0,2,1,0,0,
0,3,0,0,0,0),.Dim = c(25,6)),
X1 = c(0,0,0,1,0,0,0,1,0,0,1,0,0,0,0,1,0,0,0,0,0,0,1,0,0),
X2 = c(0,0,0,0,1,0,1,0,0,0,0,0,0,0,0,0,1,0,0,0,0,1,0,0,0),
X3 = c(1,0,0,0,0,0,0,0,1,0,0,1,0,0,1,0,0,0,0,0,1,0,0,0,0),
X4 = c(0,1,0,0,0,1,0,0,0,0,0,0,0,1,0,0,0,0,1,1,0,0,0,0,0))

list(beta = c(0,0,0,0,0,0), alpha = c(0,0,0,0,0),tau = 1)
```

TABLE 8.21

Posterior Analysis for Parkinson's Disease Study

Parameter	Mean	SD	Error	2½	Median	97½
α_1	−3.556	0.6122	0.01517	−4.761	−3.554	−2.358
α_2	−0.999	0.514	0.01503	−2.005	−1.005	0.01857
α_3	0.2185	0.5009	0.01448	−0.7569	0.2158	1.215
α_4	1.617	0.5267	0.01372	0.5942	1.618	2.657
α_5	2.755	0.6068	0.01263	1.594	2.743	3.99
β_1	−1.315	0.9916	0.0384	−3.236	−1.322	0.6969
β_2	0.05226	0.04142	0.001251	−0.02785	0.5157	0.1354
β_3	−0.9762	0.9111	0.0205	−2.807	−0.9644	0.7873
β_4	−0.199	0.9407	0.02005	−2.128	−0.1882	1.634
β_5	0.08088	0.9096	0.02089	−1.745	0.08363	1.875
β_6	0.1383	0.9228	0.02028	−1.679	0.1372	1.978
σ^2	1.097	0.9883	0.02419	0.006424	0.8789	3.601

Table 8.21 reports the posterior analysis for the Parkinson's disease study using the cumulative proportional odds model (8.95), and based on **BUGS CODE 8.9,** the analysis is executed with 150,000 observations, with a burn in of 5,000 and a refresh of 100.

Be aware that the analysis is depicted with five equations. For example, the conditional proportional odds model (8.95) for subject 7 at the third value $j = 3$ is given by $\alpha_3 - \{\beta_1 + 3\beta_2 + \beta_3 X_1[i] + \beta_4 X_2[i] + \beta_5 X_3[i] + \beta_6 X_4[i] + b_i\}$ and is estimated by $0.2158 - \{-1.322 + 3 \times 0.0488 - 0.9319 \times 0 - 0.1692 \times 1 + 0.0852$ $8 \times 0 + 0.1448 \times 0 + b_7\}$ or 1.2343 assuming $b_7 = 0$.

8.4 Comments and Conclusions

This chapter presents the Bayesian analysis of categorical responses repeated over various time points, and three approaches are considered: (1) assuming the responses follow a multinomial distribution, (2) using an estimating equations approach, and (3) with the generalized mixed linear model with random and fixed effects. In the first approach, the multinomial model is appropriate for patterns of the responses, while for the second approach, a particular model is not used, and for the third, the random effects induce a correlation between the responses over time. When the responses are categorical, the distribution of the responses is a member of the exponential class of distributions with a canonical link function that models the mean. Sometimes it is appropriate to assume the distribution for the categorical response is Poisson with a log link function, while in particular for binary data, the response is assumed to follow a Bernoulli distribution with a logistic link, and last when the response is an ordered categorical value, a cumulative proportional odds mode is often employed.

The first approach (1) is illustrated using an example from Davis,[1] where each subject responds to three drugs, and the response is binary, with either a good or a bad response to the drug; thus, there are eight possible patterns that are modeled with a multinomial distribution, and the Bayesian analysis depends on the conjugate distribution (conjugate to the multinomial). The data is reported in Table 3.14, the analysis is based on **BUGS CODE 8.1**, and the results are portrayed in Table 8.1. A second example using the multinomial distribution is a respiratory study with a binary response observed over ages 9, 10, 11, and 12, and the response is a yes if during the previous 30 days, the subject has experienced a serious wheezing incidence. It is suspected that air pollution is responsible for the wheezing, and in this case there are 16 possible patterns modeled as a multinomial. **BUGS CODE 8.2** is executed with 65,000 observations for the simulation, and the results are reported in Table 8.3. Finally, for the third example, two groups (males and females) are compared with regard to their church attendance of Cornoni-Huntley et al.,[9] where the binary response is measured at years 0, 3, and 6. The answer is yes if the subject attended church on a regular basis the previous two years. Obviously, there are eight multinomial patterns for the two sexes, and the Bayesian analysis is executed with **BUGS CODE 8.3** using 65,000 observations for the simulation. The two groups are compared on the basis of Table 8.5, which reports the Bayesian analysis. A serious limitation of the first approach is that continuous covariates cannot be employed.

GEE is the second approach and described by Equations 8.45 through 8.47, and the approach taken here depends on an accurate estimate of the variance–covariance matrix of the observations over the time points. Thall and Vail's[15] analysis of Alzheimer's disease is reanalyzed with a Bayesian version with the estimating equations approach assuming a Poisson distribution for the number of correctly recalled words observed at time points 0, 1, 2, 4, and 6 days. A log function is employed as a link and is expressed as a line function of time. See Table 8.9 for the Bayesian analysis and Figure 8.2 for a scatter plot of the response (number of correctly recalled words) over time. The plot corroborates the analysis that demonstrates a flat response over time. For the second example with a log link, the epilepsy trial of Hand and Crowder[3] is presented where the main response is the number of seizures in the previous two weeks and observed at days 0, 2, 4, 6, and 8 weeks. Again the Poisson distribution models the number of seizures and the log link function is expressed as a linear function of time with age as a covariate. Also, as required, a good estimate of the variance–covariance matrix of the response over the five time points is based on the data of Table 8.7, and the Bayesian analysis is executed with **BUGS CODE 8.5**. See Table 8.10 for the posterior analysis, which reports a very weak effect of time and age on the log of the main response.

For the third approach, namely the generalized mixed linear model, three variations are considered: (1) a Poisson distribution for the categorical response, (2) a Bernoulli distribution for a binary response, and (3) a multinomial model for an ordinal repeated measure. In the first alternative (1), a log link function is used to express the fixed and random effects, while for the second (2), the logistic link is appropriate, and for the third (3) a cumulative proportional odds mode is expressed as a linear function of fixed and random effects. It should be emphasized that the random effects take into account the heterogeneity between the subjects in the study, and their inclusion induced an association between the responses at the various time points. See Equations 8.76 and 8.77 define on general, the generalized mixed linear model.

Previous examples are revisited but now modeled as a generalized mixed linear model; therefore, consider the Parkinson's disease study using dopamine as a therapeutic agent. Recall the defining specifications for a Poisson response for the number of seizures.

The response for subject i at time j is given by Equation 8.78, where there are two random effects, one for the subjects intercept and the other for the slope with respect to time, which are assumed to be normal with mean zero and unknown variances. The four covariates $X_l, l = 1,2,3,4$, correspond to the four-dose regimens, and **BUGS CODE 8.6** is executed with 150,000 observations for the simulation and the posterior analysis is reported in Table 8.11. It shows a flat response and that the covariates (corresponding to the four doses) have minimal effect of the log clinical response, and the two random effects have very small posterior variances.

For the second example of a generalized mixed linear model, the response is binary (good and bad responses) and assumes to be Bernoulli with a logistic link function, for the Stokes, Davis, and Koch[11] study of pulmonary disease. Gender and age are covariates, and only center 1 data is employed for the example. For the Bayesian analysis, **BUGS CODE 8.7** is executed with 150,000 observations, and the posterior distributions are reported in Table 8.14, the results of which are corroborated by Figure 8.5.

The chapter is concluded with the Parkinson's disease dopamine study, which is analyzed in two ways with a multinomial response and a cumulative log odds for the link function: (1) three time points, where the responses (with values 2, 3, 4, 5, 6 for the clinical response) are averaged over the 25 subjects, and (2) 25 subjects, where the responses are averaged over the three time points using the four covariates corresponding to the four doses of dopamine.

For the first analysis, (1) the cumulative proportional odds model is expressed via Equations 8.92 through 8.94, and the posterior analysis is executed with **BUGS CODE 8.8** and reported in Table 8.18. With regard to the second scenario (2) with four discrete covariates, the multinomial distribution and the cumulative proportional odds link function is defined by

Equations 8.95 and 8.97, with 150,000 observations for the simulation and the posterior analysis is reported in Table 8.21. As before, using the effect of the covariates is minimal.

Exercises

1. Describe the three approaches of this chapter used to analyze categorical repeated data.

2. Based on the information of Table 3.14 for a subject's response (favorable or unfavorable) to three drugs, A, B, and C, confirm the posterior analysis of Table 8.1.

   a. Estimate the probability of a favorable response to drugs A, B, and C. Use the posterior median.

   b. Describe the eight multinomial patterns for the data of Table 3.14.

   c. Are the probabilities of a favorable response to the three drugs the same?

3. Refer to formula (8.8) for the Dirichlet distribution with eight categories.

   a. What do the θ_i, $i = 1,2,...,8$ represent?

   b. What do the x_i, $i = 1,2,...,8$ represent?

   c. What is the connection between the Dirichlet distribution (8,8) and the corresponding multinomial?

4. Refer to the respiratory study Davis (p. 187),[1] of Table 8.2.

   a. What are the 16 patterns of multinomial response?

   b. Show that the distribution of $X = (X_1, X_2,.., X_{16})$ is multinomial where X_i is the frequency of the ith multinomial pattern.

   c. Assuming the uninformative prior (8.19) for θ_i, $i = 1,2,...,16$, what is the posterior distribution of $\theta = (\theta_1, \theta_2,..., \theta_{16})$?

5. Based on Table 8.2 and **BUGS CODE 8.2**, execute a Bayesian analysis for the air pollution study and use 65,000 observations for the MCMC simulation with a burn in of 5,000 and a refresh of 100.

   a. Confirm the Bayesian analysis of Table 8.3.

   b. What is the probability of wheezing at age 9?

   c. Are the probabilities of wheezing at ages 9, 10, 11, and 12 the same?

   d. What is the 95% credible interval for the probability of wheezing at age 10?

   e. Display the probability density of the probability of wheezing at age 12?

6. Refer to Table 8.4 for the Iowa Church attendance study consisting of two groups, females and males. A yes or no answer was obtained from each subject at times 0, 3, and 6 years.

   a. Describe the eight multinomial patterns for females.

   b. Using an improper prior density (8.33) for the eight female probabilities, what is the posterior distribution of $\theta = (\theta_1, \theta_2, ..., \theta_8)$?

   c. Using an improper prior distribution for the male probabilities $\phi = (\phi_1, \phi_2, ..., \phi_8)$, what is the posterior distribution of ϕ?

7. Based on Table 8.4, execute a Bayesian analysis with 65,000 observations for the simulation, with a burn in of 5,000 and a refresh of 100.

   a. Confirm the Bayesian analysis reported in Table 8.5.

   b. Using the posterior median, estimate the probability of regular church attendance for males at year 3.

   c. Which of the posterior distributions of Table 8.5 are skewed?

   d. Is the probability of regular church attendance increasing or decreasing for males from year 0 to year 6?

8. Refer to Section 8.2 and describe the Bayesian version of the GEE for the unknown parameters of the model. In particular, explain the components of Equation 8.44.

9. Refer to Table 8.6, the Alzheimer study analyzed by Crowder and Hand, where the repeated measure Y_{ij} (denotes the number of correctly recalled words) is assumed to follow a Poisson distribution and the canonical link is the log function. Based on **BUGS CODE 8.4**, confirm the Bayesian posterior analysis reported in Table 8.8. Use 150,000 observations for the MCMC simulation, with a burn in of 5,000 and a refresh of 100.

   a. Confirm the Bayesian analysis of Table 8.8.

   b. Estimate the intercept term of the log link $\log(\mu_j) = \beta_1 + \beta_2 t_{ij}$.

   c. What is the 95% credible interval of $\exp(\beta_1)$?

   d. Compute the posterior density of $\exp(\beta_1)$.

   e. Is the posterior distribution of $\exp(\beta_1)$ skewed?

   f. How well does the model fit the data? Plot the predicted values Z_{ij} versus the observed Y_{ij} values. The predicted values are generated by using the following statement in **BUGS CODE 8.4**:

   ```
   For(i in 1:N){for (j in 1:M){Z_{ij}~dpos(mu[i,j])}}
   ```

10. Using **BUGS CODE 8.5**, perform a Bayesian analysis for the epileptic study, where Y_{ij} denotes the number of seizures of subject I at time j and assume it has a Poisson distribution with a log canonical link (8.65), which includes age a covariate. Use 150,000

observations for the simulation, with a burn in of 5,000 and a refresh of 100.

a. Confirm the posterior analysis of Table 8.10.

b. Using the posterior median, what is the effect of age on the link function $\log(\mu_{ij})$?

c. Does time have an effect on $\log(\mu_{ij})$?

11. Refer to Section 8.3 and describe the generalized mixed linear model-to-model categorical repeated measures.

12. a. Refer to (8.76) and explain its significance.

b. Explain the components of Equation 8.77.

13. Refer to Table 8.11, which contains the information about the Parkinson's disease study.

a. What is the covariate X_1?

b. Describe the Equation 8.78.

c. Describe the equation

$$\log\left(\mu_{ij}\,\middle|\,b_{1i},b_{2i}\right) = \beta_1 + \beta_2 t[j] + \beta_3 X_1 + \beta_4 X_2 + X_3\beta_5 + \beta_6 X_4 + b_{1i} + b_{2i} t[j]$$

d. What are the two random effects b_{1i}, b_{2i}?

14. Based on **BUGS CODE 8.6** and Table 8.11, perform a Bayesian analysis with 150,000 observations, with a burn in of 5,000 and a refresh of 100.

a. Confirm the Bayesian analysis of Table 8.12.

b. Using the posterior median, estimate the variance of the two random effects b_{1i}, b_{2i}.

c. Among the posterior distributions of Table 8.12, which are skewed? Explain your answer.

d. How well does the model (8.78) fit the data? Plot the observed Y_{ij} versus predicted values Z_{ij}? Compute the predicted values using the code

```
for(i in 1:N){for(j in 1:M){Y[i,j]~dpois(mu[i,j])}}
```

as a statement in **BUGS CODE 8.6**.

15. Explain the elements of Equations 8.85 and 8.86.

a. What does β_3 represent?

b. Describe the random effect b_{1i}.

c. What does β_1 represent?

16. Using **BUGS CODE 8.7**, execute a Bayesian analysis for the pulmonary disease study and note that the first list statement contains the data.

a. Confirm the Bayesian analysis reported in Table 8.15.

b. From the results of Table 8.15, explain how the model should be simplified.

c. How well does the model fit the data? The predicted values Z_{ij} can be generated using the code

```
for(i in 1:n){for(j in 1:M){Z[i,j]~dbern(m[i,j])}}
```

17. Describe Equations 8.89 and 8.90 for the cumulative proportional odds model.

a. In (8.90), what do Y_{ij} and X_{ij} represent?

b. Describe the fixed and random effects of (8.90).

c. What distribution is assigned to the random effects?

18. Based on **BUGS CODE 8.8**, execute a Bayesian analysis for the Parkinson's disease study with an order categorical variable for the repeated measures with values 2, 3, 4, 5, 6, 7 for the clinical response.

Use 650,000 observations for the simulation, with a burn in of 5,000 and a refresh of 100.

19. Based on Table 8.20 and **BUGS CODE 8.9**, confirm the Bayesian analysis reported in Table 8.21.

a. Describe Equations 8.95 through 8.97.

b. The cumulative proportional odds mode for this study is given by (8.97). What does the one random effect in the model represent?

c. Refer to Table 8.21, and using the posterior median, estimate the variance of the random effect.

d. How well does the model fit the data? What is the code for generating the predicted values?

20. What is the difference in the posterior analysis of Table 8.21 compared to that of Table 8.18? Explain carefully.

9

Nonlinear Models and Repeated Measures

Nonlinear models for repeated measures is the topic of this chapter and is an extension of Chapters 4 through 8. Instead of assuming linearity for the mean values, nonlinearity in the mean value function will be considered. To be more specific, when the response is continuous, the mean of the response is nonlinear in the unknown parameters, and when the response is categorical, the link function will be nonlinear in the unknown parameters.

When the response is continuous, the development will closely follow the contents of Chapter 6, whereas when the model is categorical, the development will closely follow the description of Chapter 8. Also, when the model is continuous, the choice of the variance–covariance matrix will be (1) unstructured, (2) autoregressive, (3) Toeplitz, and (4) use of random effects is crucial for the subsequent Bayesian analysis. On the other hand, when the response is categorical, there are three requirements for defining the canonical link function: (1) assume a multinomial distribution for various patterns of the data, (2) use generalized estimating equations, and (3) use as a model, the generalized linear mixed model, which includes both fixed and random effects.

In the latter case of categorical data, the distribution of the repeated measure will be assumed to follow: (1) Poisson with a log link function, when the response is discrete, (2) a binary distribution with a logistic link, and (3) for ordered categorical data, a multinomial with a cumulative proportional odds model for the link.

9.1 Nonlinear Models and a Continuous Response

When the response is continuous, normality will be assumed for the distribution of the repeated measure, and the section essentially contains two approaches, when the model contains only fixed effects and when the model contains both fixed and random effects. When the model includes only fixed effects, analyses will be performed using unstructured, autoregressive, and Toeplitz. For the mixed model, the random effects induce a correlation among the response at the various time points.

Let

$$Y_{ij} = f(X_{ij}, \beta) + e_{ij} \tag{9.1}$$

be the response for subject i at time j, where f is a known function of the 1 by p vector of covariates X_{ij}, β is an unknown k by 1 vector of unknown parameters, and e_{ij} is a normal random variable where $e_i = (e_{i1}, e_{i2}, \ldots, e_{im})$ has a m-dimensional multivariate normal distribution with mean vector 0 and variance–covariance matrix Σ for $i = 1, 2, \ldots, N$. Note that the variance–covariance matrix Σ is the same for all subjects and that it can have a special structure as well as being unstructured.

Recall that the density of e_i is the m-dimensional density

$$g(e_i | \Sigma) = (2\pi)^{-n/2} |T|^{-1/2} \exp\left(-\frac{1}{2}\right)(e_i)'T(e_i) \tag{9.2}$$

where $T = \Sigma^{-1}$ is the precision matrix, the inverse of the variance–covariance matrix. When adopting a Bayesian approach, prior distributions are assigned to the unknown parameters, and usually the components of the unknown parameter $\beta = (\beta_1, \beta_2, \ldots, \beta_q)$ are assigned independent normal distributions, and T, the precision matrix, an uninformative Wishart distribution. This applied when the covariance matrix is unstructured, but when the pattern is special, other prior distributions will be specified.

9.1.1 Unstructured Covariance Matrix

The case of the unstructured covariance matrix is introduced with an example taken from Davidian and Giltinan (p. 18)[1] and involves an intravenous injection of indomethacin into six subjects observed at time points 0.25, 0.5, 0.75, 1, 1.25, 2, 3, 4, 5, 6, 8 hours, where the repeated measure is plasma concentration measured in µg/mL (Table 9.1).

TABLE 9.1

Plasma Concentration at 11 Time Points

0.25	0.50	0.75	1	1.25	2	3	4	5	6	8
1.50	0.94	0.78	0.48	0.37	0.19	0.12	0.11	0.08	0.07	0.05
2.03	1.63	0.71	0.70	0.64	0.36	0.32	0.20	0.25	0.12	0.08
2.72	1.49	1.16	0.80	0.80	0.39	0.22	0.12	0.11	0.08	0.08
1.85	1.39	1.02	0.89	0.59	0.40	0.16	0.11	0.10	0.07	0.07
2.05	1.04	0.81	0.39	0.30	0.23	0.13	0.11	0.08	0.10	0.06
2.31	1.44	1.03	0.84	0.64	0.42	0.24	0.17	0.13	0.10	0.09

Source: Davidian, M., and Giltinan, D.M., *Nonlinear Models for Repeated Measurement Data*, Chapman & Hall, London, 1995.

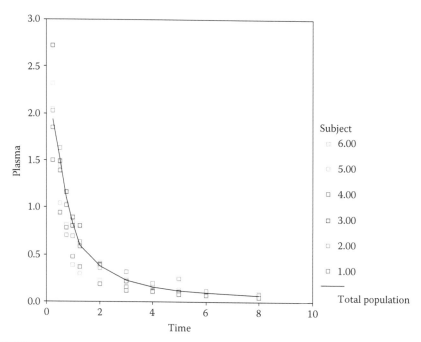

FIGURE 9.1
Plasma concentration versus time by subject.

The mean, standard deviation, and median of the plasma concentration measurements are reported at the various hourly time points and show the expected decrease in the mean and standard deviation.

Figure 9.1 is a plot of the plasma concentration versus hour by subject, where the Lowess plot is for all of the data and agrees with the descriptive statistics of Table 9.2.

The model adopted by Davidian and Giltinan[1] is based on a compartmental model for the pharmacokinetics of the log concentration and is defined as:

$$Y_{ij} = \beta_1 \exp\left(-\beta_2 t_{ij}\right) + \beta_3 \exp\left(-\beta_4 t_{ij}\right) + e_{ij} \tag{9.3}$$

where the β_i, $i = 1,2,3,4$ and the time points are given by $t_{i1} = 0.25$, $t_{i2} = 2, t_{i3} = 4, t_{i14} = 6$.

TABLE 9.2

Plasma Concentration by Hour

Hour	Mean	Standard Deviation	Median
0.25	2.0767	0.41355	2.0400
2.00	0.3317	0.09704	0.3750
4.00	0.1367	0.03882	0.1150
6.00	0.0900	0.0200	0.0900

TABLE 9.3

Computed Pearson Correlation Coefficients with
Plasma Concentrations over Four Time Points

Hour	0.25	2	4	6
0.25	1	0.627	0.202	0.249
1		0.947	0.304	−0.103
2		1	0.448	0.165
4			1	0.824
6				1

Note that the two rate constants $\exp(-\beta_2 t_{ij})$ and $\exp(-\beta_4 t_{ij})$ correspond to the two exponential phases of drug disposition.

The six vectors $e_i = (e_{i1}, e_{i2}, \ldots, e_{i11})$ are independent and follow a 11-dimensional multivariate normal distribution with mean vector 0 and unknown variance–covariance matrix Σ.

The estimated correlation matrix is portrayed in Table 9.3.

Do you perceive any pattern in the correlation matrix? I do not, and in particular do not detect a decreasing trend in the correlation with time. The Bayesian analysis is based on the data of Table 9.1, but using only four time points, the analysis is executed with **BUGS CODE 9.1**. The prior distribution for the beta coefficients for the model (9.3) is normal with variance 10, while that for the precision matrix is an uninformative Wishart distribution. The posterior analysis is somewhat sensitive to the prior distributions for the beta coefficients.

BUGS CODE 9.1

```
        model;
        {

# prior distribution for the regression coefficients.
                beta[1]~dnorm(4,.1)
                beta[2]~dnorm(-4,.1)
                beta[3]~dgamma(4,.1)
                beta[4]~dnorm(-4,.1)

for(i in 1:N){Y[i,1:M]~dmnorm(mu[],Omega[,])}
for(i in 1:N){Z[i,1:M]~dmnorm(mu[],Omega[,])}
for (j in 1:M){mu[j]<-beta[1]*exp(beta[2]*age[j])+beta[3]
*exp(beta[4]*age[j])}
# non-informative precision matrix
Omega[1:M,1:M]~dwish(R[,],4)
Sigma[1:M,1:M]<-inverse(Omega[,])
```

```
for(i in 1:4){for(j in 1:4){rho[i,j]<-Sigma[i,j]/
sqrt(Sigma[i,i]*Sigma[j,j])}}
}
list(N=6, M=4,Y = structure(.Data=c(
1.5,19,.11,.07,
2.03,.36,.20,.12,
2.72,.39,.12,.08,
1.85,.40,.11,.07,
2.05,.23,.11,.10,
2.31,.42,.17,.10),.Dim=c(6,4)),
age=c(.25,2,4,6),
R=structure(.Data=c(1,0,0,0,
                    0,1,0,0,
                    0,0,1,0,
                    0,0,0,1),.Dim=c(4,4)))
# initial values
list(beta=c(4,-4,4,-4))
```

TABLE 9.4

Posterior Analysis for Plasma Concentrations

Parameter	Mean	SD	Error	2½	Median	97½
β_1	1.913	2.82	0.0252	−3.369	1.766	7.782
β_2	−5.808	2.669	0.0265	−11.21	−5.782	−0.9568
β_3	19.72	11.03	0.1633	5.096	17.44	47.2
β_4	−9.043	2.274	0.0385	−13.4	−9.09	−4.476
ρ_{12}	−0.4636	0.3063	0.00059	−0.8892	−0.523	0.2749
ρ_{13}	−0.0606	0.362	0.000647	−0.7152	−0.0706	0.6416
ρ_{14}	−0.0423	0.3574	0.00061	−0.6962	−0.0496	0.6468
ρ_{23}	0.1078	0.3506	0.000593	−0.5937	0.1227	0.729
ρ_{24}	0.0707	0.3518	0.000559	−0.6204	0.08107	0.7081
ρ_{34}	0.0652	0.3522	0.000583	−0.6246	0.0757	0.7058

Using 370,000 observations for the simulation, with a burn in of 5,000 and a refresh of 100, the posterior analysis appears in Table 9.4.

The posterior analysis of Table 9.4 shows estimates of the correlations between the plasma concentrations at the four time points, while the estimated beta coefficients are the essential components of the model (9.3).

Using the posterior median, the estimated mean for subject i at time t_{ij} model is

$$\mu_{ij} = 1.766\exp\left(-5.782t_{ij}\right)+17.44\exp\left(-9.09t_{ij}\right) \qquad (9.4)$$

I used the posterior median instead of the posterior mean to estimate the coefficients of the model (9.3) because of the skewness of their posterior

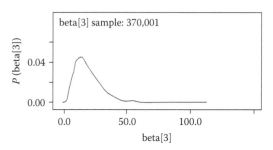

FIGURE 9.2
The posterior density of β_3.

distributions. Figure 9.2 of the posterior density of β_3 demonstrates the skewness of the posterior distribution. Also estimated are the correlation coefficients between the four time points of the plasma concentrations; thus, for example, the posterior median of the correlation between time points 3 and 4 is 0.0757.

9.1.2 Another Example Using an Unstructured Covariance Matrix

This is an important example from the pharmaceutical industry and involves the time it takes to dissolve a specified proportion of a pill. The effects of dissolving times may be affected by numerous factors including the brand, the shape of the pill, the way the pill is manufactured, the batch, and the indication for using the pill; thus, the industry is always conducting pill dissolution studies, and the information to be used for the Bayesian analysis is reported in Table 9.5. The information appearing in Table 9.5, is only a subset of the information appearing in Table A6 of Hand and Crowder.[2]

Thus, for pill 3 at 24 minutes, 70% of the pill remains undissolved.

The model adopted is a generalized version of the standard cube root form of dissolution and is generalized by Hand and Crowder,[2] whereas Hixon and Crowell[3] presented the original cube root form for the time t required until only a fraction x of the pill remains, where $t = (1 - x^{1/3})$, with $0 < x < 1$ and t is the dissolution time. Hand and Crowder[2] explain the generalization and refer to Goyan,[4] who replaced the exponent $1/3$ by an unknown parameter. Hand and Crowder[2] propose the model

$$Y_{ij} = \beta_1 + \beta_2 \left(1 - x_{ij}^{\beta_3}\right) + e_{ij} \tag{9.5}$$

where Y_{ij} is the time to a specified proportion x_{ij} remaining of the undissolved pill, at the jth specified proportion. Note that for each pill, $x_{ij} = 0.9, 0.7, 0.5, 0.3, 0.25, 0.1$; thus, $x_{i1} = 0.9, \ldots, x_{i6} = 0.1$. It will be assumed that for each i, $e_i = (e_{i1}, e_{i2}, \ldots, e_{i6})$ are independent with a multivariate normal distribution with a mean vector 0 and unknown covariance matrix

TABLE 9.5

Times at Which a Specified Proportion of Pills Remain Undissolved

Proportion		0.9	0.7	0.5	0.3	0.25	0.1
Group	Pill						
1	1	13	16	19	23	24	28
1	2	14	18	22	26	28	32
1	3	19	24	28	33	33	39
1	4	13	17	21	25	26	29
1	5	14	16	19	23	25	27
1	6	13	16	19	23	24	26
2	7	13	17	21.5	26	27.8	33.5
2	8	11.5	15.5	20.5	24.5	26.7	31.3
2	9	10.4	14.4	18.4	23.5	25.1	29.6
2	10	11.1	15.1	19	23.6	24.7	28.7
3	11	11.7	16.9	22	27.6	29.2	34.5
3	12	13.5	18.7	24.9	30	31.8	37.5
3	13	12	17	22.7	28.4	30.2	35.8
3	14	12.1	16.7	21.1	26.8	29	33.8
4	15	11	14.5	19	24.5	28	32
4	16	14.4	19	24	30	31.5	39
4	17	11	14	17.5	22	23.5	29.5

Source: Hand, D., and Crowder, M., *Practical Longitudinal Data Analysis*, Chapman & Hall, Boca Raton, FL, 1999.

Σ and that β_i, $i = 1,2,3$ are unknown parameters to be estimated using Bayesian techniques. For the Bayesian analysis, the prior distribution assigned to the beta coefficients will be informative normal distributions, the MLE values of Hand and Crowder (p. 111),[2] while that for Σ will be a noninformative Wishart distribution (Figure 9.3).

Note that the plot shows the pill-to-pill variation between the association between the times to a specified proportion remaining of the undissolved pill. **BUGS CODE 9.2** will be executed with 150,000 observations for the simulation, with a burn in of 5,000 and a refresh of 100.

Shown in Table 9.6 are the results of the Bayesian analysis and depict the estimated coefficients of the model (9.5) and the estimated correlations of the dissolution times over the six specified proportions.

For the exponent β_3 of the specified proportion, the posterior median is 0.8742 and the distribution appears to be symmetric. It seems to me that the correlations appear to decrease with larger differences between the specified proportions. The estimated beta coefficients do indeed provide a good fit to the data using model (9.5).

This example will be used assuming an autoregressive structure for the variance–covariance matrix.

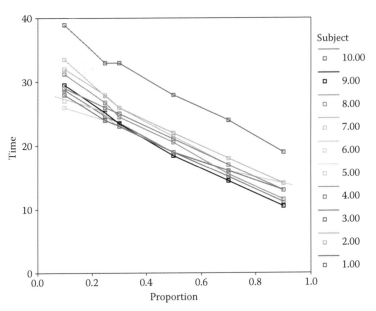

FIGURE 9.3
Times to a specified proportion remaining of the undissolved pill.

BUGS CODE 9.2

```
        model;
         {

# prior distribution for the regression coefficients.
                beta[1]~dnorm(11,1)
                beta[2]~dnorm(23,1)
                beta[3]~dunif(.7,1)

for(i in 1:N){Y[i,1:M]~dmnorm(mu[],Omega[,])}
for(i in 1:N){Z[i,1:M]~dmnorm(mu[],Omega[,])}
for (j in 1:M){mu[j]<- beta[1]+beta[2]
*(1-pow(x[j],beta[3]))}
# non-informative precision matrix
Omega[1:M,1:M]~dwish(R[,],6)
Sigma[1:M,1:M]<-inverse(Omega[,])
for(i in 1:6){for(j in 1:6){rho[i,j]<-Sigma[i,j]/
sqrt(Sigma[i,i]*Sigma[j,j])}}
}
list(N=17, M=6,

x=c(.9,.7,.5,.3,.25,.1),
Y = structure(.Data=c(13,16,19,23,24,28,
14,18,22,26,28,32,
```

```
19,24,28,33,33,39,
13,17,21,25,26,29,
14,16,19,23,24,26,
13,16,19,23,24,26,
13,17,22,26,28,34,
12,16,21,25,27,31,
10,15,18,24,25,30,
11,15,19,24,25,29,
12,17,22,28,29,35,
14,19,25,30,32,38,
12,17,23,28,30,36,
12,17,21,27,29,34,
11,15,19,25,28,32,
14,19,24,30,32,39,
11,14,18,22,24,30),.Dim=c(17,6)),

R=structure(.Data=c(1,0,0,0,0,0,
                    0,1,0,0,0,0,
                    0,0,1,0,0,0,
                    0,0,0,1,0,0,
                    0,0,0,0,1,0,
                    0,0,0,0,0,1),.Dim=c(6,6)))
# initial values

list(beta=c(11,23,.7))
```

TABLE 9.6

Posterior Analysis for the Pill Dissolution Study

Parameter	Mean	SD	Error	2½	Median	97½
β_1	12.02	1.013	0.01421	10.02	12.02	14
β_2	21.97	0.9929	0.01406	20.03	21.98	23.9
β_3	0.8652	0.0859	0.00116	0.7107	0.8724	0.9949
ρ_{12}	0.9967	0.001706	0.0000089	0.9924	0.9971	0.9989
ρ_{13}	0.9922	0.003977	0.0000188	0.9822	0.993	0.997
ρ_{14}	0.9882	0.00599	0.0000277	0.9731	0.9894	0.9959
ρ_{15}	0.985	0.007594	0.0000347	0.9659	0.9866	0.9948
ρ_{16}	0.9769	0.01151	0.000479	0.948	0.9793	0.9919
ρ_{23}	0.9976	0.001194	0.0000050	0.9947	0.9979	0.9992
ρ_{24}	0.9961	0.001969	0.0000082	0.9912	0.9965	0.9989
ρ_{25}	0.9934	0.003345	0.0000138	0.9849	0.9941	0.9977
ρ_{26}	0.9881	0.005956	0.0000231	0.973	0.9894	0.9958
ρ_{34}	0.9983	0.00088	0.0000038	0.9961	0.9985	0.9994
ρ_{35}	0.9969	0.001568	0.0000065	0.993	0.9972	0.989
ρ_{36}	0.9937	0.00317	0.0000123	0.9858	0.9944	0.9978

(Continued)

TABLE 9.6 (*Continued*)

Posterior Analysis for the Pill Dissolution Study

Parameter	Mean	SD	Error	2½	Median	97½
ρ_{45}	0.9989	0.000562	0.0000021	0.9975	0.999	0.9996
ρ_{46}	0.9965	0.001743	0.0000067	0.9922	0.9969	0.9988
ρ_{56}	0.9978	0.001133	0.0000047	0.9949	0.998	0.9992

9.1.3 Autoregressive Structure for the Covariance Matrix

Consider the nonlinear model for subject i at time j

$$Y_{ij} = f(X_{ij}, \beta) + e_{ij} \tag{9.6}$$

where:
X_{ij} is a p by 1 vector of p covariate values
β is a q by 1 vector of unknown parameters

and

$$e_{ij} = \rho e_{ij-1} + u_{ij} \tag{9.7}$$

with $i = 1,2,...,N$ and $j = 1,2,...,n$, u_{ij} are independent and normally distributed with mean 0 and variance $\sigma^2 = 1/\tau$ (τ is the precision), and the correlation ρ satisfies $-1 < \rho < 1$. Thus, for each subject, the e_{ij} follow a first-order autoregressive process with correlation ρ.

The model (9.6) can be expressed as

$$Y_{ij} = \rho Y_{ij-1} + \left[f(X_{ij}, \beta) - \rho f(X_{ij-1}, \beta) \right] + u_{ij} \tag{9.8}$$

for $i = 1,2,...,N$ and $j = 1,2,...,n$. Also assumed is that the time periods are equally spaced and each subject has the same n occasions at which the response is measured. This will later be generalized to the case where the time points are not equally spaced and the number of occasions the response is measured is not necessarily the same number n.

It can be shown that the correlation is ρ for responses one unit apart, is ρ^2 for responses two units apart, and so on. This is represented by the correlation matrix (Table 9.7).

Note that there are only two parameter σ^2 and ρ for the covariance matrix with autoregressive structure. Recall for the unstructured covariance matrix, the number of parameters for the covariance matrix is $n(n + 1)/2$, where n is the number of common time points for all subjects. I am assuming the covariance matrix is the same for all subjects.

Consider the previous example of pill dissolution, but now assuming an autoregressive structure for the correlation, corresponding to (9.7), the model is (9.8) with

$$f(X_{ij}, \beta) = \beta_1 + \beta_2 \left(1 - x_{ij}^{\beta_3}\right) \tag{9.9}$$

TABLE 9.7

Correlation Pattern for Autoregressive Structure

1	ρ	ρ^2	ρ^3	...	ρ^{n-2}	ρ^{n-1}
ρ	1	ρ	ρ^2		ρ^{n-3}	ρ^{n-2}
ρ^2		1				
ρ^3						
$\vdots$						
ρ^{n-2}					1	ρ
ρ^{n-1}	ρ^{n-2}					1

Using the data from Table 9.5 and executing the Bayesian analysis with **BUGS CODE 9.3** based on 150,000 observations for the simulation, with a burn in of 5,000 and a refresh of 100, the posterior analysis is reported in Table 9.8. Prior distributions for the coefficients of model (9.9) are noninformative normal distributions with a variance of 100, while the precision parameter τ is assigned

BUGS CODE 9.3

```
model;
      {

# prior distribution for coefficients of model
            beta[1] ~ dnorm(10,.01)
            beta[2] ~ dnorm(22,.01)
            beta[3]~dunif(.5,1)

for(i in 1:N){for (j in 2:M){Y[i,j]~dnorm(mu[i,j],tau)}}
for(i in 1:N){for (j in 2:M){mu[i,j]<-rho
*Y[i,j-1]+beta[1]+beta[2]*(1-pow(p[j],beta[3]))-rho
*(beta[1]+beta[2]*(1-pow(p[j-1],beta[3])))}}
# autoregressive structure for covariance matrix
for(i in 1:N){Y[i,1]~dnorm(mu[i,1],tau)}
for(i in 1:N){mu[i,1]<-beta[1]+beta[2]
*(1-pow(p[1],beta[3]))}

rho~dbeta(1,1)
tau~dgamma(.001,.001)
sigma<-1/tau
# powers of rho
rho12<-rho*rho
rho13<-rho*rho*rho
rho14<-rho*rho*rho*rho
rho15<-rho*rho*rho*rho*rho

}
```

```
# the time dissolution values
list(N=17, M=6,Y=structure(.Data=c(13,16,19,23,24,28,
14,18,22,26,28,32,
19,24,28,33,33,39,
13,17,21,25,26,29,
14,16,19,23,24,26,
13,16,19,23,24,26,
13,17,22,26,28,34,
12,16,21,25,27,31,
10,15,18,24,25,30,
11,15,19,24,25,29,
12,17,22,28,29,35,
14,19,25,30,32,38,
12,17,23,28,30,36,
12,17,21,27,29,34,
11,15,19,25,28,32,
14,19,24,30,32,39,
11,14,18,22,24,30),.Dim=c(17,6)),

# the specified proportions
p=c(.9,.7,.5,.3,.25,.1))
# initial values
list(beta=c(10,20,.7), tau=.5,rho=.5)
```

TABLE 9.8

Posterior Analysis for Pill Study with Autoregressive Covariance Matrix

Parameter	Mean	SD	Error	2½	Median	97½
β_1	10.93	0.3288	0.00366	10.28	10.93	11.57
β_2	26.29	1.098	0.01251	24.23	26.26	28.52
β_3	0.7137	0.05848	0.000687	0.6005	0.713	0.8299
ρ	0.9817	0.01703	0.000175	0.9367	0.9867	0.9995
σ^2	1.58	0.2312	0.000789	1.191	1.558	2.091
ρ^2	0.9664	0.03228	0.000473	0.8807	0.9735	0.999
ρ^3	0.9475	0.04692	0.000688	0.8265	0.9605	0.996
ρ^4	0.9312	0.06067	0.000889	0.7757	0.9477	0.9981
ρ^5	0.9153	0.07358	0.001078	0.7279	0.9351	0.9976

a noninformative Y distribution with first and second parameters, 0.0001 and 0.0001, respectively.

Notice that ρ, the autoregressive coefficient, estimates the correlation one unit apart, 0.2, between the specified proportions, and its posterior mean is 0.9817, with a 95% credible interval (0.9367, 0.9995), indicating a very high correlations between the dissolution times 0.2 units apart.

TABLE 9.9

Computed Pearson Correlations of Pill Dissolution Times over the Specified Proportions

Proportions	0.1	0.25	0.3	0.5	0.7	0.9
0.1		0.967	0.941	0.917	0.826	0.667
0.25			0.978	0.974	0.913	0.795
0.3				0.990	0.958	0.842
0.5					0.972	0.875
0.7						0.956

Which covariance pattern, unstructured or autoregressive, is more appropriate for the pill dissolution study? This question will be addressed in the exercises. To see if there is an autoregressive-type covariance pattern, one needs to compare the estimated (via the posterior mean) powers of ρ with the computed Pearson correlations of Table 9.9. Note Table 9.8 displays the posterior distributions of powers of the autoregressive parameter ρ; thus, for example, the posterior median of ρ^3 is 0.9605.

Note the extreme asymmetry displayed by the posterior distribution of ρ^3, as depicted in Figure 9.4. See Table 9.8 and note how the figure agrees with the posterior properties of the distribution.

9.1.4 Another Example Using an Autoregressive Covariance Structure

Hand and Crowder (p. 111)[2] describe a nonlinear example of repeated measures occurring in civil engineering, where a wood plank is inserted between concrete surfaces and fastened by four screws. The load is increased and applied to the timber, and the slippage in millimeters is recorded at two-second intervals. For our purposes, the load is the dependent variable and the slip is the independent variable, and the data exhibited is taken from Table A.15 of Hand and Crowder (p. 179)[2] (Table 9.10).

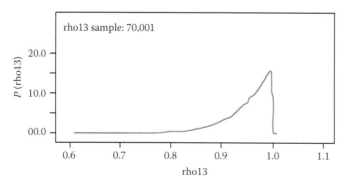

FIGURE 9.4

Posterior density of ρ^3.

TABLE 9.10

Loads for Various Slip Values

	Plank							
Slip	1	2	3	4	5	6	7	8
0	0	0	0	0	0	0	0	0
0.1	2.39	2.69	2.85	2.46	2.97	3.96	3.17	3.36
0.2	4.34	4.74	4.89	4.28	4.68	6.46	5.33	5.45
0.3	6.64	7.04	6.61	5.88	6.66	8.14	7.14	7.08
0.4	8.05	9.2	8.09	7.43	8.11	9.35	8.29	8.32
0.5	9.78	10.94	9.72	8.32	9.64	10.72	9.86	9.91
0.6	10.97	12.23	11.03	9.92	11.06	11.84	11.07	11.06
0.7	12.05	13.19	12.14	11.1	12.25	12.85	12.13	12.21
0.8	12.98	14.08	13.18	12.23	13.35	13.83	13.15	13.16
0.9	13.94	14.66	14.12	13.24	14.54	14.85	14.09	14.95
1.0	14.74	15.37	15.09	14.19	15.53	15.79	15.11	14.96

Source: Hand, D., and Crowder, M., *Practical Longitudinal Data Analysis*, Table A.15, Chapman & Hall, Boca Raton, FL, 1999.

Based on this table, the scatter plot of Figure 9.5 shows the growth of load as slip increases from 0 to 1 and reveals the heterogeneity between the eight planks.

In the field, there is a traditional equation for the association between load and slip values, and a generalized version is expressed as

$$Y_{ij} = \beta_1 + \beta_2 \exp\left(-\beta_3 x_{ij}\right) + e_{ij} \tag{9.10}$$

for plank *i* and the *j*th slip. In this case, $i = 1,2,\ldots,8$, $j = 1,2,3,4,5,6,7,8,9,10,11$ and the vector of slip values as

$$x_{ij} = (0, 0.1, 0.2, 0.3, 0.4, 0.5, 0.6, 0.7, 0.8, 0.9, 1) \tag{9.11}$$

and is the same for all eight planks.

It is assumed that the errors of (9.10) follow a first-order autoregressive process, namely

$$e_{ij} = \rho e_{ij-1} + u_{ij} \tag{9.12}$$

where ρ is the correlation coefficient, and $u_{ij} \sim nid\left(0, \sigma^2\right)$ implies that (9.10) can be expressed as

$$Y_{ij} = \rho Y_{ij-1} + \left[f\left(X_{ij}, \beta\right) - \rho f\left(X_{ij-1}, \beta\right) \right] + u_{ij} \tag{9.13}$$

where

$$f\left(x_{ij}, \beta\right) = \beta_1 + \beta_2 \exp\left(-\beta_3 x_{ij}\right) \tag{9.14}$$

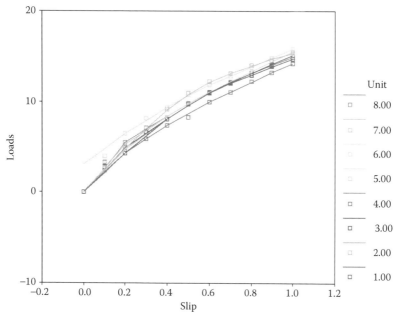

FIGURE 9.5
Load versus slip.

Priors for the Bayesian analysis are as follows: (1) for the three beta parameters of the model (9.14) normal (0, 0.0001) are employed, (2) for the precision tau of the autoregressive errors and uninformative gamma distribution is used, and (3) for the auto correlation parameter rho, a uniform distribution is assigned. **BUGS CODE 9.4** is executed with 150,000 observations for the simulation, with a burn in of 5,000 and a refresh of 100.

BUGS CODE 9.4

```
model;
      {
                beta[1]  ~  dnorm(0,.001)
                beta[2]  ~  dnorm(0,.001)
      beta[3]  ~  dnorm(0,.001)

for(i in 1:N){for (j in 2:M){Y[i,j]~dnorm(mu[i,j],tau)}}
for(i in 1:N){for (j in 2:M){mu[i,j]<-rho
*Y[i,j-1]+beta[1]+beta[2]*exp(-beta[3]*s[j])-rho
*(beta[1]+beta[2]*exp(-beta[3]*s[j-1]))}}

for(i in 1:N){Y[i,1]~dnorm(mu[i,1],tau)}
for(i in 1:N){mu[i,1]<-beta[1]+beta[2]
*exp(-beta[3]*s[1])}}
```

```
rho~dbeta(1,1)
tau~dgamma(.001,.001)
sigma<-1/tau

rho12<-rho*rho
rho13<-rho*rho*rho
rho14<-rho*rho*rho*rho
rho15<-rho*rho*rho*rho*rho


}

list(N=8, M=11,Y=structure(.Data=c(

0,2.38,4.34,6.64,8.05,9.78,10.97,12.05,12.98,13.94,14.74,
0,2.69,4.74,7.04,9.20,10.94,12.23,13.19,14.08,14.66,15.37,
0,2.85,4.89,6.61,8.09,9.72,11.03,12.14,13.18,14.12,15.09,
0,2.46,4.28,5.88,7.43,8.32,9.92,11.10,12.23,13.24,14.19,
0,2.97,4.68,6.66,8.11,9.64,11.06,12.25,13.35,14.54,15.53,
0,3.96,6.46,8.14,9.35,10.72,11.84,12.85,13.83,14.85,15.79,
0,3.17,5.33,7.14,8.29,9.86,11.07,12.13,13.15,14.09,15.11,
0,3.36,5.45,7.08,8.32,9.91,11.06,12.21,13.16,14.05,14.96),.
Dim=c(8,11)),
s=c(0,.1,.2,.3,.4,.5,.6,.7,.8,.9,1))
# initial values
list(beta=c(0,0,0), tau=.5,rho=.5)
```

The Bayesian analysis of Table 9.11 agrees very well with the analysis of Hand and Crowder (p. 112),[2] and the analysis implies that $\beta_1 = -\beta_2$, which in turn implies that the revised model has mean value

$$f(x_{ij}, \beta) = \beta \left[1 - \exp(-\beta_3 x_{ij}) \right] \tag{9.15}$$

TABLE 9.11

Posterior Analysis of Timber Load Study

Parameter	Mean	SD	Error	2½	Median	97½
β_1	19.64	1.647	0.07889	18.36	19.76	21.26
β_2	−19.64	1.629	0.07809	−21.28	−19.74	−18.34
β_3	1.619	2.775	0.1243	1.26	1.433	1.613
ρ	0.9107	0.04258	0.002102	0.8226	0.9098	0.9906
ρ_{12}	0.8312	0.07733	0.003816	0.6767	0.8278	0.9813
ρ_{13}	0.7602	0.1056	0.005211	0.5567	0.7531	0.9721
ρ_{14}	0.6968	0.1287	0.006345	0.458	0.6852	0.9629
ρ_{15}	0.6401	0.1474	0.007262	0.3768	0.6264	0.9539
σ^2	0.09408	0.1331	0.006631	0.0609	0.0813	0.1147

TABLE 9.12

Posterior Analysis for Revised Model with Timber Slip Study

Parameter	Mean	SD	Error	2½	Median	97½
β	18.49	12.03	0.732	18.19	19.68	21.3
β_3	1.415	0.1875	0.01039	1.233	1.435	1.603
ρ	0.9199	0.04688	0.002724	0.8125	0.9275	0.9929
σ^2	0.0862	0.04331	0.002468	0.06052	0.08079	0.1149

which is the original formulation of the model. Refer to Hand and Crowder (p. 111)[2] for additional details.

The analysis is easily revised as is **BUGS CODE 9.4**, and it can be shown that posterior analysis for the revised model is reported in Table 9.12.

Revised estimates of ρ and σ^2 differ very little from the corresponding parameters of Table 9.11, for the original model, and the posterior median of β_3 is the same for both models.

9.1.5 Models with Random Effects

For nonlinear models, the mixed (random and fixed effects) model is expressed as

$$Y_{ij} = f\left(X_{ij}, \beta, Z_{ij}, b\right) + e_{ij} \qquad (9.16)$$

where:

f is a known function

X_{ij} is a p by 1 vector of p covariate values for subject i

β is a q by 1 vector of unknown parameters

Z_{ij} is a r by 1 vector of known covariate values

b is a s by 1 random vector with 0 mean vector and sth order dispersion matrix Σ, which is the same for all subjects

In addition, the components of the independent random vectors $e_i = (e_{i1}, e_{i2}, ..., e_{in})$ are normal random variables with mean 0 and common variance σ^2.

Recall that the linear mixed model analogous to (9.16) is expressed as

$$Y_{ij} = X_{ij}\beta + Z_{ij}b + e_{ij} \qquad (9.17)$$

Because there are numerous ways to introduce random effects into the nonlinear model, the nonlinear version is much more complex than the linear version and presents some challenges for the Bayesian analysis. Also recall that the random effects in the model (9.17) induce a correlation structure for the repeated measure among the various time points.

Consider the previous timber slip example with the model

$$Y_{ij} = \beta_1 + \beta_2 \exp\left(-\beta_3 x_{ij}\right) + e_{ij} \qquad (9.18)$$

where Y_{ij} is the load for timber with the jth slippage value. Notice the variety of ways that random effects can be added to the model. For example, (1) the simplest situation is to add one random effect b_i to the right-hand side of (9.18), then (2) replace β_2 by $\beta_2 + b_{2i}$, and last in addition, (3) replace β_3 by $\beta_3 + b_{3i}$, where the three random effects are independent normal random variables with mean zero and unknown variances. Of course, there are many more scenarios to the addition of random effects cases of these three cases. In choosing the addition of random effects, one is inducing a correlation structure between the repeated measures of the various time points. For example, to induce compound symmetry, the model

$$Y_{ij} = \beta_1 + b_i + \beta_2 \exp\left(-\beta_3 x_{ij}\right) + e_{ij} \tag{9.19}$$

where e_{ij} are independent normal random variables with mean 0 and variance σ^2 and independent of the n random effects b_i, which are normal random variables with mean 0 and variance σ_b^2. It can be shown that the covariance between Y_{ij} and Y_{ik} is σ_b^2 and that the variance of Y_{ij} is $\sigma_b^2 + \sigma^2$, which completely determines the dispersion matrix of the observations.

The Bayesian analysis is executed with **BUGS CODE 9.5**

BUGS CODE 9.5

```
model;
     {
             beta[1] ~ dnorm(0,.001)
             beta[2] ~ dnorm(0,.001)
     beta[3] ~ dnorm(0,.001)

for(i in 1:N){for (j in 1:M){Y[i,j]~dnorm(mu[i,j],tau)}}
for(i in 1:N){for (j in 1:M){mu[i,j]<-b[i]+
beta[1]+beta[2]*exp(-beta[3]*s[j])}}

for(i in 1:N){b[i]~dnorm(0,tau1)}

tau~dgamma(.001,.001)

# sigma is the variance of the errors
sigma<-1/tau
tau1~dgamma(.001,.001)
# sigma1 is the variance of the random effects
sigma1<-1/tau1

# rho is the common correlation
rho<-sigma1/(sigma+sigma1)
var<-sigma+sigma1

}

list(N=8, M=11,Y=structure(.Data=c(
```

```
0,2.38,4.34,6.64,8.05,9.78,10.97,12.05,12.98,13.94,14.74,
0,2.69,4.74,7.04,9.20,10.94,12.23,13.19,14.08,14.66,15.37,
0,2.85,4.89,6.61,8.09,9.72,11.03,12.14,13.18,14.12,15.09,
0,2.46,4.28,5.88,7.43,8.32,9.92,11.10,12.23,13.24,14.19,
0,2.97,4.68,6.66,8.11,9.64,11.06,12.25,13.35,14.54,15.53,
0,3.96,6.46,8.14,9.35,10.72,11.84,12.85,13.83,14.85,15.79,
0,3.17,5.33,7.14,8.29,9.86,11.07,12.13,13.15,14.09,15.11,
0,3.36,5.45,7.08,8.32,9.91,11.06,12.21,13.16,14.05,14.96),.
Dim=c(8,11)),

s=c(0,.1,.2,.3,.4,.5,.6,.7,.8,.9,1))

# initial values
list(beta=c(0,0,0), tau=.5,tau1=1)
```

Based on **BUGS CODE 9.5**, the posterior analysis is conducted.

The estimates of the beta coefficients of the nonlinear mean are almost the same as those assuming an autoregressive process for the errors (see Table 9.11) of the common correlation is 0.6865 based on the posterior mean. Also, the variance of the errors in (9.19) is estimated as 0.119 with the posterior median, and that of the variance between the various planks is estimated as 0.3445 with the posterior mean. Again it is obvious that $\beta_1 = -\beta_2$; thus, the model can be revised in the same way as was done with the autoregressive version of the model.

Consider an expansion of the nonlinear mixed model (9.19) to

$$Y_{ij} = \beta_1 + b_{1i} + (\beta_2 + b_{2i})\exp(-\beta_3 x_{ij}) + e_{ij} \tag{9.20}$$

where there are two independent random effects b_{1i} and b_{2i} with a normal distribution with means 0 and variances σ_1^2 and σ_2^2, respectively. This would imply a covariance structure that is not compound symmetry as it is for (9.19) version with one random effect. What is the covariance between two observations, say, at time points j and k, where j is not equal to k? It can be shown that

$$\mathrm{Cov}(Y_{ij}, Y_{ik}) = \sigma_1^2 + \sigma_2^2 \exp{-\beta_3(x_{ij} + x_{ik})} \tag{9.21}$$

and

$$\mathrm{Var}(Y_{ij}) = \sigma_1^2 + \sigma_2^2 \exp(-2\beta_3 x_{ij}) \tag{9.22}$$

where $i = 1,2,...,N$ and $j = 1,2,...,n$, with N the number of subjects and n the number of time points.

The analysis for the timber load data is executed with **BUGS CODE 9.5a** using 150,000 observations for the simulation, with a burn in of 5,000 and a refresh of 100. The # symbol indicates remarks about the various program statement. In particular, I have remarked about the main formula (9.20) for the model with two random effects and also about the formulas for the covariances (9.21) and variances (9.22). Recall this is the expansion of the model for the timber load data, and the model now includes two random

BUGS CODE 9.5a

```
model;
     {
               beta[1] ~ dnorm(0,.001)
               beta[2] ~ dnorm(0,.001)
    beta[3] ~ dnorm(0,.001)
# the model formula (9.20)
for(i in 1:N){for (j in 1:M){Y[i,j]~dnorm(mu[i,j],tau)}}
for(i in 1:N){for (j in 1:M){mu[i,j]<-b1[i]+beta[1]+
(beta[2]+b2[i])*exp(-beta[3]*s[j])}}

for(i in 1:N){b1[i]~dnorm(0,tau1)}
for(i in 1:N){b2[i]~dnorm(0,tau2)}
# the variances (9.22)
for(j in 1:M){v[j]<-sigma1+sigma2*exp(-2*beta[3]*s[j])}
# the covariances (9.21)
for (j in 1:M){for(k in 1:M){c[j,k]<-sigma1+sigma2
*exp(-beta[3]*(s[j]+s[k]))}}
# the correlations
for(j in 1:M){for(k in 1:M){rho[j,k]<-c[j,k]/
sqrt(v[j]*v[k])}}

tau~dgamma(.01,.01)
sigma<-1/tau
tau1~dgamma(.01,.01)
tau2~dgamma(.01,.01)
sigma1<-1/tau1
sigma2<-1/tau2

}

list(N=8, M=11,Y=structure(.Data=c(

0,2.38,4.34,6.64,8.05,9.78,10.97,12.05,12.98,13.94,14.74,
0,2.69,4.74,7.04,9.20,10.94,12.23,13.19,14.08,14.66,15.37,
0,2.85,4.89,6.61,8.09,9.72,11.03,12.14,13.18,14.12,15.09,
0,2.46,4.28,5.88,7.43,8.32,9.92,11.10,12.23,13.24,14.19,
0,2.97,4.68,6.66,8.11,9.64,11.06,12.25,13.35,14.54,15.53,
0,3.96,6.46,8.14,9.35,10.72,11.84,12.85,13.83,14.85,15.79,
0,3.17,5.33,7.14,8.29,9.86,11.07,12.13,13.15,14.09,15.11,
0,3.36,5.45,7.08,8.32,9.91,11.06,12.21,13.16,14.05,14.96),.
Dim=c(8,11)),

s=c(0,.1,.2,.3,.4,.5,.6,.7,.8,.9,1))

# initial values
list(beta=c(0,0,0),
```

TABLE 9.13

Posterior Analysis for Timber Load Study with Compound Symmetry Dispersion Matrix

Parameter	Mean	SD	Error	2½	Median	97½
β_1	20.23	0.5253	0.02854	19.2	20.25	21.22
β_2	−20.0	0.4425	0.02438	−20.83	−20.01	−19.13
β_3	1.33	0.05869	0.003427	1.23	1.324	1.451
ρ	0.6865	0.1216	0.001271	0.4371	0.6924	0.9014
σ^2	0.121	0.01983	0.000146	0.08836	0.119	0.1658
σ_b^2	0.3445	0.2835	0.003392	0.09886	0.2677	1.058
Var	0.4655	0.284	0.003389	0.2161	0.3893	1.177

TABLE 9.14

Posterior Analysis for Timber Load Model with Two Random Effects

Parameter	Mean	SD	Error	2½	Median	97½
β_1	20.23	0.5874	0.02689	19.15	20.19	21.49
β_2	−19.99	0.5134	0.02262	−21.12	−19.96	−19.06
β_3	1.331	0.0626	0.003027	1.207	1.333	1.449
σ^2	0.107	0.0197	0.000196	0.07494	0.1048	0.1515
σ_1^2	0.5466	0.4945	0.007586	0.1293	0.4142	1.762
σ_2^2	0.2978	0.3884	0.005661	0.01024	0.186	1.276

effects, which induce the variances (9.22) and covariances (9.21) producing a 11th order correlation matrix with entries $\rho[i,j]$ (Table 9.15).

Tables 9.14 and 9.15 report the posterior analysis for the timber slip information, where the former table reveals estimates of the main parameters of the model (9.20), while the latter table displays estimates of the correlation matrix of the repeated measures (load data) among 11 slip values.

It appears that the posterior means of Table 9.14 for the coefficients of the model (9.20) are almost the same as the corresponding values for a model with one variance component (see Table 9.13). Also, note that the estimates imply $\beta_1 = -\beta_2$, which was also apparent from Table 9.13.

Refer to the code relevant to estimating the correlations, where the following program statements appear:

1. Code for the variances, see (9.22)

```
for(j in 1:M){v[j]<-sigma1+sigma2*exp(-2*beta[3]*s[j])}
```

2. Code for the covariances, see (9.21)

```
for (j in 1:M){for(k in 1:M){c[j,k]<-sigma1+sigma2
*exp(-beta[3]*(s[j]+s[k]))}}
```

3. Code for the correlations

```
for(j in 1:M){for(k in 1:M){rho[j,k]<-c[j,k]/sqrt(v[j]*v[k])}}
```

TABLE 9.15

Posterior Means of the Correlation of Timber Loads among 11 Slip Values

Slip	0	0.1	0.2	0.3	0.4	0.5	0.6	0.7	0.8	0.9	1
0		0.998	0.994	0.987	0.979	0.969	0.958	0.947	0.936	0.925	0.915
0.1		1	0.998	0.994	0.988	0.981	0.972	0.963	0.954	0.944	0.935
0.2			1	0.998	0.995	0.990	0.983	0.976	0.968	0.960	0.952
0.3				1	0.998	0.995	0.991	0.985	0.979	0.973	0.966
0.4					1	0.999	0.996	0.992	0.988	0.982	0.977
0.5						1	0.999	0.997	0.993	0.99	0.985
0.6							1	0.999	0.997	0.994	0.991
0.7								1	0.999	0.997	0.995
0.8									1	0.999	0.998
0.9										1	0.999
1											1

To one decimal place, the estimated correlations are the same, implying a compound symmetry type of structure.

How well does the model (9.20) fit the data? This is explored in Figure 9.6, which is a scatter plot of predicted versus observed load values, and it appears that the fit is exceptionally good. It would be interesting to compare the fit of the model with two random effects (9.20) with the model (9.19) that includes one random effect. This will be left as an exercise.

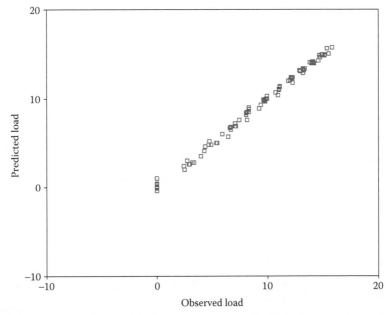

FIGURE 9.6
Predicted versus observed loads of timber load study with two random effects.

9.1.6 Mixed Model for Energy Expenditure

The next example is explained by Crowder and Hand (p. 116)[2] and is based on data from Crowder[5] and involves female students whose energy expenditure (calories per minute) and heart rate (beats per minute) is measured four times, corresponding to four tasks: (1) lying, (2) sitting, (3) walking, and (4) skipping. As one would expect, the heart rate as well as the energy expenditure increases with successive tasks. Also measured is the weight in kilograms of each of the seven student volunteers, and Table 9.16 reports the data from the study, where W means weight in kilograms, E designates energy expenditure, and HR denotes heart rate (Figure 9.7).

TABLE 9.16

Energy Expenditure and Task

Task	W	E	HR	Subject
1	51.10	1.57	50	1
2	51.10	1.31	60	1
3	51.10	4.67	125	1
4	51.10	4.23	135	1
1	58.10	1.46	880	2
2	58.10	1.41	78	2
3	58.10	4.49	120	2
4	58.10	5.56	131	2
1	61.50	1.20	85	3
2	61.50	1.22	81	3
3	61.50	3.65	127	3
4	61.50	7.83	135	3
1	60.30	1.20	65	4
2	60.30	0.94	74	4
3	60.30	3.31	100	4
4	60.30	7.85	122	4
1	61.50	1.16	70	5
2	61.50	1.35	74	5
3	61.50	4.13	84	5
4	61.50	10.60	125	5
1	50.90	1.44	86	6
2	50.90	2.78	92	6
3	50.90	4.18	121	6
4	50.90	6.45	130	6
1	50.10	1.29	78	7
2	50.10	1.33	95	7
3	50.10	3.03	110	7
4	50.10	5.06	128	7

Source: Crowder, M.J., On concurrent regression lines, *Applied Statistics*, 27, 310–318, 1978.

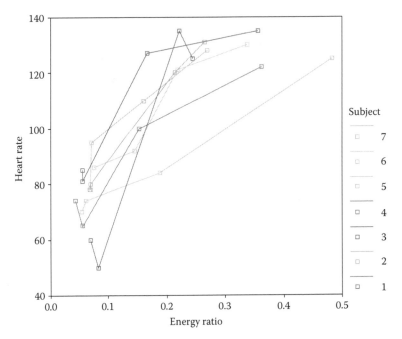

FIGURE 9.7
Heart rate versus energy expenditure by subject.

As expected, the heart rate, energy expenditure, and energy ratio increase with increasing physical exertion (Table 9.17).

The model proposed by Hand and Crowder (p. 117)[2] is the mixed nonlinear model

$$Y_{ij} = (\beta_1 + b_{1i}) + (\beta_2 + b_{2i})(X_{ij} - \eta_2) + e_{ij} \tag{9.23}$$

where

$$X_{ij} = \frac{E_{ij}}{W_i^{\eta_1}} \tag{9.24}$$

where:
E_{ij} is the energy expenditure of subject i for task j
W_i is the weight of the ith subject for $i = 1,2,\ldots,7$ and $j = 1,2,3,4$

Note that i is the ith subject and j is the jth task.

The unknown parameters are $\beta_1, \beta_2, \eta_1, \eta_2, \sigma^2, \sigma_1^2$, and σ_2^2, where e_{ij} are independent normal random variables with mean 0 and variance σ^2, and the two random effects b_{i1}, b_{21} are independent normal random variables with mean 0 and variances σ_1^2 and σ_2^2, respectively.

The Bayesian analysis is executed with **BUGS CODE 9.6**, and the initial values for the simulation are taken from Hand and Crowder (p. 118).[2] These

TABLE 9.17

Descriptive Statistics: Energy Expenditure, Heart Rate, Energy Ratio by Task

Task	Mean	SD	Median
Energy Expenditure			
1	1.3314	0.1585	1.29
2	1.4771	0.5947	1.33
3	3.9229	0.6104	4.130
4	6.7971	2.5163	6.45
Heart Rate			
1	73.429	12.8304	78
2	79.143	11.8382	78
3	112.429	15.6570	120
4	129.429	4.8599	130
Energy Ratio			
1	0.0655	0.01442	0.0685
2	0.0732	0.03338	0.0670
3	0.1922	0.03444	0.1881
4	0.3278	0.08662	0.3385

are informative distributions, and for example, the prior distribution for β_1 is normal with mean −76 and variance 10. In a similar way, the prior distribution of η_1 is normal with mean 24 and variance 10.

The Bayesian analysis is executed with 150,000 observations, with a burn in of 5,000 and a refresh of 100 and appears in Table 9.18.

BUGS CODE 9.6

```
model;
    {
            beta[1] ~ dnorm(-76,.1)
            beta[2] ~ dnorm(24,.1)
    eta[1] ~ dnorm(19,.1)
    eta[2]~ dnorm(-33,.1)

for(i in 2:N){for (j in 1:M){Y[i,j]~dnorm(mu[i,j],tau)}}
for(i in 2:N){for (j in 1:M){Z[i,j]~dnorm(mu[i,j],tau)}}
for(i in 2:N){for (j in 1:M){mu[i,j]<-(beta[1]+b1[i])+
(beta[2]+b2[i])*(x[i,j]-eta[2])}}
for(i in 2:N){for(j in 1:M){x[i,j]<-E[i,j]/
(pow(w[i],eta[1]))}}

for(i in 1:N){b1[i]~dnorm(0,tau1)}
for(i in 1:N){b2[i]~dnorm(0,tau2)}
```

```
tau~dgamma(1.25,2)
sigma<-1/tau
tau1~dgamma(.92,2)
sigma1<-1/tau1
tau2~dgamma(100,2)
sigma2<-1/tau2
}

list(N=7, M=4,Y=structure(.Data=c(50.0,60.0,125.0,135.0,
80.0,78.0,120.0,131.0,
85.0,81.0,127.0,135.0,
65.0,74.0,100.0,122.0,
70.0,74.0,84.0,125.0,
86.0,92.0,121.0,130.0,
78.0,95.0,110.0,128.0),.Dim=c(7,4)),

E=structure(.Data=c(1.57,1.31,4.67,4.23,
1.46,1.41,4.49,5.56,
1.20,1.22,3.65,7.83,
1.20,.94,3.31,7.85,
1.16,1.35,4.13,10.60,
1.44,2.78,4.18,6.45,
1.29,1.33,3.03,5.06),.Dim=c(7,4)),

w=c(51,58,62,60,62,51,50))

# initial values
list(beta=c(-70,24), eta=c(20,-33),tau=1.25,tau1=.934,
tau2=100)
```

TABLE 9.18

Posterior Analysis for Energy Expenditure

Parameter	Mean	SD	Error	2½	Median	97½
β_1	−76.62	3.162	0.011090	−82.81	−76.62	−70.42
β_1	6.391	1.239	0.0462	4.932	6.217	8.596
η_1	18.89	3.132	0.0176	12.83	18.97	25.11
η_2	−28.6	3.652	0.06751	−35.63	−28.66	−213
σ^2	539.3	167.1	0.5205	303.1	510.2	945.1
σ_1^2	802.4	15730	622	0.5586	2.927	51.76
σ_2^2	0.02017	0.002041	0.0000054	0.01655	0.02004	0.02457

Skewness of the posterior distribution is apparent for the three variances and is quite apparent for σ_1^2 with a posterior mean of 802 but a posterior median of 2.927. In a similar way, the posterior median of σ^2 is 510 compared to a posterior mean of 593. The reader should compare the results in Table 9.18 with those via maximum likelihood as performed by Hand and Crowder (p. 118).[2]

9.1.7 Blood Glucose Levels

Blood glucose levels for seven volunteers were checked, where each person took alcohol at time 0 and provided a blood sample at 14 time points over five hours. Times are displayed as minutes divided by 10. Crowder and Tredger[6] analyzed these results, which are also explained by Hand and Crowder (p. 118).[2] The procedure was repeated with the same subjects, but were given a dietary supplement.

Table 9.19 reports the descriptive statistics for the glucose values. As expected, the mean blood glucose value increases with time, reaches a peak at approximately 60 minutes, and then decreases. The time −1 serves as a baseline, while alcohol was taken at time 1. The first entry is for the first data, while the second is for date 2.

Figure 9.8 is a plot of blood glucose levels at the various time points and corroborates the descriptive statistics and show an increase to a maximum at 60 minutes but also displays the intersubject heterogeneity. A lowess curve is used to exhibit the outcome for each of the seven subjects by session and appears to show higher glucose values for session 1, where the subjects received alcohol.

Based on the basic data, the descriptive statistics, and the scatter plot, what is the most appropriate model? Crowder and Tredger[6] propose the following model for the blood glucose value Y_{ij} with two random effects. Below is the model for session 1.

$$Y_{1ij} = \left(\beta_1 + b_{1i}\right) + \left(\beta_2 + b_{2i}\right)x_{ij}^3 \exp\left(-\gamma_1 x_{ij}\right) + \varepsilon_{1ij} \tag{9.25}$$

TABLE 9.19

Descriptive Statistics of Blood Glucose Levels

Time	Mean	SD	Median
−1	3.50, 3.48	0.3829, 0.6176	3.5, 3.6
0	3.41, 3.51	0.4220, 0.5047	3.50, 3.6
2	4.91, 4.75	1.004, 0.7322	5.00, 4.40
4	6.385, 6.085	1.5388, 0.6694	6.20, 5.8
6	6.842, 6.842	1.4339, 0.9466	6.80, 7.0
8	5.585, 5.587	1.162, 1.1625	5.9, 5.90
10	4.314, 4.257	0.8532, 0.8599	4.6, 4.40
12	4.507, 3.885	1.142, 0.6517	4.1, 3.9
15	4.357, 3.742	0.3909, 0.3735	4.3, 3.8
18	4.341, 3.671	0.467, 0.3199	4.1, 3.5
21	3.628, 3.442	0.7087, 0.2636	3.4, 3.5
24	3.428, 3.5	0.660, 0.2582	3.5, 3.5
27	3.342, 0.3728	0.5287, 0.3545	3.5, 3.5
30	3.571, 3.671	0.5314, 0.2811	3.8, 3.7

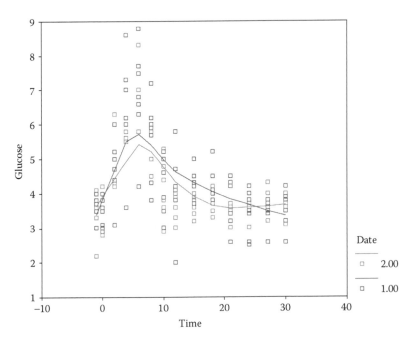

FIGURE 9.8
Glucose levels versus time by date.

and for the second date, the model is defined by

$$Y_{2ij} = \left(\beta_3 + b_{3i}\right) + \left(\beta_4 + b_{4i}\right)x_{ij}^3 \exp\left(-\gamma_2 x_{ij}\right) + \varepsilon_{2ij} \tag{9.26}$$

where x_{ij} is the time for subject i at the jth time point, β_1 and β_2 are unknown parameters, and the two random effects b_{1i} and b_{2i} are independent normal random variables with mean 0 and variances σ_{b1}^2 and σ_{b2}^2, respectively. In addition, the error terms ε_{1ij} are independent normal random variables with mean 0 and variance σ_1^2. The blood glucose values expressed as (9.25) imply a cubic response damped by the exponential term $\exp\left(-\gamma_1 x_{ij}\right)$, where γ_1 is an unknown positive parameter. Let the 14 by 1 vector of times be denoted by

$$x_{ij} = (-1,0,2,4,6,8,10,12,15,18,21,24,27,30) \tag{9.27}$$

for all subjects i. Of course, a similar interpretation is given to the parameters of the second session model (9.26). The information from the study for the two sessions is given in the list statements of **BUGS CODE 9.7**, and the prior information for the various coefficients are specified as noninformative distributions.

BUGS CODE 9.7

```
model;
    {

# prior distribution for the beta coefficients
            beta[1] ~ dnorm(0,.01)
            beta[2] ~ dnorm(0,.01)
            beta[3] ~ dnorm(0,.01)
            beta[4] ~ dnorm(0,.01)

# prior distribution for the eta coefficients

  eta[1] ~ dnorm(3,.1)
  eta[2] ~ dnorm(3,.1)

for(i in 1:N){for (j in 1:M){Y1[i,j]~dnorm(mu1[i,j],tau1)}}
for(i in 1:N){for(j in 1:M){Y2[i,j]~dnorm(mu2[i,j],tau2)}}
for(i in 1:N){for (j in 1:M){Z1[i,j]~dnorm(mu1[i,j],tau1)}}
for(i in 1:N){for (j in 1:M){Z2[i,j]~dnorm(mu2[i,j],tau2)}}
for(i in 1:N){for (j in 1:M){mu1[i,j]<-(beta[1]+b1[i])+
(beta[2]+b2[i])*exp(-eta[1]*x[j])}}
for(i in 1:N){for (j in 1:M){mu2[i,j]<-(beta[3]+b3[i])+
(beta[4]+b4[i])*exp(-eta[2]*x[j])}}

# the distribution for the four random effects

for(i in 1:N){b1[i]~dnorm(0,taub1)}
for(i in 1:N){b2[i]~dnorm(0,taub2)}
for(i in 1:N){b3[i]~dnorm(0,taub3)}
for(i in 1:N){b4[i]~dnorm(0,taub4)}

# the prior distribution for the various variances

tau1~dgamma(.001,.001)
sigma1<-1/tau1
taub1~dgamma(.001,.001)
sigmab1<-1/taub1
taub2~dgamma(.001,.001)
sigmab2<-1/taub2

tau2~dgamma(.001,.001)
sigma2<-1/tau2
taub3~dgamma(.001,.001)
sigmab3<-1/taub3
taub4~dgamma(.001,.001)
sigmab4<-1/taub4

}

# the glucose values from session 1

list(N=7, M=14,Y1=structure(.Data=c(
```

```
3.00,3.00,4.70,6.00,6.30,4.30,3.00,2.00,4.50,3.80,3.20,
2.60,2.60,2.60,
4.00,3.60,6.00,8.60,8.80,7.20,5.00,3.80,4.20,4.00,2.60,
2.50,2.6,3.80,
3.50,3.50,6.00,7.30,7.50,6.20,5.00,4.10,4.30,4.10,3.40,
3.80,3.80,3.90,
3.80,3.80,4.40,6.00,6.80,5.70,4.60,3.80,4.30,4.50,4.50,
4.20,3.80,4.2,

3.70,4.00,5.20,7.00,6.60,6.00,5.20,4.70,4.50,4.50,4.20,
3.50,3.70,3.80,
3.50,3.10,3.10,3.60,4.20,3.80,3.50,4.20,3.70,4.10,3.20,
3.4,3.40,3.2,

3.00,2.90,5.00,6.20,7.70,5.90,3.90,5.80,5.00,5.20,4.30,
4.00,3.50,3.50),.Dim=c(7,14)),

# the glucose values for the second data

Y2=structure(.Data=c(
2.20,2.80,4.40,5.60,5.80,4.50,3.60,3.30,3.20,3.30,3.00,
3.00,3.20,3.10,
4.10,4.20,6.30,7.00,8.30,5.70,2.90,3.00,3.40,3.50,3.20,
3.80,4.30,3.70,
3.80,3.80,5.00,5.50,7.00,5.00,3.80,3.60,3.60,3.50,3.50,
3.50,3.50,4.00,
3.60,3.60,4.30,5.50,6.30,5.70,5.30,4.70,4.00,3.50,3.60,
3.70,4.00,3.70,
3.80,3.80,4.70,7.00,7.70,6.00,5.00,4.70,4.30,4.20,3.70,
3.40,3.70,3.80,
3.60,3.50,4.40,6.20,7.00,5.90,4.80,3.90,3.90,4.00,3.70,
3.50,3.80,3.80,
3.30,2.90,4.20,5.80,5.80,5.80,4.40,4.00,3.80,3.70,3.40,
3.60,3.60,3.60),.Dim=c(7,14)),

# the time points

x=c(-1.00,.00,2.00,4.00,6.00,8.00,10.00,12.00,
15.00,18.00,21.00,24.00,27.00,30.00))

# initial values
list(beta=c(0,0,0,0),eta=c(10,10),tau1=1,taub1=1,taub2=1,
tau2=1,taub3=1,taub4=1)
```

Based on **BUGS CODE 9.7**, the Bayesian analysis is executed with 750,000 observations, with a burn in of 5,000 and a refresh of 100.

The two model have almost the same estimated coefficients, except for β_2 and β_4. Also, $\sigma_{b1}^2 \neq \sigma_{b3}^2$, and $\sigma_{b2}^2 \neq \sigma_{b4}^2$, however more formal analysis, might not imply the same thing. The equality of the two parameters is left as an exercise (Table 9.20).

TABLE 9.20

Posterior Analysis for the Glucose Study

Parameter	Mean	SD	Error	2½	Median	97½
β_1	4.585	5.279	0.1723	−8.106	5.406	14.72
β_2	0.3516	5.289	0.1726	−9.719	−0.551	13.07
β_3	4.978	5.451	0.1786	−5.526	5.238	17.06
β_4	−0.1843	5.461	0.1789	−12.24	−0.5351	10.4
γ_1	0.03706	0.2356	0.005773	−0.06807	−0.00687	0.8555
γ_2	0.03508	0.2196	0.004963	−0.05435	−0.00368	0.7707
σ_1^2	1.576	0.2473	0.001529	1.164	1.551	2.129
σ_2^2	1.223	0.1868	0.000996	0.9114	1.205	1.641
σ_{b1}^2	0.1629	0.3345	0.001642	0.00099	0.06363	0.8813
σ_{b2}^2	0.1245	0.3318	0.003145	0.000752	0.0265	0.8236
σ_{b3}^2	0.05157	0.1151	0.000408	0.000702	0.01618	0.3095
σ_{b4}^2	0.04996	0.1821	0.001383	0.000646	0.0122	0.324

9.1.8 Michaelis–Menten Equation

In the biology literature, one may find many examples of nonlinear models and especially for modeling the pharmacokinetics of drug studies. Ruppert, Cressie, and Carroll[7] is a good reference of the general area of the use of nonlinear models in pharmacokinetics, and Draper and Smith[8] provide information about curve fitting with nonlinear models. Our presentation of the Michaelis-Menten model is based on the book by Jones (p. 139)[9] who define the mixed model as

$$Y_{ij} = \frac{x_{ij}\exp(\beta_1 + b_{1i})}{\left[x_{ij} + \exp(\beta_2 + b_{2i})\right]} + e_{ij} \tag{9.28}$$

where the response for subject i at the dose j is Y_{ij}, β_1 and β_2 are unknown parameters, and the two random effects b_{1i}, b_{2i} are independent normal random variables with mean 0 and variances σ_1^2 and σ_2^2 and are independent of the residuals e_{ij} that are independent normal random variables with mean 0 and variance σ^2. An alternative form of the model given by Bates and Watts (p. 270)[10] as

$$Y_{ij} = \frac{x_{ij}\beta_1}{\left(x_{ij} + \beta_2\right)} + e_{ij} \tag{9.29}$$

without the random effects and without the exponential operations of (9.28). The exponential functions are used to put on positive constraint on the parameters, and of course without the random effects, the interpretation based on (9.29) is simpler. In this form, according to Jones,[9] the velocity of an enzymatic reaction is the dependent variable Y_{ij} as a function of substrate concentration x_{ij}. Of course with the mixed model version (9.28), a correlation matrix for the response is induced, whereas with the simpler version (9.29), a correlation pattern, must be specified.

For the first data set using this model, the study by Treloar[11] involves the velocity of the reaction expressed as counts per minute of a radioactive product from the reaction measured as a function of substrate concentration in parts per million, and the velocity is computed in counts per minute. The experiment is conducted once with the enzyme treated with Puromycin and once without the enzyme being treated. The velocity is assumed to depend on the substrate concentration according to the Michaelis-Menten equation, and it thought that the final velocity β_1 should be affected by Puromycin but not the half velocity β_2. The information from the study is reported in Table 9.21.

It is left as an exercise to analyze this data.

The second example is based on an example provided by Jones and is simulated data set based on the model (9.28), and there are 30 subjects and the response is observed at dose levels. The descriptive statistics are shown in Table 9.22. Simulation using the Michaelis-Menten model assumes that $\beta_1 = \beta_2 = 0$ (Figure 9.9).

Upon inspection of the posterior analysis, the estimated values are compatible with the simulation parameters of the model, where $\beta_1 = \beta_2 = 0$. Note that the model includes two random effects, but it is difficult to determine the correlation structure between the nine dose levels; thus, consider the

TABLE 9.21

Reaction Velocity versus Substrate Concentration

Concentration	Treated	Untreated
0.02	76	67
0.02	47	51
0.06	97	84
0.06	107	86
0.11	123	98
0.11	139	115
0.22	159	131
0.22	152	124
0.56	191	144
0.56	201	158
1.1	207	160
1.1	200	

TABLE 9.22

Descriptive Statistics with Simulated Data

Dose	Mean	SD	Median
0.0625	0.0585	0.0135	0.0575
0.125	0.1129	0.0232	0.1185
0.25	0.1833	0.0405	0.1985
0.5	0.3289	0.0560	0.330
1	0.5026	0.1037	0.5310
2	0.6753	0.1385	0.672
4	0.7785	0.1826	0.7375
8	0.8867	0.2112	0.8550
16	0.9446	0.2469	0.9060

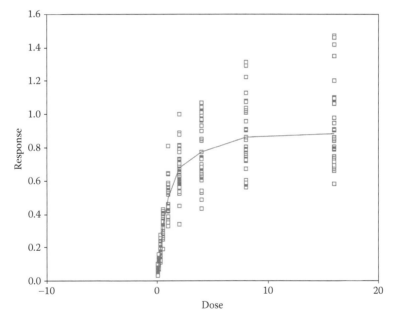

FIGURE 9.9

Simulated data with 30 subjects and 9 dose levels. (Data from Jones, R.H., *Longitudinal Data with Serial Correlation: A State Space Approach*, Chapman and Hall, London, p. 148, 1993.)

Michaelis-Menten model with one random effect, which induces a compound symmetry pattern for the covariance matrix. Thus, consider the version

$$Y_{ij} = b_i + \frac{x_{ij}\exp(\beta_1)}{x_{ij} + \exp(\beta_2)} + e_{ij} \tag{9.30}$$

where $b_i \sim ind(0,\sigma^2)$, $i = 1,2,\ldots,N$, and the other terms in the model are defined as in (9.28).

BUGS CODE 9.8 will be the foundation for executing the posterior analysis:

BUGS CODE 9.8

```
model;
{
            beta[1] ~ dnorm(0,.1)
            beta[2] ~ dnorm(0,.1)

for(i in 1:N){for (j in 1:M){Y[i,j]~dnorm(mu[i,j],tau)}}

for(i in 1:N){for (j in 1:M){Z[i,j]~dnorm(mu[i,j],tau)}}

for(i in 1:N){for (j in 1:M){mu[i,j]<-(x[j]*exp(-
beta[1]+b1[i]))/(x[j]+exp(-beta[2]+b2[i])*x[j])}}

for(i in 1:N){b1[i]~dnorm(0,taub1)}
for(i in 1:N){b2[i]~dnorm(0,taub2)}

tau~dgamma(.01,.01)
sigma1<-1/tau
taub1~dgamma(.01,.01)
sigmab1<-1/taub1
taub2~dgamma(.01,.01)
sigmab2<-1/taub2
}

list(N=30, M=9,Y=structure(.Data=c(
.032,.072,.124,.286,.349,.449,.541,.590,.748,
.100,.161,.274,.420,.808,1.001,1.043,1.316,1.097,
.071,.129,.192,.405,.559,.879,1.010,1.023,1.353,
.064,.130,.197,.409,.554,.811,.967,1.010,1.204,
.069,.135,.220,.316,.443,.536,.732,.727,.738,
.057,.117,.238,.297,.544,.815,1.070,1.222,1.460,
.065,.129,.240,.403,.576,.774,.932,1.077,1.066,
.063,.120,.252,.355,.537,.704,.682,.864,.912,
.059,.079,.173,.257,.362,.537,.487,.602,.664,
.045,.075,.112,.249,.330,.340,.436,.576,.660,
.074,.139,.243,.373,.643,.657,.772,.959,.979,
.053,.108,.193,.312,.462,.631,.734,.754,.866,
.056,.116,.200,.329,.519,.730,.843,1.038,.910,
.068,.137,.217,.428,.586,.728,.997,1.130,1.103,
.044,.085,.151,.191,.350,.601,.532,.563,.581,
.067,.097,.194,.331,.554,.669,.964,.790,1.089,
.053,.087,.175,.251,.421,.590,.664,.675,.792,
.070,.147,.235,.306,.530,.694,.698,.913,.906,
.043,.098,.162,.340,.385,.675,.633,.803,.853,
.057,.121,.213,.323,.451,.527,.605,.720,.806,
.042,.104,.183,.298,.432,.618,.741,.904,.843,
.053,.128,.265,.341,.533,.713,.841,.846,.946,
.038,.069,.128,.300,.434,.605,.680,.723,.674,
.070,.120,.222,.347,.557,.776,.902,1.073,.906,
```

```
.069,.125,.200,.381,.645,.790,.851,1.293,1.471,
.050,.108,.156,.257,.537,.560,.674,.763,.795,
.066,.127,.196,.354,.572,.774,1.016,1.040,1.089,
.058,.126,.209,.359,.532,.890,1.036,1.081,1.421,
.055,.106,.221,.334,.423,.592,.654,.784,.714,
.046,.094,.169,.316,.450,.595,.620,.744,.692),.
Dim=c(30,9)),
x=c(.0625,.125,.25,.5,1,2,4,8,16))
# initial values
list(beta=c(0,0), tau=1,ta
```

TABLE 9.23

Posterior Analysis for Simulated Michaelis–Menten Model

Parameter	Mean	SD	Error	2½	Median	97½
β_1	0.3819	0.7409	0.02436	−2.215	0.6239	0.7734
β_2	2.738	2.469	0.06727	−2.872	2.767	7.498
σ^2	0.1214	0.0106	0.0000295	0.1023	0.1208	0.144
σ_{b1}^2	0.0137	0.01007	0.000163	0.00309	0.01095	0.04021
σ_{b2}^2	0.447	1.406	0.0699	0.00541	0.0759	0.3297

TABLE 9.24

Posterior Analysis for Simulated Michaelis–Menten Model with Compound Symmetry

Parameter	Mean	SD	Error	2½	Median	97½
β_1	−1.731	3.632	0.2016	−11.69	0.4366	0.7781
β_2	2.033	8.313	0.4427	−12.43	1.199	20.05
σ^2	0.1213	0.01074	0.0000462	0.1019	0.1207	0.1441
σ_{b1}^2	0.0061	0.00323	0.0000312	0.002104	0.00544	0.01435

Based on **BUGS CODE 9.8**, the posterior analysis and the Bayesian analysis for the simulated Michaelis-Menten data with compound symmetry are reported in Tables 9.23 and 9.24. One's estimate of β_1 and β_2 is 0 based on the 95% credible interval and the compound symmetry pattern has correlation $\rho = \sigma_b^2/(\sigma^2 + \sigma_b^2)$ estimated as 0.04314 via the posterior median.

9.2 Nonlinear Repeated Measures with Categorical Data

Recall from Chapter 8 that the Bayesian analysis for categorical data was presented, but assuming linearity for the response, and the presentation was divided into three parts: (1) a multinomial pattern for the repeated measures outcome,

(2) a Bayesian version of generalized estimating equation, and (3) when random effects were incorporated into the model. With regard to part (1), the correlation structure was determined by Dirichlet posterior distribution of the parameters, and regard to (3), the correlation structure was determined by the random effects in the model. In the latter part (3), the distribution of the repeated measures was assumed to be binary, categorical with several possible values, and ordinal.

For binary response, the probability of success was transformed to the logit scale, which in turn was expressed as a linear function of fixed and random effects. For categorical responses with a few possible values, the responses were assumed to have a Poisson distribution, with a mean transformed to the log scale, which in turn was expressed as a linear function of fixed and random effects. Last, for ordinal categorical data, the outcomes were assumed to have a multinomial distribution whose parameters (probabilities) are transformed to cumulative log odds ratios, and in turn, the log odds ratios were expressed as a linear function of fixed and random effects.

For nonlinear model for categorical data, the Bayesian analysis will follow the following alternatives: (1) for binary data, the probability of success will be transformed to the logit scale, which will then be expressed as a nonlinear function of fixed and random effects, (2) for categorical responses with a few possible values, the responses are assumed to have a Poisson distribution, with a mean transformed to the log scale, which in turn will be expressed as a nonlinear function of fixed and random effects, and (3) lastly, for ordinal categorical data, the outcomes are assumed to have a multinomial distribution whose parameters (probabilities) are transformed to cumulative log odds ratios, and in turn, the log odds ratios will be expressed as a nonlinear function of fixed and random effects.

9.2.1 Pill Dissolution Study with a Binary Response

Recall this example from the pharmaceutical industry, which involves the time it takes to dissolve a specified proportion of a pill. Also recall for this example that the response is the time it takes the pill to dissolve to a specified proportion (left undissolved). I have replaced the time values (measured in seconds) by a binary response, where 0 represents values less than the median value of 23 seconds, while 1 represents values greater than 23 seconds. The proposed model is given by a revised form for (9.5) that includes a random effect for the logit that measures the pip-to-pill variation in the factor that scales the specified proportions. Recall for a binary response with success probability, the canonical link is given by

$$\operatorname{logit}\left(\theta_{ij}\right) = \beta_1 + \beta_2\left(1 - x_{ij}^{\beta_3 + b_i}\right) \tag{9.31}$$

where:
$\theta_{ij} = P\left[Y_{ij} = 1/b_i\right]$
x_{ij} is the specified proportion, that is, the proportion left of the undissolved pill.

Of course, I have replaced the original continuous data with binary values Y_{ij}; thus, there is loss of information, but nevertheless, it serves to illustrate a nonlinear model for categorical data. Prior information for the parameters of the model (9.31) is as follows: For β, the distribution is normal with a mean of 11, and a precision (1/variance) is 0.1, and for $β_2$ the distribution is normal with a mean of 23 and a precision of 0.1, while for $β_3$ the prior is uniform over (0.7, 1). The Bayesian analysis is executed with **BUGS CODE 9.9**.

The Bayesian analysis is executed with 75,000 observations, with a burn in of 5,000 and a refresh of 100, and the results are shown in Table 9.25. Uninformative prior distributions were assigned to all parameters of the model (9.31).

BUGS CODE 9.9

```
model;
{

# prior distribution for the regression coefficients
            beta[1]~dnorm(11,.1)
            beta[2]~dnorm(-23,.1)
            beta[3]~dunif(.7,1)

for(i in 1:N){for(j in 1:M){Y[i,j]~dpois(theta[i,j])}}

for(i in 1:N){for (j in 1:M)
{logit(theta[i,j])<-beta[1]+beta[2]*(1-
pow(x[j],beta[3]+b1[i]))}}
for (i in 1:N){b1[i]~dnorm(0,tau)}
# prior distribution for the precision of the random
effect
tau~dgamma(3,2)
sigma<-1/tau

}

list(N=10, M=6,Y = structure(.Data=c(1,1,0,0,0,0,
1,1,1,0,0,0,
1,1,1,1,1,0,
1,1,1,0,0,0,
1,1,0,0,0,0,
1,1,0,0,0,0,
1,1,1,0,0,0,
1,1,1,0,0,0,
1,1,1,0,0,0,
1,1,1,0,0,0),.Dim=c(10,6)),
# x is the vector of specified proportions
x=c(.1,.25,.3,.5,.7,.9))

# initial values

list(beta=c(11,-23,.7)
```

TABLE 9.25

Posterior Analysis for Pill Dissolution Study with Binary Response

Parameter	Mean	SD	Error	2½	Median	97½
β_1	16.87	2.375	0.0367	12.29	16.85	21.56
β_2	−16.3	2.467	0.03691	−21.21	−16.3	−11.49
β_3	0.8429	0.0865	0.001273	0.7065	0.8385	0.9911
σ^2	0.9611	0.9768	0.0164	0.264	0.7085	30.205

It appears that the variance of the random effect is estimated as 0.7085 with the posterior median and represents the pip-to-pill variation in the effect on the specified proportion. Also, the MCMC errors are reasonable, for example, the error for the posterior mean 16.87 of β_1 is 0.0367. How do these compare to the Bayesian analysis using the full data set (see Table 9.6), but remember the covariance structure for that analysis was unstructured and the binary response assumes a model with one random effect. Also, the present model, a canonical link logit function, is used to model the mean; thus, the effects of the coefficients of the present model are not interpreted the same way as the coefficients for the original data.

Using the results based on model (9.31), what is the correlation pattern for the repeated time measurements over the eight proportions? It is difficult to answer this question; thus, consider a version of the model (9.5) assuming a compound symmetry–type pattern, namely,

$$\mathrm{logit}\left(\theta_{ij}\right) = b_i + \beta_1 + \beta_2\left(1 - x_{ij}^{\beta_3}\right) \qquad (9.32)$$

where the random effects $b_i \sim nid\left(0, \sigma^2\right)$ for $i = 1,2,\ldots,N$. What will be the variation in the intercept on the logit scale?

Based on a small revision of **BUGS CODE 9.9**, one may show the Bayesian analysis is provided by Table 9.26. Uninformative prior distributions are employed for all parameters.

The Bayesian analysis is executed with 65,000 observations for the simulation, with a burn in of 5,000 and a refresh of 100. Note the skewness in the posterior distribution of the variance of the random effect; thus, the pill-pill variation in the intercept b_i is estimated as 0.7232 using the posterior median. Also note very little difference in the estimates of the model coefficients with the two random effect models (9.31) and (9.32). Beware that the two random

TABLE 9.26

Posterior Analysis for Pill Study with One Random Effect

Parameter	Mean	SD	Error	2½	Median	97½
β_1	−6.836	1.986	0.01843	−11.21	−6.652	−3.474
β_2	12.89	3.769	0.03507	6.642	12.51	21.29
β_3	0.8448	0.08631	0.000803	0.7068	0.8425	0.9915
σ^2	0.9238	0.7324	0.006702	0.2751	0.7232	2.787

TABLE 9.27

Bayesian Analysis for Pill Dissolution Study and Three Random Effects

Parameter	Mean	SD	Error	2½	Median	97½
β_1	−7.432	2.612	0.02506	−13.38	−7.105	−3.246
β_1	13.72	4.752	0.04399	6.242	13.1	24.6
β_3	0.8826	0.0817	0.000733	0.7165	0.896	0.9959
σ_1^2	0.8794	0.6552	0.005419	0.2697	0.6994	2.578
σ_2^2	0.5106	0.2903	0.002951	0.2014	0.4377	1.255
σ_3^2	1.003	1.305	0.002438	0.2767	0.7495	3.228

effects of the two models are measuring different characteristics of the logit. This will be explored further in the exercises.

For the last scenario of the pill dissolution study, consider a model with three random effects

$$\log it\left(\theta_{ij}\right) = b_{1i} + \beta_1 + \left(\beta_2 + b_{2i}\right)\left(1 - x_{ij}^{\beta_3 + b_{3i}}\right) \tag{9.33}$$

one for the intercept, b_{2i} for the slope, and b_{3i} for the scaling factor.

The analysis is executed with 150,000 observations, with a burn in of 5,000 and a refresh of 100, and the output is reported in Table 9.27. Noninformative prior distributions are used for all parameters.

Of the six posterior distributions, the three corresponding to the variances of the three random effects are skewness with the posterior means larger than the corresponding posterior medians; thus, based on the medians, the random effect of the slope appears be the smallest among the three.

Three versions of a generalized nonlinear mixed model for the pill dissolution study are presented, and which one gives the best fit? This will be investigated by the student in the exercises.

9.2.2 Categorical Data with Load Slip Data

Hand and Crowder (p. 111)[2] describe a nonlinear example of repeated measures occurring in civil engineering, where a wood plank is inserted between concrete surfaces and fastened by four screws. The load is increased and applied to the timber, and the slippage in millimeters is recorded at two-second intervals. For our purposes, the load is the dependent variable and the slip the independent variable, and the data exhibited is taken from Table A.15 of Hand and Crowder (p. 179).[2]

A revision of model (9.10) is considered; thus let Y_{ij} be the load of slip j for plank i, where Y_{ij} has a Poisson distribution with conditional mean and variance given one random effect as $E\left(Y_{ij} | b_i\right) = Var\left(Y_{ij} | b_i\right)$, where

$$\log\left[E\left(Y_{ij} | b_i\right)\right] = \beta_1 + b_{1i} + \beta_2 \exp\left(-\beta_3 x_{ij}\right) \tag{9.34}$$

where $Y_{ij} = 1$ when the original load data is between 0 and 3.24 mm. In addition, let $Y_{ij} = 2$ when the original load data is between 3.25 and 6.48 mm. Also let $Y_{ij} = 3$ when the original load data is between 6.49 and 9.72 mm, and last let $Y_{ij} = 4$ for values greater than 9.72 mm. The random effects $b_i \sim nid\left(0, \sigma^2\right)$ and the beta parameters are unknown coefficients.

Model (9.34) is a generalized nonlinear mixed model with a log link function for the Poisson distribution of the categorical load data. The Bayesian analysis will be executed with **BUGS CODE 9.10** using 150,000 observations for the simulation, with a burn in of 5,000 and a refresh of 100. Uninformative prior distributions were used for all parameters: (1) for the beta coefficients, a normal distribution with mean 0 and precision 0.001, while for the precision of the random effect, a gamma with first and second parameters 0.001.

BUGS CODE 9.10

```
model;
      {
             beta[1] ~ dnorm(0,.001)
             beta[2] ~ dnorm(0,.001)
             beta[3] ~ dnorm(0,.001)

for(i in 1:N){for (j in 1:M){Y[i,j]~dpois(mu[i,j])}}
for(i in 1:N){for (j in 1:M){log(mu[i,j])<-b[i]+beta[1]+
beta[2]*exp(-beta[3]*s[j])}}

for(i in 1:N){b[i]~dnorm(0,tau1)}


tau1~dgamma(.001,.001)
# sigma1 is the variance of the random effects
sigma1<-1/tau1

}

list(N=8, M=11,Y=structure(.Data=
c(1,1,2,3,3,4,4,4,4,4,4,
1,1,2,3,3,4,4,4,4,4,4,
1,1,2,3,3,3,4,4,4,4,4,
1,1,2,2,3,3,4,4,4,4,4,
1,1,2,3,3,3,4,4,4,4,4,
1,2,2,3,3,4,4,4,4,4,4,
1,1,2,3,3,4,4,4,4,4,4,
1,2,2,3,3,4,4,4,4,4,4),.Dim=c(8,11)),

s=c(0,.1,.2,.3,.4,.5,.6,.7,.8,.9,1))

# initial values
list(beta=c(0,0,0), tau1=.5)
```

TABLE 9.28

Posterior Analysis for Timber Slip Data with Nonlinear Mixed Model
for Categorical Data

Parameter	Mean	SD	Error	2½	Median	97½
β_1	1.449	0.1419	0.002531	1.227	1.431	1.792
β_2	−1.663	0.3206	0.001686	−2.335	−1.648	−1.076
β_3	4.469	1.778	0.02712	1.914	4.57	8.96
σ^2	0.006419	0.01025	0.0000597	0.0004609	0.003297	0.03122

Table 9.28 reveals the posterior distribution of the parameters of the model
(9.33) and demonstrates an extremely small estimate of the variance of the
random effect.

The simulation errors are reasonable, for example, the simulation error for
the posterior mean of β_3 is 0.02712, that is, the estimate 4.469 is within 0.027
units of the "true" posterior mean of β_3.

Now consider an extension of (9.34) to include two random factors b_{1i} and
b_{2i}, where $b_{2i} \sim nid(0, \sigma_2^2)$, then the link function is the canonical link

$$\log\left[E\left(Y_{ij}|b_i\right)\right] = \beta_1 + b_{1i} + (\beta_2 + b_{2i})\exp(-\beta_3 x_{ij}), \qquad (9.35)$$

Thus, using the program statement

```
for(i in 1:N){for (j in 1:M){log(mu[i,j])<-b1[i]+beta[1]+(bet
a[2]+b2[i])*exp(-beta[3]*s[j])}}
```

with **BUGS CODE 9.10** to execute the Bayesian analysis with 150,000 observations for the simulation gives the following results for the posterior analysis reported in Table 9.29. It is interesting to observe that the estimates of
the beta coefficients and the variance of the random factor b_{1i} are almost the
same as those reported in Table 9.28, and that the plank-to-plank variation
for the intercept and slope are quite small. Also note that the posterior distributions for the two variances are skewed; thus, I recommend using the
posterior medians as estimates. For example, the posterior median of σ_1^2 is
0.003392 compared to a posterior mean of 0.006563.

TABLE 9.29

Bayesian Analysis for the Timber Slip Study with Two Random Effects

Parameter	Mean	SD	Error	2½	Median	97½
β_1	1.45	0.1742	0.005539	1.225	1.427	1.794
β_2	−1.685	0.3388	0.003354	−2.041	−1.666	−1.074
β_3	4.585	1.809	0.03168	1.923	4.648	9.03
σ_1^2	0.006563	0.01087	0.000059	0.000470	0.003392	0.03166
σ_2^2	0.054	0.1388	0.001297	0.000658	0.01299	0.3694

TABLE 9.30

Posterior Analysis for Timber Slip Study with Three Random Factors

Parameter	Mean	SD	Error	2½	Median	97½
β_1	1.425	0.1353	0.002712	1.214	1.41	1.726
β_2	−1.682	0.3396	0.002159	−2.399	−1.666	−1.066
β_3	5.228	1.935	0.03912	2.168	4.979	9.834
σ_1^2	0.00689	0.01166	0.0000727	0.000472	0.00345	0.03419
σ_2^2	0.05459	0.1377	0.001156	0.000651	0.01339	0.3712
σ_3^2	0.6223	3.518	0.06036	0.000823	0.04551	4.637

Last, consider the version with three random factors; thus, let

$$\log\left[E\left(Y_{ij}\,\middle|\,b_i\right)\right] = \beta_1 + b_{1i} + \left(\beta_2 + b_{2i}\right)\exp\left[-\left(\beta_3 + b_{3i}\right)x_{ij}\right] \qquad (9.36)$$

where the random effects $b_{3i} \sim nid\left(0,\sigma_3^2\right)$ are independent of the other two random factors. Remember to use the code

```
for(i in 1:N){for (j in 1:M){log(mu[i,j])<-b1[i]+beta[1]+(bet
a[2]+b2[i])*exp(-(beta[3]+b3[i])*s[j])}}
```

in **BUGS CODE 9.10**.

I executed the analysis with the revised version of **BUGS CODE 9.10**, using 150,000 observations for the simulation, with a burn in of 5,000 and a refresh of 100, and the Bayesian analysis is portrayed in Table 9.30. Uninformative prior distributions are used for all parameters.

Comparing Table 9.30 with 9.29 reveals that the estimates of the common parameters are almost the same, and that the variance σ_3^2 of the third random factor has an extremely skewed distribution with a posterior mean of 0.6223 compared to a posterior median of 0.04551.

9.2.3 Heart Rate of a Nonlinear Mixed Model with a Log Canonical Link

The next example is on data from Crowder[5] and involves female students whose energy expenditure (calories per minute) and heart rate (beats per minute) is measured four times. As in the previous example, the heart rate data is transformed into five categories, as shown in Table 9.31.

The relationship between heart rate and task will be explored with the mixed nonlinear model with a log link function

$$\log\left(\mu_{ij}\right) = \left(\beta_1 + b_{1i}\right) + \left(\beta_2 + b_{2i}\right)\left(X_{ij} - \beta_3\right) \qquad (9.37)$$

with X_{ij} the task number for subject i at task j, for $i = 1,\ldots,7$, and $j = 1,2,3,4$.

TABLE 9.31

Categories for Heart Rate

Range	Category
50–67	1
68–84	2
85–101	3
102–118	4
>119	5

Note that μ_{ij} is the conditional mean of Y_{ij} given b_{1i} and b_{2i}. Also, the Y_{ij} (the heart rate for subject i at task j) are assumed to follow a Poisson distribution, where the two random factors are independent, and in addition, $b_{1i} \sim nid\left(0, \sigma_1^2\right)$ and $b_{2i} \sim nid\left(0, \sigma_2^2\right)$.

Prior distributions for the parameters of the model (9.37) are uninformative. For example, the three beta coefficients are assigned normal distributions with mean 0 and variance 100, while the variances of the two random factors are designated as gamma with parameters 3 and 2. The first list statement of **BUGS CODE 9.11** contains the categories of the heart rate data.

BUGS CODE 9.11

```
model;
    {
            beta[1] ~ dnorm(0,.01)
            beta[2] ~ dnorm(0,.01)
    beta[3] ~ dnorm(0,.01)

for(i in 2:N){for (j in 1:M){Y[i,j]~dpois(mu[i,j])}}
for(i in 2:N){for (j in 1:M){log(mu[i,j])<-(beta[1]+b1[i])+
(beta[2]+b2[i])*(x[j]-beta[3])}}

for(i in 1:N){b1[i]~dnorm(0,tau1)}
for(i in 1:N){b2[i]~dnorm(0,tau2)}

tau~dgamma(3,2)
sigma<-1/tau
tau1~dgamma(3,2)
sigma1<-1/tau1
tau2~dgamma(3,2)
sigma2<-1/tau2

}

list(N=7, M=4,Y=structure(.Data=c(1,1,5,5,
2,2,5,5,
3,2,5,5,
```

```
1,2,3,5,
2,2,2,5,
3,3,5,5,
2,3,4,5),.Dim=c(7,4)),

x=c(1,2,3,4))

# initial values
list(beta=c(0,0,0), tau1=.934,tau2=100)
```

TABLE 9.32

Posterior Analysis for Heart Rate Study with Two Random Factors

Parameter	Mean	SD	Error	2½	Median	97½
β_1	1.229	0.7698	0.02593	−0.387	1.225	2.814
β_2	0.3388	0.305	0.00976	−0.2753	0.3392	0.9556
β_3	2.917	2.001	0.06711	−1.343	2.969	6.874
σ_1^2	0.5431	0.3187	0.00287	0.2101	0.4645	1.337
σ_2^2	0.4756	0.2566	0.002359	0.1939	0.4131	1.13

Table 9.32 reports the results of the Bayesian analysis, which is executed with 150,000 observations, with a burn in of 5,000 and a refresh of 100.

MCMC errors are quite small, and the posterior distributions for the two variance components are slightly skewed. Using the posterior median 0.4645 for σ_1^2 estimates, the subject to subject variation for the intercept on the log scale.

9.2.4 Blood Glucose Study of Poisson Distribution for Glucose Values

Recall Section 9.1.8 and the blood glucose levels for seven volunteers, where each person took alcohol at time 0 and provided a blood sample at 14 time points over five hours. Times are displayed as minutes divided by 10. Crowder and Tredger[6] analyzed these results that are also explained by Hand and Crowder (p. 118)[2] The procedure was repeated with the same subjects, but they were given a dietary supplement.

Table 9.19 reports the descriptive statistics for the glucose values. As expected, the mean blood glucose value increases with time, reaches a peak at approximately 60 minutes, and then decreases. The time −1 serves as a baseline, while alcohol was taken at time 1.

Based on the basic data, the descriptive statistics, and the scatter plot, what is the most appropriate model? Crowder and Tredger[6] propose the following model for the blood glucose value Y_{ij} with two random effects. The following is the model for session 1:

$$\log(\mu_{1ij}) = (\beta_1 + b_{1i}) + (\beta_2 + b_{21})x_{ij}^3 \exp(-\gamma_1 x_{ij}) \qquad (9.38)$$

while for the second date, the model is defined by

$$\log\left(\mu_{2ij}\right) = \left(\beta_3 + b_{3i}\right) + \left(\beta_4 + b_{4i}\right)x_{ij}^3\exp\left(-\gamma_2 x_{ij}\right) \tag{9.39}$$

where:

x_{ij} is the time for subject i at the jth time point

β_1 and β_2 are unknown parameters

the two random effects b_{1i} and b_{1i} are independent normal random variables with mean 0 and variances σ_1^2 and σ_2^2, respectively

The blood glucose values imply a cubic response damped by the exponential term $\exp\left(-\gamma_1 x_{ij}\right)$, where γ_1 is an unknown positive parameter. Let the 14 by 1 vector of times be denoted by

$$x_{ij} = (-1, 0, 2, 4, 6, 8, 10, 12, 15, 18, 21, 24, 27, 30)$$

for all subjects i. Of course, a similar interpretation is given to the parameters of the second session model (9.26). The information from the study for the two sessions is given in the list statements of **BUGS CODE 9.12**, and the prior information for the various coefficients is specified as noninformative distributions. It is important that the model (9.37) for categorical data is not the same as that for the continuous glucose values of model (9.25). Although there are similarities, there are differences. What are those differences? For the continuous model, the expected glucose value Y_{ij} is expressed directly by $\left(\beta_1 + b_{1i}\right) + \left(\beta_2 + b_{21}\right)x_{ij}^3\,e^{-\gamma_1 x_{ij}}$, but on the other hand for the categorical

BUGS CODE 9.12

```
model;
    {
            beta[1] ~ dnorm(0,.01)
            beta[2] ~ dnorm(0,.01)
            beta[3] ~ dnorm(0,.01)
            beta[4] ~ dnorm(0,.01)

        eta[1] ~ dunif(.5,2)
        eta[2] ~ dunif(.5,2)

for(i in 1:N){for (j in 1:M){Y1[i,j]~dpois(mu1[i,j])}}
for(i in 1:N){for(j in 1:M){Y2[i,j]~dpois(mu2[i,j])}}

for(i in 1:N){for (j in 1:M){log(mu1[i,j])<-(beta[1]+
b1[i])+(beta[2]+b2[i])*exp(-eta[1]*x[j])}}
for(i in 1:N){for (j in 1:M){log(mu2[i,j])<-(beta[3]+
b3[i])+(beta[4]+b4[i])*exp(-eta[2]*x[j])}}
```

```
for(i in 1:N){b1[i]~dnorm(0,taub1)}
for(i in 1:N){b2[i]~dnorm(0,taub2)}
for(i in 1:N){b3[i]~dnorm(0,taub3)}
for(i in 1:N){b4[i]~dnorm(0,taub4)}


taub1~dgamma(.001,.001)
sigmab1<-1/taub1
taub2~dgamma(.001,.001)
sigmab2<-1/taub2

taub3~dgamma(.001,.001)
sigmab3<-1/taub3
taub4~dgamma(.001,.001)
sigmab4<-1/taub4


}

list(N=7, M=14,Y1=structure(.Dat
a=c(1,1,2,3,3,2,1,1,2,2,1,1,1,1,
2,1,3,4,4,4,2,2,2,2,1,1,1,2,
1,1,3,4,4,3,2,2,2,2,1,2,2,2,
2,2,2,3,3,3,2,2,2,2,2,2,2,2,
1,2,2,3,3,3,2,2,2,2,1,1,1,2,
1,1,1,1,2,2,1,2,1,2,1,1,1,1,
1,1,2,3,4,3,2,3,2,2,2,2,1,1),.Dim=c(7,14)),

Y2=structure(.Data=c(1,1,2,3,3,2,1,1,1,1,1,1,1,1,
2,2,3,3,4,3,1,1,1,1,1,2,2,1,
2,2,2,3,3,2,2,1,1,1,1,1,1,2,
1,1,2,3,3,3,2,2,2,1,1,1,2,1,
2,2,2,3,4,3,2,2,2,2,1,1,1,2,
1,1,2,3,3,3,2,2,2,2,1,1,2,2,
1,1,2,3,3,3,2,2,2,1,1,1,1,1),.Dim=c(7,14)),

x=c(-1.00,.00,2.00,4.00,6.00,8.00,10.00,
12.00,15.00,18.00,21.00,24.00,27.00,30.00))

# initial values
list(beta=c(0,0,0,0), eta=c(.5,.5),taub1=1,taub2=1,taub3=
1,taub4=1)
```

values, the log of the expected value of Y_{ij} is expressed by the same model $(\beta_1 + b_{1i}) + (\beta_2 + b_{21})x_{ij}^3 e^{-\gamma_1 x_{ij}}$.

The analysis is executed with **BUGS CODE 9.12** using 300,000 observations, with a burn in of 5,000 and a refresh of 100, and uninformative priors are assigned to the parameters of the models (9.37) and (9.38). Note the uniform (0.5, 2) prior assigned to γ_1 and γ_2. The blood glucose values have been categorized into four classes and labeled 1, 2, 3, and 4, and the values are

TABLE 9.33

Posterior Analysis Blood Glucose Values with Four Random Factors

Parameter	Mean	SD	Error	2½	Median	97½
β_1	0.7026	0.02098	0.000559	0.5159	0.704	0.8798
β_2	−0.2395	0.1601	0.000920	−0.5963	−0.2216	0.02436
β_3	0.6064	0.09052	0.000447	0.4246	0.6082	0.7789
β_4	−0.1326	0.1431	0.000844	−0.4511	−0.1184	0.1181
γ_1	0.9786	0.3545	0.002245	0.5207	0.8998	1.82
γ_2	1.001	0.3665	0.002281	0.5201	0.9239	1.845
σ_{b1}^2	0.0198	0.03822	0.00035	0.000603	0.00789	0.1099
σ_{b2}^2	0.02324	0.0575	0.000601	0.000569	0.007138	0.1473
σ_{b3}^2	0.01191	0.02543	0.000279	0.000512	0.005021	0.06459
σ_{b4}^2	0.02333	0.06321	0.000686	0.000580	0.006958	0.1444

assumed to follow a Poisson distribution with a log link. The prior distributions are uninformative.

The main focus for the analysis reported in Table 9.33 is to compare the coefficients of the two dates (session 1 versus session 2).

Based on the results of the posterior analysis, are the two models the same? In order to answer this question, one would have to compare the corresponding coefficients in (9.37) and (9.38). Of course, one should also plot the glucose values versus time and use a lowess plot to delineate the two groups. This will provide important information (along with the posterior analysis) for comparing the glucose response between the two sessions. It appears to me that the two models are the same. Do you agree? This is left as an exercise,

9.2.5 Categorical Data and the Michaelis–Menten Model

The last example for nonlinear regression is in the area of pharmacokinetics.

In the biology literature one, may find many examples of nonlinear models and especially for modeling the pharmacokinetics of drug studies. Ruppert, Cressie, and Carroll[7] is a good reference of the general area of the use of nonlinear models in pharmacokinetics, and Draper and Smith[8] provide information about curve fitting with nonlinear models. Our presentation of the Michaelis-Menten model is based on the book by Jones (p. 139),[9] who defines the mixed model as

$$Y_{ij} = \frac{x_{ij} \exp(\beta_1 + b_{1i})}{x_{ij} + \exp(\beta_2 + b_{2i})} + e_{ij}$$

where the response for subject i at the dose j is Y_{ij}, β_1 and β_2 are unknown parameters, and the two random effects b_{1i}, b_{2i} are independent normal random variables with mean 0 and variances σ_1^2 and σ_2^2, and are independent of the residuals e_{ij} are independent normal random variables with mean 0 and variance σ^2.

The exponential functions are used to put on positive constraint on the parameters, and of course without the random effects, the interpretation based on (9.29) is simpler. In this form, according to Treloar,[11] the velocity of an enzymatic reaction is the dependent variable Y_{ij} as a function of substrate concentration x_{ij}. Of course with the mixed model version (9.28), a correlation matrix for the response is induced, whereas with the simpler version (9.29) a correlation pattern must be specified. Refer to Section 9.19 for an explanation of using the model with simulate data, and Figure 9.9 portrays the response as a function of dose. The simulated data follows the model (9.28) with $\beta_1 = \beta_2 = 0$. For our purposes, the simulated data is classified into five categories:

1 0.032–0.318

2. 0.319–0.604

3. 0.605–0.890

4. 0.8981–1.176

5. >1.176

then, the data analyzed with the model

$$\log \mu_{ij} = \frac{x_{ij} \exp(\beta_1 + b_{1i})}{x_{ij} + \exp(\beta_2 + b_{2i})} \tag{9.40}$$

where $\mu_{ij} = E(Y_{ij} \mid b_{1i}, b_{2i}) = \mathrm{Var}(Y_{ij} \mid b_{1i}, b_{2i})$ and $Y_{ij} \sim \mathrm{Poisson}(\mu_{ij})$ and is the response for subject i at dose j. Also, the two random effects are assumed to be independent, where $b_{1i} \sim nid(0, \sigma_1^2)$ and $b_{2i} \sim nid(0, \sigma_2^2)$.

Based on **BUGS CODE 9.13**, the Bayesian analysis is executed with 65,000 observations for the simulation, with a burn in of 5,000 and a refresh of 100, and the prior distributions for the parameters of the model (9.39) are assigned uninformative prior distributions.

BUGS CODE 9.13

```
model;

{

for (i in 1:N){for(j in 1:M){Y[i,j]~dpois(mu[i,j])}}
for(i in 1:N){for(j in 1:M){log(mu[i,j])<-
x[j]*(beta[1]+b1[i])/(x[j]+exp(beta[2]+b2[i]))}}

for(i in 1:2){beta[i]~dnorm(0,.001)}

for (i in 1:N){b1[i]~dnorm(0,taub1)}
taub1~dgamma(.001,.001)
sigmab1<-1/taub1
```

```
for (i in 1:N){b2[i]~dnorm(0,taub2)}
taub2~dgamma(.001,.001)
sigmab2<-1/taub2


}

list(N=30,M=9,

Y=structure(.Data=c(1,1,1,1,2,2,2,2,3,
1,1,1,2,3,4,4,5,4,
1,1,1,2,2,3,4,4,5,
1,1,1,2,2,3,4,4,5,
1,1,1,1,2,2,3,3,3,
1,1,1,1,2,3,4,5,5,
1,1,1,2,2,3,4,4,4,
1,1,1,2,2,3,3,3,4,
1,1,1,1,2,2,2,2,3,
1,1,1,1,2,2,2,2,3,
1,1,1,2,3,3,3,4,4,
1,1,1,1,2,3,3,3,3,
1,1,1,2,2,3,3,4,4,
1,1,1,2,2,3,4,4,4,
1,1,1,1,2,2,2,2,2,
1,1,1,2,2,3,4,3,4,
1,1,1,1,2,2,3,3,3,
1,1,1,1,2,3,3,4,4,
1,1,1,2,2,3,3,3,3,
1,1,1,2,2,2,3,3,3,
1,1,1,1,2,3,3,4,3,
1,1,1,2,2,3,3,3,4,
1,1,1,1,2,3,3,3,3,
1,1,1,2,2,3,4,4,4,
1,1,1,2,3,3,3,5,5,
1,1,1,1,2,2,3,3,3,
1,1,1,2,2,3,4,4,4,
1,1,1,2,2,3,4,4,5,
1,1,1,2,2,2,3,3,3,
1,1,1,1,2,2,3,3,3),.Dim=c(30,9)),

x=c(.0625,.1250,.2500,.5000,1.0000,2.0000,4.0000,8.0000,1
6.0000))

# initial values
list(beta=c(0,0), taub1=1,taub2=1)
```

TABLE 9.34

Posterior Analysis with Michaelis–Menten Model for Categorical Data

Parameter	Mean	SD	Error	2½	Median	97½
β_1	1.42	0.08992	0.001275	1.252	1.418	1.604
β_2	0.06273	0.2639	0.004768	−0.4609	0.06399	0.5752
σ_1^2	0.007071	0.008951	0.000187	0.000499	0.003949	0.03199
σ_2^2	0.02624	0.04659	0.001854	0.000625	0.009229	0.1608

Recall that the estimates of the coefficients of the model (9.39) are interpreted on the log scale; thus, as Table 9.34 indicates, the effect of the dose on the log mean is estimated as 1.42 with a 95% credible interval of (1.252, 1.604). Also, on the log scale, the 95% credible interval (0.4609, 0.5752) for β_2 suggests it is 0.

9.3 Comments and Conclusion

Chapter 5 presents the Bayesian analysis of repeated measures using nonlinear models with continuous and categorical data. There are two parts: (1) continuous data and (2) categorical data.

Section 9.1 begins with nonlinear models with an unstructured covariance pattern and the first example deals with plasma concentration versus time with the basic data displayed in Table 9.1 and the associated descriptive statistics reported in Table 9.2. Corroborating the descriptive statistics, Figure 9.1 plots the plasma concentration versus time by subject. The specific nonlinear model is represented by (9.3) and the Bayesian analysis is executed with **BUGS CODE 9.1** with the results displayed in Table 9.4.

A pill dissolution study is the second example assuming an unstructured covariance matrix and the information is given in Table 9.5. The specific model is defined by (9.5) and the association between times to dissolve a specified proportion of the pill left not dissolved. Bayesian analysis is executed with **BUGS CODE 9.2** and the results are displayed in Table 9.6.

The third example is again based on the pill dissolution study but with an autoregressive correlation structure for the responses. Table 9.8 reports the Bayesian analysis, which is executed with **BUGS CODE 9.3** and based on the model (9.9). As one would expect, the estimates of the model coefficients are the same as those (via Table 9.6) with an unstructured covariance pattern, but for the present case one parameter, the autoregressive correlation, is the main focus.

Section 9.1.4 is another example assuming an autoregressive pattern for the correlation structure using the timber load slip data, portrayed in Table 9.10, and the association (between load and slip) depicted in Figure 9.5. For the Bayesian analysis **BUGS CODE 9.3** is executed and the posterior analysis reported in Table 9.11, and the analysis implies that one coefficient is the negative of another; thus, a revised model is proposed and the Bayesian analysis is repeated as reported in Table 9.12.

Our first encounter with a mixed nonlinear model with continuous data is presented in Section 9.1.5. Formula (9.18) defines the specific model for the load slip data and includes one random effect as an intercept term. See the posterior analysis reported in Table 9.13 and executed with **BUGSCODE 9.5**. After the initial analysis, the model is expanded to include more random factors, and **BUGS CODE 9.5a** is executed for the expanded analysis with two random factors and reported in Table 9.14 and for three random factors in Table 9.15.

Section 9.1.6 is an example involving energy expenditure and heart rate measured with an increasing level of exercise, with the model including two random effects. Figure 9.7 depicts heart rate versus energy expenditure, while Table 9.16 describes energy expenditure versus task, and heart rate versus task. Formula (9.23) is the specific model that expresses energy expenditure versus task including two random effects.

The blood glucose values are revisited in Section 9.1.7 and the descriptive statistics are shown in Table 9.19, while models defined by (9.28) and (9.29) express the blood glucose values versus time separately for two groups. Table 9.20 reports the Bayesian analysis, which is executed with **BUGS CODE 9.7**, and there are two random factors for each group, and the variance of those four factors is of primary interest.

Finally, for the first part of the chapter dealing with continuous data, simulated data for repeated measures at various doses is investigated with the Michaelis-Menten model, and Table 9.22 displays the simulated data. There are 30 subjects and nine dose levels, and Figure 9.9 depicts the association between the response and dose. Last, Table 9.23 reveals the posterior analysis based on **BUGS CODE 9.8**, where the model contained one random factor.

For the second part of this chapter, a categorical repeated measure is expressed by a nonlinear model. The approach is to specify a distribution (Bernoulli or Poisson) for the main response and a canonical link, which in turn expresses the nonlinear association. For example, if the main response has a Bernoulli distribution, the canonical link is the logit, while if the main response has a Poisson distribution, the log is the canonical link.

For the examples employed in the first part of the chapter, I will partition the continuous data into binary and ordinal categories. Section 9.2.1 returns to the pill dissolution study, where the time to a specific proportion is dichotomized. Formula (9.31) defines the model, using a logistic link, and includes one random factor, and **BUGS CODE 9.9** executes the analysis with results reported in Table 9.25. As a result of the analysis, the model is revised, where one random factor is included as an intercept on the logit scale, and the posterior analysis is reported in Table 9.26. A more complex model including three random factors is executed with a revision of **BUGS CODE 9.9** and Table 9.27 explains the posterior analysis. Of course, a pertinent question is which model (one, two, or three random factors) is the most appropriate?

The timber load slip data is reanalyzed as a categorical response in Section 9.1.4 with four levels (1,2,3,4) for the load data, and formula (9.35) defines the nonlinear model (with two random factors) as a log link. The categorized data is assumed to follow a Poisson distribution, and the analysis executed with **BUG CODE 9.10**, while Table 9.28 displays the posterior analysis. Three random factors are included in an expanded version of the model, where Table 9.30 shows the posterior analysis.

Section 9.2.3 uses a categorized form of the energy expenditure data, and Section 9.2.4 revisits the blood glucose example using various scenarios for the inclusion of the random factors. In both cases, the main response is assigned a Poisson distribution, and the link is the log function, which is a nonlinear function of the independent variable and random factors.

Using the cumulative log odds as the canonical link for ordinal categorical data with a multinomial distribution is not considered but will be part of the exercises for the student.

Exercises

1. Describe the content of this chapter.
    a. What types of variables are used for the repeated measures of this chapter?
    b. For repeated continuous data, with a normal distribution, define a nonlinear model with fixed effects and covariates.
    c. For repeated continuous data with a normal distribution, define a nonlinear model with fixed and random effects.
    d. For binary repeated data with a Bernoulli distribution, define a nonlinear model with mixed effects using the logistic link.
    e. For categorical repeated data with a Poisson distribution, define a nonlinear model with mixed effects using the log link.
2. Describe carefully the nonlinear model (9.1).
3. Based on Table 9.1, verify the descriptive statistics of Table 9.2.
4. Based on the model (9.3) for the plasma concentration data of Table 9.1, and **BUGS CODE 9.1**, verify the posterior analysis of Table 9.6.
    a. What is the 95% credible interval for β_3?
    b. Identify the posterior distributions are symmetric about the posterior mean.
    c. What is your estimate of ρ_{12} based on the posterior median?
5. For the pill dissolution study,
    a. Describe Equation 9.5.

b. What are the dependent and independent variables?

c. Refer to Figure 9.3 and describe the pill–pill heterogeneity and the overtrend of time over the specified proportions.

d. Using **BUGS CODE 9.4**, verify the posterior analysis of Table 9.6.

e. Do the estimated correlations of Table 9.6 suggest a pattern for the dissolution times?

f. Identify the posterior distributions that are symmetric about the posterior mean.

6. a. Define the components of the nonlinear model (9.6) with an autoregressive error structure.

b. Based on Equations 9.6 and 9.7, derive Equation 9.8.

c. Verify the correlation pattern depicted in Table 9.7.

d. Describe the nonlinear model (9.9) for the time to pill dissolution.

e. Using 150,000 observations for the MCMC simulation, with a burn in of 5,000 and a refresh of 100, verify the posterior analysis of Table 9.8.

f. What prior distributions are assigned to the parameters of the model (9.8)?

7. a. Refer to Figure 9.5 and describe the overall trend of the load over the slip values.

b. Describe the nonlinear model (9.10) for the load data as a function of the slip values.

c. Using 150,000 observations for the simulation and a burn in of 5,000, execute **BUGS CODE 9.4** and verify the posterior analysis of Table 9.11.

d. What are the posterior medians of the beta parameters?

e. What is the 95% credible interval for the correlation ρ?

f. Which posterior distributions are skewed?

g. Based on Table 9.11, is $\beta_1 = -\beta_2$? Explain your answer.

8. Equation 9.15 is the model for load as a function of slip, assuming $\beta_1 = -\beta_2$.

a. From Equation 9.14, derive Equation 9.15.

b. Compare the estimated parameters of Table 9.11 to the corresponding estimates of Table 9.12.

c. Do the estimates of β_3, ρ, σ^2 change from the estimates of Table 9.11 to those of Table 9.12?

d. Does the model (9.15) provide a good fit to the load slip data? Plot the observed load values to the corresponding predicted load values using **BUGS CODE 9.4**.

9. a.   Describe the nonlinear mixed model (9.19) and explain the random effects b_i.

   b.   Based on (9.19), what is $E(Y_{ij} \mid b_i)$?

   c.   Perform a Bayesian analysis using 150,000 observations and a burn in of 5,000 and verify the posterior analysis of Table 9.13.

   d.   What are the prior distributions employed for the Bayesian analysis?

   e.   What is the 95% credible interval for the variance σ_1^2 of the random factor?

   f.   Is $\beta_1 = -\beta_2$? Justify your answer.

   g.   Based on the MCMC simulation errors of Table 9.13, is 150,000 observations sufficient for the simulation?

10. a. Describe the mixed nonlinear model (9.20) with two random factors b_{1i}, b_{2i}.

   b.   Perform a Bayesian analysis using **BUGS CODE 9.5a** with 150,000 observations for the simulation and a burn in of 5,000.

   c.   Verify Table 9.14.

   d.   What prior distributions are assigned for the parameters of model (9.20)?

   e.   Is $\beta_1 = -\beta_2$?

   f.   What are your estimates of σ_1^2 and σ_2^2?

   g.   Verify Table 9.15, the estimated correlations. Can you identify any pattern in the correlation structure?

   h.   In the Bayesian analysis reported in Table 9.14, the two random factors b_{1i}, b_{2i} are assumed to be independent. Perform a Bayesian analysis but assume b_{1i} and b_{2i} have a bivariate normal distribution with a zero mean vector and unknown variance–covariance matrix Σ. You will have to revise **BUGS CODE 9.5a**.

11. a. Refer to Figure 9.7 and describe the subject-subject heterogeneity of the energy expenditure study.

   b.   Based on Table 9.16, verify the descriptive statistics of Table 9.17.

   c.   Based on Table 9.17, does the average heart rate increase with task?

   d.   Perform a Bayesian analysis using **BUGS CODE 9.6** with 150,000 observations for the simulation and verify the posterior analysis of Table 9.18.

   e.   Refer to model (9.23) and Table 9.18 and identify those distributions that are not symmetric about the posterior mean.

   f.   Estimate $E(Y_{ij} \mid b_{1i}, b_{2i})$ for $i = 3$ and $j = 3$.

12. a. Refer to Figure 9.8 and describe the association between blood glucose values and time for the two groups.

   b. Define the components of Equations 9.25 and 9.26.

   c. Execute **BUGS CODE 9.7** with 150,000 observations for the simulation and 5,000 for the burn in and verify the Bayesian analysis for the blood glucose study.

   d. What prior distributions are assigned to the parameters of the Bayesian analysis?

   e. Compare the coefficients of the two models. Are the models the same? Refer to Tables 9.27 and 9.28.

   f. Based on the results of part e, revise the Bayesian analysis using a revision of **BUGS CODE 9.7**.

13. a. When the repeated measure is binary with a Bernoulli distribution, what is the canonical link function?

   b. When the repeated measure is categorical with a Poisson distribution, what is the canonical link?

14. Equation 9.31 is the canonical link (which is logit) expressing a nonlinear function with three coefficients plus one random factor. The repeated measure is the time to a specified proportion (which gives the proportion of the pill not dissolved) and the independent variable is the vector of specified proportions. The time variable is dichotomized.

   a. Perform a Bayesian analysis with **BUGS CODE 9.9** using 150,000 observations and a burn in of 5,000 and verify the posterior analysis reported in Table 9.25.

   b. What is the posterior median of the variance σ^2 of the random effect of Equation 9.31.

   c. What prior distributions are assigned to the parameters of model (9.31)?

   d. Identify the posterior distributions that are symmetric about the posterior mean.

   e. The 95% credible interval for β_1 is (−12.96, −2.918) and a posterior mean of −6.994. Interpret this value relative to Equation 9.31.

15. a. Describe Equation 9.32.

   b. What is the role of the random effect in Equation 9.32?

   c. Using a revision of **BUGS CODE 9.9**, perform a Bayesian analysis with 65,000 observations for the simulation and a burn in of 5,000, and validate the posterior analysis revealed in Table 9.26.

   d. Compare the posterior means of $\beta_1, \beta_2, \beta_3$ of Table 9.25 with the corresponding parameters of Table 9.26. Identify the differences (if any) in the posterior means.

e. What are the prior distributions for the parameters of the model (9.23)?

16. a. Explain the components of Equation 9.33. What is the link function and what is the independent variable?

b. The logistic link function is expressed as a nonlinear function of the independent variable and also includes three random factors.

c. Revise **BUGS CODE 9.9** and perform a Bayesian analysis with 150,000 observations for the simulation, with a burn in of 5,000, and verify Table 9.27, the posterior analysis.

d. Identify the posterior distributions that are skewed (not symmetric about the posterior mean).

e. What prior distributions did you use in the Bayesian analysis?

f. Does model (9.33) provide a good fit to the data? Explain your answer.

17. a. The load data for the load slip timber study is assumed to have Poisson distribution and (9.34) the log link is a nonlinear model with three "fixed" beta coefficients and one random effect. Describe Equation 9.34 and identify the independent variable.

b. What is the distribution of the repeated measure, the load values (which is categorized into four values)?

c. Using **BUGS CODE 9.10**, perform a Bayesian analysis with 150,000 observations and a burn in of 5,000. Verify the posterior analysis shown in Table 9.29.

d. What prior distribution is assigned for the three beta coefficients and the variance of the random effect?

18. a. Describe Equation 9.35, a generalization of (9.34) to two random factors b_{1i}, b_{2i}.

b. Perform a Bayesian analysis using **BUGS CODE 9.10** with 150,000 observations and a burn in of 4,000, and validate the posterior analysis reported in Table 9.29.

c. Identify the posterior distribution of Table 9.29 that are skewed.

d. The posterior median of σ_2^2 is 0.012299, a very small value. Does this imply $b_{2i} = 0$?

e. What prior distribution did you use for the parameters of the model (9.35)?

19. The heart rate data is categorized into five values, as shown in Table 9.31.

a. Describe the log-link function (9.31). What is the independent variable?

b. What distribution is assigned to the two random factors b_{1i}, b_{2i}?

c. Perform a Bayesian analysis with **BUGS CODE 9.11** using 150,000 observations for the simulation and with a burn in of 5,000. Verify the posterior analysis of Table 9.32.

d. What prior distributions did you use for the parameters of (9.31)?

e. Refer to the estimates of $\beta_1, \beta_2, \beta_3$. Should model (9.31) be revised?

20. The blood glucose values are categorized into four values assumed to follow a Poisson distribution.

a. Describe Equation 9.38.

b. Describe Equation 9.39.

c. What do the x_{ij} represent?

d. Verify the posterior analysis of Table 9.33. Using 150,000 observations for the simulation, perform the Bayesian analysis with **BUGS CODE 9.12**.

10

Bayesian Techniques for Missing Data

10.1 Introduction

Missing data arise in almost all serious statistical analyses. This chapter presents the Bayesian approach to missing data. To begin the chapter, various missing data mechanisms are defined, and then the chapter continues with four topics: (1) linear models for repeated measures with continuous data, (2) liner models with categorical data, (3) nonlinear models with continuous data, and (4) nonlinear models with categorical data. Of course, these four topics have been presented in Chapters 7 through 9, but with complete data for all subjects.

Missing data often occurs when conducting a research study, which presents a problem in that what should one do to handle those cases when missing data occurs. Up to this point, I have not considered the case of missing data, and one has to consider some alternatives for missing data. For a good reference for the problem with missing data, the book by Little and Rubin[1] should be consulted, and another reference recommended is Schafer.[2] Several situations will be outlined for dealing with missing data, but first definition of various missing data mechanisms will be explained.

In order to define the various mechanisms of missing data, consider a repeated measures response Y_{ij} for subject i at the jth time, where $i = 1,2,\ldots,N$ and $j = 1,2,\ldots,n$, and let R be the corresponding N by n matrix with ijth element $R_{ij} = 0$ if Y_{ij} is missing, otherwise it is 1. Let the probability model for R be

$$P\left(R|Y_{\text{obs}}, Y_{\text{miss}}, \eta\right) \tag{10.1}$$

where:
  Y_{obs} are the components of the N by n matrix Y that are not missing
  Y_{miss} are the components of Y that are missing or not observed
  η is an unknown parameter associated with the probability of a missing observation

Note Y is the N by n matrix with ijth element Y_{ij}.

The first mechanism to be defined is the case of missing completely at random (MCAR). For example, a value may be missing because equipment malfunctioned, the weather was terrible, people got sick, or the data was not entered correctly. Here the data is MCAR. When we say that data is MCAR, we mean that the probability that an observation is missing is unrelated to the value of or to the value of any other variables. Thus, data on family income would not be considered MCAR if people with low incomes were less likely to report their family income than people with higher incomes. Similarly, if Whites were more likely to omit reporting income than African Americans, we again would not have data that was MCAR because missing pattern would be correlated with ethnicity. However, if a participant's data were missing because he or she was stopped for a traffic violation and missed the data collection session, his or her data would presumably be MCAR. Another way to think of MCAR is to note that in that case any piece of data is just as likely to be missing as any other piece of data.

More formally, the MCAR mechanism is expressed as

$$P\left(R|Y_{\text{obs}}, Y_{\text{miss}}, \eta\right) = P\left(R|\eta\right) \qquad (10.2)$$

That is to say, the probability an observation will be missing is independent of the observed and of the missing information.

Notice that it is the value of the observation, and not its "missingness," that is important. If people who refused to report personal income were also likely to refuse to report family income, the data could still be considered MCAR, so long as neither of these had any relation to the income value itself. This is an important consideration, because when a data set consists of responses to several survey instruments, someone who did not complete the Beck Depression Inventory (BDI) would be missing all BDI subscores, but that would not affect whether the data can be classified as MCAR.

This nice feature of data that is MCAR is that the analysis remains unbiased. We may lose power for our design, but the estimated parameters are not biased by the absence of data.

Often data is not MCAR, but it may be classifiable as missing at random (MAR). (MAR is not really a good name for this condition because most people would take it to be synonymous with MCAR, which it is not. However, the label has stuck.) The phraseology is a bit awkward here because we tend to think of randomness as not producing bias, and thus might well think that MAR is not a problem. Unfortunately, it is a problem, although in this case we have ways of dealing with the issue so as to produce meaningful and relatively unbiased estimates. But just because a variable is MAR does not mean that you can just forget about the problem. But nor does it mean that you have to throw up your hands and declare that there is nothing to be done. Consider the formal definition of MAR, namely

$$P(R|Y_{obs}, Y_{miss}, \eta) = P(R|Y_{obs}, \eta) \qquad (10.3)$$

that is, the probability an observation is missing depends only on the observed data and not on the data that is missing. This has important implications for data imputation by regression techniques, an important topic to be covered. The scenario in which the data is MAR is sometimes referred to as ignorable missingness. Regression techniques among other are available to impute data for the missing values without introducing bias. MCAR is a special case of MAR.

A third alternative for the missing mechanism is missing not at random (NMAR). For example, if we are examining the mental health of subjects who have been diagnosed as depressed are less likely than others to report their mental status, the data is NMAR. Clearly, the mean mental status score for the available data will not be an unbiased estimate of the mean that we would have obtained with complete data. The same thing happens when people with low income are less likely to report their income on a data collection form.

When one is considering an analysis with data that is NMAR, the problem is challenging. The only way to obtain an unbiased estimate of parameters is to model the mechanism of missing. That is to say, one would need to determine a model that takes into account the missing data, and the model could then be included into a more elaborate model that estimates the missing values. The NMAR will not be considered for analysis.

According to Schafer (p. 12),[2] Little and Rubin[1] show that under the case of MAR, inferences can be based on a model that does not need to include the model for R (10.3) and the parameter η of the missing mechanism.

Note that

$$P(R|Y_{obs}\theta, \eta) = P(R|Y_{obs}, \eta)P(Y_{obs}|\theta) \qquad (10.4)$$

where:
θ is the vector of parameters for the distribution of Y

This in turn implies that inferences for θ can be based on the likelihood function that depend only on the observed data, that is,

$$L(\theta|Y_{obs}) \propto P(Y_{obs}|\theta) \qquad (10.5)$$

Of course, for a Bayesian, the likelihood is multiplied by the prior for θ, which by Bayes's theorem is the posterior density of θ.

What methods are best for accounting for missing data?

A common approach is to delete those cases (subjects for repeated measures) with missing data and to run our analyses on what remains. Thus, for example, in the repeated measures study of Davidian and Giltinan[3] measuring the plasma concentration at time points 0.25, 0.5, 0.75, 1, 1.25, 2, 3, 4, 5, 6, and 8 hours, if for some reason measurement at 1, 2, and 4 hours are missing, then one could delete those subjects from the analysis. This approach is usually called list-wise deletion, but it is also known as complete case analysis, because only those subjects that have be measured at all time points have been included in the analysis.

Although list-wise deletion often results in a substantial decrease in the sample size available for the analysis, it does have important advantages. In particular, under the assumption that data is MCAR, it leads to unbiased parameter estimates. Unfortunately, even when the data is MCAR, there is a loss in power using this approach, especially if we have to rule out a large number of subjects. And when the data is not MCAR, bias results. (For example, when low-income individuals are less likely to report their income level, the resulting mean is biased in favor of higher incomes.) The alternative approaches discussed later should be considered as a replacement for list-wise deletion, though in some cases we may be not.

An old procedure that should certainly be relegated to the past was the idea of substituting a mean for the missing data. For example, in the plasma concentration study, if the observations are missing at times 4, 6, and 8 hours, calculate the mean and the plasma concentration at the previous times, and substitute the mean at times 4, 6, and 8. There are a couple of problems with this approach. In the first place, it adds no new information. The overall mean, with or without replacing my missing data, will be the same. In addition, such a process leads to an underestimate of error. In this case of plasma concentration if there is a known mean profile that is decreasing, the mean substitutions method will bias the analysis.

If we don't like mean substitution, why not try using regression to predict what the missing score should be on the basis of other variables that are present? We use existing variables to make a prediction, and then substitute that predicted value as if it were an actual obtained value. This approach has been around for a long time and has at least one advantage over mean substitution. At least the imputed value is in some way conditional on other information we have about the person. With mean substitution, if we were missing a person's weight, we assign him or her the average weight. Put somewhat incorrectly, with regression substitution we would assign him or her the weight of males or females of around the same age.

That has to be an improvement. But the problem of error variance remains. By substituting a value that is perfectly predictable from other variables, we have not really added more information but we have increased the sample size and reduced the standard error. This will generally be the approach taken here, but with using the Bayesian predictive density in the MCMC simulation, which will impute thousands of values (based on the model) for the missing data. Additional information about missing data can be found in Howell,[4] Allison,[5] Fitzmaurice, Laird, and Ware,[6] and Daniels and Hogan,[7] and the last reference is from a Bayesian viewpoint.

As mentioned earlier, MAR is fairly easy to address by regression techniques, but unfortunately one cannot be sure that the data is actually MAR or whether the missing mechanism depends on unobserved variables or on the missing data themselves. Of course, the problem is that the unobserved factors that might influence the probability of an observation being missing will probably never be known with certainty. In practice, we will try and include all covariates that will possibly affect the observed repeated measures and hope for the best.

In WinBUGS, missing outcomes in a regression can easily be addressed by simply including the data vector (the repeated measure) by designating a missing value as NA, and then the software explicitly models the outcome variable, in effect imputing missing values at each iteration of the MCMC simulation. As we have seen in Chapters 5 through 9, 50,000 observations for the simulation is not unusual, thus if missing data is present, there would be 50,000 imputed values, from which the predictive mean, standard deviation, and 95% credible interval are computed. When there are missing values in the predictors (covariates), one must employ a different approach than just simply using an NA for the missing value. One needs to assume that the predictor follows a particular distribution, and then a NA symbol can be used for each missing value. Of course, this presents another problem, namely, determining the appropriate distribution for the covariate.

Suppose the posterior density of θ is

$$\pi(\theta|Y) = cL(Y|\theta)\pi(\theta) \tag{10.6}$$

where c is the normalizing constant and the likelihood function is

$$L(Y|\theta) = f(Y|\theta) \tag{10.7}$$

Also, note that f is the density of the observations. The predictive density of a future observation Z is

$$g(z|y) \propto \int_{\Omega} f(z|\theta)\pi(\theta|y)d\theta \tag{10.8}$$

and the integration is taken with respect to the parameter space Ω.

It is the predictive density (10.8) of a future (missing) observation Z that Bugs uses for data imputation.

10.2 Missing Data and Linear Models of Repeated Measures

The first example to be considered is taken from Schafer (p. 195),[2] which is a study based on an investigation by Ryan and Joiner[8] and consists of 28 heart attack subjects, where the cholesterol levels are measured at 2, 4, and 14 days after the attack (Table 10.1). I assume the MAR mechanism is valid.

TABLE 10.1

Cholesterol Levels of Heart Attack Subjects

Subject	Day 2	Day 4	Day 14
1	270	218	156
2	236	234	
3	210	214	242
4	142	116	
5	280	200	
6	272	276	256
7	160	146	142
8	220	182	216
9	226	238	248
10	242	288	
11	186	190	168
12	266	236	236
13	206	244	
14	318	258	200
15	294	240	264
16	282	294	
17	234	220	264
18	224	200	
19	276	220	188
20	282	186	182
21	360	352	294
22	310	202	214
23	280	218	
24	278	248	198
25	288	278	
26	288	248	256
27	244	270	280
28	236	242	204

Source: Ryan, T.A., and Joiner, B.L., Normal probability plots and tests for normality. Technical Report, Department of Statistics Pennsylvania State University, University Park, PA, 1976.

One may verify the descriptive statistics of Table 10.2, which reveals a downward trend in the cholesterol levels from day 2 to day 14. Notice that there are nine missing values that occur at day 14.

The model for the cholesterol level Y_{ij} for subject i and the jth time point

$$Y_{ij} = \beta_1 + \beta_2 t_j + e_{ij} \tag{10.9}$$

where the independent errors

TABLE 10.2

Descriptive Statistics Cholesterol Study

Day	Mean	Median	SD	N
2	253.92	268	47.71	28
4	230.64	235	46.96	28
14	221.47	216	43.18	19

$$e_i = (e_{i1}, e_{i2}, e_{i3}) \tag{10.10}$$

have a multivariate normal distribution with mean vector 0 and variance-covariance matrix Σ, for $i = 1,2,...,N \ (= 28)$ and $j = 1,2,3$. Also the time points are $t_1 = 2, t_2 = 4, t_3 = 14$. Figure 10.1 confirms the downward trend via the linear regression line. The unknown parameters are the beta coefficients of (10.9) and the elements of the variance–covariance matrix Σ. For this example, the covariance pattern is unstructured, but in what is to follow other patterns will be considered.

The Bayesian analysis is performed with **BUGS CODE 10.1** and uses a linear regression with an unstructured covariance matrix, and the prior distributions are uninformative normal distributions for the regression coefficients and an uninformative Wishart for the 3 by 3 precision matrix of the

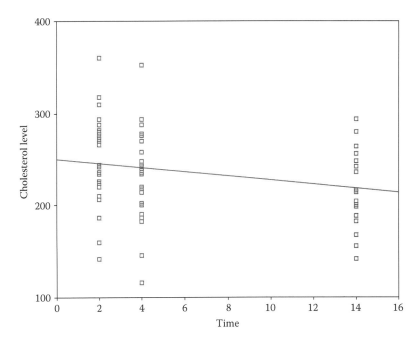

FIGURE 10.1
Cholesterol levels over three days.

BUGS CODE 10.1

```
        model;
         {
# prior distribution for the regression coefficients.
            beta1 ~ dnorm(0.0, 0.001)
            beta2 ~ dnorm(0.0, 0.001)

for(i in 1:N1){Y[i,1:M1]~dmnorm(mu1[],Omega[,])}
for (j in 1:M1){mu1[j]<-beta1+beta2*age[j]}
# non-informative precision matrix
Omega[1:M1,1:M1]~dwish(R[,],3)
Sigma[1:M1,1:M1]<-inverse(Omega[,])
rho[1,2]<-Sigma[1,2]/sqrt(Sigma[1,1]*Sigma[2,2])
rho[1,3]<-Sigma[1,3]/sqrt(Sigma[1,1]*Sigma[3,3])
rho[2,3]<-Sigma[2,3]/sqrt(Sigma[2,2]*Sigma[3,3])
}

list(N1 = 28,M1 = 3,Y = structure(.Data = c(270.00
218.00 156.00
236.00,       234.00,       NA,
210.00,       214.00,       242.00,
142.00,       116.00,       NA,
280.00,       200.00,       NA,
272.00,       276.00,       256.00,
160.00,       146.00,       142.00,
220.00,       182.00,       216.00,
226.00,       238.00,       248.00,
242.00,       288.00,       NA,
186.00,       190.00,       168.00,
266.00,       236.00,       236.00,
206.00,       244.00,       NA,
318.00,       258.00,       200.00,
294.00,       240.00,       264.00,
282.00,       294.00,       NA,
234.00,       220.00,       264.00,
224.00,       200.00,       NA,
276.00,       220.00,       188.00,
282.00,       186.00,       182.00,
360.00,       352.00,       294.00,
310.00,       202.00,       214.00,
280.00,       218.00,       NA,
278.00,       248.00,       198.00,
288.00,       278.00,       NA,
244.00,       270.00,       280.00,
236.00,       242.00,       204.00,
288.00,       248.00,       256.00),.Dim = c(28,3)),
```

```
age = c(2,4,14),

R = structure(.Data = c(1,0,0,
                        0,1,0,
                        0,0,1),.Dim = c(3,3)))

# initial values

list(beta1 = 0,beta2 = 0)
```

cholesterol levels over three days. WinBUGS uses multiple imputation via the Bayesian predictive density and the simulation generated 65,000 observations, with a burn in of 5,000 and a refresh of 100. Thus, 65,000 observations are imputed for the missing values.

Note that the missing values are denoted by the NA symbol. The Bayesian analysis is reported in Table 10.3. Also notice that posterior mean of Y[2,3] is 229, with a posterior median of 229, and a 95% credible interval (169, 288) based on 65,000 imputations. WinBUGS executes the analysis explicitly with these imputed values. The estimated mean of the slope is positive, but the 95% credible interval (−1.773, 2.623) contains 0. The missing value Y[2,3] is considered another parameter in the Bayesian analysis.

There are three estimated correlations and two regression parameters, plus the missing value for Y[2,3] depicted in Table 10.3. Of the 84 observations, nine are missing or about 11%.

What is an autoregressive structure for the covariance matrix? Consider the model

$$Y_{ij} = X'_{ij}\beta + e_{ij} \tag{10.11}$$

where

$$e_{ij} = \rho e_{ij-1} + u_{ij}$$

TABLE 10.3

Posterior Analysis for Cholesterol Study

Parameter	Mean	SD	Error	2½	Median	97½
β_1	214.9	15.7	0.1363	179.3	216.6	240.6
β_2	0.2002	1.114	0.01016	−1.773	0.1249	2.623
ρ_{12}	0.6845	0.1223	0.000672	0.4057	0.6982	0.8854
ρ_{13}	0.3704	0.2207	0.001165	−0.1068	0.389	0.747
ρ_{23}	0.7222	0.1217	0.000656	0.4231	0.7466	0.8937
Y[2,3]	229	30.04	0.1392	169.6	229	288.4

with $i = 1,2,...,N$ and $j = 1,2,...,n$, u_{ij} are independent and normally distributed with mean 0 and variance $\sigma^2 = 1/\tau$, τ is the precision, and the correlation ρ satisfied $-1 < \rho < 1$. Thus, for each subject, the e_{ij} follow a first-order autoregressive process with correlation ρ. The model (10.11) is rearranged to give

$$Y_{ij} = \rho Y_{ij-1} + \left(X'_{ij} - \rho X'_{ij-1} \right)\beta + u_{ij} \tag{10.12}$$

for $i = 1,2,...,N$ and $j = 1,2,...,n$. Also assumed is that the time periods are equally spaced and each subject has n occasions at which the response is measured. This will later be generalized to the case where the time points are not equally spaced and the number of occasions the response is measured is not necessarily the same number n.

It can be shown that the correlation is ρ for responses one unit apart, is ρ^2 for responses two units apart, and so on. This is represented by the correlation matrix.

Consider the exercise therapy trial assuming an autoregressive structure for the correlation, then corresponding to (10.12), the model is

$$Y_{ij} = \rho Y_{ij-1} + \beta_1 (1-\rho) + \beta_2 \left(t_{ij} - \rho t_{ij-1} \right) + u_{ij} \tag{10.13}$$

where for each i, $t_{ij} = 0,2,4,6,8,10,12$ and $j = 1,2,3,4,5,6,7$.

This example is considered in Section 6.2.2 and the data set had 5 missing values out of 84. I modified this data set so that there are 18 missing values or about 21%. See **BUGS CODE 6.2** for the original data with five missing values. This data set is modified to have 18 missing values or about 21%, and the data with 18 missing values is included in the first list statement of **BUGS CODE 10.2**.

The Bayesian analysis is executed with 75,000 observations for the simulation, with a burn in of 5,000 and a refresh of 100.

BUGS CODE 10.2

```
model;
     {
          beta1 ~ dnorm(0.0, 0.001)
          beta2 ~ dnorm(0.0, 0.001)

for(i in 1:N1){for (j in 2:M1){Y[i,j]~dnorm(mu[i,j],tau)}}
for(i in 1:N1){for (j in 2:M1){mu[i,j]<-beta1*(1-rho)
+beta2*(age[j]-rho*age[j-1])+rho*Y[i,j-1]}}
for(i in 1:N1){Y[i,1]~dnorm(mu[i,1],tau)}

rho~dbeta(1,1)
```

```
for(i in 1:N1){mu[i,1]<-beta1+beta2*age[1]}
tau~dgamma(.01,.01)
sigma<-1/tau
}


# weight = 1

list(N1 = 16,M1 = 7,Y = structure(.Data = c(79,NA,79,
80,    NA,    78,    80,
83,    NA,    85,    85,    86,    87,    87,
81,    83,    82,    82,    NA,    83,    82,
81,    81,    81,    82,    82,    NA,    81,
80,    81,    82,    82,    82,    NA, 86,
76, NA, 76,    76,    76,    76,    75,
81, 84,        83,    NA,    85,    85,    85,
77,    78,    79,    79,    81,    82,    NA,
84,    85,    87,    89, NA, NA, 86,
74,    75,    NA,    78,    79,    78,    78,
76,    NA,    77,    77,    77,    76,    76,
84,    84,    86,    NA,    86,    86,    86,
79,    NA,    79,    80,    80,    82,    82,
78,    78,    77, NA, 75,    75,    76,
78,    80,    77,    77,    75, NA,     75,
84,    85,    85,    85,    85,    83,    NA),.Dim = c(16,7)),

age = c(0,2,4,6,8,10,12))

# initial values

list(beta1 = 0,beta2 = 0,rho =.5,tau = 1)
```

Table 10.4 reports the Bayesian analysis for the exercise therapy trial with five missing values. I assume MAR is the mechanism and that the Bayesian predictive distribution will not introduce bias. This will be investigated in more depth in the exercises.

Table 10.5 reports the Bayesian analysis for the exercise therapy trial with 18 missing values.

TABLE 10.4

Bayesian Analysis for Exercise Therapy Trial with Five Missing Values

Parameter	Mean	SD	Error	2½	Median	97½
β_1	79.68	0.4127	0.004817	78.87	79.68	80.5
β_2	0.1269	0.07604	0.001163	−0.02113	0.1255	0.279
ρ	0.9633	0.02721	0.001163	0.8998	0.9683	0.9985
σ^2	2.653	0.3792	0.002313	2.014	2.618	3.495

TABLE 10.5

Bayesian Analysis for Exercise Therapy Trial with 18 Missing Values

Parameter	Mean	SD	Error	2½	Median	97½
β_1	79.68	0.4155	0.005091	78.87	79.68	80.51
β_2	0.134	0.07827	0.001603	−0.0214	0.1336	0.2845
ρ	0.9634	0.02827	0.002276	0.8968	0.9694	0.9983
σ^2	2.749	0.4202	0.003152	2.049	2.708	3.69

Comparing Table 10.4 with 10.5, one concludes that the missing data patterns (one with 5 missing values, the other with 18) produce about the same estimates for the parameters of the model. How many additional missing values will produce quite different estimates for the model parameters?

Our next example is the previously analyzed hematocrit study for hip-replacement studies, which is presented in Chapter 6.

In order to do this, an example introduced in Chapter 5 will be utilized.

For this example, a quadratic is proposed to model the mean profile of two groups, which was analyzed by Crowder and Hand (p. 79),[9] where the hematocrit of hip-replacement patients is measured on four occasions. Also measured is the age of each subject at the beginning of the study, and there are two groups, male and female patients. Hematocrit is a measure of how much volume the red blood cells take up in the blood and is measured as a percentage, thus, an observed value of 33 means red blood cells comprise 33% by volume of the blood. For our study, one follows the hematocrit of hip-replacement patients in order to see if they are becoming anemic. The data for Crowder and Hand is found in Table 5.5, where the four times are denoted by t_1, t_2, t_3, and t_4, and missing values are denoted by periods.

Recall from Figure 5.2 that a quadratic mean response is estimated for the mean profile of the combined group. The expectation for the mean profile of 30 subjects at four time points is

$$E\left(Y_{ij}\right) = \beta_1 + \beta_2 t_{ij} + \beta_3 t_{ij}^2 + \beta_4 \text{age}_i \tag{10.14}$$

where:
Y_{ij} is the time for subject i at time t_{ij}
β_4 is the effect of age on the average hematocrit values

Note that there are 13 male and 17 female hip-replacement patients and that not all patients have a complete set of four measurements. Indeed some have four, some three, and some two, but all have at least two repeated values.

The present analysis will focus on the combined group of 30 patients.

When the pattern of the covariance matrix is first-order autoregressive, it can be shown that the expectation of the hematocrit of subject i at time t_j is

$$E\left[Y_{ij}\right] = \rho Y_{i,j-1} + \beta_1(1-\rho) + \beta_2\left(t_j - \rho t_{j-1}\right)$$
$$+ \beta_3\left(t_j^2 - \rho t_{j-1}^2\right) + \beta_4 age_i(1-\rho)$$

(10.15)

where the first-order autoregressive coefficient is ρ and satisfies the constraint $-1 < \rho < 1$. It can also be shown that the variance σ^2 of the Y_{ij} is constant for all occasions.

Missing data accounts for 17% of the 120 observations, and the Bayesian analysis will be based on the model (10.15), which is similar to the analysis reported in Table 6.18, except the present analysis includes age as a covariate.

The prior distributions for the beta coefficients are uninformative normal distributions with mean 0 and variance 1000, while that for the autoregressive parameter is uniform (0,1). Based on the model (10.15) and **BUGS CODE 10.3**,

BUGS CODE 10.3

```
model;
{
        for (i in 1:4){beta[i] ~ dnorm(0.0, 0.001)}

for(i in 1:N){for (j in 2:M){Y[i,j]~dnorm(mu[i,j],tau)}}
# future values for hematocrit denoted by Z
for(i in 1:N){for (j in 1:M){Z[i,j]~dnorm(mu[i,j],tau)}}
for(i in 1:N){for (j in 2:M){mu[i,j]<-beta[1]*(1-rho)+
beta[2]*(time[j]-rho*time[j-1])+beta[3]*(time[j]*time[j]-rho*
time[j-1]*time[j-1])+rho*Y[i,j-1]+beta[4]*age[i]*(1-rho)}}
for(i in 1:N){Y[i,1]~dnorm(mu[i,1],tau)}
for(i in 1:N){mu[i,1]<-beta[1]+beta[2]*time[1]+beta[3]*
time[1]*time[1]+beta[4]*age[i]}
rho~dbeta(1,1)
tau~dgamma(.001,.001)
sigma<-1/tau


}

list(N = 30,M = 4,

age = c(66,70,44,70,74,65,54,63,71,68,69,64,70,60,52,52,75,
72,54,71,58,77,66,
53,74,78,74,79,71,68),

Y = structure(.Data =
```

```
c(47.1,31.1,NA,32.8,
44.1,31.5,NA,37.0,
39.7,33.7,NA,24.5,
43.3,18.4,NA,36.6,
37.4,32.3,NA,29.1,
45.7,35.5,NA,39.8,
44.9,34.1,NA,32.1,
42.9,32.1,NA,NA,
46.1,28.8,NA,37.8,
42.1,34.4,34.0,36.1,
38.3,29.4,32.9,30.5,
43.0,33.7,34.1,36.7,
37.8,26.6,26.7,30.6,
37.3,26.5,NA,38.5,
NA,28.0,NA,33.9,
27.0,32.5,NA,32.0,
38.4,32.3,NA,37.9,
38.8,32.6,NA,26.9,
44.7,32.2,NA,34.2,
38.0,27.1,NA,37.9,
34.0,23.2,NA,26.0,
44.8,37.2,NA,29.7,
46.0,29.1,NA,26.7,
41.9,32.0,37.1,37.6,
38.0,31.7,38.4,35.7,
42.2,34.0,32.9,33.3,
39.7,33.5,26.6,32.7,
37.5,28.2,28.8,30.3,
34.6,31.0,30.1,28.7,
35.5,24.7,28.1,29.8),.Dim = c(30,4)),

time = c(1,2,3,4))
# initial values
list(beta = c(0,0,0,0),rho =.5, tau = 1)
```

the posterior analysis is executed with 130,000 observations, with a burn in of 5,000 and a refresh of 100.

For subject 1 at time 3, the observation is missing and is estimated as 28.97 via the posterior median and using 130,000 imputations. Also, based on the 95% credible interval (−0.0690, 0.1623), the effect of age on average hematocrit is minimal. Note that the correlations between hematocrit values one unit apart are estimated as 0.2731, a somewhat modest amount (Table 10.6).

Our analysis of the hip-replacement study will be examined to study the effect of an increasing number of missing values on the subsequent posterior estimates. Missing values will be increased for the hematocrit values on for

TABLE 10.6

Posterior Analysis of Hip-Replacement Study

Parameter	Mean	SD	Error	2½	Median	97½
Y[1,3]	28.97	4.159	0.01473	20.79	28.97	37.12
Y[2,3]	30.19	4.172	0.01419	21.9	30.22	38.34
β_1	52.09	4.444	0.03505	43.32	52.12	60.71
β_2	−18.12	2.083	0.02028	−22.2	−18.13	−14.01
β_3	3.158	0.4076	0.004027	2.356	3.16	3.962
β_4	0.0448	0.05911	0.000549	−0.069030	0.0443	0.1623
ρ	0.2732	0.1207	0.001497	0.04456	0.2731	0.5112
σ^2	18.49	2.749	0.008787	13.88	18.23	24.61

the ages of the 30 subjects. Using WinBUGS, missing values are imputed for a covariate by assuming it follows a probability distribution, which presents a problem. What is a reasonable distribution for age of the 30 subjects? Figure 10.2 reveals that a normal distribution for age is reasonable via the Q–Q plot.

The model defined by (10.15) is amended by assuming a normal distribution for age, that is

$$\text{Age[i]} \sim \text{dnorm(vu, taua)} \tag{10.16}$$

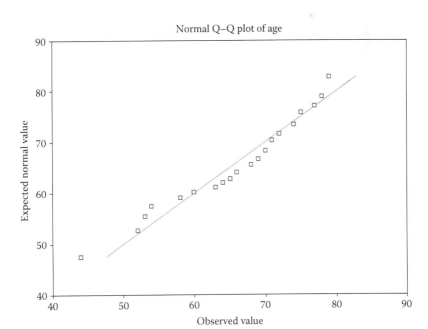

FIGURE 10.2
Test for normal distribution of age.

where the prior distribution for vu is

$$vu \sim dnorm(0, .001) \tag{10.17}$$

and for taua is

$$taua \sim dgamma(.001, .001) \tag{10.18}$$

BUGS CODE 10.3 is amended to **BUGS CODE 10.3a** and includes the statements for the prior distributions given by (10.16) through (10.18).

BUGS CODE 10.3a

```
        model;
        {
                for (i in 1:4){beta[i] ~ dnorm(0.0, 0.001)}

                for (i in 1:N){age[i]~dnorm(vu, taua)}
                vu~dnorm(.001,.001)
                taua~dgamma(.001,.001)
                sigmaa<-1/taua


for(i in 1:N){for (j in 2:M){Y[i,j]~dnorm(mu[i,j],tau)}}
# future values for hematocrit denoted by Z
for(i in 1:N){for (j in 1:M){Z[i,j]~dnorm(mu[i,j],tau)}}
for(i in 1:N){for (j in 2:M){mu[i,j]<-beta[1]*(1-rho)+
beta[2]*(time[j]-rho*time[j-1])+beta[3]*(time[j]*time[j]-rho*
time[j-1]*time[j-1])+rho*Y[i,j-1]+beta[4]*age[i]*(1-rho)}}
for(i in 1:N){Y[i,1]~dnorm(mu[i,1],tau)}
for(i in 1:N){mu[i,1]<-beta[1]+beta[2]*time[1]+beta[3]*
time[1]*time[1]+beta[4]*age[i]}
rho~dbeta(1,1)
tau~dgamma(.001,.001)
sigma<-1/tau
}

list(N = 30,M = 4,

age = c(66,NA,44,70,NA,65,54,63,71,NA,69,NA,70,60,NA,52,75,
72,54,71,58,NA,66,
NA,74,78,74,NA,71,68),

Y = structure(.Data =
c(47.1,31.1,NA,32.8,
44.1,31.5,NA,37.0,
39.7,33.7,NA,24.5,
43.3,18.4,NA,36.6,
37.4,32.3,NA,29.1,
```

```
45.7,35.5,NA,39.8,
44.9,34.1,NA,32.1,
42.9,32.1,NA,NA,
46.1,28.8,NA,37.8,
42.1,34.4,34.0,36.1,
38.3,29.4,32.9,30.5,
43.0,33.7,34.1,36.7,
37.8,26.6,26.7,30.6,
37.3,26.5,NA,38.5,
NA,28.0,NA,33.9,
27.0,32.5,NA,32.0,
38.4,32.3,NA,37.9,
38.8,32.6,NA,26.9,
44.7,32.2,NA,34.2,
38.0,27.1,NA,37.9,
34.0,23.2,NA,26.0,
44.8,37.2,NA,29.7,
46.0,29.1,NA,26.7,
41.9,32.0,37.1,37.6,
38.0,31.7,38.4,35.7,
42.2,34.0,32.9,33.3,
39.7,33.5,26.6,32.7,
37.5,28.2,28.8,30.3,
34.6,31.0,30.1,28.7,

35.5,24.7,28.1,29.8),.Dim = c(30,4)),
time = c(1,2,3,4))
# initial values
list(beta = c(0,0,0,0),rho =.5, tau = 1,taua = 1,tauvu = 1,
vu = 0)
```

The posterior analysis is executed via **BUG CODE 10.3a**, using 65,000 observations for the simulation, with a burn in of 5,000 and a refresh of 100. There are missing age values for subjects 2, 5, 10, 12, 15, 22, 24, and 28, comprising $8/30 = 26\%$. The number of missing hematocrit values is 18, the same as in the previous analysis, and the posterior analysis is reported in Table 10.7.

One should compare Table 10.6 to 10.7. One sees that the estimate of the autoregressive parameter is approximately the same for the two analyses, one with 18 missing data in the hematocrit values compared to the case with 8 missing values for age. The largest difference is with β_1 with a posterior mean of 52.09 for the analysis reported in Table 10.6 compared to a posterior mean of 48.43 for that reported in Table 10.7.

Also, note the estimated missing value of 68.53 (via the posterior mean) for the age of subject 2 and 63.95 for subject 5. It is interesting to observe that the estimates of the two hematocrit values for subjects 1 and 2 at time 3 are almost the same for the two analyses.

TABLE 10.7

Posterior Analysis for Hematocrit Study with Missing Observations for Age and Hematocrit

Parameter	Mean	SD	Error	2½	Median	97½
Age[2]	68.53	8.935	0.08428	50.86	68.51	86.16
Age[5]	63.95	8.721	0.0473	46.71	64.95	81.09
β_1	48.43	4.764	0.2151	39.18	48.29	58.86
β_2	−18.23	1.948	0.09025	−22.01	−18.25	−14.38
β_3	3.178	0.3807	0.01748	2.427	3.182	3.921
β_4	0.1016	0.06742	0.001729	−0.04108	0.09389	0.2247
σ^2	18.09	2.702	0.01916	13.53	17.84	24.06
σ_a^2	83.24	28.18	0.162	44.6	77.72	152.8
ρ	0.2542	0.1187	0.00059	0.0355	0.2519	0.4923
Y[1,3]	28.94	4.128	0.0292	20.89	28.93	37.06
Y[2,3]	30.15	4.174	0.0272	21.89	30.15	38.26

One more analysis will be performed with more missing values for the hematocrit and age values. There are 11 missing age values and 31 missing hematocrit values. The missing values are indicated by NA in the first list statement of **BUGS CODE 10.3b**, which is executed with 130,000 observations for the simulation.

BUGS CODE 10.3b

```
model;
{
        for (i in 1:4){beta[i] ~ dnorm(0.0, 0.001)}

        for (i in 1:N){age[i]~dnorm(vu, taua)}
        vu~dnorm(.001,.001)
        taua~dgamma(.001,.001)
        sigmaa<-1/taua

for(i in 1:N){for (j in 2:M){Y[i,j]~dnorm(mu[i,j],tau)}}
# future values for hematocrit denoted by Z
for(i in 1:N){for (j in 1:M){Z[i,j]~dnorm(mu[i,j],tau)}}
for(i in 1:N){for (j in 2:M){mu[i,j]<-beta[1]*(1-rho)+
beta[2]*(time[j]-rho*time[j-1])+beta[3]*(time[j]*time[j]-rho
*time[j-1]*time[j-1])+rho*Y[i,j-1]+beta[4]*age[i]*(1-rho)}}
for(i in 1:N){Y[i,1]~dnorm(mu[i,1],tau)}
for(i in 1:N){mu[i,1]<-beta[1]+beta[2]*time[1]+beta[3]
*time[1]*time[1]+beta[4]*age[i]}
rho~dbeta(1,1)
```

```
tau~dgamma(.001,.001)
sigma<-1/tau


}

list(N = 30,M = 4,


age = c(66,NA,44,70,NA,65,54,NA,71,NA,69,NA,70,60,NA,52,
75,NA,54,71,58,NA,66,
NA,74,NA,74,NA,71,68),

Y = structure(.Data =

c(47.1,31.1,NA,32.8,
NA,31.5,NA,37.0,
39.7,33.7,NA,24.5,
43.3,18.4,NA,36.6,
37.4,32.3,NA,29.1,
NA,35.5,NA,39.8,
44.9,34.1,NA,32.1,
42.9,32.1,NA,NA,
46.1,28.8,NA,37.8,
NA,34.4,34.0,36.1,
38.3,NA,32.9,30.5,
43.0,33.7,34.1,NA,
37.8,NA,26.7,30.6,
37.3,26.5,NA,38.5,
NA,28.0,NA,33.9,
27.0,32.5,NA,32.0,
38.4,32.3,NA,37.9,
38.8,32.6,NA,26.9,
44.7,32.2,NA,34.2,
38.0,27.1,NA,37.9,
34.0,23.2,NA,26.0,
44.8,37.2,NA,29.7,
46.0,29.1,NA,26.7,
41.9,32.0,37.1,37.6,
38.0,31.7,38.4,35.7,
42.2,NA,32.9,33.3,
39.7,33.5,NA,32.7,
37.5,NA,28.8,30.3,
34.6,31.0,NA,28.7,

35.5,24.7,28.1,29.8),.Dim = c(30,4)),
time = c(1,2,3,4))
# initial values
list(beta = c(0,0,0,0),rho =.5, tau = 1,taua = 1,vu = 0)
```

TABLE 10.8

Posterior Analysis of Hip-Replacement Data with Missing Values: 31 Hematocrit and 11 Age Values

Parameter	Mean	SD	Error	2½	Median	97½
Age[2]	66.1	9.066	0.06388	20.83	29.35	37.92
Age[5]	63.45	8.93	0.04802	45.73	63.46	80.93
β_1	47.87	5	0.2322	38.04	47.86	57.72
β_2	−16.61	2.41	0.121	−21.32	−16.66	−11.74
β_3	2.878	0.4675	0.02323	1.933	2.885	3.787
β_4	0.08049	0.06818	0.002943	−0.04672	0.08508	0.2201
σ^2	19.24	3.065	0.02474	14.15	18.94	26.14
σ_a^2	84.12	31.19	0.2203	43	77.86	161.4
ρ	0.2007	0.1152	0.001036	0.01495	0.1924	0.4438
$Y[1,3]$	29.36	4.338	0.03293	20.83	29.35	37.92
$Y[2,3]$	30.28	4.37	0.0314	21.73	30.29	38.86

Compared to Table 10.6 (the analysis with 18 missing hematocrit values only), the estimate of β_1 is further decreased to 47.87 compared to a posterior mean of 52.09 reported in Table 10.6. In addition, the posterior mean of ρ is decreased from a value of 0.2732 in Table 10.6 to an estimate of 0.2007 reported in Table 10.8. Also, the 95% credible interval for β_4 includes 0 for the three analyses reported in Tables 10.6 through 10.8. This implies that age has little effect on the mean of the hematocrit values and the model (10.14) revised by setting $\beta_4 = 0$.

Consider the timber load slip example with the model

$$Y_{ij} = \beta_1 + \beta_2 \exp\left(-\beta_3 x_{ij}\right) + e_{ij} \tag{10.19}$$

where:
Y_{ij} is the load for timber with the *j*th slippage value

Notice the variety of ways that random effects can be added to the model. For example, (1) the simplest situation is to add one random effect b_i to the right-hand side of (9.19), then (2) replace β_2 by $\beta_2 + b_{2i}$, and last in addition (3) replace β_3 by $\beta_3 + b_3$, where the three random effects are independent normal random variables with mean zero and unknown variances. Of course, there are many more scenarios to the addition of random effects cases of these three cases. In choosing the addition of random effects, one is inducing a correlation structure between the repeated measures of the various time points. For example, to induce compound symmetry, the model

$$Y_{ij} = \beta_1 + b_i + \beta_2 \exp\left(-\beta_3 x_{ij}\right) + e_{ij} \tag{10.20}$$

where e_{ij} are independent normal random variables with mean 0 and variance σ^2 and independent of the *n* random effects b_i, which are normal random variables with mean 0 and variance σ_b^2. It can be shown that the covariance between Y_{ij} and Y_{ik} is σ_b^2 and that the variance of Y_{ij} is $\sigma_b^2 + \sigma^2$, which completely determines the dispersion matrix of the observations.

The model is a linear model with one random effect and **BUGS CODE 10.4** closely follows formulas (10.19) and (10.20). I have modified the original data set included in **BUGS CODE 9.5** by inserting missing 24 missing values or $24/88 = 27\%$, which are designated by NA included in the list statement of **BUGS CODE 10.4**. The one random effect of the model induces a compound symmetry correlation structure with intraclass correlation coefficient

$$\rho = \frac{\sigma_b^2}{\sigma_b^2 + \sigma^2} \tag{10.21}$$

where σ_b^2 and σ^2 are defined by (10.20).

I specified uninformative prior distributions for all the unknown parameters of the model and performed a Bayesian analysis with **BUGS CODE 10.4** using 65,000 observations for the simulation, with a burn in of 5,000 and a refresh of 100.

The posterior analysis of Table 10.9 should be compared to Table 9.13, which reports the posterior analysis with no missing data.

BUGS CODE 10.4

```
model;
    {
            beta[1] ~ dnorm(0,.001)
            beta[2] ~ dnorm(0,.001)
        beta[3] ~ dnorm(0,.001)

for(i in 1:N){for (j in 1:M){Y[i,j]~dnorm(mu[i,j],tau)}}
for(i in 1:N){for (j in 1:M)
{mu[i,j]<-b[i]+beta[1]+beta[2]*exp(-beta[3]*s[j])}}

for(i in 1:N){b[i]~dnorm(0,tau1)}

tau~dgamma(.001,.001)
sigma<-1/tau
tau1~dgamma(.001,.001)
sigma1<-1/tau1
rho<-sigma1/(sigma+sigma1)
var<-sigma+sigma1


}

list(N = 8, M = 11,Y = structure(.Data = c(

0,2.38,4.34,NA,8.05,9.78,NA,12.05,12.98,NA,14.74,
```

```
0,2.69,NA,7.04,NA,10.94,12.23,NA,14.08,14.66,15.37,
0,2.85,NA,6.61,8.09,NA,11.03,12.14,13.18,14.12,NA,
NA,2.46,4.28,5.88,7.43,NA,9.92,11.10,NA,13.24,14.19,
0,2.97,NA,6.66,8.11,NA,11.06,12.25,13.35,NA,15.53,
0,NA,6.46,8.14,9.35,NA,11.84,12.85,NA,14.85,15.79,
0,3.17,NA,7.14,8.29,9.86,NA,12.13,13.15,NA,15.11,
0,NA,5.45,7.08,NA,9.91,11.06,12.21,NA,14.05,14.96),.Dim =
c(8,11)),

s=c(0,.1,.2,.3,.4,.5,.6,.7,.8,.9,1))

# initial values
list(beta = c(0,0,0), tau =.5,tau1 = 1)
```

TABLE 10.9

Posterior Analysis of Timber Load Slip Data with 24 Missing Load Values

Parameter	Mean	SD	Error	2½	Median	97½
β_1	19.81	0.538	0.03026	18.79	19.8	20.85
β_2	−19.71	0.4552	0.02545	−20.6	−19.7	−18.85
β_3	1.393	0.06422	0.00382	1.274	1.392	1.521
ρ	0.6834	0.126	0.001295	0.4198	0.6911	0.9023
σ^2	0.1174	0.02333	0.000170	0.08029	0.1144	0.1709
σ_b^2	0.3297	0.2791	0.003836	0.09192	0.2551	1.028

The only slight difference that I see is with β_1, which has a posterior mean of 20.23 with no missing values, compared to a posterior mean of 19.81 with missing values. Posterior mean of the other parameters are almost the same with or without missing values.

For the next version, there are five observations missing for each plank, for a total of 40% or 45.4%, compared to the previous version with 32 missing observations. An amended version of **BUGS CODE 10.4** is executed with 65,000 observations for the simulation, and the results are reported in Table 10.10. The following list statement displays the data set with the missing values indicated by NA.

```
list(N = 8, M = 11,Y = structure(.Data = c(
0,2.38,4.34,NA,8.05,9.78,NA,NA,NA,NA,14.74,
0,2.69,NA,7.04,NA,10.94,12.23,NA,NA,14.66,NA,
NA,2.85,NA,6.61,8.09,NA,11.03,NA,13.18,14.12,NA,
NA,2.46,4.28,NA,7.43,NA,9.92,11.10,NA,13.24,NA,
0,2.97,NA,6.66,8.11,NA,11.06,NA,13.35,NA,NA,
NA,NA,6.46,8.14,9.35,NA,11.84,12.85,NA,14.85,NA,
0,3.17,NA,7.14,NA,9.86,NA,12.13,13.15,NA,NA,
0,NA,5.45,7.08,NA,9.91,NA,12.21,NA,14.05,NA),.Dim = c(8,11)),
```

TABLE 10.10

Posterior Analysis for the Timber Load Slip Study with 40 Missing Load Values

Parameter	Mean	SD	Error	2½	Median	97½
β_1	19.2	0.6383	0.03656	17.98	19.17	20.58
β_2	−20.03	0.5617	0.03166	−20.26	−19.0	−19.95
β_3	1.453	0.08399	0.005008	1.282	1.452	1.635
ρ	0.7044	0.1247	0.001089	0.4356	0.7153	0.9115
σ^2	0.1163	0.02819	0.00026	0.07366	0.1123	0.1827
σ_b^2	0.3594	0.2978	0.002573	0.09974	0.2814	1.083

The addition of missing values to 40 from 32 makes very little difference in the posterior means of the parameters of the model. To see this, compare Table 10.9 with 10.10.

10.3 Missing Data and Categorical Repeated Measures

There are two parts to this section, the first concentrates on linear models and the second on nonlinear.

In Chapter 8, where the Bayesian analysis of categorical data is presented, three cases were considered: (1) the multinomial distribution for various patterns in the data, (2) generalized estimating equations, and (3) generalized mixed linear models. Part (1) is appropriate if the number of responses (possible values of the repeated measures) is not too large and covariates are not considered. I favor part (3) generalized mixed linear model, where the response can have a Bernoulli distribution for binary data and a Poisson distribution for more than two responses. Thus, consider the case where the repeated measure has a Poisson distribution.

1. Conditional on the random effects b_i, the distribution of Y_{ij} is Poisson and the observations over n time points are independent. The conditional variance is

$$\text{Var}\left(Y_{ij}|b_i\right) = E\left(Y_{ij}|b_i\right) \tag{10.22}$$

2. The canonical link is the log function

$$\log\left[E\left(Y_{ij}|b_i\right)\right] = v_{ij} = X_{ij}\beta + Z_{ij}b_i \tag{10.23}$$

3. The random effects are independent multivariate normal with mean vector 0 and variance–covariance matrix Σ. In addition, the random effects are independent of the covariates.

Also consider the experimental use of dopamine in Chapter 8. Table 8.11 revised the data set so that there are missing observations, which appear in the list statement of **BUGS CODE 10.5**.

Refer to Table 8.11, and note that $X_1 = 1$ if dose = 8.4 mg, otherwise 0; $X_2 = 1$ if dose = 16.8 mg, otherwise 0; $X_3 = 1$ if dose = 33.5 mg, otherwise 0; and $X_4 = 1$ if dose = 67 mg, otherwise 0. Also, when $X_1 = X_2 = X_3 = X_4 = 0$, the subject is assigned to placebo. Based on this information, the proposed model is

$$Y_{ij} \sim \text{Poisson}(\mu_{ij}) \tag{10.24}$$

$$\log(\mu_{ij}|b_{1i}, b_{2i}) = \beta_1 + \beta_2 t[j] + \beta_3 X_1 + \beta_4 X_2 + X_3\beta_5 + \beta_6 X_4 + b_{1i} + b_{2i} t[j]$$

where the (b_{1i}, b_{2i}), $i = 1,2,\ldots,N$ are independent and have a two-dimensional normal distribution with a 2 by 1 mean vector 0 and unknown 2 by 2 covariance matrix Σ. Therefore, the proposed model is a generalized linear model where the fixed effects designate a linear response with respect to time and dose effects, and with two random effects, one corresponding to the constant term, and the other b_{2i} representing the subject-to-subject variation of the time effect. The effect of the first dose 8.5 mg of the log response is β_3, while the effect of the highest dose 67 mg is β_6, thus, it is of interest to compare the various doses using the contrasts

$$d_{34} = \beta_3 - \beta_4$$

$$d_{35} = \beta_3 - \beta_5$$

$$\vdots \tag{10.25}$$

$$d_{56} = \beta_5 - \beta_6$$

For example, d_{34} compares the effect of dose 8.4 mg on the log response with the effect of dose 33.5 mg on the clinical response.

For the first analysis, **BUGS CODE 10.5** is executed with 65,000 observations and a burn in of 5,000. Note the model (10.25) has two independent random factors with variances σ_1^2 and σ_2^2. I revised the data so that there are now 25 missing values (or 33.3%) included in the list statement. The two random factors b_{1i}, b_{2i} are assumed to be independent, and the case where they have a bivariate normal distribution is left as an exercise.

The analysis implies no difference in the effect of dose on the log clinical response. Compare the results of Table 9.11 (with no missing values) with those of Table 10.11 (with five missing values for clinical response). Also it is apparent that the two random factors have extremely small variances 0.005294 for b_{1i} and 0.000537 for b_{2i}. Obviously, the model can be modified so that it does not take into consideration the four doses, which is also implied

BUGS CODE 10.5

```
model;

{

for (i in 1:N){for (j in 1:M){Y[i,j]~dpois(mu[i,j])

log(mu[i,j])<-beta[1]+beta[2]*tim[j]+beta[3]*X1[i]+beta[4]
*X2[i]+beta[5]*X3[i]+beta[6]*X4[i]+b1[i]+b2[i]*tim[j]}}

for(i in 1:6){beta[i]~dnorm(0,.0001)}

for (i in 1:N){b1[i]~dnorm(0,tau1)}

for (i in 1:N){b2[i]~dnorm(0,tau2)}

tau1~dgamma(0.001,.001)
tau2~dgamma(0.001,.001)

sigma1<-1/tau1
sigma2<-1/tau2

d34<-beta[3]-beta[4]
d35<-beta[3]-beta[5]
d36<-beta[3]-beta[6]
d45<-beta[4]-beta[5]
d46<-beta[4]-beta[6]
d56<-beta[5]-beta[6]

}


# the data having missing values denoted by NA
list(N = 25,M = 3,

Y = structure(.Data =
c(2,NA,4,
2,3,NA,
NA,3,4,
3,NA,2,
3,3,NA,
5,NA,5,
NA,4,4,
4,NA,2,
NA,3,3,
4,4,NA,
4,NA,4,
NA,4,3,
NA,3,4,
```

```
3,7,NA,
5,NA,5,
NA,6,4,
NA,3,3,
6,5,NA,
5,NA,3,
NA,3,4,
3,5,NA,
6,NA,5,
3,3,3,
4,NA,5,
NA,3,3)),.Dim = c(25,3)),

X1 = c(0,0,0,1,0,0,0,1,0,0,1,0,0,0,0,1,0,0,0,0,0,0,1,0,0),

X2 = c(0,0,0,0,1,0,1,0,0,0,0,0,0,0,0,0,0,1,0,0,0,0,1,0,0,0),

X3 = c(1,0,0,0,0,0,0,0,0,1,0,0,1,0,0,1,0,0,0,0,0,1,0,0,0,0),

X4 = c(0,1,0,0,0,1,0,0,0,0,0,0,0,1,0,0,0,0,1,1,0,0,0,0,0),

tim = c(2,7,14))

list(beta = c(0,0,0,0,0,0), tau1 = 1,tau2 = 1)
```

TABLE 10.11

Bayesian Analysis of Parkinson's Disease Study with 25 Missing Values

Parameter	Mean	SD	Error	2½	Median	97½
β_1	1.41	0.204	0.001602	1	1.412	1.805
β_2	−0.00567	0.01589	0.000122	−0.03708	−0.005594	0.02541
β_3	−0.1524	0.2567	0.002	−0.06614	−0.1508	0.3433
β_4	−0.0384	0.2713	0.002259	−0.5707	−0.0367	0.4915
β_5	−0.09342	0.2606	0.002068	−0.6109	−0.09195	0.4091
β_6	−0.01928	0.249	0.001932	−0.5109	−0.01802	0.47
d_{34}	−0.1139	0.2846	0.002206	−0.6744	−0.1141	0.4517
d_{35}	−0.0589	0.2721	0.002126	−0.5865	−0.05999	0.4749
d_{36}	−0.133	0.2652	0.00297	−0.6608	−0.1312	0.3849
d_{45}	0.05495	0.2875	0.002304	−0.5095	0.0548	0.619
d_{46}	−0.01919	0.2785	0.002269	−0.5738	−0.0195	0.5252
d_{56}	−0.014140	0.2651	0.002069	−0.5929	−0.07413	0.4449
σ_1^2	0.01118	0.01631	0.000318	0.000526	0.005294	0.057
σ_2^2	0.000628	0.000369	0.0000035	0.0002036	0.000537	0.00159

by the fact that all the differences d_{ij} have 95% credible intervals that contain 0, thus, the modified model is

$$\log\left(\mu_{ij} \big| b_{1i}, b_{2i}\right) = \beta_1 + \beta_2 t[j] + b_{1i} + b_{2i} t[j] \tag{10.26}$$

and the Bayesian analysis repeated with a revision of **BUGS CODE 10.5** namely, **BUGS CODE 10.5a**.

Again 65,000 observations are generated for the simulation and 5,000 for the burn in.

BUGS CODE 10.5a

```
model;

{

for (i in 1:N){for (j in 1:M){Y[i,j]~dpois(mu[i,j])

log(mu[i,j])<-beta[1]+beta[2]*tim[j]+b1[i]+b2[i]*tim[j]}}

for(i in 1:2){beta[i]~dnorm(0,.0001)}

for (i in 1:N){b1[i]~dnorm(0,tau1)}

for (i in 1:N){b2[i]~dnorm(0,tau2)}

tau1~dgamma(0.001,.001)
tau2~dgamma(0.001,.001)

sigma1<-1/tau1
sigma2<-1/tau2
}

list(N = 25,M = 3,

Y = structure(.Data =
c(2,NA,4,
  2,3,NA,
  NA,3,4,
  3,NA,2,
  3,3,NA,
  5,NA,5,
  NA,4,4,
  4,NA,2,
  NA,3,3,
  4,4,NA,
  4,NA,4,
  NA,4,3,
```

```
NA,3,4,
3,7,NA,
5,NA,5,
NA,6,4,
NA,3,3,
6,5,NA,
5,NA,3,
NA,3,4,
3,5,NA,
6,NA,5,
3,3,3,
4,NA,5,
NA,3,3),.Dim = c(25,3)),

tim = c(2,7,14))

list(beta = c(0,0), tau1 = 1,tau2 = 1)
```

The posterior analysis for the modified model (10.26) is reported in Table 10.12 and implies that the slope on the log scale is 0.

Based on the posterior median, the estimates of the other parameters are almost the same as those reported in Table 10.11. Overall it appears that the clinical scores are the same across the three time points and that the random effects corresponding to the intercept and slope have extremely small variances. Under the modified model and the addition of 10 missing values for a total of 35 out of 75, the Bayesian analysis is repeated with 65,000 observations for the simulation and 5,000 for the burn in.

The missing values are in the following list statement from the WinBUGS code.

```
Y = structure(.Data =
c(2,NA,NA,
NA,3,NA,
NA,3,4,
3,NA,2,
```

TABLE 10.12

Posterior Analysis for Parkinson's Disease with 25 Missing Values and a Modified Model

Parameter	Mean	SD	Error	2½	Median	97½
β_1	1.368	0.1385	0.000743	1.09	1.369	1.634
β_2	−0.006557	0.01573	0.0000989	−0.0376	−0.00647	0.0243
σ_1^2	0.009481	0.01346	0.000251	0.000514	0.004761	0.04643
σ_2^2	0.000573	0.000322	0.0000029	0.0001932	0.000494	0.002402

```
NA,3,NA,
5,NA,5,
NA,4,NA,
4,NA,2,
NA,3,NA,
NA,4,NA,
4,NA,4,
NA,4,NA,
NA,3,NA,
3,7,NA,
5,NA,5,
NA,6,4,
NA,3,NA,
NA,5,NA,
5,NA,3,
NA,3,NA,
3,5,NA,
6,NA,5,
NA,3,NA,
4,NA,5,
NA,3,3),.Dim = c(25,3)),
```

Even with an additional 10 missing values, the estimated parameters reported in Table 10.13 are approximately the same as those reported in Tables 10.11 and 10.12.

The posterior distributions are well behaved and the MCMC simulation errors are quite small, indicating that the posterior means are very accurate estimates.

The next example has a binary repeated measure, thus, let $Y_{ij} = 0, 1$ be a binary response for subject i at time j, then the logistic generalized mixed model is defined as:

1. Conditional on the random effect b_i, Y_{ij} are independent and are Bernoulli distributed with mean $E(Y_{ij} | b_i)$ and variance

$$\text{Var}\left(Y_{ij} | b_i\right) = E\left(Y_{ij} | b_i\right)\left[1 - E\left(Y_{ij} | b_i\right)\right] \tag{10.27}$$

TABLE 10.13

Posterior Analysis for the Parkinson's Disease Study with 35 Missing Values

Parameter	Mean	SD	Error	2½	Median	97½
β_1	1.335	0.09511	0.000581	1.143	1.337	1.517
σ_1^2	0.01087	0.01604	0.000315	0.000537	0.005274	0.05446
σ_2^2	0.000724	0.000451	0.0000045	0.000217	0.000611	0.001892

2. The logistic link function is given by

$$\nu_{ij} = X_{ij}\beta + b_i \qquad (10.28)$$

where

$$\nu_{ij} = \frac{\log\left[\Pr\left(Y_{ij} = 1 \middle| b_i\right)\right]}{\Pr\left(Y_{ij} = 0 \middle| b_i\right)} \qquad (10.29)$$

3. The random effects $b_i(i = 1,2,\dots,n)$ are independent and have a normal distribution with mean 0 and variance σ_1^2.

An example is based on the study of Stokes, Davis, and Koch[10] of a clinical trial for subjects with pulmonary disease and consists of 111 patients recruited from two clinics and are randomized to receive placebo or an active treatment. Responses were recorded as 1 (good) or 0 (poor) and measured at baseline and visits 1, 2, 3, and 4. Also known are the gender of subject, their age, and if they have received placebo or not.

Suppose that the response at time j ($= 0,1,2,3,4$) for subject i is given by

$$Y_{ij} \sim \text{Bernoulla}\left(\mu_{ij}\right) \qquad (10.30)$$

where

$$\text{logit}\left(\mu_{ij}\right) = \beta_1 + \beta_2 t[j] + \beta_3 t^2[j] + \beta_4 \text{age}[i] + \beta_5 \text{gen}[i] + \beta_6 \text{grp}[i] + b_{1i} \qquad (10.31)$$

and $i = 1,2,\dots,N$, $j = 1,2,\dots,n$. N is the number of subjects and n the number of time points per subject. A quadratic response is proposed for the proportion at each time point $t[j]$ and there is one random effect, which accounts for the subject-to-subject variation of the intercept on the logistic scale. The Bayesian analysis is based on **BUGS CODE 10.6** and the data from Stokes, Davis, and Koch[10] is included in the first list statement of the code. For this example, only 56 patients from center 1 are included. I coded the 34 missing values out of a total of 280 as NA in the list statement of **BUGS CODE 10.6**. Recall that the data set with no missing values was analyzed in Chapter 8 using **BUGS CODE 8.6**. Note from (10.31) the linear effect of time is given by β_2 and the quadratic effect by β_3, β_4 is the age effect, and β_5 the gender effect, and last the treatment effect is given by β_6 on the logistic scale.

For the present study with 34 missing values, the analysis is performed with 75,000 observations and 5,000 for the burn in.

Table 10.14 reports the posterior analysis for the pulmonary study, and it is seen that the 95% credible interval for the intercept includes 0 as does the 95% credible interval for β_4, β_5, and β_6, the effects for age, gender, and group, respectively.

BUGS CODE 10.6

```
model;
{

for(i in 1:N){for (j in 1:M){Y[i,j]~dbern(mu[i,j])}}
for(i in 1:N){for(j in 1:M){logit(mu[i,j])<-beta[1]+beta[2]
*t[j]+beta[3]*t[j]*t[j]+beta[4]*age[i]+beta[5]*gen[i]+
beta[6]*grp[i]+b1[i]}}
for(i in 1:N){for (j in 1:M){p[i,j]<-(exp(beta[1]+beta[2]
*t[j]+beta[3]*t[j]*t[j]+beta[4]*age[i]+beta[5]*gen[i]+
beta[6]*grp[i]+b1[i]))/(1+exp(beta[1]+beta[2]*t[j]+beta[3]
*t[j]*t[j]+beta[4]*age[i]+beta[5]*gen[i]+beta[6]*grp[i]
+b1[i]))}}
for(i in 1:6){beta[i]~dnorm(0,.0001)}
for(i in 1:N){b1[i]~dnorm(0,tau1)}
tau1~dgamma(.0001,.0001)
sig1<-1/tau1
}


list(N = 56,M = 5,

Y = structure(.Data = c(0, 0,      NA,    0,     0,
0,      0,      NA,    0,     0,
1,      1,      NA,    1,     1,
1,      1,      NA,    1,     0,
1,      1,      NA,    1,     1,
0,      0,      NA,    0,     0,
0,      1,      NA,    1,     1,
0,      0,      NA,    0,     0,
1,      1,      NA,    1,     1,
1,      0,      NA,    1,     0,
1,      1,      1,     1,     1,
0,      1,      NA,    1,     0,
1,      1,      0,     0,     0,
0,      0,      NA,    0,     0,
1,      1,      1,     1,     1,
0,      0,      NA,    0,     1,
0,      0,      0,     0,     0,
0,      0,      NA,    1,     1,
0,      0,      0,     0,     0,
1,      1,      NA,    1,     0,
0,      1,      0,     1,     0,
1,      1,      NA,    1,     1,
1,      1,      1,     1,     1,
0,      1,      NA,    0,     0,
0,      0,      0,     0,     0,
0,      0,      NA,    0,     0,
```

```
1,      1,      0,      1,      1,
0,      1,      NA,     0,      0,
0,      1,      1,      0,      0,
0,      0,      NA,     0,      0,
0,      0,      0,      0,      0,
0,      1,      NA,     1,      1,
1,      1,      1,      1,      0,
0,      0,      NA,     1,      0,
0,      0,      0,      1,      0,
0,      0,      NA,     0,      0,
0,      0,      1,      0,      0,
0,      0,      NA,     0,      0,
0,      0,      0,      0,      0,
0,      0,      NA,     0,      0,
0,      0,      0,      0,      1,
0,      0,      NA,     0,      0,
1,      1,      1,      1,      1,
0,      0,      NA,     0,      0,
0,      1,      0,      1,      1,
0,      0,      NA,     0,      0,
0,      0,      0,      0,      0,
1,      1,      NA,     1,      1,
1,      1,      1,      1,      1,
0,      0,      NA,     1,      1,
0,      0,      NA,     1,      1,
0,      0,      NA,     1,      0,
0,      0,      NA,     1,      0,
1,      1,      NA,     1,      0,
1,      1,      NA,     1,      1,
0,      1,      NA,     0,      1
),.Dim = c(56,5)),

age = c(4,4,6,4,2,8,2,3,4,4,1,3,3,4,4,3,2,8,3,1,3,7,3,0,1,4,
2,3,3,0,2,0,2,2,2,5,4,7,3,1,2,0,2,6,4,6,3,2,4,8,3,5,2,6,2,3,
3,6,1,9,2,8,3,7,2,3,3,0,1,5,2,6,4,5,3,1,5,0,2,8,2,6,1,4,3,1,
1,3,2,7,2,6,4,9,6,3,5,7,2,7,2,2,1,5,4,3,3,2,1,1,2,4,2,5),

grp = c(0,0,1,0,0,1,0,1,1,0,1,1,0,0,0,1,0,1,0,1,1,1,1,1,0,
1,0,0,0,1,0,1,1,0,1,0,1,1,0,0,0,1,0,0,0,0,0,1,0,1,1,0,1,1,
0,1),

gen = c(0,0,0,0,1,0,0,0,0,0,0,0,0,0,0,0,0,1,1,0,0,0,0,0,1,
0,0,1,0,0,0,0,0,0,0,1,0,0,0,0,0,0,0,0,0,0,0,0,0,0,0,0,1,0,0,0),

t = c(0,1,2,3,4))

list(beta = c(0,0,0,0,0,0), tau1 = 1)
```

TABLE 10.14

Bayesian Analysis for Pulmonary Study with 34 Missing Values

Parameter	Mean	SD	Error	2½	Median	97½
β_1	−2.076	1.242	0.02013	−4.682	−2.03	0.2568
β_2	1.545	0.5568	0.00363	0.4878	1.534	2.68
β_3	−0.347	0.1328	0.000843	−0.6168	−0.3439	−0.09571
β_4	−0.06087	0.2762	0.004269	−0.6223	−0.0586	0.4888
β_5	−1.146	1.685	0.02615	−4.616	−1.109	2.115
β_6	1.43	1.114	0.01878	−0.671	1.391	3.759
σ_1^2	12.31	5.861	0.08518	4.839	11.06	27.12

The implication of the posterior analysis of Table 10.14 is that the model can be simplified to

$$\text{logit}\left(\mu_{ij}\right) = \beta_1 + \beta_2 t[j] + \beta_3 t^2[j] + b_{1i} \qquad (10.32)$$

with the linear and quadratic time effects, plus the one random factor b_{1i}. The Bayesian analysis for the modified model (10.32) is implemented with **BUGS CODE 10.6a** using 75,000 observations for the simulation, with a burn in of 5,000. Note the data set for **BUGS CODE 10.6a** is in the first list statement of **BUGS CODE 10.6**.

The posterior analysis for the modified model (10.32) is reported in Table 10.15 and shows some similarity with the analysis for the full model reported in Table 10.15; however, there are some differences.

BUGS CODE 10.6a

```
model;
{

for(i in 1:N){for (j in 1:M){Y[i,j]~dbern(mu[i,j])}}
for(i in 1:N){for(j in 1:M){logit(mu[i,j])<-
beta[1]+beta[2]*t[j]+beta[3]*t[j]*t[j]+b1[i]}}

for(i in 1:N){for (j in 1:M){p[i,j]<-
(exp(beta[1]+beta[2]*t[j]+beta[3]*t[j]*t[j]+b1[i]))/(1+exp(
beta[1]+beta[2]*t[j]+beta[3]*t[j]*t[j]+b1[i]))}}
for(i in 1:3){beta[i]~dnorm(0,.0001)}
for(i in 1:N){b1[i]~dnorm(0,tau1)}
tau1~dgamma(.0001,.0001)
sig1<-1/tau1

list(beta = c(0,0,0), tau1 = 1)
```

TABLE 10.15

Posterior Analysis for Respiratory Study with 34 Missing Values and Modified Model

Parameter	Mean	SD	Error	2½	Median	97½
β_1	−1.606	0.6356	0.007828	−2.923	−1.58	−0.4341
β_2	1.455	0.5417	0.003183	0.4265	1.444	2.549
β_3	−0.3262	0.1294	0.000736	−0.5865	−0.3238	−0.0791
σ_1^2	10.12	4.543	0.05162	4.133	9.215	21.45

By comparing Table 10.14 with 10.15, there are some differences in the estimates of the parameters, most notably for β_1, which is estimated as −2.076 for the full model and as −1.606 for the modified model. Also apparent is the difference in the posterior median of 9.215 for the variance of the random effect for the modified model, compared to a posterior median of 11.06 for σ_1^2 of the full model. I increased the number of missing values to 56, all occurring at time 2, and repeated the Bayesian analysis with **BUGS CODE 10.6a** using again 75,000 observations for the simulation and a burn in of 5,000, and the results are revealed in Table 10.16.

Comparing Table 10.16 with 10.15, one sees a difference in the posterior mean of σ_1^2 and that there is still a strong quadratic time effect on the logistic scale.

Our next example for a generalized linear mixed model for ordered categorical data is the Parkinson's disease study examined in Chapter 8.

Consider the response for subject i at time j, $Y_{ij} = 1, 2, ..., K$, the generalized linear mixed model for ordinal categorical data is specified by:

1. The conditional distribution of the Y_{ij} given a q by 1 vector of random effects b_i is multinomial.

2. The link function is the cumulative response probabilities given as

$$\log \frac{\Pr\left(Y_{ij} \le k|b_i\right)}{\Pr\left(Y_{ij} > k|b_i\right)} = v_{ij} \tag{10.33}$$

where

$$v_{ij} = \alpha_k + X_{ij}\beta + Z_{ij}b_i \tag{10.34}$$

TABLE 10.16

Posterior Analysis for Respiratory Study with 56 Missing Values and Modified Model

Parameter	Mean	SD	Error	2½	Median	97½
β_1	−1.836	0.6991	0.009514	−3.314	−1.8	−0.5519
β_2	2.046	0.6416	0.004605	0.8409	2.025	3.363
β_3	−0.4705	0.1539	0.001077	−0.7855	−0.4657	−0.1814
σ_1^2	13.01	6.296	0.07925	5.038	11.66	28.8

where:

    X_{ij} is a 1 by p known vector of covariates
    β is a p by 1 unknown vector of regression parameters
    Z_{ij} is a 1 by q vector of known covariate values
    b_i is a q by 1 vector of random effects
    α_k is an unknown scalar for the ordinal value k

3. The random effects b_i are independent and have a multivariate normal distribution with mean vector 0 and unknown covariance matrix Σ.

As an example, we revisited the experimental use of dopamine. Table 10.17 contains the information about this study, where the first three columns give the clinical rating at each time point ($t = 2, 7,$ and 14 days) for each of 25 patients. When employing the proportional odds model of (8.89) and (8.90), it is useful to express the information in the following format.

Thus, at day 2, 2 subjects scored 2, 11 scored 3, 4 scored 4. 5 scored 5, 2 scored 6, and last 0 scored 7. There were no 1 scores, thus, it will not be included in the analysis.

Consider the model

$$Y_{ij} \sim \text{multinomial}\left(p_{i1}, p_{i2}, \ldots, p_{i6}; n_i\right) \tag{10.35}$$

where $p_{ij} = \Pr(Y_{ij} = j)$ with $i = 2,7,14$ and $j = 1,2,3,4,5,6$. That is for time i and score j, the probability a subject scores j is p_{ij}. Thus, for each time point, the various scores are assumed to follow a multinomial distribution. Note n_i is the total number of scores at time I, summed over six scores.

The proportional odds assumptions is expressed by

$$\text{logit}\left(\gamma_{ij}\right) = \alpha_j - \left(\beta_1 + \beta_2 i + b_i\right) \tag{10.36}$$

where:

    i denotes time (= 2, 7, and 14 days)
    $j = 1, 2, \ldots, K - 1$
    the ordinal scores are 1, 2, ..., K

TABLE 10.17

Multinomial Outcomes for Parkinson's Disease Study

Time (days)	2 Rating	3 Rating	4 Rating	5 Rating	6 Rating	7 Rating
2	2	11	4	5	2	0
7	0	12	6	5	1	1
14	2	8	8	5	2	0
Total	4	31	18	15	5	1

$$\gamma_{ij} = \Pr\left(Y_{ij} \leq j | b_i\right) \tag{10.37}$$

and the

$$b_i \sim nid(0, \tau) \tag{10.38}$$

where $\tau > 0$.

Note that the proportional odds assumption (8.92) assumes that $\beta_1 + \beta_2 i + b_i$ and does not depend on the ordinal score j.

The random effects measure the heterogeneity over the three time points with variance $\sigma^2 = 1/\tau$.

For the Bayesian analysis, β_1 and β_2 are given noninformative normal distributions $n(0, 0.0001)$ and τ is specified a noninformative gamma distribution $(0.001, 0.001)$. Note $\beta_1 + b_i$ denotes the constant term at time point i. Consider two time points i and $i + 5$, then the difference of the logits at ordinal score j is

$$\log it\left(\gamma_{i+5,j}\right) - \log it\left(\gamma_{i,j}\right) = 5\beta_2 \tag{10.39}$$

which measures the change in the logit over a period of five days.

BUGS CODE 10.7 will execute the analysis for the Parkinson's disease study using the proportional odds model and is based on the code found in Congdon (p. 102).[11] For additional information about the analysis of categorical data including the proportional odds model, see Agresti (p. 164),[12] and for information from a Bayesian view using the proportional odds model for ordered categorical data, see Broemeling.[13] The Bayesian analysis will be performed with **BUGS CODE 10.7** using 65,000 observations for the simulation and a burn in of 5,000. There are 38 missing values for a total of 36 used in the analysis compared to the original data set of 74 used for Bayesian analysis with **BUGS CODE 8.8**. I reduced the count by approximately one half. The original data set is given by the list statement of **BUGS CODE 8.8**, namely

```
Y = structure(.Data =
c(2,11,4,5,2,0,
0,12,6,5,1,1,
2,8,8,5,2,0),.Dim = c(3,6)),
```

The Bayesian analysis with the missing values is reported in Table 10.18. Note that this posterior analysis is based on 38 observations, instead of the full data set with 74 cell count total.

The random effects have a variance estimated at 2.696, but 0.03947 with the posterior median, thus, I recommend using the latter. Note that there are five models for this model (10.36) one for each of the five ordinal values, but that the three beta coefficients remain the same for each cumulative proportional odds. Comparing Table 8.19 (the posterior analysis with no missing data) to

BUGS CODE 10.7

```
model;
{

for (i in 1:M){Y[i,1:K]~dmulti(p[i,1:K],N[i])
N[i]<-sum(Y[i,])}

for(i in 1:M){for(j in 1:K-1)
{logit(gamma[i,j])<-alpha[j]-mu[i]}}

for(i in 1:M){p[i,1]<-gamma[i,1]}

for(i in 1:M){for(j in 2:K-1)
{p[i,j]<-gamma[i,j]-gamma[i,j-1]}}

for(i in 1:M){p[i,K]<-1-gamma[i,K-1]}

for(i in 1:M){mu[i]<-beta[1]+beta[2]*tim[i]+b[i]}

for(i in 1:2){beta[i]~dnorm(0,.0001)}
alpha[1]~dnorm(0,1)
alpha[2]~dnorm(0,1)
alpha[3]~dnorm(0,1)
alpha[4]~dnorm(0,1)
alpha[5]~dnorm(0,1)
for (i in 1:M){b[i]~dnorm(0,tau)}
tau~dgamma(.001,.001)
sigma<-1/tau
}

list(K = 6,M = 3,

Y = structure(.Data =
c(1,8,2,3,1,0,
  0,7,3,3,1,1,
  2,6,5,4,1,0),.Dim = c(3,6)),

tim = c(2,7,14))

list(beta = c(0,0), alpha = c(0,0,0,0,0),tau = 1)
```

Table 10.18 (the one with approximately 50% missing data), one sees differences in the estimates of all the parameters. For example, based on Table 10.18, the posterior median for α_1 is -2.691, but based on Table 8.19, α_1 has a posterior median of -3.107. It is difficult to draw any conclusions about the effect of missing values on the estimates of the parameters. I reduced each cell entry by one half, thus, the empirical multinomial proportions are approximately the same.

TABLE 10.18

Bayesian Analysis for the Parkinson's Disease Study with 36 Missing Values and the Proportional Cumulative Odds Model

Parameter	Mean	SD	Error	2½	Median	97½
β_1	−0.4832	1.276	0.07035	−2.965	−0.5321	2.296
β_2	−0.009231	0.15	0.008654	−0.4794	0.005556	0.2395
σ^2	2.696	26.79	0.7803	0.000805	0.03947	18.56
α_1	−2.696	0.5806	0.01613	−3.851	−2.691	−1.589
α_2	−0.7009	0.4983	0.01826	−1.692	−0.6909	0.235
α_3	0.166	0.486	0.0179	−0.7984	0.1765	1.09
α_4	1.202	0.5019	0.0165	0.1903	1.208	2.162
α_5	2.14	0.5864	0.01488	1.01	2.134	3.323

For the last scenario for missing data, there are now 54 fewer cell count totals as follows:

```
Y = structure(.Data =
c(1,2,1,1,1,0,
0,3,1,1,1,1,
1,2,2,1,1,0),.Dim = c(3,6)),
```

which is the list statement in a modified version of **BUGS CODE 10.8**.

As before, the analysis is executed with 65,000 observations and a burn in of 5,000, and the posterior analysis is reported in Table 10.19.

The results of Table 10.19 (the case with 54 missing cell counts) should be compared to Table 10.18 (the case with 36 cell counts). Thus, there are three scenarios for the Parkinson's disease analysis: (1) no missing data, (2) 36 missing cell count totals, and (3) 54 missing cell count totals. A comparison among the three scenarios is left as an exercise for the student.

TABLE 10.19

Bayesian Analysis for the Parkinson's Disease Study with 54 Missing Cell Totals and the Proportional Cumulative Odds Model

Parameter	Mean	SD	Error	2½	Median	97½
β_1	−0.3749	1.264	0.05893	−2.764	−0.3789	2.315
β_2	−0.01024	0.1433	0.007141	−0.3146	−0.003394	0.2373
σ^2	2.407	38.89	0.4621	0.001012	0.08377	15.77
α_1	−2.105	0.5849	0.01455	−3.293	−2.902	−0.9789
α_2	−0.6017	0.5078	0.01667	−1.592	−0.6011	0.4094
α_3	0.1081	0.501	0.017	−0.8779	0.1148	1.119
α_4	0.7585	0.5118	0.01623	−0.2432	0.7527	1.787
α_5	1.822	0.6108	0.01445	0.6717	1.805	3.071

BUGS CODE 10.8

```
model;
{

# prior distribution for the regression coefficients.
            beta[1]~dnorm(11,.1)
            beta[2]~dnorm(-23,.1)
            beta[3]~dunif(.7,1)

for(i in 1:N){for(j in 1:M){Y[i,j]~dpois(theta[i,j])}}

for(i in 1:N){for (j in 1:M)
{logit(theta[i,j])<-beta[1]+beta[2]*(1-
pow(x[j],beta[3]+b1[i]))}}
for (i in 1:N){b1[i]~dnorm(0,tau)}
# prior distribution for the precision of the random
effect
tau~dgamma(3,2)
sigma<-1/tau

}

list(N = 10, M = 6,Y = structure(.Data = c(1,NA,0,NA,0,0,
1,NA,NA,0,NA,0,
1,NA,1,NA,1,0,
NA,1,NA,0,NA,0,
1,1,NA,0,NA,0,
1,NA,0,NA,NA,0,
1,NA,1,NA,NA,0,
1,NA,1,NA,NA,0,
NA,NA,1,NA,0,0,
1,NA,NA,0,NA,0),.Dim = c(10,6)),
# x is the vector of specified proportions
x = c(.1,.25,.3,.5,.7,.9))

# initial values

list(beta = c(11,-23,.7)
```

The second part of this section is focused on nonlinear categorical data. Recall that Chapter 9 dealt with nonlinear models, for both continuous and discrete data, and the approach taken here is to introduce missing data in the data sets considered for nonlinear categorical data of Chapter 9.

For nonlinear model for categorical data, the Bayesian analysis will follow the following alternatives: (1) for binary data, the probability of success will be transformed to the logit scale, which will then be expressed as a nonlinear function of fixed and random effects, (2) for categorical responses with a few possible values, the responses are assumed to have a Poisson distribution, with a mean transformed to the log scale, which in turn will be expressed as a nonlinear function of fixed and random effects, and (3) last, for ordinal categorical data, the outcomes are assumed to have a multinomial distribution whose parameters (probabilities) are transformed to cumulative log odds ratios, and in turn, the log odds ratios will be expressed as a nonlinear function of fixed and random effects.

Recall this example from the pharmaceutical industry, which involves the time it takes to dissolve a specified proportion of a pill. The effects of dissolving times may be affected by numerous factors including the brand, the shape of the pill, the way the pill is manufactured, the batch, and the indication for using the pill, thus the industry is always conducting pill dissolution studies, and the information to be used for the Bayesian analysis is reported in Table 9.5. The information appearing in the table is only a subset of the information appearing in Table A6 of Hand and Crowder.[14] Recall for this example, the response is the time it takes the pill to dissolve to a specified proportion (left undissolved). I have replaced the time values (measured in seconds) by a binary response, where 0 represents values less than the median value of 23 seconds, while a 1 represents the values greater than 23 seconds. The proposed model is given by a revised form for (9.5) that includes a random effect for the logit that measures the pip-to-pill variation in the factor that scales the specified proportions. Recall for a binary response with success probability, the canonical link is given by

$$\log \mathrm{it}\left(\theta_{ij}\right) = \beta_1 + \beta_2 \left(1 - x_{ij}^{\beta_3 + b_i}\right) \tag{10.40}$$

where:

$\theta_{ij} = P\left[Y_{ij} = 1 / b_i\right]$

x_{ij} is the specified proportion, that is the proportion left of the undissolved pill

Of course, I have replaced the original continuous data with binary values Y_{ij}, thus there is loss of information, but nevertheless, it serves to illustrate a nonlinear model for categorical data. Prior information for the parameters of the model (10.40) is as follows: for β_1, the distribution is normal with a mean of 11, and the variance (1/variance) is 10, and for β_2 the distribution is normal with mean −23 and variance 10, while for β_3, the prior is uniform over (0.7, 1). The Bayesian analysis is executed with **BUGS CODE 10.8**, using 65,000 observations to generate the posterior distribution.

The Bayesian analysis is revealed in Table 10.20 and is quite similar to that reported in Table 9.23, the analysis with no missing values.

TABLE 10.20

Pill Dissolution Study with Binary Responses and 30 Missing Values

Parameter	Mean	SD	Error	2½	Median	97½
β_1	17.05	2.346	0.03589	12.43	17.07	21.56
β_2	−17.11	2.462	0.0405	−21.91	−17.11	−12.29
β_3	0.8391	0.08648	0.001345	0.706	0.8341	0.9907
σ^2	0.9401	0.8662	0.01391	0.266	0.7066	3.042

How does the posterior analysis of Table 10.20 (with 30 missing values) compare to the analysis of Table 9.23 (with no missing values)? The simulation size is 65,000 for both, and the prior distributions are the same.

One can see there is very little difference in the two analyses. Perhaps the prior distributions are too informative and consequently the sample sizes do not make a difference.

Recall the example using a nonlinear generalized mixed model for categorical data of Hand and Crowder (p. 111),[14] who describe a nonlinear example of repeated measures occurring in civil engineering, where a wood plank is inserted between concrete surfaces and fastened by four screws. The load is increased and applied to the timber, and the slippage in millimeters is recorded at two-second intervals. For our purposes, the load is the dependent variable and the slip the independent variable, and the data exhibited is taken from Table A.15 of Hand and Crowder (p. 179).[14]

A revision of model (9.10) is considered, thus let Y_{ij} be the load of slip j for plank i, where Y_{ij} has a Poisson distribution with conditional mean and variance given one random effect as $E\left(Y_{ij}|b_i\right) = \mathrm{Var}\left(Y_{ij}|b_i\right)$, where

$$\log\left[E\left(Y_{ij}|b_i\right)\right] = \beta_1 + b_{1i} + \beta_2 \exp\left(-\beta_3 x_{ij}\right) \tag{10.41}$$

where $Y_{ij} = 1$ when the original load data is between 0 and 3.24 mm. In addition, let $Y_{ij} = 2$ when the original load data is between 3.25 and 6.48 mm. Also let $Y_{ij} = 3$ when the original load data is between 6.49 and 9.72 mm, and last let $Y_{ij} = 4$ for values greater than 9.72 mm. The random effects $b_i \sim nid\left(0, \sigma^2\right)$ and the beta parameters are unknown coefficients.

Model (9.34) is a generalized nonlinear mixed model with a log link function for the Poisson distribution of the categorical load data. Bayesian analysis will be executed with **BUGS CODE 10.9** using 300,000 observations for the simulation, with a burn in of 5,000 and a refresh of 100. Uninformative prior distributions were used for all parameters: (1) for the beta coefficients, a normal distribution with mean 0 and precision 0.001, while for the precision of the random effect a gamma with first and second parameters 0.001. For the load values, there are three missing load values for each of the eight planks, that is, 25% of the original load values are missing.

See the list statement of **BUGS CODE 10.9**.

BUGS CODE 10.9

```
model;
    {
            beta[1] ~ dnorm(0,.001)
            beta[2] ~ dnorm(0,.001)
            beta[3] ~ dnorm(0,.001)

for(i in 1:N){for (j in 1:M){Y[i,j]~dpois(mu[i,j])}}
for(i in 1:N){for (j in 1:M){log(mu[i,j])<-b[i]+beta[1]
+beta[2]*exp(-beta[3]*s[j])}}

for(i in 1:N){b[i]~dnorm(0,tau1)}

tau1~dgamma(.001,.001)
# sigma1 is the variance of the random effects
sigma1<-1/tau1

}

list(N = 8, M = 11,Y = structure(.Data =
c(1,1,2,NA,3,NA,4,4,NA,4,4,
1,NA,2,3,NA,4,4,4,4,NA,4,
1,NA,2,3,NA,3,4,4,NA,4,4,
1,NA,2,2,3,3,NA,4,4,NA,4,
1,NA,2,3,NA,3,4,4,NA,4,4,
1,NA,2,NA,3,4,4,4,NA,4,4,
1,1,NA,3,3,NA,4,4,4,NA,4,
1,NA,2,3,NA,4,4,4,4,NA,4),.Dim = c(8,11)),

s = c(0,.1,.2,.3,.4,.5,.6,.7,.8,.9,1))

# initial values
list(beta = c(0,0,0), tau1 =.5)
```

As reported in Table 10.21, the posterior median of the imputed missing value $Y[1,4]$ is 3, and none of the 95% credible intervals for the beta coefficients contain 0. Skewness of the posterior distribution for the variance σ_1^2 of the random effect is evident, with a median of 0.003905 compared to a posterior mean of 0.008168. Also note that the MCMC errors are extremely small, thus, I am very confident the posterior means are quite close to the actual posterior means. After referring to the list statement of **BUGS CODE 10.9**, the median of 3 for $Y[1,4]$ is reasonable.

How does the posterior analysis of Table 10.21 (based on 24 missing values) compare to the posterior analysis with no missing values as reported in Table 9.26? Using the posterior median, there is very little difference in the two analyses.

TABLE 10.21

Posterior Analysis for the Study with 24 Missing Load Values with One Random Effect

Parameter	Mean	SD	Error	2½	Median	97½
$Y[1,4]$	2.776	1.724	0.005173	0	3	7
β_1	1.456	0.196	0.004408	1.183	1.429	1.981
β_2	−1.6	0.3585	0.001812	−2.348	−1.59	−0.9389
β_3	5.304	4.155	0.04963	1.483	4.592	13.06
σ_1^2	0.008015	0.01368	0.000058	0.000484	0.003859	0.04054

Figure 10.3 displays the posterior distribution of the missing observation $Y[1,4]$ and demonstrates the skewness of the 75,000 imputed values.

Now consider and extension of (9.34) to include two random factors b_{1i} and b_{2i}, where $b_{2i} \sim nid(0, \sigma_2^2)$, then the link function is the canonical link

$$\log\left[E\left(Y_{ij}|b_i\right)\right] = \beta_1 + b_{1i} + \left(\beta_2 + b_{2i}\right)\exp\left(-\beta_3 x_{ij}\right) \qquad (10.42)$$

thus, using the program statement

```
for(i in 1:N){for (j in 1:M){log(mu[i,j])<-
b1[i]+beta[1]+(beta[2]+b2[i])*exp(-beta[3]*s[j])}}
```

with **BUGS CODE 10.9** to execute the Bayesian analysis with 300,000 observations for the simulation gives the following results for the posterior analysis reported in Table 10.22.

Upon comparing the results of Table 10.21 with 10.22, the addition of one random effect makes a large difference in the posterior estimates of the parameters of the model. One should introduce random effects if there is a good scientific reason for doing so. Remember when interpreting the estimates of the model (10.42), it is with respect to the log scale, not the original scale of the observations. To compare the effect of 24 missing values, one

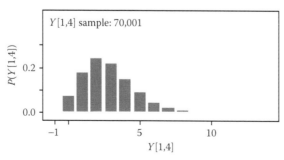

FIGURE 10.3
Posterior mass function of $Y[1,4]$.

TABLE 10.22

Posterior Analysis for Timber Study with 24 Missing Load Values and Two Random Factors

Parameter	Mean	SD	Error	2½	Median	97½
β_1	−11.26	10.72	0.4089	−39.42	−7.716	−0.2724
β_2	11.82	10.69	0.4091	0.9739	8.267	39.97
β_3	−0.1768	0.1737	0.006235	−0.6399	−0.1137	−0.02336
σ_1^2	0.009167	0.01636	0.000075	0.000493	0.004216	0.04817
σ_2^2	0.008038	0.0151	0.0000823	0.000476	0.003768	0.04142
$Y[1,4]$	2.331	1.56	0.003111	0	2	6

needs to compare Table 10.22 with 9.27. Remember the model contains two random effects. One observes that the estimates for the beta coefficients are quite different but those for the two variance components σ_1^2 and σ_2^2 are similar. This is an interesting result that needs to be explored and will be left as an exercise for the student.

The next example is explained by Crowder and Hand (p. 116)[9] and is based on data from Crowder[15] and involves female students whose energy expenditure (calories per minute) and heart rate (beats per minute) is measured four times, corresponding to four tasks: (1) lying, (2) sitting, (3) walking, and (4) skipping. As one would expect, the heart rate as well as the energy expenditure increases with successive tasks. Also measured is the weight in kilograms of each of the seven student volunteers, and Table 10.23 reports how the heart rate in beats per minutes is categorized into five values.

The relationship between heart rate and task will be explored with the mixed nonlinear model with a log link function

$$\log(\mu_{ij}) = (\beta_1 + b_{1i}) + (\beta_2 + b_{2i})(X_{ij} - \beta_3) \tag{10.43}$$

with X_{ij} the task number for subject i at task j, for $i = 1,\ldots,7$, and $j = 1,2,3,4$.

Note that μ_{ij} is the conditional mean of Y_{ij} given b_{1i} and b_{2i}. Also, the Y_{ij} (the heart rate for subject i at task j) are assumed to follow a Poisson distribution, where the two random factors are independent, and in addition, $b_{1i} \sim nid(0,\sigma_1^2)$ and $b_{2i} \sim nid(0,\sigma_2^2)$.

TABLE 10.23

Categories for Heart Rate

Range	Category
50–67	1
68–84	2
85–101	3
102–118	4
>119	5

Prior distributions for the parameters of the model (10.43) are uninformative. For example, the three beta coefficients are assigned normal distributions with mean 0 and variance 100, while the variances of the two random
factors are designated as gamma with parameters 3 and 2. The first list statement of **BUGS CODE 10.10** contains the categories of the heart rate data
where the missing values are denoted by NA, and there are 14 missing values and 14 observed values.

A Bayesian analysis is executed with 150,000 observations generated from
the joint posterior distribution of the model, thus for each missing value,
150,000 observations are imputed.

BUGS CODE 10.10

```
model;
    {
                    beta[1] ~ dnorm(0,.01)
                    beta[2] ~ dnorm(0,.01)


  beta[3] ~ dnorm(0,.01)


for(i in 2:N){for (j in 1:M){Y[i,j]~dpois(mu[i,j])}}
for(i in 2:N){for (j in 1:M){Z[i,j]~dpois(mu[i,j])}}
for(i in 2:N){for (j in 1:M){log(mu[i,j])<-(beta[1]+b1[i])+
(beta[2]+b2[i])*(x[j]-beta[3])}}


for(i in 1:N){b1[i]~dnorm(0,tau1)}
for(i in 1:N){b2[i]~dnorm(0,tau2)}


tau1~dgamma(3,2)
sigma1<-1/tau1tau2~dgamma(3,2)
sigma2<-1/tau2

}
list(N = 7, M = 4,Y = structure(.Data = c(1,NA,NA,5,
NA,2,NA,5,
NA,2,NA,5,
1,NA,NA,5,
NA,2,NA,5,
NA,3,NA,5,
NA,3,NA,5),.Dim = c(7,4)),
x = c(1,2,3,4))
# initial values
list(beta = c(0,0,0), tau1 =.934,tau2 = 100)
```

TABLE 10.24

Posterior Analysis of Heart Rate Study with Two Random Factors and 14 Missing Values

Parameter	Mean	SD	Error	2½	Median	97½
β_1	1.278	0.8255	0.02124	−0.3408	1.264	2.955
β_2	0.4482	0.3411	0.004903	−0.2139	0.4456	1.135
β_3	3.387	1.651	0.04204	−0.0119	3.406	6.605
σ_1^2	0.5568	0.323	0.002082	0.2156	0.4754	1.383
σ_2^2	0.5073	0.2774	0.001534	0.2036	0.4396	1.213
$Y[7,1]$	2.308	2.474	0.01101	0	2	9

The posterior analysis reveals small simulation errors and a posterior median of 2 for the missing value for patient 7 at task 1. As expected, the 95% credible interval for the beta coefficients contains 0 (Table 10.24).

How does the posterior analysis of Table 10.24 (with 14 missing values) compare to the corresponding analysis with no missing data reported in Table 9.30? I found very little difference in the two analyses, even though the number of missing values account for 50% of the non-missing information. This next example also assumed a Poisson distribution for the categorical repeated measure with the log as the link function, expressed as a nonlinear mixed model.

Recall Section 9.1.8 and the blood glucose levels for seven volunteers, where each person took alcohol at time 0 and provided a blood sample at 14 time points over five hours. Times are displayed as minutes divided by 10. Crowder and Tredger[16] analyzed these results, which are also explained by Hand and Crowder (p. 118).[14] The procedure was repeated with the same subjects, but they were given a dietary supplement.

Table 9.18 reports the descriptive statistics for the glucose values. As expected, the mean blood glucose value increases with time, reaches a peak at approximately 60 minutes, and then decreases. The time −1 serves as a baseline, while alcohol was taken at time 1.

Based on the basic data, the descriptive statistics, and the scatter plot, what is the most appropriate model? Crowder and Tredger[16] propose the following model for the blood glucose value Y_{ij} with two random effects. The following is the model for session 1.

$$\log\left(\mu_{1ij}\right) = \left(\beta_1 + b_{1i}\right) + \left(\beta_2 + b_{21}\right)x_{ij}^3 \exp\left(-\gamma_1 x_{ij}\right) \qquad (10.44)$$

while for the second date, the model is defined by

$$\log\left(\mu_{2ij}\right) = \left(\beta_3 + b_{3i}\right) + \left(\beta_4 + b_{4i}\right)x_{ij}^3 \exp\left(-\gamma_2 x_{ij}\right) \qquad (10.45)$$

where:

x_{ij} is the time for subject i at the jth time point

β_1 and β_2 are unknown parameters

the two random effects b_{1i} and b_{1i} are independent normal random variables with mean 0 and variances σ_1^2 and σ_2^2, respectively

The blood glucose values imply a cubic response damped by the exponential term $\exp(-\gamma_1 x_{ij})$, where γ_1 is an unknown positive parameter. Let the 14 by 1 vector of times be denoted by

$$x_{ij} = (-1, 0, 2, 4, 6, 8, 10, 12, 15, 18, 21, 24, 27, 30)$$

for all subjects i. The information from the study for the two sessions is given in the list statements of **BUGS CODE 9.12**, and the prior information for the various coefficients is specified as noninformative distributions. It is important that the model (9.37) for categorical data is not the same as that for the continuous glucose values of model (9.25). Although there are similarities, there are differences. What are those differences? For the continuous model, the expected glucose value Y_{ij} is expressed directly by $(\beta_1 + b_{1i}) + (\beta_2 + b_{21})x_{ij}^3 \, e^{-\gamma_1 x_{ij}}$, but on the other hand for the categorical values, the log of the expected value of Y_{ij} is expressed by the same model $(\beta_1 + b_{1i}) + (\beta_2 + b_{21})x_{ij}^3 \, e^{-\gamma_1 x_{ij}}$. The Bayesian analysis is performed with 75,000 observations with a burn in of 5,000. The data is included in the list statement of **BUGS CODE 10.11**, where

BUGS CODE 10.11

```
model;
{

            beta[1]  ~ dnorm(0,.01)
            beta[2]  ~ dnorm(0,.01)
            beta[3]  ~ dnorm(0,.01)
            beta[4]  ~ dnorm(0,.01)


        eta[1]  ~ dunif(.5,2)
        eta[2]  ~ dunif(.5,2)

for(i in 1:N){for (j in 1:M){Y1[i,j]~dpois(mu1[i,j])}}
for(i in 1:N){for(j in 1:M){Y2[i,j]~dpois(mu2[i,j])}}

for(i in 1:N){for (j in 1:M){log(mu1[i,j])<-(beta[1]+b1[i])+
(beta[2]+b2[i])*exp(-eta[1]*x[j])}}
for(i in 1:N){for (j in 1:M){log(mu2[i,j])<-(beta[3]+b3[i])+
(beta[4]+b4[i])*exp(-eta[2]*x[j])}}
```

```
for(i in 1:N){b1[i]~dnorm(0,taub1)}
for(i in 1:N){b2[i]~dnorm(0,taub2)}
for(i in 1:N){b3[i]~dnorm(0,taub3)}
for(i in 1:N){b4[i]~dnorm(0,taub4)}
taub1~dgamma(.001,.001)
sigmab1<-1/taub1
taub2~dgamma(.001,.001)
sigmab2<-1/taub2
taub3~dgamma(.001,.001)
sigmab3<-1/taub3
taub4~dgamma(.001,.001)
sigmab4<-1/taub4


}


list(N = 7, M = 14,Y1 = structure(.Data = c(1,1,NA,3,3,2,
1,1,2,NA,1,1,1,1,
2,1,3,4,NA,4,2,2,2,2,1,NA,1,2,
1,1,3,4,4,3,NA,2,2,2,1,2,NA,2,
2,2,NA,3,3,3,2,2,2,2,2,NA,2,2,
1,2,2,NA,3,3,2,2,2,NA,2,1,1,2,
1,1,1,1,2,2,1,NA,1,2,1,1,NA,1,
1,1,NA,3,4,3,2,3,2,2,2,NA,1,1),.Dim = c(7,14)),


Y2 = structure(.Data = c(NA,1,2,3,3,2,NA,1,1,1,1,1,1,1,
2,2,3,3,4,3,1,NA,1,1,1,NA,2,1,
2,2,2,3,3,2,2,1,NA,1,1,1,NA,2,
1,1,2,3,3,3,2,2,2,1,1,1,2,1,
2,2,2,3,NA,3,2,2,2,2,1,NA,1,2,
NA,1,2,3,3,NA,2,2,2,2,1,1,2,2,
1,NA,2,3,3,3,2,2,2,1,NA,1,1,1),.Dim = c(7,14)),


x = c(-1.00,.00,2.00,4.00,6.00,8.00,10.00,12.00,15.00,18.
00,21.00,24.00,27.00,30.00))

# initial values
list(beta = c(0,0,0,0), eta = c(.5,.5),taub1 = 1,taub2 =
1,taub3 = 1,taub4 = 1)
```

a missing value is denoted by NA, and there are 28 missing values with 14 for each group.

The posterior analysis is reported in Table 10.25, where the main focus is to compare the two groups. Notice that $Y_1[2,5]$ denotes the missing value for subject 2 at the fifth time point, and the Bayesian analysis generated 75,000 imputed values with a median of 2.

TABLE 10.25

Posterior Analysis Blood Glucose Values with Four Random Factors and 28 Missing Values

Parameter	Mean	SD	Error	2½	Median	97½
β_1	0.6963	0.1017	0.000704	0.4913	0.6981	0.8917
β_2	−0.2405	0.1626	0.001269	−0.6069	−0.2224	0.02488
β_3	0.6187	0.09387	0.000583	0.4302	0.62	0.799
β_4	−0.1019	0.1589	0.001428	−0.4475	−0.08895	0.1844
γ_1	0.9707	0.3514	0.002194	0.519	0.8921	1.814
γ_2	1.014	0.3756	0.002241	0.519	0.9366	1.862
σ^2_{b1}	0.02233	0.04586	0.000436	0.000616	0.008543	0.1283
σ^2_{b2}	0.02349	0.06165	0.000698	0.000587	0.00714	0.1477
σ^2_{b3}	0.01163	0.02322	0.000204	0.000511	0.004923	0.06358
σ^2_{b4}	0.02727	0.07458	0.000922	0.000591	0.007679	0.06358
$Y_1[2,5]$	2.087	1.475	0.005732	0	2	5
$Y_2[2,8]$	1.891	1.394	0.005064	0	2	5

It appears that β_2 is the same as β_4, and both have 95% credible intervals that contain 0; therefore, the model can be modified. It also appears that the two groups have the same parameters.

10.4 Comments and Conclusions

Introductory remarks outline the main topics with missing data to be covered in this chapter: (1) linear mean with continuous repeated measures, (2) linear link function with categorical repeated measure, (3) nonlinear model with continuous response, and (4) nonlinear model with categorical response.

Various missing value mechanisms are explained, including MCAR, MAR, and NMAR. It is shown that if the mechanism is MAR, then inferences can be based on the likelihood function of the unknown parameters given the observed data, thus for the Bayesian analysis, the posterior density of the unknown parameters is proportional to the product of the likelihood function and the prior distribution of the unknown parameters. For the case MAR using WinBUGS, the missing value is treated as an unknown parameter and the imputed values are generated with the MCMC simulation. One can then use either the posterior mean or the median for the missing value, which is automatically done with the software.

The first example to be considered is the cholesterol example of Ryan and Joiner 8 where there are nine missing values occurring at the third time point.

The missing values are placed by the author, thus the analysis with missing data can be compared to that with no missing data. The model is linear in time with two beta coefficients for intercept and slope, and the cholesterol levels are assumed to have a trivariate normal distribution for each time point with unstructured 3 by 3 covariance matrix. Posterior inferences for the parameters are implemented with **BUGS CODE 10.1** and reported in Table 10.3. The posterior means and medians are computed for the two regression coefficients and the three correlations. Also reported is the posterior distribution of the missing observations $Y[2,3]$. This example is reanalyzed with an autoregressive error structure and the posterior distribution generated with **BUGS CODE 10.2**, and the results are reported in Table 10.4. To see the effect of the number of missing values, the number of which is increased to 18 values and the posterior analysis reported in Table 10.4, those results are compared to the analysis with only 5 missing values.

As the second example, recall the hematocrit values of the hip-replacement study. In addition to the hematocrit response as the dependent variable, age is a covariate and the assumed model (10.14) uses an autoregressive error structure and 17% of the values are missing. With **BUGS CODE 10.3**, the Bayesian analysis is reported in Table 10.6. A very interesting feature of this example is that some of the age values are also missing. With WinBUGS, the missing values for a covariate are not automatically imputed, thus I assumed a normal distribution for age (after viewing the P-P plot), and then the software will automatically impute values for the missing data. With **BUGS CODE 10.3a**, 130,000 are used for the situation where there are missing values for both the dependent response and the covariate and the posterior analysis is reported in Table 10.7. To see the effect of the number of missing values, the number missing for age is 11 and 31 for hematocrit. Refer to **BUGS CODE 10.3b** and Table 10.8 for the posterior distribution of the parameters of the model. One can now assess the effect of the number of missing values on the posterior distribution of the parameters.

For the third example, recall the timber load slip study where the model is defined by (10.20). For the analysis, there are 24 missing values or 27% for the load data, and the results are revealed in Table 10.9. A comparison of Table 10.9 with 9.13 shows one the effect of missing data on the posterior analysis, and a main difference appears to be in the posterior median of β_1. In this next phase for the analysis of this study, 40 missing load values are used, and **BUGS CODE 10.4** implements the analysis with results revealed in Table 10.9. In order to see the effect of increasing the number of missing values, one should compare the posterior analysis of Table 10.9 with that of 10.8.

The remaining part of this chapter deals with a categorical response for the repeated measure, the main subject of Chapter 8. The Parkinson's disease study is now examined, where the repeated measure is a continuous clinical score, which is categorized into seven values. It is assumed that the number of missing values is 25 or 33.3% of the full data set. The model (10.24) assumes a Poisson distribution for the categorized scores and a log link, expressed as a linear function of time values and four covariates corresponding to four doses. There are also two random factors and the Bayesian analysis is explained in Table 10.11, and the posterior results compared to the case of no missing data are revealed by Table 8.11. As a result of the analysis (which showed that the four dose levels did not affect the expected response on the log scale), the model is modified to the model defined in 10.26, which now includes only two beta coefficients and the two random factors. With the modified model, the number of missing value is 35 of 75, and the Bayesian analysis is explained by Table 10.13.

Continuing with categorical data, an example with a binary response with a Bernoulli distribution is assumed for a pulmonary disease with a logistic link function. Gender and age are covariates and 34 of 280 missing values for the binary response (good or bad pulmonary function). See **BUGS CODE 10.6** and Table 10.14 for the Bayesian analysis, where it is determined that age and gender have no effect on the mean response on the logistic scale. With the implied simplified model (10.32) with one random factor and the number of missing values at 56 of 280, the Bayesian analysis is repeated with **BUGS CODE 10.6a** and revealed by Table 10.16.

For the last case of a linear link function with categorical data, the Parkinson's disease is reexamined with a generalized mixed model (10.36) with a cumulative proportional odds for the link. See Table 10.18 for the report of the posterior analysis. The cell totals were reduced by approximately 50% to see the effect of missing data on the posterior distribution.

A nonlinear link function is now used to analyze studies with a categorical response. Recall that Chapter 9 emphasizes the nonlinear model for both continuous data, thus I have used examples from Chapter 9 and various missing value patterns. For example, the pill dissolution study is reanalyzed with 30 missing binary values and the analysis implemented with **BUGS CODE 10.8**, and the results are reported in Table 10.20. Also reanalyzed is the timber load slip data with four categories assuming a Poisson distribution for the four categories and a nonlinear log link function (10.41) with three beta coefficients and two random factors. I put 24 missing load values and implemented the Bayesian analysis with **BUGS CODE 10.9** and the report of the analysis displayed by Table 10.21 and Figure 10.3. The latter displays the posterior mass function of the missing value $Y[1,4]$. As a result of the analysis, the model is modified to (10.42).

Now consider the heart rate study defined by (10.43) and a nonlinear link function with two random factors. The heart rate is categorized with five categories assumed to follow a Poisson distribution with 14 missing values. The last example is a reexamination of the blood glucose study with two groups and 28 missing values. See Table 10.24 for the analysis, which is implemented by **BUGS CODE 10.11**. The effect of the number of missing values on the posterior analysis is studied by referring to the corresponding tables of Chapter 9.

Exercises

1. Refer to Section 10.1, the introduction, and write a brief essay on the following topics:

   a. What are the three missing values mechanisms?

   b. Define mathematically the MAR assumption.

   c. How does WinBUGS implement the Bayesian analysis with missing data?

2. For the cholesterol study, conduct a Bayesian analysis using **BUGS CODE 10.1** with 75,000 observations and a burn in of 5,000.

   a. Verify the posterior analysis of Table 10.3.

   b. What is the 95% credible interval for β_2?

   c. What is your estimate of $Y[2,3]$?

   d. Plot the posterior density of ρ_{13}.

   e. Which posterior distributions are not symmetric about the mean?

3. Based on Equation 10.11, derive Equation 10.12, the linear model with autoregressive errors.

4. Explain Equation 10.13, the model for the exercise therapy trial.

   a. What is Y_{ij}?

   b. What is P?

   c. What do β_1 and β_2 measure?

5. Using **BUGS CODE 10.2** with 75,000 observations and a burn in of 4,000,

   a. Verify the posterior analysis of Table 10.4.

   b. Compare the estimates of the parameters of Table 10.4 with that of Table 10.5.

   c. What is the effect of missing values on the estimates of the parameters?

6. a. Explain Equation 10.14, the mean for hematocrit of the hip-replacement study.

   b. What is Y_{ij}?

   c. What parameter measures the effect of age on average hematocrit?

   d. What do β_1, β_2 and β_3 measure?

   e. Does the model (10.14) provide a good fit to the hematocrit data?

7. a. Derive Equation 10.15, the mean for hematocrit, with autoregressive error.

   b. What do β_1, β_2 and β_3 measure?

   c. What does ρ measure?

8. Using **BUGS CODE 10.3**, verify the posterior analysis of Table 10.6 using 130,000 observations for the simulation and a burn in of 5,000.

   a. What is the MCMC error for estimating the missing value $Y[1,3]$?

   b. What are the prior distributions for the parameters that are involved in **BUGS CODE 10.3**?

   c. What posterior distributions are symmetric about the posterior mean?

   d. Explain the difference in the posterior analyses of Table 10.7 versus that of Table 10.6.

9. Using **BUGS CODE 10.3b** with 130,000 observations, verify the posterior analysis of Table 10.8.

   a. What is the posterior median of age[2]?

   b. Is the posterior density of P symmetric about the posterior mean?

   c. Compare the posterior mean of β_1 of Table 10.6 (the analysis with 18 missing hematocrit values) to the posterior mean of β_1, reported in Table 10.8 (31 missing hematocrit values and 11 missing age values).

10. Describe Equation 10.19, the model for the timber load data.

    a. What is Y_{ij}?

    b. What do β_1, β_2 and β_3 measure?

    c. Explain why (10.19) is a nonlinear model in the unknown parameters.

    d. Describe Equation 10.20.

    e. How does model (10.20) differ from model (10.19)? Explain carefully.

    f. Does the addition of the random effect in model (10.20) change the estimates of β_1, β_2 and β_3 of the posterior analysis of Table 10.19?

11. Analyze the timber load study using **BUGS CODE 10.4** with 65,000 observations for the simulation and a burn in of 5,000, and verify the posterior analysis of Table 10.9.

    a. What is your estimate of the variance of the random effect b_{1i}?

    b. What does P represent in (10.20)?

    c. How many missing load values are there?

    d. Compare the estimate of the parameters of Table 10.9 with those of Table 9.13 (with no missing load values).

    e. What prior distribution was specified in **BUGS CODE 10.4**?

12. Using an amended version of **BUGS CODE 10.4**, perform a Bayesian analysis with 65,000 observations for the simulation and a burn in of 5,000 and verify Table 10.10.

    a. How many missing load values are there?

    b. What is the 95% credible interval for β_2?

    c. Compare Tables 10.9 and 10.10 and explain the difference in the estimates of the parameters.

    d. What prior distributions are used with the analysis using **BUGS CODE 10.4**?

13. Refer to Equation 10.24, the model for the Parkinson's disease study.

    a. What is Y_{ij}?

    b. What do the β_i, $i = 3,4,5,6$, measure?

    c. Is the log link linear in the parameters?

    d. What do the two random factors b_{1i}, b_{2i} represent? Interpret the random effects in the log link function.

    e. Use **BUGS CODE 10.5** and verify the posterior analysis of Table 10.11.

    f. Identify the statements in the code that specify the prior distributions for the Bayesian analysis.

14. Refer to Table 10.11.

    a. Using the estimates of Table 10.11, estimate the log of the mean of Y_{ij}.

    b. What is the effect of the various dose levels on the log of the mean of the response?

    c. In light of Table 10.11, should the model (10.14) be modified? How?

    d. How many missing values are there in this data set?

15. Refer to the modified model (10.26) and use **BUGS CODE 10.5** with 65,000 observations for the simulation and a burn in of 5,000.

    a. Verify Table 10.12.

    b. Compare the estimates of the parameters of Table 10.12 with those of Table 10.11.

    c. Are the estimates of β_1 and β_2 of Table 10.11 different form those of Table 10.12?

    d. What are the prior distributions specified in the modified model (10.26)?

16. Describe models (10.30) and (10.31) for the pulmonary study.

    a. What does Y_{ij} represent?

    b. What do the betas represent in the model (10.31)?

    c. Based on Table 10.31, estimate the mean of Y_{ij}.

    d. What are the three covariates in the model (10.31)?

17. Using **BUGS CODE 10.6** with 75,000 observations and a burn in of 5,000, verify the posterior analysis of Table 10.14.

    a. Refer to **BUGS CODE 10.6** and identify the code that specifies the prior distribution for the parameters.

    b. How many missing values are there?

    c. Explain how the posterior estimates of Table 10.14 imply the simplified model (10.32).

    d. Is the posterior distribution of σ_1^2 skewed?

18. Using **BUGS CODE 10.6a** with 75,000 observations for the simulation, verify the posterior analysis of Table 10.15.

    a. Based on Table 10.15, should the model be further simplified?

    b. Do the estimates of Table 10.15 imply a definite quadratic response over time on the log scale?

    c. Based on the estimates of the parameters of Table 10.15, estimate the log of the mean of Y_{ij}?

19. Explain the cumulative proportional odds model defined by (10.33) and (10.34).

20. Refer to (10.35) and (10.36) as the model for the Parkinson's disease study.

    a. What is Y_{ij}?

    b. What do α_j represent?

    c. Based on (10.36), derive (10.39).

    d. What do β_1 and β_2 measure?

21. Using **BUGS CODE 10.7** with 65,000 observations for the simulation, verify the posterior analysis of Table 10.18.

    a. The posterior means of the α_j, $j = 1, 2,..., 5$ are increasing. Is this reasonable?

    b. Identify the code statements that specify the prior distributions of the parameters of the Bayesian analysis.

    c. Based on the posterior estimates of β_1 and β_2, should the model (10.36) be simplified?

References

Chapter 1

1. Fitzmaurice, G.M., Laird, N.M., and Ware, J.H., *Applied Longitudinal Analysis*, 2nd edition, John Wiley & Sons, New York, 2011.
2. Aerts, M., Geys, H., Molenberghs, G., and Ryan, L.M., *Topics in Modeling of Clustered Data*, Chapman & Hall, Boca Raton, FL, 2002.
3. Price, C.J., Kimmel, C.A., Tyl, R.W., and Marr, M.C., The developmental toxicity of ethylene glycol in mice, *Toxicology and Applied Pharmacology*, 8, 115–127, 1988.
4. Davis, C.S., *Statistical Methods for the Analysis of Repeated Measurements*, Springer-Verlag, New York, 2002.
5. Little, R.J.A., and Rubin, D.B., *Statistical Analysis with Missing Data*, John Wiley & Sons, New York, 1987.
6. Schafer, J.L., *Analysis of Incomplete Multivariate Data*, Chapman & Hall, London, 1997.
7. Daniels, M.J., and Hogan, J.W., *Missing Data in Longitudinal Studies: Strategies for Bayesian Modeling and Sensitivity Analysis*, Chapman & Hall, Boca Raton, FL, 2008.
8. Bayes, T., An essay towards solving a problem in the doctrine of chances, *Philosophical Transactions of the Royal Society of London*, 53, 370, 1778.
9. Laplace, P.S., Memorie des les probabilities, Memories de l'Academie des sciences de Paris, 227, 1778.
10. Stigler, M., *The History of Statistics: The Measurement of Uncertainty Before 1900*, The Belknap Press of Harvard University Press, Cambridge, MA, 1986.
11. Hald, A., *A History of Mathematical Statistics: From 1750–1930*, Wiley Interscience, London, 1990.
12. Lhoste, E., Le calcul des probabilities appliqué a l'artillirie, lois de probabilite a prior, *Revu d'artillirie*, Mai, 405, 1923.
13. Jeffreys, H., *An Introduction to Probability*, Clarendon Press, Oxford, 1939.
14. Savage, L.J., *The Foundation of Statistics*, John Wiley & Sons, New York, 1954.
15. Lindley, D.V., *Introduction to Probability and Statistics from a Bayesian Viewpoint, Volumes I and II*, Cambridge University Press, Cambridge, 1965.
16. Broemeling, L.D., and Broemeling, A.L., Studies in the history of probability and statistics XLVIII: The Bayesian contributions of Ernest Lhoste, *Biometrika*, 90(3), 728, 2003.
17. Box, G.E.P., and Tiao, G.C., *Bayesian Inference in Statistical Analysis*, Addison Wesley, Reading, MA, 1973.
18. Zellner, A., *An Introduction to Bayesian Inference in Econometrics*, John Wiley & Sons, New York, 1971.
19. Dale, A.I., *A History of Inverse Probability from Thomas Bayes to Karl Pearson*, Springer-Verlag, Berlin, Germany, 1991.
20. Leonard, T., and Hsu, J.S.J., *Bayesian Methods: An Analysis for Statisticians and Interdisciplinary Researchers*, Cambridge University Press, Cambridge, 1999.

21. Gelman, A., Carlin, J.B., Stern, H.S., and Rubin, D.B., *Bayesian Data Analysis*, Chapman & Hall, New York, 1997.
22. Congdon, P., *Applied Bayesian Modeling*, John Wiley & Sons, Hoboken, NJ, 2003.
23. Carlin, B.P., and Louis, T.A., *Bayes and Empirical Bayes for Data Analysis*, Chapman & Hall, 1996, New York.
24. Gilks, W.R., Richardson, S., and Spiegelhalter, D.J., *Markov Chain Monte Carlo in Practice*, Chapman & Hall, Boca Raton, FL, 1996.
25. Broemeling, L.D., *Bayesian Analysis of Linear Models*, Marcel Dekker, New York, 1985.
26. Lehmann, E.L., *Testing Statistical Hypotheses*, John Wiley & Sons, New York, 1959.
27. Lee, P.M., *Bayesian Statistics: An Introduction*, 2nd edition, Arnold, London, 1997.
28. Mielke, C.H., Shields, J.P., and Broemeling, L.D., Coronary artery calcium scores for men and women of a large asymptomatic population, *CVD Prevention*, 2, 194–198, 1999.
29. Casella, G., and George, E.I., Explaining the Gibbs sampler, *The American Statistician*, 46, 167–174, 2004.
30. Gregurich, M.A., and Broemeling, L.D., A Bayesian analysis for estimating the common mean of independent normal populations using the Gibbs sampler, *Communications in Statistics*, 26(1), 25–31, 1997.
31. Ntzoufras, I., *Bayesian Modeling Using WinBUGS*, John Wiley & Sons, Hoboken, NJ, 2009.
32. Broemeling, L.D., *Bayesian Biostatistics and Diagnostic Medicine*, Chapman & Hall, Boca Raton, FL, 1996.
33. Woodworth, G.G., *Biostatistics: A Bayesian Introduction*, John Wiley & Sons, Hoboken, NJ, 2004.
34. The BUGS Project Resource. http://www.mrc-bsu.cam.ac.uk/bugs/winbugs/contents.shtml.

Chapter 2

1. Chatterjee, S., and Price, B., *Regression Analysis by Example*, John Wiley & Sons, New York, 1991.
2. Ntzoufras, I., *Bayesian Modeling Using WinBUGS*, John Wiley & Sons, New York, 2009.
3. Hosmer, D.W., and Lemeshow, S., *Applied Logistic Regression*, John Wiley & Sons, New York, 1989.
4. Agresti, A., *Categorical Data Analysis*, John Wiley & Sons, New York, 1990.
5. Congdon, P., *Bayesian Models for Categorical Data*, John Wiley & Sons, New York, 2005.
6. Woolson, R.F., *Statistical Methods for the Analysis of Biomedical Data*, John Wiley & Sons, New York, 1987.
7. Montgomery, D.C., Peck, E.A., and Vining, G.G., *Introduction to Linear Regression Analysis*, John Wiley & Son, New York, 2001.

8. Bache, C.A., Serum, J.W., Youngs, D.W., and Lisk, D.J., Polychlorinated biphenyl residues: Accumulation in Lake Cayuga trout with age, *Science*, 117, 1192–1193, 1972.
9. Hand, D.J., and Taylor, C.C., *Multivariate Analysis of Variance and Repeated Measures,* Chapman & Hall, London, 1987.
10. Broemeling, L.D., *Advanced Bayesian Methods for Medical Test Accuracy*, CRC Press, Boca Raton, FL, 2012.
11. Bates, D.M., and Watts, D.G., *Nonlinear Regression Analysis and Its Application*, John Wiley & Sons, New York, 1988.
12. Denison, T., Holmes, C.C., Mallick, B.K., and Smith, A.F.M., *Bayesian Methods for Nonlinear Classification and Regression*, John Wiley & Sons, New York, 2002.
13. Davidian, M., and Giltinan, D.M., *Nonlinear Models for Repeated Measures Data*, Chapman & Hall, London, 1995.

Chapter 3

1. Lippil, I.E., A double-blind crossover evaluation of progabide in partial seizures, *Neurology*, 35, 285, 1985.
2. Thall, P.F., and Vail, S.C., Some covariance models for longitudinal count data with over dispersion, *Biometrics*, 46, 657–671, 1990.
3. Fitzmaurice, G.M., Laird, N.M., and Ware, J.H., *Applied Longitudinal Analysis*, 2nd edition, John Wiley & Sons, New York, 2011.
4. Fuller, W.A., *Measurement Error Models*, John Wiley & Sons, New York, 1987.
5. Crawford, P., and Chadwick, D., A comparative study of progabide and placebo as add-on therapy in patients with refractory epilepsy, *Journal of Neurology, Neurosurgery, and Psychiatry*, 49, 1251–1267, 1986.
6. Chadwick, D., Prospects for new drug treatment in epilepsy: A review, *Journal of the Royal Society of Medicine*, 83(363), 383–386, 1990.
7. Hand, D.J., and Taylor, C.C., *Multivariate Analysis of Variance and Repeated Measures*, Chapman & Hall, London, 1987.
8. Hand, D., and Crowder, M., *Practical Longitudinal Data Analysis,* Chapman & Hall, Boca Raton, FL, 1999.
9. Little, A., Levy, R., Chuaqui-Kidd, P., and Hand, D., A double-blind, placebo controlled trial of high-dose lecithin in Alzheimer's disease, *Journal of Neurology, Neurosurgery, and Psychiatry*, 48, 736–742, 1985.
10. Davis, C.S., *Statistical Methods for the Analysis of Repeated Measurements*, Springer-Verlag, New York, 2002.
11. Agresti, A., *Categorical Data Analysis*, John Wiley & Sons, New York, 1990.
12. Cornoni-Huntley, J., Brock, D.B., Ostfeld, A., Taylor, J.O., and Wallace, R.B., *Established Populations for Epidemiologic Studies of the Elderly: Resource Data Book*, National Institutes of Health (NIH Publ. No. 86-2443), Bethesda, MD, 1986.

Chapter 4

1. Potthoff, R.F., and Roy, S.N., A generalized multivariate analysis variance model useful especially for growth curve models, *Biometrika*, 51, 313–326, 1964.
2. Hand, D., and Crowder, M., *Practical Longitudinal Data Analysis*, Chapman & Hall, Boca Raton, FL, 1996.
3. Fitzmaurice, G.M., Laird, N.M., and Ware, J.H., *Applied Longitudinal Analysis*, 2nd edition, John Wiley & Sons, New York, 2011.
4. Miller, R.G., *Beyond ANOVA: Basics of Applied Statistics*, John Wiley & Sons, New York, 1986.
5. Broemeling, L.D., *Bayesian Methods for Measures of Agreement*, Taylor & Francis, Boca Raton, FL, 2009.
6. Leonard, T., and Hsu, J.S., *Bayesian Methods: An Analysis for Statisticians and Interdisciplinary Researchers*, Cambridge University Press, Cambridge, 1999.
7. Crowder, M.J., and Hand, D.J., *Analysis of Repeated Measures*, Chapman & Hall, New York, 1990.
8. Box, G.E.P., and Tiao, G.C., *Bayesian Inference in Statistical Analysis*, Addison Wesley, Reading, MA, 1973.
9. Broemeling, L.D., *Bayesian Analysis of Linear Models*, Marcel Dekker, New York, 1985.
10. Carroll, R.J., and Ruppert, D., *Transformation and Weighting in Regression*, Chapman & Hall, New York, 1988.
11. Davidian, M., and Giltinan, D.M., *Nonlinear Models for Repeated Measurement Data*, Chapman & Hall, New York, 1995.
12. Aerts, M., Geys, H., Molenberghs, G., and Ryan, L.M., *Topics in Modeling of Clustered Data*, Chapman & Hall, Boca Raton, FL, 2002.

Chapter 5

1. Fitzmaurice, G.M., Laird, N.M., and Ware, N.M., *Applied Longitudinal Analysis*, 2nd edition, John Wiley & Sons, New York, 2011.
2. Van der Lende, R., Kok, T.J., Peset, R., Quanjer, P.H., Schouten, J.P., and Orle, N.G.M., Decreases in the VC and FEV1 with time: Indicators of the effects of smoking and air pollution, *Bulletin of European Physiopathology and Respiration*, 17, 775–792, 1981.
3. Crowder, M.J., and Hand, D.J., *Analysis of Repeated Measures*, Chapman & Hall, London, 1990.
4. Rhoads, G.C., Treatment of lead-exposed trial group, safety and efficacy of succimer in toddlers with blood lead levels of 20–40 µg/dL, *Pediatric Research*, 48, 593–599, 2000.
5. Cornoni-Huntley, J., Brock, D.B., Ostfeld, A., Taylor, J.O., and Wallace, R.B., *Established Populations for Epidemiologic Studies of the Elderly: Resource Data Book*, National Institutes of Health (NIH Publ. No. 86-2443), Bethesda, MD, 1986.
6. Davis, C.S., *Statistical Methods for the Analysis of Repeated Measurements*, Springer-Verlag, New York, 2002.

7. DeGroot, M.H., *Optimal Statistical Decisions*, McGraw-Hill, New York, 1970.
8. Broemeling, L.D., *Bayesian Methods for Measures of Agreement*, CRC Press, Boca Raton, FL, 2009.
9. Zerbe, G.O., Randomization analysis for the completely randomized blocks extended to growth and response curves, *Journal of the American Statistical Association*, 57, 348–368, 1979.

Chapter 6

1. Fitzmaurice, G.M., Laird, N.M., and Ware, J.H., *Applied Longitudinal Analysis*, 2nd edition, John Wiley & Sons, New York, 2011.
2. Freund, R.J., Littell, R.C., and Spector, P.C., *SAS Systems for Linear Models*, SAS Institute, Cary, NC, 1986.
3. Gelman, A., Carlin, J.B., Stern, H.S., and Rubin, D.B., *Bayesian Data Analysis*, Chapman & Hall, Boca Raton, FL, 1997.
4. Ntzoufras, I., *Bayesian Modeling Using WinBUGS*, John Wiley & Sons, New York, 2009.
5. Crowder, M.J., and Hand, D.J., *Analysis of Repeated Measures*, Chapman & Hall, New York, 1990.
6. Leonard, T., and Hsu, J.S.J., *Bayesian Methods: An Analysis for Statisticians and Interdisciplinary Researchers*, Cambridge University Press, Cambridge, 1999.
7. Congdon, P., *Applied Bayesian Modeling*, John Wiley & Sons, Hoboken, NJ, 2003.
8. Jones, R.H., *Longitudinal Data with Serial Correlation: A State-Space Approach*, Chapman & Hall, London, 1993.
9. Carroll, R.J., and Ruppert, D., *Transformation and Weighting in Regression*, Chapman & Hall, New York, 1988.
10. Jones, B.J., and Kenward, M.G., *Design and Analysis of Cross-Over Trials*, Chapman & Hall, London, 1988.
11. Daniels, M.J., and Hogan, J.W., *Missing Data in Longitudinal Studies: Strategies for Bayesian Modeling and Sensitivity Analysis*, Chapman & Hall, Boca Raton, FL, 2008.

Chapter 7

1. Fitzmaurice, G.M., Laird, N.M., and Ware, J.H., *Applied Logistic Regression*, 2nd edition, John Wiley & Sons, New York, 2011.
2. Van der Lende, R., Kok, T.J., Peset, R., Quanjer, P.H., Schouten, J.F., and Orle, N.G.M., Decreases in the VC and FEV1 with time: Indicators for effects of smoking and air pollution, *Bulletin of European Physiopathology and Respiration*, 17, 775–792, 1981.
3. Crowder, M.J., and Hand, D.J., *Analysis of Repeated Measures*, Chapman & Hall, New York, 1990.

4. Rhoads, G.G., Treatment of lead-exposed children (TLC) trial group. Safety and efficacy of Succimer in toddlers with blood lead levels of 20–44 µg/dL, *Pediatric Research*, 48, 593–599, 2000.

5. Golub, G.H., and Van Loan, C.F., *Matrix Computations*, Johns Hopkins University Press, 4th edition, 2012.

Chapter 8

1. Davis, C.S., *Statistical Methods for the Analysis of Repeated Measures*, Springer-Verlag, New York, 2002.

2. Fitzmaurice, G.M., Laird, N.M., and Ware, J.H., *Applied Longitudinal Analysis*, 2nd edition, John Wiley & Son, New York, 2011.

3. Hand, D., and Crowder, M., *Practical Longitudinal Analysis*, Chapman & Hall, Boca Raton, FL, 1999.

4. Crowder, M.J., and Hand, D.J., *Analysis of Repeated Measures*, Chapman & Hall, New York, 1990.

5. Jones, B., and Kenward, M.G., *Design and Analysis of Cross-Over Trials*, Chapman & Hall, London, 1988.

6. Aerts, M., Geys, H., Molenberghs, G., and Ryan, L.M., *Topics in Modelling of Clustered Data*, Chapman & Hall, Boca Raton, FL, 2002.

7. Ware, J.H., Linear models for the analysis longitudinal studies, *The American Statistician*, 39, 95–101, 1985.

8. Agresti, A., *Categorical Data Analysis*, John Wiley & Sons, New York, 1990.

9. Cornoni-Huntley, J., Bock, D.B., Ostfeld, A., Taylor, J.O., and Wallace, R.B., *Established Populations for Epidemiologic Studies of the Elderly: Resource Data Book*. National Institutes of Health (NIH Publ. No. 86-2443), Bethesda, MD, 1986.

10. Hand, D.J., and Taylor, C.C., *Multivariate Analysis of Variance and Repeated Measures*, Chapman & Hall, London, 1987.

11. Stokes, D.O., Davis, C.S., and Koch, G.G., *Categorical Data Analysis Using the SAS System*, SAS Institute, Cary, NC, 1995.

12. Congdon, P., *Applied Bayesian Modelling*, John Wiley & Son, Chichester, 2003.

13. Agresti, A., *Analysis of Ordered Categorical Data*, 2nd Edition, John Wiley & Sons, New York, 2010.

14. Broemeling, L.D., *Bayesian Methods in Epidemiology*, Chapman & Hall, Boca Raton, FL, 2014.

15. Thall, P.F., and Vail, S.C., Some covariance models for the longitudinal count data with over dispersion, *Biometrics*, 46, 657–671, 1988.

Chapter 9

1. Davidian, M., and Giltinan, D.M., *Nonlinear Models for Repeated Measurement Data*, Chapman & Hall, London, 1995.

2. Hand, D., and Crowder, M., *Practical Longitudinal Data Analysis*, Chapman & Hall, Boca Raton, FL, 1999.
3. Hixon, A.W., and Crowell, J.H., Dependence of reaction velocity upon surface and agitation, *Journal of Industrial and Engineering Chemistry*, 23, 923–931, 1931.
4. Goyan, J.E., Dissolution rate studies. III. Penetration models for describing dissolution of a multiparticulate system, *Journal of Pharmaceutical Sciences*, 54, 645–647, 1965.
5. Crowder, M.J., On concurrent regression lines, *Applied Statistics*, 27, 310–318, 1978.
6. Crowder, M.J., and Tredger, J.A., The use of exponentially damped polynomials for biological recovery data, *Applied Statistics*, 32, 15–18, 1981.
7. Ruppert, D., Cressie, N., and Carroll, R.J., A transformation-weighting model for estimating Michaelis Menten parameters, *Biometrics*, 45, 637–656, 1989.
8. Draper, N.R., and Smith, H., *Applied Regression Analysis*, 2nd edition, John Wiley & Sons, New York, 1981.
9. Jones, R.H., *Longitudinal Data with Serial Correlation: A State Space Approach*, Chapman & Hall, London, 1993.
10. Bates, D.M., and Watts, D.G., *Nonlinear Regression Analysis and Its Applications*, John Wiley & Sons, New York, 1988.

Chapter 10

1. Little, R.J.A., and Rubin, D.B., *Statistical Analysis with Missing Data*, John Wiley & Sons, New York, 1987.
2. Schafer, J.J., *Analysis of Incomplete Multivariate Data*, Chapman & Hall, New York, 1997.
3. Davidian, M., and Giltinan, D.M., *Nonlinear Models for Repeated Measurement Data*, Chapman & Hall, New York, 1995.
4. Howell, D.C., Treatment of missing data: Part I, 2001. http://www.uvm.edu/StatPages/More_Stuff/Missing-Data/Missing.html.
5. Allison, P.D., *Missing Data*, Sage Publications, Thousand Oaks, CA, 2001.
6. Fitzmaurice, G.M., Laird, N.M., and Ware, J.H., *Applied Longitudinal Analysis*, 2nd edition, John Wiley & Sons, New York, 2011.
7. Daniels, M.J., and Hogan, J.W., *Missing Data in Longitudinal Studies: Strategies in Bayesian Modeling and Sensitivity Analysis*, Chapman & Hall, New York, 2008.
8. Ryan, T.A., and Joiner, B.L., Normal probability plots and tests for normality. Technical Report, Department of Statistics, Pennsylvania State University, University Park, PA, 1976.
9. Crowder, M.J., and Hand, D.J., *Analysis of Repeated Measures*, Chapman & Hall, New York, 1990.
10. Stokes, M.E., Davis, C.S., and Koch, G.G., *Categorical Data Analysis Using the SAS System*, SAS Institute, Cary, NC, 1995.
11. Congdon, P., *Applied Bayesian Modeling*, John Wiley & Sons, Hoboken, NY, 2003.
12. Agresti, A., *Categorical Data Analysis*, John Wiley & Sons, New York, 1990.

13. Broemeling, L.D., *Advanced Bayesian Modeling for Medical Test Accuracy*, CRC Press, Boca Raton, FL, 2012.
14. Hand, D., and Crowder, M., *Practical Longitudinal Data Analysis*, Chapman & Hall, Boca Raton, FL, 1996.
15. Crowder, M.J., Gaussian estimation for correlated binary data, *Journal of the Royal Statistical Society B*, 47, 229–237, 1985.
16. Crowder, M.J., and Tredger, J.A., The use of exponentially damped polynomials for biological recovery, *Applied Statistics*, 30, 147–152, 1981.

Index

Note: Locators followed by "*f*" and "*t*" denote figures and tables in the text

Printed and bound by CPI Group (UK) Ltd, Croydon, CR0 4YY

22/10/2024

01777615-0018